OXFORD MATHEMATICAL MONOGRAPHS

OXFORD MATHEMATICAL MONOGRAPHS

A. Belleni-Moranti: *Applied semigroups and evolution equations*
A. M. Arthurs: *Complementary variational principles* 2nd edition
M. Rosenblum and J. Rovnyak: *Hardy classes and operator theory*
J. W. P. Hirschfeld: *Finite projective spaces of three dimensions*
A. Pressley and G. Segal: *Loop groups*
D. E. Edmunds and W. D. Evans: *Spectral theory and differential operators*
Wang Jianhua: *The theory of games*
S. Omatu and J. H. Seinfeld: *Distributed parameter systems: theory and applications*
J. Hilgert, K. H. Hofmann, and J. D. Lawson: *Lie groups, convex cones, and semigroups*
S. Dineen: *The Schwarz lemma*
S. K. Donaldson and P. B. Kronheimer: *The geometry of four-manifolds*
D. W. Robinson: *Elliptic operators and Lie groups*
A. G. Werschulz: *The computational complexity of differential and integral equations*
L. Evens: *Cohomology of groups*
G. Effinger and D. R. Hayes: *Additive number theory of polynomials*
J. W. P. Hirschfeld and J. A. Thas: *General Galois geometries*
P. N. Hoffman and J. F. Humphreys: *Projective representations of the symmetric groups*
I. Györi and G. Ladas: *The oscillation theory of delay differential equations*
J. Heinonen, T. Kilpelainen, and O. Martio: *Non-linear potential theory*
B. Amberg, S. Franciosi, and F. de Giovanni: *Products of groups*
M. E. Gurtin: *Thermomechanics of evolving phase boundaries in the plane*
I. Ionescu and M. F. Sofonea: *Functional and numerical methods in viscoplasticity*
N. Woodhouse: *Geometric quantization* 2nd edition
U. Grenander: *General pattern theory*
J. Faraut and A. Koranyi: *Analysis on symmetric cones*
I. G. Macdonald: *Symmetric functions and Hall polynomials* 2nd edition
B. L. R. Shawyer and B. B. Watson: *Borel's methods of summability*
D. McDuff and D. Salamon: *Introduction to symplectic topology*
M. Holschneider: *Wavelets: an analysis tool*
Jacques Thévenaz: *G-algebras and modular representation theory*
Hans-Joachim Baues: *Homotopy type and homology*
P. D. D'Eath: *Black holes: gravitational interactions*
R. Lowen: *Approach spaces: the missing link in the topology–uniformity–metric triad*
Nguyen Dinh Cong: *Topological dynamics of random dynamical systems*
J. W. P. Hirschfeld: *Projective geometries over finite fields* 2nd edition
K. Matsuzaki and M. Taniguchi: *Hyperbolic manifolds and Kleinian groups*
David E. Evans and Yasuyuki Kawahigashi: *Quantum symmetries on operator algebras*
Norbert Klingen: *Arithmetical similarities: prime decomposition and finite group theory*
Isabelle Catto, Claude Le Bris, and Pierre-Louis Lions: *The mathematical theory of thermodynamic limits: Thomas–Fermi type models*

Symmetric Functions and Hall Polynomials

Second Edition

I. G. MACDONALD

Queen Mary, University of London

with contribution by

A. V. ZELEVINSKY

CLARENDON PRESS • OXFORD

OXFORD
UNIVERSITY PRESS

Great Clarendon Street, Oxford, OX2 6DP,
United Kingdom

Oxford University Press is a department of the University of Oxford.
It furthers the University's objective of excellence in research, scholarship,
and education by publishing worldwide. Oxford is a registered trade mark of
Oxford University Press in the UK and in certain other countries

First published 1998
First published in paperback 2008
Reprinted in paperback 2015

Published in the United States of America by Oxford University Press
198 Madison Avenue, New York, NY 10016, United States of America

British Library Cataloguing in Publication Data
Data available

Library of Congress Cataloging in Publication Data
Data available

ISBN 978–0–19–873912–8

Foreword

The theory of symmetric functions has shown phenomenal development in recent years, in no small part due to Ian Macdonald's masterful exposition of this subject. Most of the material outside the first chapter is unavailable else-where in a unified form and thus provides an invaluable service to researchers whose work involves symmetric functions. I myself have been entranced on innumerable occasions by the beauty and richness of this subject. Worth special mention are the "examples" (really exercises with solutions) at the end of each section, which are essential additions to the main text.

Chapter VI especially stands out. This chapter introduces the symmetric functions now known as *Macdonald symmetric functions* or *Macdonald polynomials*. They were discovered by Macdonald around 1988 and received their most comprehensive treatment in the first printing (1995) of the second (current) edition of this book. Macdonald polynomials have subsequently found a bewildering number of unexpected connections to such topics as Hilbert schemes, diagonal harmonics, affine Hecke algebras, Gromov-Witten invariants, integrable probability, k-Schur functions, affine Grassmannians, rational Cherednik algebras, quantum groups, Yang-Baxter equations, cluster algebras, ...

Macdonald's book remains as timely now as when it was first published, so Oxford University Press has performed a real service to the mathematical community by coming out with this new paperback printing. At the time of this writing, Google Scholar gives an astonishing 6720 citations to the first printing. A new generation of mathematicians with suitable algebraic background await enthrallment by the pleasures within.

Richard Stanley
February 7, 2015

Preface to the second edition

The first edition of this book was translated into Russian by A. Zelevinsky in 1984, and for the Russian version both the translator and the author furnished additional material, both text and examples. Thus the original purpose of this second edition was to make this additional material accessible to Western readers. However, in the intervening years other developments in this area of mathematics have occurred, some of which I have attempted to take account of: the result, I am afraid, is a much longer book than its predecessor. Much of this extra bulk is due to two new chapters (VI and VII) about which I shall say something below.

For readers acquainted with the first edition, it may be of use to indicate briefly the main additions and new features of this second edition. The text of Chapter I remains largely unchanged, except for a discussion of transition matrices involving the power-sums in §6, and of the internal (or inner) product of symmetric functions in §7. On the other hand, there are more examples at the ends of the various sections than there were before. To the appendix on polynomial functors I have added an account of the related theory of polynomial representations of the general linear groups (always in characteristic zero), partly for its own sake and partly with the aim of rendering the account of zonal polynomials in Chapter VII self-contained. I have also included, as Appendix B to Chapter I, an account, following Specht's thesis, of the characters of wreath products $G \sim S_n$ (G any finite group), along the same lines as the account of the characters of the symmetric groups in Chapter I, §7: this may serve the reader as a sort of preparation for the more difficult Chapter IV on the characters of the finite general linear groups.

In Chapter II, one new feature is that the formula for the Hall polynomial (or, more precisely, for the polynomial $g_S(t)$ (4.1)) is now made completely explicit in (4.11). The chapter is also enhanced by the appendix, written by A. Zelevinsky for the Russian edition.

The main addition to Chapter III is a section (§8) on Schur's Q-functions, which are the case $t = -1$ of the Hall–Littlewood symmetric functions. In this context I have stopped short of Schur's theory of the projective representations of the symmetric groups, for which he introduced these symmetric functions, since (a) there are now several recent accounts of this theory available, among them the monograph of P. Hoffman and J. F. Humphreys in this series, and (b) this book is already long enough.

Chapters IV and V are unchanged, and require no comment.

Chapter VI is new, and contains an extended account of a family of symmetric functions $P_\lambda(x; q, t)$, indexed as usual by partitions λ, and depending rationally on two parameters q and t. These symmetric functions include as particular cases many of those encountered earlier in the book: for example, when $q = 0$ they are the Hall–Littlewood functions of Chapter III, and when $q = t$ they are the Schur functions of Chapter I. They also include, as a limiting case, Jack's symmetric functions depending on a parameter α. Many of the properties of the Schur functions generalize to these two-parameter symmetric functions, but the proofs (at present) are usually more elaborate.

Finally, Chapter VII (which was originally intended as an appendix to Chapter VI, but outgrew that format) is devoted to a study of the zonal polynomials, long familiar to statisticians. From one point of view, they are a special case of Jack's symmetric functions (the parameter α being equal to 2), but their combinatorial and group-theoretic connections make them worthy of study in their own right. This chapter can be read independently of Chapter VI.

London, 1995 I. G. M.

Preface to the first edition

This monograph is the belated fulfilment of an undertaking made some years ago to publish a self-contained account of Hall polynomials and related topics.

These polynomials were defined by Philip Hall in the 1950s, originally as follows. If M is a finite abelian p-group, it is a direct sum of cyclic subgroups, of orders $p^{\lambda_1}, p^{\lambda_2}, \ldots, p^{\lambda_r}$ say, where we may suppose that $\lambda_1 \geqslant \lambda_2 \geqslant \ldots \geqslant \lambda_r$. The sequence of exponents $\lambda = (\lambda_1, \ldots, \lambda_r)$ is a partition, called the *type* of M, which describes M up to isomorphism. If now μ and ν are partitions, let $g^{\lambda}_{\mu\nu}(p)$ denote the number of subgroups N of M such that N has type μ and M/N has type ν. Hall showed that $g^{\lambda}_{\mu\nu}(p)$ is a polynomial function of p, with integer coefficients, and was able to determine its degree and leading coefficient. These polynomials are the Hall polynomials.

More generally, in place of finite abelian p-groups we may consider modules of finite length over a discrete valuation ring $\mathfrak{o}$ with finite residue field: in place of $g^{\lambda}_{\mu\nu}(p)$ we have $g^{\lambda}_{\mu\nu}(q)$ where q is the number of elements in the residue field.

Next, Hall used these polynomials to construct an algebra which reflects the lattice structure of the finite $\mathfrak{o}$-modules. Let $H(q)$ be a free $\mathbf{Z}$-module with basis (u_λ) indexed by the set of all partitions λ, and define a multiplication in $H(q)$ by using the $g^{\lambda}_{\mu\nu}(q)$ as structure constants, i.e.

$$u_\mu u_\nu = \sum_\lambda g^{\lambda}_{\mu\nu}(q) u_\lambda.$$

It is not difficult to show (see Chapter II for the details) that $H(q)$ is a commutative, associative ring with identity, which is freely generated (as $\mathbf{Z}$-algebra) by the generators $u_{(1^r)}$ corresponding to the elementary $\mathfrak{o}$-modules.

Symmetric functions now come into the picture in the following way. The ring of symmetric polynomials in n independent variables is a polynomial ring $\mathbf{Z}[e_1, \ldots, e_n]$ generated by the elementary symmetric functions $e_1, \ldots, e_n$. By passing to the limit with respect to n, we obtain a ring $\Lambda = \mathbf{Z}[e_1, e_2, \ldots]$ of symmetric functions in infinitely many variables. We might therefore map $H(q)$ isomorphically onto Λ by sending each generator $u_{(1^r)}$ to the elementary symmetric function e_r. However, it turns out that a better choice is to define a homomorphism $\psi\colon H(q) \to \Lambda \otimes \mathbf{Q}$ by $\psi(u_{(1^r)}) = q^{-r(r-1)/2} e_r$ for each $r \geqslant 1$. In this way we obtain a family of symmetric functions $\psi(u_\lambda)$, indexed by partitions. These symmetric functions are essentially the Hall–Littlewood functions, which are the subject of Chapter III. Thus the combinatorial lattice properties of finite $\mathfrak{o}$-modules are reflected in the multiplication of Hall–Littlewood functions.

The formalism of symmetric functions therefore underlies Hall's theory, and Chapter I is an account of this formalism—the various types of symmetric functions, especially the Schur functions (S-functions), and the relations between them. The character theory of the symmetric groups, as originally developed by Frobenius, enters naturally in this context. In an appendix we show how the S-functions arise 'in nature' as the traces of polynomial functors on the category of finite-dimensional vector spaces over a field of characteristic 0.

In the past few years, the combinatorial substructure, based on the 'jeu de taquin', which underlies the formalism of S-functions and in particular the Littlewood–Richardson rule (Chapter I, §9), has become much better understood. I have not included an account of this, partly from a desire to keep the size of this monograph within reasonable bounds, but also because Schützenberger, the main architect of this theory, has recently published a complete exposition [S7].

The properties of the Hall polynomials and the Hall algebra are developed in Chapter II, and of the Hall–Littlewood symmetric functions in Chapter III. These are symmetric functions involving a parameter t, which reduce to S-functions when $t = 0$ and to monomial symmetric functions when $t = 1$. Many of their properties generalize known properties of S-functions.

Finally, Chapters IV and V apply the formalism developed in the previous chapters. Chapter IV is an account of J. A. Green's work [G11] on the characters of the general linear groups over a finite field, and we have sought to bring out, as in the case of the character theory of the symmetric groups, the role played by symmetric functions. Chapter V is also about general linear groups, but this time over a non-archimedean local field rather than a finite field, and instead of computing characters we compute spherical functions. In both these contexts Hall's theory plays a decisive part.

Queen Mary College,
London 1979

I. G. M.

Contents

I. SYMMETRIC FUNCTIONS

1. Partitions 1
2. The ring of symmetric functions 17
3. Schur functions 40
4. Orthogonality 62
5. Skew Schur functions 69
6. Transition matrices 99
7. The characters of the symmetric groups 112
8. Plethysm 135
9. The Littlewood–Richardson rule 142

Appendix A: Polynomial functors and polynomial representations 149
Appendix B: Characters of wreath products 169

II. HALL POLYNOMIALS

1. Finite $\mathfrak{o}$-modules 179
2. The Hall algebra 182
3. The *LR*-sequence of a submodule 184
4. The Hall polynomial 187

Appendix (by A. Zelevinsky): Another proof of Hall's theorem 199

III. HALL–LITTLEWOOD SYMMETRIC FUNCTIONS

1. The symmetric polynomials R_λ 204
2. Hall–Littlewood functions 208
3. The Hall algebra again 215
4. Orthogonality 222
5. Skew Hall–Littlewood functions 226
6. Transition matrices 238
7. Green's polynomials 246
8. Schur's Q-functions 250

IV. THE CHARACTERS OF GL_n OVER A FINITE FIELD

1. The groups L and M 269
2. Conjugacy classes 270
3. Induction from parabolic subgroups 273
4. The characteristic map 276
5. Construction of the characters 280

6. The irreducible characters 284
Appendix: proof of (5.1) 291

V. THE HECKE RING OF GL_n OVER A LOCAL FIELD

1. Local fields 292
2. The Hecke ring $H(G,K)$ 293
3. Spherical functions 298
4. Hecke series and zeta functions for $GL_n(F)$ 300
5. Hecke series and zeta functions for $GSp_{2n}(F)$ 302

VI. SYMMETRIC FUNCTIONS WITH TWO PARAMETERS

1. Introduction 305
2. Orthogonality 309
3. The operators D_n^r 315
4. The symmetric functions $P_\lambda(x;q,t)$ 321
5. Duality 327
6. Pieri formulas 331
7. The skew functions $P_{\lambda/\mu}$, $Q_{\lambda/\mu}$ 343
8. Integral forms 352
9. Another scalar product 368
10. Jack's symmetric functions 376

VII. ZONAL POLYNOMIALS

1. Gelfand pairs and zonal spherical functions 388
2. The Gelfand pair (S_{2n}, H_n) 401
3. The Gelfand pair $(GL_n(\mathbf{R}), O(n))$ 414
4. Integral formulas 424
5. The complex case 440
6. The quaternionic case 446

BIBLIOGRAPHY 457

NOTATION 467

INDEX 473

I

SYMMETRIC FUNCTIONS

1. Partitions

Many of the objects we shall consider in this book will turn out to be parametrized by partitions. The purpose of this section is to lay down some notation and terminology which will be used throughout, and to collect together some elementary results on orderings of partitions which will be used later.

Partitions

A *partition* is any (finite or infinite) sequence

$$\lambda = (\lambda_1, \lambda_2, \ldots, \lambda_r, \ldots) \tag{1.1}$$

of non-negative integers in decreasing order:

$$\lambda_1 \geqslant \lambda_2 \geqslant \ldots \geqslant \lambda_r \geqslant \ldots$$

and containing only finitely many non-zero terms. We shall find it convenient not to distinguish between two such sequences which differ only by a string of zeros at the end. Thus, for example, we regard $(2,1)$, $(2,1,0)$, $(2,1,0,0,\ldots)$ as the same partition.

The non-zero λ_i in (1.1) are called the *parts* of λ. The number of parts is the *length* of λ, denoted by $l(\lambda)$; and the sum of the parts is the *weight* of λ, denoted by $|\lambda|$:

$$|\lambda| = \lambda_1 + \lambda_2 + \ldots.$$

If $|\lambda| = n$ we say that λ is a *partition of* n. The set of all partitions of n is denoted by $\mathscr{P}_n$, and the set of all partitions by $\mathscr{P}$. In particular, $\mathscr{P}_0$ consists of a single element, the unique partition of zero, which we denote by 0.

Sometimes it is convenient to use a notation which indicates the number of times each integer occurs as a part:

$$\lambda = (1^{m_1} 2^{m_2} \ldots r^{m_r} \ldots)$$

means that exactly m_i of the parts of λ are equal to i. The number

$$m_i = m_i(\lambda) = \mathrm{Card}\{j \colon \lambda_j = i\} \tag{1.2}$$

is called the *multiplicity* of i in λ.

Diagrams

The *diagram* of a partition λ may be formally defined as the set of points $(i,j) \in \mathbf{Z}^2$ such that $1 \leqslant j \leqslant \lambda_i$. In drawing such diagrams we shall adopt the convention, as with matrices, that the first coordinate i (the row index) increases as one goes downwards, and the second coordinate j (the column index) increases as one goes from left to right.† For example, the diagram of the partition (5441) is

consisting of 5 points or nodes in the top row, 4 in the second row, 4 in the third row, and 1 in the fourth row. More often it is convenient to replace the nodes by squares, in which case the diagram is

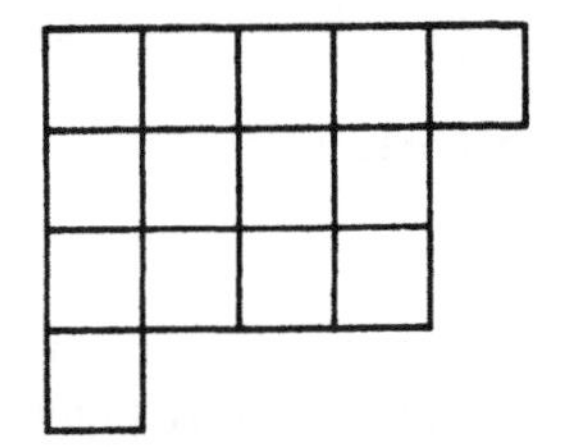

We shall usually denote the diagram of a partition λ by the same symbol λ.

The *conjugate* of a partition λ is the partition λ' whose diagram is the transpose of the diagram λ, i.e. the diagram obtained by reflection in the main diagonal. Hence λ'_i is the number of nodes in the ith column of λ, or equivalently

(1.3) $$\lambda'_i = \operatorname{Card}\{j : \lambda_j \geqslant i\}.$$

In particular, $\lambda'_1 = l(\lambda)$ and $\lambda_1 = l(\lambda')$. Obviously $\lambda'' = \lambda$.

For example, the conjugate of (5441) is (43331).

From (1.2) and (1.3) we have

(1.4) $$m_i(\lambda) = \lambda'_i - \lambda'_{i+1}.$$

† Some authors (especially Francophones) prefer the convention of coordinate geometry (in which the first coordinate increases from left to right and the second coordinate from bottom to top) and define the diagram of λ to be the set of $(i,j) \in \mathbf{Z}^2$ such that $1 \leqslant i \leqslant \lambda_j$. Readers who prefer this convention should read this book upside down in a mirror.

For each partition λ we define

(1.5) $$n(\lambda) = \sum_{i \geqslant 1} (i-1)\lambda_i,$$

so that n(λ) is the sum of the numbers obtained by attaching a zero to node in the top row of the diagram of λ, a 1 to each node in the second row, and so on. Adding up the numbers in each column, we see that

(1.6) $$n(\lambda) = \sum_{i \geqslant 1} \binom{\lambda'_i}{2}.$$

Another notation for partitions which is occasionally useful is the following, due to Frobenius. Suppose that the main diagonal of the diagram of λ consists of r nodes (i,i) $(1 \leqslant i \leqslant r)$. Let $\alpha_i = \lambda_i - i$ be the number of nodes in the ith row of λ to the right of (i,i), for $1 \leqslant i \leqslant r$, and let $\beta_i = \lambda'_i - i$ be the number of nodes in the ith column of λ below (i,i), for $1 \leqslant i \leqslant r$. We have $\alpha_1 > \alpha_2 > \ldots > \alpha_r \geqslant 0$ and $\beta_1 > \beta_2 > \ldots > \beta_r \geqslant 0$, and we denote the partition λ by

$$\lambda = (\alpha_1, \ldots, \alpha_r \mid \beta_1, \ldots, \beta_r) = (\alpha \mid \beta).$$

Clearly the conjugate of $(\alpha \mid \beta)$ is $(\beta \mid \alpha)$.

For example, if $\lambda = (5441)$ we have $\alpha = (421)$ and $\beta = (310)$.

(1.7) *Let λ be a partition and let $m \geqslant \lambda_1$, $n \geqslant \lambda'_1$. Then the $m+n$ numbers*

$$\lambda_i + n - i \quad (1 \leqslant i \leqslant n), \qquad n - 1 + j - \lambda'_j \quad (1 \leqslant j \leqslant m)$$

are a permutation of $\{0, 1, 2, \ldots, m+n-1\}$.

Proof. The diagram of λ is contained in the diagram of (m^n), which is an $n \times m$ rectangle. Number the successive segments of the boundary line between λ and its complement in (m^n) (marked thickly in the picture) with the numbers $0, 1, \ldots, m+n-1$, starting at the bottom. The numbers attached to the vertical segments are $\lambda_i + n - i$ $(1 \leqslant i \leqslant n)$, and by transposition those attached to the horizontal segments are

$$(m+n-1) - (\lambda'_j + m - j) = n - 1 + j - \lambda'_j \qquad (1 \leqslant j \leqslant m). \quad \blacksquare$$

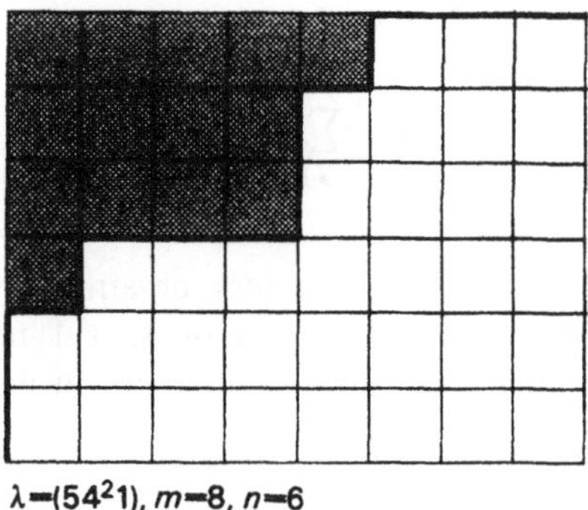

$\lambda=(54^21)$, $m=8$, $n=6$

Let

$$f_{\lambda,n}(t)=\sum_{i=1}^{n} t^{\lambda_i+n-i}.$$

Then (1.7) is equivalent to the identity

(1.7′) $$f_{\lambda,n}(t)+t^{m+n-1}f_{\lambda',m}(t^{-1})=(1-t^{m+n})/(1-t).$$

Skew diagrams and tableaux

If λ, μ are partitions, we shall write $\lambda \supset \mu$ to mean that the diagram of λ contains the diagram of μ, i.e. that $\lambda_i \geqslant \mu_i$ for all $i \geqslant 1$. The set-theoretic difference $\theta = \lambda - \mu$ is called a *skew diagram*. For example, if $\lambda = (5441)$ and $\mu = (432)$, the skew diagram $\lambda - \mu$ is the shaded region in the picture below:

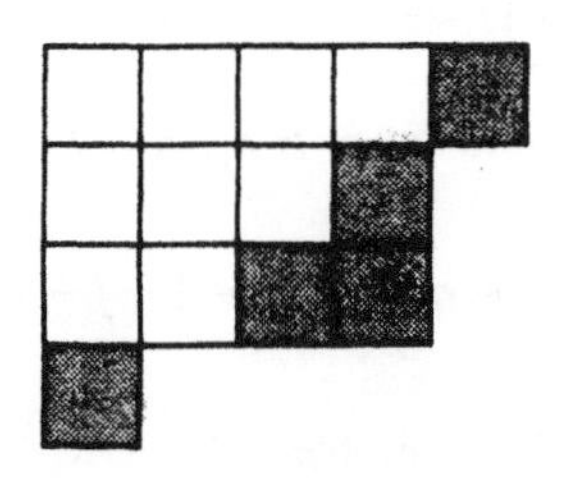

A *path* in a skew diagram θ is a sequence $x_0, x_1, \ldots, x_m$ of squares in θ such that x_{i-1} and x_i have a common side, for $1 \leqslant i \leqslant m$. A subset φ of θ is said to be *connected* if any two squares in φ can be connected by a path in φ. The maximal connected subsets of θ are themselves skew diagrams, called the *connected components* of θ. In the example above, there are three connected components.

The *conjugate* of a skew diagram $\theta = \lambda - \mu$ is $\theta' = \lambda' - \mu'$. Let $\theta_i = \lambda_i - \mu_i$, $\theta_i' = \lambda_i' - \mu_i'$, and

$$|\theta| = \sum \theta_i = |\lambda| - |\mu|.$$

A skew diagram θ is a *horizontal m-strip* (resp. a *vertical m-strip*) if $|\theta| = m$ and $\theta'_i \leqslant 1$ (resp. $\theta_i \leqslant 1$) for each $i \geqslant 1$. In other words, a horizontal (resp. vertical) strip has at most one square in each column (resp. row).

If $\theta = \lambda - \mu$, a necessary and sufficient condition for θ to be a horizontal strip is that the sequences λ and μ are interlaced, in the sense that $\lambda_1 \geqslant \mu_1 \geqslant \lambda_2 \geqslant \mu_2 \geqslant \ldots$.

A skew diagram θ is a *border strip* (also called a *skew hook* by some authors, and a *ribbon* by others) if θ is connected and contains no 2×2 block of squares, so that successive rows (or columns) of θ overlap by exactly one square. The *length* of a border strip θ is the total number $|\theta|$ of squares it contains, and its *height* is defined to be one less than the number of rows it occupies. If we think of a border strip θ as a set of nodes rather than squares, then by joining contiguous nodes by horizontal or vertical line segments of unit length we obtain a sort of staircase, and the height of θ is the number of vertical line segments or 'risers' in the staircase.

If $\lambda = (\alpha_1, \ldots, \alpha_r \mid \beta_1, \ldots, \beta_r)$ and $\mu = (\alpha_2, \ldots, \alpha_r \mid \beta_2, \ldots, \beta_r)$, then $\lambda - \mu$ is a border strip, called the *border* (or *rim*) of λ.

A (column-strict) *tableau* T is a sequence of partitions

$$\mu = \lambda^{(0)} \subset \lambda^{(1)} \subset \ldots \subset \lambda^{(r)} = \lambda$$

such that each skew diagram $\theta^{(i)} = \lambda^{(i)} - \lambda^{(i-1)}$ $(1 \leqslant i \leqslant r)$ is a horizontal strip. Graphically, T may be described by numbering each square of the skew diagram $\theta^{(i)}$ with the number i, for $1 \leqslant i \leqslant r$, and we shall often think of a tableau as a numbered skew diagram in this way. The numbers inserted in $\lambda - \mu$ must increase strictly down each column (which explains the adjective 'column-strict') and weakly from left to right along each row. The skew diagram $\lambda - \mu$ is called the *shape* of the tableau T, and the sequence $(|\theta^{(1)}|, \ldots, |\theta^{(r)}|)$ is the *weight* of T.

We might also define row-strict tableaux by requiring strict increase along the rows and weak increase down the columns, but we shall have no use for them; and throughout this book a *tableau* (unqualified) will mean a column-strict tableau, as defined above.

A *standard tableau* is a tableau T which contains each number $1, 2, \ldots, r$ exactly once, so that its weight is $(1, 1, \ldots, 1)$.

Addition and multiplication of partitions

Let λ, μ be partitions. We define $\lambda + \mu$ to be the sum of the sequences λ and μ:

$$(\lambda + \mu)_i = \lambda_i + \mu_i.$$

Also we define $\lambda \cup \mu$ to be the partition whose parts are those of λ and μ, arranged in descending order. For example, if $\lambda = (321)$ and $\mu = (22)$, then $\lambda + \mu = (541)$ and $\lambda \cup \mu = (32221)$.

Next, we define $\lambda\mu$ to be the componentwise product of the sequences λ, μ:

$$(\lambda\mu)_i = \lambda_i \mu_i.$$

Also we define $\lambda \times \mu$ to be the partition whose parts are $\min(\lambda_i, \mu_j)$ for all $i \leqslant l(\lambda)$ and $j \leqslant l(\mu)$, arranged in descending order.

The operations $+$ and $\cup$ are dual to each other, and so are the two multiplications:

$$(1.8) \qquad \begin{aligned} (\lambda \cup \mu)' &= \lambda' + \mu', \\ (\lambda \times \mu)' &= \lambda'\mu'. \end{aligned}$$

Proof. The diagram of $\lambda \cup \mu$ is obtained by taking the rows of the diagrams of λ and μ and reassembling them in order of decreasing length. Hence the length of the kth column of $\lambda \cup \mu$ is the sum of the lengths of the kth columns of λ and of μ, i.e. $(\lambda \cup \mu)'_k = \lambda'_k + \mu'_k$.

Next, the length of the kth column of $\lambda \times \mu$ is equal to the number of pairs (i, j) such that $\lambda_i \geqslant k$ and $\mu_j \geqslant k$, hence it is equal to $\lambda'_k \mu'_k$. Consequently $(\lambda \times \mu)'_k = \lambda'_k \mu'_k$. ∎

Orderings

Let L_n denote the *reverse lexicographic ordering* on the set $\mathscr{P}_n$ of partitions of n: that is to say, L_n is the subset of $\mathscr{P}_n \times \mathscr{P}_n$ consisting of all (λ, μ) such that either $\lambda = \mu$ or the first non-vanishing difference $\lambda_i - \mu_i$ is *positive*. L_n is a total ordering. For example, when $n = 5$, L_5 arranges $\mathscr{P}_5$ in the sequence

$$(5), (41), (32), (31^2), (2^21), (21^3), (1^5).$$

Another total ordering on $\mathscr{P}_n$ is L'_n, the set of all (λ, μ) such that either $\lambda = \mu$ or else the first non-vanishing difference $\lambda^*_i - \mu^*_i$ is *negative*, where $\lambda^*_i = \lambda_{n+1-i}$. The orderings L_n, L'_n are distinct as soon as $n \geqslant 6$. For example, if $\lambda = (31^3)$ and $\mu = (2^3)$ we have $(\lambda, \mu) \in L_6$ and $(\mu, \lambda) \in L'_6$.

(1.9) *Let $\lambda, \mu \in \mathscr{P}_n$. Then*

$$(\lambda, \mu) \in L'_n \Leftrightarrow (\mu', \lambda') \in L_n.$$

Proof. Suppose that $(\lambda, \mu) \in L'_n$ and $\lambda \neq \mu$. Then for some integer $i \geqslant 1$ we have $\lambda_i < \mu_i$, and $\lambda_j = \mu_j$ for $j > i$. If we put $k = \lambda_i$ and consider the diagrams of λ and μ, we see immediately that $\lambda'_j = \mu'_j$ for $1 \leqslant j \leqslant k$, and that $\lambda'_{k+1} < \mu'_{k+1}$, so that $(\mu', \lambda') \in L_n$. The converse is proved similarly. ∎

An ordering which is more important than either L_n or L'_n is the *natural* (partial) *ordering* N_n on $\mathscr{P}_n$ (also called the *dominance* partial ordering by some authors), which is defined as follows:

$$(\lambda, \mu) \in N_n \Leftrightarrow \lambda_1 + \ldots + \lambda_i \geqslant \mu_1 + \ldots + \mu_i \quad \text{for all } i \geqslant 1.$$

As soon as $n \geqslant 6$, N_n is not a total ordering. For example, the partitions (31^3) and (2^3) are incomparable with respect to N_6.

We shall write $\lambda \geqslant \mu$ in place of $(\lambda, \mu) \in N_n$.

(1.10) *Let* $\lambda, \mu \in \mathscr{P}_n$. *Then*

$$\lambda \geqslant \mu \Rightarrow (\lambda, \mu) \in L_n \cap L'_n.$$

Proof. Suppose that $\lambda \geqslant \mu$. Then either $\lambda_1 > \mu_1$, in which case $(\lambda, \mu) \in L_n$, or else $\lambda_1 = \mu_1$. In that case either $\lambda_2 > \mu_2$, in which case again $(\lambda, \mu) \in L_n$, or else $\lambda_2 = \mu_2$. Continuing in this way, we see that $(\lambda, \mu) \in L_n$.

Also, for each $i \geqslant 1$, we have

$$\begin{aligned} \lambda_{i+1} + \lambda_{i+2} + \ldots &= n - (\lambda_1 + \ldots + \lambda_i) \\ &\leqslant n - (\mu_1 + \ldots + \mu_i) \\ &= \mu_{i+1} + \mu_{i+2} + \ldots \end{aligned}$$

Hence the same reasoning as before shows that $(\lambda, \mu) \in L'_n$. |

Remark. It is not true in general that $N_n = L_n \cap L'_n$. For example, when $n = 12$ and $\lambda = (63^2)$, $\mu = (5^2 1^2)$ we have $(\lambda, \mu) \in L_{12} \cap L'_{12}$, but $(\lambda, \mu) \notin N_{12}$.

(1.11) *Let* $\lambda, \mu \in \mathscr{P}_n$. *Then*

$$\lambda \geqslant \mu \Leftrightarrow \mu' \geqslant \lambda'.$$

Proof. Clearly it is enough to prove one implication. Suppose then that $\mu' \not\geqslant \lambda'$. Then for some $i \geqslant 1$ we have

$$\lambda'_1 + \ldots + \lambda'_j \leqslant \mu'_1 + \ldots + \mu'_j \qquad (1 \leqslant j \leqslant i-1)$$

and

(1) $$\lambda'_1 + \ldots + \lambda'_i > \mu'_1 + \ldots + \mu'_i$$

from which it follows that $\lambda'_i > \mu'_i$.

Let $l = \lambda'_i$, $m = \mu'_i$. From (1) it follows that

(2) $$\lambda'_{i+1} + \lambda'_{i+2} + \ldots < \mu'_{i+1} + \mu'_{i+2} + \ldots$$

Now $\lambda'_{i+1}+\lambda'_{i+2}+\ldots$ is equal to the number of nodes in the diagram of λ which lie to the right of the ith column, and therefore

$$\lambda'_{i+1}+\lambda'_{i+2}+\ldots=\sum_{j=1}^{l}(\lambda_j-i).$$

Likewise

$$\mu'_{i+1}+\mu'_{i+2}+\ldots=\sum_{j=1}^{m}(\mu_j-i).$$

Hence from (2) we have

$$(3)\qquad \sum_{j=1}^{m}(\mu_j-i)>\sum_{j=1}^{l}(\lambda_j-i)\geqslant\sum_{j=1}^{m}(\lambda_j-i)$$

in which the right-hand inequality holds because $l>m$ and $\lambda_j\geqslant i$ for $1\leqslant j\leqslant l$. From (3) we have

$$\mu_1+\ldots+\mu_m>\lambda_1+\ldots+\lambda_m$$

and therefore $\lambda\ngeqslant\mu$. |

Raising operators

In this subsection we shall work not with partitions but with integer vectors $a=(a_1,\ldots,a_n)\in\mathbf{Z}^n$. The symmetric group S_n acts on $\mathbf{Z}^n$ by permuting the coordinates, and the set

$$P_n=\{b\in\mathbf{Z}^n:b_1\geqslant b_2\geqslant\ldots\geqslant b_n\}$$

is a fundamental domain for this action, i.e. the S_n-orbit of each $a\in\mathbf{Z}^n$ meets P_n in exactly one point, which we denote by a^+. Thus a^+ is obtained by rearranging $a_1,\ldots,a_n$ in descending order of magnitude.

For $a,b\in\mathbf{Z}^n$ we define $a\geqslant b$ as before to mean

$$a_1+\ldots+a_i\geqslant b_1+\ldots+b_i\qquad(1\leqslant i\leqslant n).$$

(1.12) *Let* $a\in\mathbf{Z}^n$. *Then*

$$a\in P_n\Leftrightarrow a\geqslant wa\quad\textit{for all } w\in S_n.$$

Proof. Suppose that $a\in P_n$, i.e. $a_1\geqslant\ldots\geqslant a_n$. If $wa=b$, then $(b_1,\ldots,b_n)$ is a permutation of $(a_1,\ldots,a_n)$, and therefore

$$a_1+\ldots+a_i\geqslant b_1+\ldots+b_i\qquad(1\leqslant i\leqslant n)$$

so that $a\geqslant b$.

Conversely, if $a \geqslant wa$ for all $w \in S_n$ we have in particular

$$(a_1, \dots, a_n) \geqslant (a_1, \dots, a_{i-1}, a_{i+1}, a_i, a_{i+2}, \dots, a_n)$$

for $1 \leqslant i \leqslant n-1$, from which it follows that

$$a_1 + \dots + a_{i-1} + a_i \geqslant a_1 + \dots + a_{i-1} + a_{i+1},$$

i.e. $a_i \geqslant a_{i+1}$. Hence $a \in P_n$. |

Let $\delta = (n-1, n-2, \dots, 1, 0) \in P_n$.

(1.13) *Let $a \in P_n$. Then for each $w \in S_n$ we have $(a + \delta - w\delta)^+ \geqslant a$.*

Proof. Since $\delta \in P_n$ we have $\delta \geqslant w\delta$ by (1.12), hence

$$a + \delta - w\delta \geqslant a.$$

Let $b = (a + \delta - w\delta)^+$. Then again by (1.12) we have

$$b \geqslant a + \delta - w\delta.$$

Hence $b \geqslant a$. |

For each pair of integers i, j such that $1 \leqslant i < j \leqslant n$ define $R_{ij}: \mathbf{Z}^n \to \mathbf{Z}^n$ by

$$R_{ij}(a) = (a_1, \dots, a_i + 1, \dots, a_j - 1, \dots, a_n).$$

Any product $R = \prod_{i<j} R_{ij}^{r_{ij}}$ is called a *raising operator*. The order of the terms in the product is immaterial, since they commute with each other.

(1.14) *Let $a \in \mathbf{Z}^n$ and let R be a raising operator. Then*

$$Ra \geqslant a.$$

For we may assume that $R = R_{ij}$, in which case the result is obvious.

Conversely:

(1.15) *Let $a, b \in \mathbf{Z}^n$ be such that $a \leqslant b$ and $a_1 + \dots + a_n = b_1 + \dots + b_n$. Then there exists a raising operator R such that $b = Ra$.*

Proof. We may take

$$R = \prod_{k=1}^{n-1} R_{k,k+1}^{r_k}$$

where

$$r_k = \sum_{i=1}^{k} (b_i - a_i) \geqslant 0. \quad |$$

(1.16) *If λ, μ are partitions of n such that $\lambda > \mu$, and if λ, μ are adjacent for the natural ordering (so that $\lambda \geqslant \nu \geqslant \mu$ implies either $\nu = \lambda$ or $\nu = \mu$), then $\lambda = R_{ij}\mu$ for some $i < j$.*

Proof. Suppose first that $\lambda_1 > \mu_1$, and let $i \geqslant 2$ be the least integer for which $\lambda_1 + \ldots + \lambda_i = \mu_1 + \ldots + \mu_i$. Then we have $\mu_i > \lambda_i \geqslant \lambda_{i+1} \geqslant \mu_{i+1}$, so that $\mu_i > \mu_{i+1}$. Consequently $\nu = R_{1i}\mu$ is a partition, and one sees immediately that $\lambda \geqslant \nu$. Hence $\lambda = \nu = R_{1i}\mu$.

If now $\lambda_1 = \mu_1$, then for some $j > 1$ we have $\lambda_k = \mu_k$ for $k < j$ and $\lambda_j > \mu_j$. The preceding argument may now be applied to the partitions $(\lambda_j, \lambda_{j+1}, \ldots)$ and $(\mu_j, \mu_{j+1}, \ldots)$. |

Remark. This proposition leads directly to an alternative proof of (1.11); for it shows that it is enough to prove (1.11) in the case that $\lambda = R_{ij}\mu$, in which case it is obvious.

Examples

1. Let λ be a partition. The *hook-length* of λ at $x = (i,j) \in \lambda$ is defined to be

$$h(x) = h(i,j) = \lambda_i + \lambda'_j - i - j + 1.$$

From (1.7'), with λ and λ' interchanged, and $m = \lambda_1$, we have

$$\sum_{j=1}^{\lambda_1} t^{\lambda'_j + \lambda_1 - j} + \sum_{j=1}^{n} t^{\lambda_1 - 1 + j - \lambda_j} = \sum_{j=0}^{\lambda_1 + n - 1} t^j$$

or, putting $\mu_i = \lambda_i + n - i$ $(1 \leqslant i \leqslant n)$,

$$\sum_{j=1}^{\lambda_1} t^{h(1,j)} + \sum_{j=2}^{n} t^{\mu_1 - \mu_j} = \sum_{j=1}^{\mu_1} t^j. \tag{1}$$

By writing down this identity for the partition $(\lambda_i, \lambda_{i+1}, \ldots)$ and then summing over $i = 1, 2, \ldots, l(\lambda)$ we obtain

$$\sum_{x \in \lambda} t^{h(x)} + \sum_{i<j} t^{\mu_i - \mu_j} = \sum_{i \geqslant 1} \sum_{j=1}^{\mu_i} t^j. \tag{2}$$

From (2) it follows that

$$\prod_{x \in \lambda} (1 - t^{h(x)}) = \frac{\prod_{i \geqslant 1} \prod_{j=1}^{\mu_i} (1 - t^j)}{\prod_{i<j} (1 - t^{\mu_i - \mu_j})} \tag{3}$$

and in particular, by dividing both sides of (3) by $(1-t)^{|\lambda|}$ and then setting $t=1$, that

$$\prod_{x\in\lambda} h(x) = \frac{\prod_{i\geqslant 1}\mu_i!}{\prod_{i<j}(\mu_i-\mu_j)}. \tag{4}$$

2. The sum of the hook-lengths of λ is

$$\sum_{x\in\lambda} h(x) = n(\lambda)+n(\lambda')+|\lambda|.$$

3. For each $x=(i,j)\in\lambda$, the *content* of x is defined to be $c(x)=j-i$. We have

$$\sum_{x\in\lambda} c(x) = n(\lambda')-n(\lambda).$$

If n is any integer $\geqslant l(\lambda)$, the numbers $n+c(x)$ for x in the ith row of λ are $n-i+1,\dots,n-i+\lambda_i$, and therefore

$$\prod_{x\in\lambda}(1-t^{n+c(x)}) = \prod_{i\geqslant 1}\frac{\varphi_{\lambda_i+n-i}(t)}{\varphi_{n-i}(t)}$$

where $\varphi_r(t)=(1-t)(1-t^2)\dots(1-t^r)$.

4. If $\lambda=(\lambda_1,\dots,\lambda_n)=(\alpha_1,\dots,\alpha_r\mid\beta_1,\dots,\beta_r)$ in Frobenius notation, then

$$\sum_{i=1}^{n} t^i(1-t^{-\lambda_i}) = \sum_{j=1}^{r}(t^{\beta_j+1}-t^{-\alpha_j}).$$

5. For any partition λ,

$$\sum_{x\in\lambda}(h(x)^2-c(x)^2) = |\lambda|^2.$$

6. Let λ be a partition and let r,s be positive integers. Then $\lambda_i-\lambda_{i+r}\geqslant s$ for all $i\leqslant l(\lambda)$ if and only if $\lambda'_j-\lambda'_{j+s}\leqslant r$ for all $j\leqslant l(\lambda')$.

7. The set $\mathscr{P}_n$ of partitions of n is a lattice with respect to the natural ordering: in other words, each pair λ,μ of partitions of n has a least upper bound $\sigma=\sup(\lambda,\mu)$ and a greatest lower bound $\tau=\inf(\lambda,\mu)$. (Show that τ defined by

$$\sum_{i=1}^{r}\tau_i = \min\left(\sum_{i=1}^{r}\lambda_i,\ \sum_{i=1}^{r}\mu_i\right)$$

for all $r\geqslant 1$ is indeed a partition; this establishes the existence of $\inf(\lambda,\mu)$. Then define $\sigma=\sup(\lambda,\mu)$ by $\sigma'=\inf(\lambda',\mu')$. The example $\lambda=(31^3)$, $\mu=(2^3)$, $\sigma=(321)$ shows that it is not always true that

$$\sum_{i=1}^{r}\sigma_i = \max\left(\sum_{i=1}^{r}\lambda_i,\ \sum_{i=1}^{r}\mu_i\right).\Bigg)$$

8. Let p be an integer $\geqslant 2$.

(a) Let λ, μ be partitions of length $\leqslant m$ such that $\lambda \supset \mu$, and such that $\lambda - \mu$ is a border strip of length p. Let $\delta_m = (m-1, m-2, \ldots, 1, 0)$ and let $\xi = \lambda + \delta_m$, $\eta = \mu + \delta_m$. Show that η is obtained from ξ by subtracting p from some part ξ_i of ξ and rearranging in descending order. (Consider the diagrams of ξ and η.)

(b) With the same notation, suppose that ξ has m_r parts ξ_i congruent to r modulo p, for each $r = 0, 1, \ldots, p-1$. These ξ_i may be written in the form $p\xi_k^{(r)} + r$ $(1 \leqslant k \leqslant m_r)$, where $\xi_1^{(r)} > \xi_2^{(r)} > \ldots > \xi_{m_r}^{(r)} \geqslant 0$. Let $\lambda_k^{(r)} = \xi_k^{(r)} - m_r + k$, so that $\lambda^{(r)} = (\lambda_1^{(r)}, \ldots, \lambda_{m_r}^{(r)})$ is a partition. The collection $\lambda^* = (\lambda^{(0)}, \lambda^{(1)}, \ldots, \lambda^{(p-1)})$ is called the *p-quotient* of the partition λ. The effect of changing $m \geqslant l(\lambda)$ is to permute the $\lambda^{(r)}$ cyclically, so that λ^* should perhaps be thought of as a 'necklace' of partitions.

The m numbers $ps + r$, where $0 \leqslant s \leqslant m_r - 1$ and $0 \leqslant r \leqslant p - 1$, are all distinct. Let us arrange them in descending order, say $\tilde{\xi}_1 > \ldots > \tilde{\xi}_m$, and define a partition $\tilde{\lambda}$ by $\tilde{\lambda}_i = \tilde{\xi}_i - m + i$ $(1 \leqslant i \leqslant m)$. This partition $\tilde{\lambda}$ is called the *p-core* (or *p-residue*) of λ. Both $\tilde{\lambda}$ and λ^* (up to cyclic permutation) are independent of m, provided that $m \geqslant l(\lambda)$.

If $\lambda = \tilde{\lambda}$ (i.e. if λ^* is empty), the partition λ is called a *p-core*. For example, the only 2-cores are the 'staircase' partitions $\delta_m = (m-1, m-2, \ldots, 1)$.

Following G. D. James, we may conveniently visualize this construction in terms of an abacus. The runners of the abacus are the half-lines $x \geqslant 0$, $y = r$ in the plane $\mathbf{R}^2$, where $r = 0, 1, 2, \ldots, p-1$, and λ is represented by the set of beads at the points with coordinates $(\xi_k^{(r)}, r)$ in the notation used above. The removal of a border strip of length p from λ is recorded on the abacus by moving some bead one unit to the left on its runner, and hence the passage from λ to its p-core corresponds to moving all the beads on the abacus as far left as they will go.

This arithmetical construction of the p-quotient and p-core is an analogue for partitions of the division algorithm for integers (to which it reduces if the partition has only one part).

(c) The p-core of a partition λ may be obtained graphically as follows. Remove a border strip of length p from the diagram of λ in such a way that what remains is the diagram of a partition, and continue removing border strips of length p in this way as long as possible. What remains at the end of this process is the p-core $\tilde{\lambda}$ of λ, and it is independent of the sequence of border strips removed. For by (a) above, the removal of a border strip of length p from λ corresponds to subtracting p from some part of ξ and then rearranging the resulting sequence in descending order; the only restriction is that the resulting set of numbers should be all distinct and non-negative.

(d) The p-quotient of λ can also be read off from the diagram of λ, as follows. For $s, t = 0, 1, \ldots, p-1$ let

$$R_s = \{(i,j) \in \lambda \colon \lambda_i - i \equiv s (\operatorname{mod} p)\},$$

$$C_t = \{(i,j) \in \lambda \colon j - \lambda'_j \equiv t (\operatorname{mod} p)\},$$

so that R_s consists of the rows of λ whose right-hand node has content (Example 3) congruent to s modulo p, and likewise for C_t. If now $(i,j)\in R_s\cap C_t$, the hook-length at (i,j) is

$$h(i,j)=\lambda_i+\lambda'_j-i-j+1\equiv s-t+1(\bmod p)$$

and therefore p divides $h(i,j)$ if and only if $t\equiv s+1(\bmod p)$.

On the other hand, if $\xi_i=p\xi_k^{(r)}+r$ as in (b) above, the hook lengths of λ in the ith row are the elements of the sequence $(1,2,\dots,\xi_i)$ after deletion of $\xi_i-\xi_{i+1},\dots,\xi_i-\xi_m$. Hence those divisible by p are the elements of the sequence $(p,2p,\dots,p\xi_k^{(r)})$ after deletion of $p(\xi_k^{(r)}-\xi_{k+1}^{(r)}),\dots,p(\xi_k^{(r)}-\xi_{m_r}^{(r)})$. They are therefore p times the hook lengths in the kth row of $\lambda^{(r)}$, and in particular there are $\lambda_k^{(r)}$ of them.

It follows that each $\lambda^{(r)}$ is embedded in λ as $R_s\cap C_{s+1}$, where $s\equiv r-m(\bmod p)$, and that the hook lengths in $\lambda^{(r)}$ are those of the corresponding nodes in $R_s\cap C_{s+1}$, divided by p. In particular, if m is a multiple of p (which we may assume without loss of generality) then $\lambda^{(r)}=\lambda\cap R_r\cap C_{r+1}$ for each r (where $C_p=C_0$).

(e) From (c) and (d) it follows that the p-core (resp. p-quotient) of the conjugate partition λ' is the conjugate of the p-core (resp. p-quotient) of λ.

(f) For any two partitions λ,μ we shall write

$$\lambda\sim_p\mu$$

to mean that $\tilde{\lambda}=\tilde{\mu}$, i.e. that λ and μ have the same p-core. As above, let $\xi=\lambda+\delta_m$, $\eta=\mu+\delta_m$, where $m\geqslant\max(l(\lambda),l(\mu))$. Then it follows from (a) and (b) that $\lambda\sim_p\mu$ if and only if $\eta\equiv w\xi(\bmod p)$ for some permutation $w\in S_m$. Also, from (e) above it follows that $\lambda\sim_p\mu$ if and only if $\lambda'\sim_p\mu'$.

(g) From the definitions in (b) it follows that a partition λ is uniquely determined by its p-core $\tilde{\lambda}$ and its p-quotient λ^*, together with the integer m (modulo p). Since $|\lambda|=|\tilde{\lambda}|+p|\lambda^*|$, the generating function for partitions with a given p-core $\tilde{\lambda}$ is

$$\sum_{\tilde{\mu}=\tilde{\lambda}}t^{|\mu|}=t^{|\tilde{\lambda}|}P(t^p)^p$$

where $P(t)=\prod_{n\geqslant 1}(1-t^n)^{-1}$ is the partition generating function. Hence the generating function for p-cores is

$$\sum t^{|\tilde{\lambda}|}=P(t)/P(t^p)^p$$

$$=\prod_{n\geqslant 1}\frac{(1-t^{np})^p}{1-t^n}.$$

In particular, when $p=2$ we obtain the identity

$$(*)\qquad \sum_{m\geqslant 1}t^{m(m-1)/2}=\prod_{n\geqslant 1}\frac{1-t^{2n}}{1-t^{2n-1}}.$$

We shall leave it to the interested reader to write down the corresponding identity for $p > 2$. It turns out to be a specialization of the 'denominator formula' for the affine Lie algebra of type $A_{p-1}^{(1)}$ [K1]. Thus in particular $(*)$ is a specialization of Jacobi's triple product identity.

9. (a) A partition is *strict* if all its parts are distinct. If $\mu = (\mu_1, \dots, \mu_r)$ is a strict partition of length r (so that $\mu_1 > \mu_2 > \dots > \mu_r > 0$), the *double* of μ is the partition $\lambda = (\mu_1, \dots, \mu_r \mid \mu_1 - 1, \dots, \mu_r - 1)$ in Frobenius notation, and the diagram of λ is called the *double diagram* $D(\mu)$ of μ. The part of $D(\mu)$ that lies above the main diagonal is called the *shifted diagram* $S(\mu)$ of μ; it is obtained from the usual diagram of μ by moving the ith row $(i-1)$ squares to the right, for each $i > 1$. Thus $D(\mu)$ consists of $S(\mu)$ dovetailed into its reflection in the diagonal.

Let $m \geqslant l(\lambda) = \mu_1$, and let $\xi = \lambda + \delta_m$ where $\delta_m = (m-1, m-2, \dots, 1, 0)$ as in Example 8. The first r parts of ξ are $\mu_1 + m, \dots, \mu_r + m$, and since the partition $(\lambda_{r+1}, \dots, \lambda_m)$ is the conjugate of $(\mu_1 - r, \dots, \mu_r - 1)$, it follows from (1.7) that ξ is obtained from the sequence $(\mu_1 + m, \mu_2 + m, \dots, \mu_r + m, m-1, m-2, \dots, 1, 0)$ by deleting the numbers $m - \mu_r, \dots, m - \mu_1$. Hence ξ satisfies the following condition:

$(*)$ an integer j between 0 and $2m$ occurs in ξ if and only if $2m - j$ does not.

Conversely, if ξ satisfies this condition, then λ is the double of a strict partition.

(b) Let p be an integer $\geqslant 2$, and consider the p-quotient and p-core (Example 8) of λ. Without loss of generality we may assume that m is a multiple of p, so that $2m = (n+1)p$ with n odd. As in Example 8, suppose that for each $r = 0, 1, \dots, p-1$ the ξ_i congruent to r modulo p are $p\xi_k^{(r)} + r$ $(1 \leqslant k \leqslant m_r)$, where $\xi_1^{(r)} > \dots > \xi_{m_r}^{(r)} \geqslant 0$. Since $p\xi_k^{(r)} + r \leqslant 2m = (n+1)p$, it follows that $\xi_k^{(r)} \leqslant n$ if $r \neq 0$, and that $\xi_k^{(0)} \leqslant n+1$.

Suppose first that $r \neq 0$, and let $s = p - r$. From above, for each $k = 1, 2, \dots, m_r$ the number $2m - (p\xi_k^{(r)} + r) = p(n - \xi_k^{(r)}) + s$ does not occur in ξ. Hence the numbers $n - \xi_k^{(r)}$ $(1 \leqslant k \leqslant m_r)$ and $\xi_k^{(s)}$ $(1 \leqslant k \leqslant m_s)$ fill the interval $[0, n]$ of $\mathbf{Z}$, from which it follows first that $m_r + m_s = n + 1$, and second (by (1.7)) that the components $\lambda^{(r)}$ and $\lambda^{(s)}$ of the p-quotient of λ are conjugate partitions. In particular, if p is even, $\lambda^{(p/2)}$ is self-conjugate.

Next, if $r = 0$ we have $2m - p\xi_k^{(0)} = p(n + 1 - \xi_k^{(0)})$ and therefore the sequence $\xi^{(0)} = (\xi_1^{(0)}, \dots, \xi_{m_0}^{(0)})$ satisfies the condition $(*)$, so that $2m_0 = n + 1$ and $\lambda^{(0)}$ is the double of a strict partition.

Finally, it follows from the definition of the p-core in Example 8(b) and the fact that $m_r + m_s = n + 1$ when $r + s \equiv 0 \pmod p$ that the sequence $\tilde{\xi}$ (*loc. cit.*) satisfies the condition $(*)$, and hence the p-core $\tilde{\lambda}$ is the double of a strict partition.

(c) Let μ as before be a strict partition of length r. If $(i,j) \in \mu$ and $k \geqslant i$, an *$(i,j)_k$-bar* of μ consists of the squares $(a,b) \in \mu$ such that $a = i$ or k and $b \geqslant j$, and is defined only when the diagram obtained by the removal of these squares has no two rows of equal length; so that (i) if $k > i$ there is an $(i,j)_k$-bar only when $j = 1$, in which case it consists of the ith and kth rows of μ; and (ii) when $k = i$ there is an $(i,j)_k$-bar only when $j - 1$ is not equal to any of $\mu_{i+1}, \dots, \mu_r$. The *length* of a bar is the number of squares it contains.

Show that for each $i \geqslant 1$ the lengths of the $(i,j)_k$-bars of μ are the hook lengths in the double diagram $D(\mu)$ at the squares (i,k) with $k > i$, that is to say at the squares in the ith row of the shifted diagram $S(\mu)$.

Let ν be the strict partition whose diagram is obtained from that of μ by removing a bar of length $p > 1$ and then rearranging the rows in descending order of length. Thus ν is obtained from μ in case (i) by deleting μ_i and μ_k, where $\mu_i + \mu_k = p$, and in case (ii) by replacing μ_i by $\mu_i - p$ and rearranging. Show that the double diagram of ν is obtained from that of μ by removing two border strips of length p, one of which lies in rows $i, i+1, \dots$, and the other in columns $i, i+1, \dots$.

(d) A strict partition is a *p-bar core* if it contains no bars of length p. By starting with a strict partition μ, removing a bar of length p, rearranging the rows if necessary, then repeating the process as often as necessary, we shall end up with a strict partition $\bar{\mu}$ called the *p-bar core* of μ. It follows from above that the double of $\bar{\mu}$ is the p-core (Example 8) of the double of μ, and that $\bar{\mu}$ is independent of the sequence of moves described above to reach it.

10. For any partition λ, let

$$h(\lambda) = \prod_{x \in \lambda} h(x)$$

denote the product of the hook-lengths of λ (Example 1). With the notation of Example 8, we have

$$h(\lambda) = p^{|\lambda^*|} h(\lambda^*) h'(\lambda),$$

where $h(\lambda^*) = \prod_{r=0}^{p-1} h(\lambda^{(r)})$, and $h'(\lambda)$ is the product of the hook-lengths $h(x)$ that are not multiples of p. (Use formula (4) of Example 1.)

If, moreover, p is prime, then

$$h'(\lambda) \equiv \sigma_\lambda h(\tilde{\lambda}) \pmod p$$

where $\sigma_\lambda = \pm 1$. Hence in particular, when p is prime, λ is a p-core if and only if $h(\lambda)$ is prime to p.

11. Let λ be a partition. The *content polynomial* of λ is the polynomial

$$c_\lambda(X) = \prod_{x \in \lambda} (X + c(x))$$

(see §3, Example 4) where X is an indeterminate.

(a) Let $m \geqslant l(\lambda)$ and let $\xi_i = \lambda_i + m - i$ $(1 \leqslant i \leqslant m)$ as in Example 8. Then

$$\frac{c_\lambda(X+m)}{c_\lambda(X+m-1)} = \prod_{i=1}^{m} \frac{X+\xi_i}{X+m-i}.$$

(b) Let p be a prime number. If θ is a border strip of length p, the contents $c(x)$, $x \in \theta$ are p consecutive integers, hence are congruent modulo p to $0, 1, \dots, p-1$ in some order. If $\theta = \lambda - \mu$, it follows that

$$c_\lambda(X) \equiv c_\mu(X)(X^p - X) \pmod p.$$

Hence, for any partition λ,

$$c_\lambda(X) \equiv c_{\tilde{\lambda}}(X)(X^p - X)^{|\lambda^*|} \pmod p$$

where (Example 8) $\tilde{\lambda}$ and λ^* are respectively the p-core and p-quotient of λ.

(c) From (a) and (b) it follows that if p is prime and $|\lambda| = |\mu|$, then

$$\lambda \sim_p \mu \Leftrightarrow c_\lambda(X) \equiv c_\mu(X) \pmod p.$$

12. Let $\mathscr{P}_n$ denote the set of partitions of n, and $\mathbf{N}^+$ the set of positive integers. For each $r \geqslant 1$ let

$$a(r,n) = \operatorname{Card}\{(\lambda, i) \in \mathscr{P}_n \times \mathbf{N}^+ : \lambda_i = r\},$$

$$b(r,n) = \operatorname{Card}\{(\lambda, i) \in \mathscr{P}_n \times \mathbf{N}^+ : m_i(\lambda) \geqslant r\}.$$

Show that

$$a(r,n) = b(r,n) = p(n-r) + p(n-2r) + \dots$$

where $p(m)$ is the number of partitions of m.

Deduce that

$$\prod_{\lambda \in \mathscr{P}_n} \Big(\prod_{i \geqslant 1} i^{m_i(\lambda)} \Big) = \prod_{\lambda \in \mathscr{P}_n} \Big(\prod_{i \geqslant 1} m_i(\lambda)! \Big).$$

Let $h(r,n) = \operatorname{Card}\{(\lambda, x) : |\lambda| = n,\ x \in \lambda \text{ and } h(x) = r\}$, where (Example 1) $h(x)$ is the hook-length of λ at x. Show that

$$h(r,n) = ra(r,n).$$

13. A matrix of non-negative real numbers is said to be *doubly stochastic* if its row and column sums are all equal to 1.

Let λ, μ be partitions of n. Show that $\lambda \geqslant \mu$ if and only if there exists a doubly stochastic $n \times n$ matrix M such that $M\lambda = \mu$ (where λ, μ are regarded as column vectors of length n). (If $\lambda \geqslant \mu$ we may by (1.16) assume that $\lambda = R_{ij}\mu$. Now define $M = (m_{rs})$ by

$$m_{ii} = m_{jj} = (\lambda_i - \lambda_j - 1)/(\lambda_i - \lambda_j), \qquad m_{ij} = m_{ji} = 1/(\lambda_i - \lambda_j)$$

and $m_{rs} = \delta_{rs}$ otherwise. Then M is doubly stochastic and $M\lambda = \mu$.)

14. Let λ be a partition. If $s = (i,j)$, with $i,j \geqslant 1$, is any square in the first quadrant, we define the *hook-length* of λ at s to be $h(s) = \lambda_i + \lambda'_j - i - j + 1$. When $s \in \lambda$ this agrees with the previous definition, and when $s \notin \lambda$ it is negative. For each $r \in \mathbf{Z}$ let $u_r(\lambda)$ denote the number of squares s in the first quadrant such that $h(s) = r$. Show that $u_{-r}(\lambda) = u_r(\lambda) + r$ for all $r \in \mathbf{Z}$.

If $u_r(\lambda) = u_r(\mu)$ for all $r \in \mathbf{Z}$, does it follow that $\lambda = \mu$ or $\lambda = \mu'$?

15. (a) Let λ be a partition, thought of as an infinite sequence, and let σ be the sequence $(\frac{1}{2}-i)_{i\geqslant 1}$. Show that $\lambda+\sigma$ and $-(\lambda'+\sigma)$ are complementary subsequences of $\mathbf{Z}+\frac{1}{2}$.

(b) Let $g_\lambda(t)=(1-t)\Sigma_{i\geqslant 1}t^{\lambda_i-i}$, which is a polynomial in t and t^{-1}. Show that $g_\lambda(t)=g_{\lambda'}(t^{-1})$.

16. Let λ,μ,ν,π be partitions such that $\lambda\geqslant\nu$ and $\mu\geqslant\pi$. Show that $\lambda+\mu\geqslant\nu+\pi$, $\lambda\cup\mu\geqslant\nu\cup\pi$, $\lambda\mu\geqslant\nu\pi$, and $\lambda\times\mu\geqslant\nu\times\pi$.

17. Let $\lambda=(\lambda_1,\dots,\lambda_n)$ be a partition of length $\leqslant n$ with $\lambda_1\leqslant n$, so that $\lambda\subset(n^n)$. The *complement* of λ in (n^n) is the partition $\hat\lambda=(\hat\lambda_1,\dots,\hat\lambda_n)$ defined by $\hat\lambda_i=n-\lambda_{n+1-i}$, so that the diagram of $\hat\lambda$ is obtained by giving the complement of the diagram of λ in (n^n) a half-turn.

Suppose now that $\lambda=(\alpha\mid\beta)$ in Frobenius notation. Show that $\hat\lambda=(\hat\beta\mid\hat\alpha)$, where $\hat\alpha$ (resp. $\hat\beta$) is the complement in $[0,n-1]$ of the sequence α (resp. β).

18. For λ,μ partitions of n, let π_n be the probability that $\lambda\geqslant\mu$. Does $\pi_n\to 0$ as $n\to\infty$?

Notes and references

The idea of representing a partition by its diagram goes back to Ferrers and Sylvester, and the diagram of a partition is called by some authors the Ferrers diagram or graph, and by others the Young diagram. Tableaux and raising operators were introduced by Alfred Young in his series of papers on quantitative substitutional analysis [Y2].

Example 8. The notion of the p-core of a partition was introduced by Nakayama [N1], and the p-quotient by Robinson [R6] and Littlewood [L10].

Example 9. The notions of a bar and of the p-bar core are due to A. O. Morris [M14]. See also Morris and Yaseen [M16] and Humphreys [H12].

2. The ring of symmetric functions

Consider the ring $\mathbf{Z}[x_1,\dots,x_n]$ of polynomials in n independent variables $x_1,\dots,x_n$ with rational integer coefficients. The symmetric group S_n acts on this ring by permuting the variables, and a polynomial is *symmetric* if it is invariant under this action. The symmetric polynomials form a subring

$$\Lambda_n=\mathbf{Z}[x_1,\dots,x_n]^{S_n}.$$

Λ_n is a graded ring: we have

$$\Lambda_n=\bigoplus_{k\geqslant 0}\Lambda_n^k$$

where Λ_n^k consists of the homogeneous symmetric polynomials of degree k, together with the zero polynomial.

For each $\alpha = (\alpha_1, \dots, \alpha_n) \in \mathbf{N}^n$ we denote by x^α the monomial

$$x^\alpha = x_1^{\alpha_1} \dots x_n^{\alpha_n}.$$

Let λ be any partition of length $\leqslant n$. The polynomial

$$m_\lambda(x_1, \dots, x_n) = \sum x^\alpha \tag{2.1}$$

summed over all distinct permutations α of $\lambda = (\lambda_1, \dots, \lambda_n)$, is clearly symmetric, and the m_λ (as λ runs through all partitions of length $\leqslant n$) form a $\mathbf{Z}$-basis of Λ_n. Hence the m_λ such that $l(\lambda) \leqslant n$ and $|\lambda| = k$ form a $\mathbf{Z}$-basis of Λ_n^k; in particular, as soon as $n \geqslant k$, the m_λ such that $|\lambda| = k$ form a $\mathbf{Z}$-basis of Λ_n^k.

In the theory of symmetric functions, the number of variables is usually irrelevant, provided only that it is large enough, and it is often more convenient to work with symmetric functions in infinitely many variables. To make this idea precise, let $m \geqslant n$ and consider the homomorphism

$$\mathbf{Z}[x_1, \dots, x_m] \to \mathbf{Z}[x_1, \dots, x_n]$$

which sends each of $x_{n+1}, \dots, x_m$ to zero and the other x_i to themselves. On restriction to Λ_m this gives a homomorphism

$$\rho_{m,n} : \Lambda_m \to \Lambda_n$$

whose effect on the basis (m_λ) is easily described; it sends $m_\lambda(x_1, \dots, x_m)$ to $m_\lambda(x_1, \dots, x_n)$ if $l(\lambda) \leqslant n$, and to 0 if $l(\lambda) > n$. It follows that $\rho_{m,n}$ is surjective. On restriction to Λ_m^k we have homomorphisms

$$\rho_{m,n}^k : \Lambda_m^k \to \Lambda_n^k$$

for all $k \geqslant 0$ and $m \geqslant n$, which are always surjective, and are bijective for $m \geqslant n \geqslant k$.

We now form the inverse limit

$$\Lambda^k = \varprojlim_n \Lambda_n^k$$

of the $\mathbf{Z}$-modules Λ_n^k relative to the homomorphisms $\rho_{m,n}^k$: an element of Λ^k is by definition a sequence $f = (f_n)_{n \geqslant 0}$, where each $f_n = f_n(x_1, \dots, x_n)$ is a homogeneous symmetric polynomial of degree k in $x_1, \dots, x_n$, and $f_m(x_1, \dots, x_n, 0, \dots, 0) = f_n(x_1, \dots, x_n)$ whenever $m \geqslant n$. Since $\rho_{m,n}^k$ is an isomorphism for $m \geqslant n \geqslant k$, it follows that the projection

$$\rho_n^k : \Lambda^k \to \Lambda_n^k,$$

which sends f to f_n, is an isomorphism for all $n \geqslant k$, and hence that Λ^k has a **Z**-basis consisting of the *monomial symmetric functions* m_λ (for all partitions λ of k) defined by

$$\rho_n^k(m_\lambda) = m_\lambda(x_1, \ldots, x_n)$$

for all $n \geqslant k$. Hence Λ^k is a free **Z**-module of rank $p(k)$, the number of partitions of k.

Now let

$$\Lambda = \bigoplus_{k \geqslant 0} \Lambda^k,$$

so that Λ is the free **Z**-module generated by the m_λ for *all* partitions λ. We have surjective homomorphisms

$$\rho_n = \bigoplus_{k \geqslant 0} \rho_n^k : \Lambda \to \Lambda_n$$

for each $n \geqslant 0$, and ρ_n is an isomorphism in degrees $k \leqslant n$.

It is clear that Λ has a structure of a graded ring such that the ρ_n are ring homomorphisms. The graded ring Λ thus defined is called the *ring of symmetric functions*† in countably many independent variables $x_1, x_2, \ldots$.

Remarks. 1. Λ is *not* the inverse limit (in the category of rings) of the rings Λ_n relative to the homomorphisms $\rho_{m,n}$. This inverse limit, $\hat{\Lambda}$ say, contains for example the infinite product $\prod_{i=1}^{\infty}(1 + x_i)$, which does not belong to Λ, since the elements of Λ are by definition finite sums of monomial symmetric functions m_λ. However, Λ is the inverse limit of the Λ_n in the category of *graded* rings.

2. We could use any commutative ring A in place of **Z** as coefficient ring; in place of Λ we should obtain $\Lambda_A \cong \Lambda \otimes_{\mathbf{Z}} A$.

Elementary symmetric functions

For each integer $r \geqslant 0$ the rth *elementary symmetric function* e_r is the sum of all products of r distinct variables x_i, so that $e_0 = 1$ and

$$e_r = \sum_{i_1 < i_2 < \ldots < i_r} x_{i_1} x_{i_2} \ldots x_{i_r} = m_{(1^r)}$$

for $r \geqslant 1$. The generating function for the e_r is

$$E(t) = \sum_{r \geqslant 0} e_r t^r = \prod_{i \geqslant 1} (1 + x_i t) \tag{2.2}$$

† The elements of Λ (unlike those of Λ_n) are no longer polynomials: they are formal infinite sums of monomials. We have therefore reverted to the older terminology of 'symmetric functions'.

(t being another variable), as one sees by multiplying out the product on the right. (If the number of variables is finite, say n, then e_r (i.e. $\rho_n(e_r)$) is zero for all $r > n$, and (2.2) then takes the form

$$\sum_{r=0}^{n} e_r t^r = \prod_{i=1}^{n} (1 + x_i t),$$

both sides now being elements of $\Lambda_n[t]$. Similar remarks will apply to many subsequent formulas, and we shall usually leave it to the reader to make the necessary (and obvious) adjustments.)

For each partition $\lambda = (\lambda_1, \lambda_2, \dots)$ define

$$e_\lambda = e_{\lambda_1} e_{\lambda_2} \dots .$$

(2.3) *Let λ be a partition, λ' its conjugate. Then*

$$e_{\lambda'} = m_\lambda + \sum_\mu a_{\lambda\mu} m_\mu$$

where the $a_{\lambda\mu}$ are non-negative integers, and the sum is over partitions $\mu < \lambda$ in the natural ordering.

Proof. When we multiply out the product $e_{\lambda'} = e_{\lambda'_1} e_{\lambda'_2} \dots$, we shall obtain a sum of monomials, each of which is of the form

$$(x_{i_1} x_{i_2} \dots)(x_{j_1} x_{j_2} \dots) \dots = x^\alpha,$$

say, where $i_1 < i_2 < \dots < i_{\lambda'_1}$, $j_1 < j_2 < \dots < j_{\lambda'_2}$, and so on. If we now enter the numbers $i_1, i_2, \dots, i_{\lambda'_1}$ in order down the first column of the diagram of λ, then the numbers $j_1, j_2, \dots, j_{\lambda'_2}$ in order down the second column, and so on, it is clear that for each $r \geqslant 1$ all the symbols $\leqslant r$ so entered in the diagram of λ must occur in the top r rows. Hence $\alpha_1 + \dots + \alpha_r \leqslant \lambda_1 + \dots + \lambda_r$ for each $r \geqslant 1$, i.e. we have $\alpha \leqslant \lambda$. By (1.12) it follows that

$$e_{\lambda'} = \sum_{\mu \leqslant \lambda} a_{\lambda\mu} m_\mu$$

with $a_{\lambda\mu} \geqslant 0$ for each $\mu \geqslant \lambda$, and the argument above also shows that the monomial x^λ occurs exactly once, so that $a_{\lambda\lambda} = 1$. |

(2.4) *We have*

$$\Lambda = \mathbf{Z}[e_1, e_2, \dots]$$

and the e_r are algebraically independent over $\mathbf{Z}$.

Proof. The m_λ form a **Z**-basis of Λ, and (2.3) shows that the e_λ form another **Z**-basis: in other words, every element of Λ is uniquely expressible as a polynomial in the e_r. ∎

Remark. When there are only finitely many variables $x_1, \ldots, x_n$, (2.4) states that $\Lambda_n = \mathbf{Z}[e_1, \ldots, e_n]$, and that $e_1, \ldots, e_n$ are algebraically independent. This is the usual statement of the 'fundamental theorem on symmetric functions'.

Complete symmetric functions

For each $r \geqslant 0$ the rth *complete symmetric function* h_r is the sum of all monomials of total degree r in the variables $x_1, x_2, \ldots$, so that

$$h_r = \sum_{|\lambda| = r} m_\lambda.$$

In particular, $h_0 = 1$ and $h_1 = e_1$. It is convenient to define h_r and e_r to be zero for $r < 0$.

The generating function for the h_r is

(2.5) $$H(t) = \sum_{r \geqslant 0} h_r t^r = \prod_{i \geqslant 1} (1 - x_i t)^{-1}.$$

To see this, observe that

$$(1 - x_i t)^{-1} = \sum_{k \geqslant 0} x_i^k t^k,$$

and multiply these geometric series together.

From (2.2) and (2.5) we have

(2.6) $$H(t)E(-t) = 1$$

or, equivalently,

(2.6′) $$\sum_{r=0}^{n} (-1)^r e_r h_{n-r} = 0$$

for all $n \geqslant 1$.

Since the e_r are algebraically independent (2.4), we may define a homomorphism of graded rings

$$\omega : \Lambda \to \Lambda$$

by

$$\omega(e_r) = h_r$$

for all $r \geqslant 0$. The symmetry of the relations (2.6′) as between the e's and the h's shows that

(2.7) *ω is an involution, i.e. ω^2 is the identity map.* |

It follows that ω is an automorphism of Λ, and hence from (2.4) that

(2.8) *We have*

$$\Lambda = \mathbf{Z}[h_1, h_2, \ldots]$$

and the h_r are algebraically independent over $\mathbf{Z}$. |

Remark. If the number of variables is finite, say n (so that $e_r = 0$ for $r > n$) the mapping $\omega: \Lambda_n \to \Lambda_n$ is defined by $\omega(e_r) = h_r$ for $1 \leqslant r \leqslant n$, and is still an involution by reason of (2.6′); we have $\Lambda_n = \mathbf{Z}[h_1, \ldots, h_n]$ with $h_1, \ldots, h_n$ algebraically independent, but $h_{n+1}, h_{n+2}, \ldots$ are non-zero polynomials in $h_1, \ldots, h_n$ (or in $e_1, \ldots, e_n$).

As in the case of the e's, we define

$$h_\lambda = h_{\lambda_1} h_{\lambda_2} \ldots$$

for any partition $\lambda = (\lambda_1, \lambda_2, \ldots)$. By (2.8), the h_λ form a $\mathbf{Z}$-basis of Λ. We now have three $\mathbf{Z}$-bases, all indexed by partitions: the m_λ, the e_λ, and the h_λ, the last two of which correspond under the involution ω. If we define

$$f_\lambda = \omega(m_\lambda)$$

for each partition λ, the f_λ form a fourth $\mathbf{Z}$-basis of Λ. (The f_λ are the 'forgotten' symmetric functions: they have no particularly simple direct description.)

The relations (2.6′) lead to a determinant identity which we shall make use of later. Let N be a positive integer and consider the matrices of $N+1$ rows and columns

$$H = (h_{i-j})_{0 \leqslant i,j \leqslant N}, \qquad E = \left((-1)^{i-j} e_{i-j}\right)_{0 \leqslant i,j \leqslant N}$$

with the convention mentioned earlier that $h_r = e_r = 0$ for $r < 0$. Both H and E are lower triangular, with 1's down the diagonal, so that $\det H = \det E = 1$; moreover the relations (2.6′) show that they are inverses of each other. It follows that each minor of H is equal to the complementary cofactor of E', the transpose of E.

Let λ, μ be two partitions of length $\leqslant p$, such that λ' and μ' have length $\leqslant q$, where $p + q = N + 1$. Consider the minor of H with row indices $\lambda_i + p - i$ $(1 \leqslant i \leqslant p)$ and column indices $\mu_i + p - i$ $(1 \leqslant i \leqslant p)$. By

(1.7) the complementary cofactor of E' has row indices $p-1+j-\lambda'_j$ $(1\leqslant j\leqslant q)$ and column indices $p-1+j-\mu'_j$ $(1\leqslant j\leqslant q)$. Hence we have

$$\det\left(h_{\lambda_i-\mu_j-i+j}\right)_{1\leqslant i,j\leqslant p}=(-1)^{|\lambda|+|\mu|}\det\left((-1)^{\lambda'_i-\mu'_j-i+j}e_{\lambda'_i-\mu'_j-i+j}\right)_{1\leqslant i,j\leqslant q}.$$

The minus signs cancel out, and therefore we have (Aitken [A1])

(2.9) $$\det\left(h_{\lambda_i-\mu_j-i+j}\right)_{1\leqslant i,j\leqslant p}=\det\left(e_{\lambda'_i-\mu'_j-i+j}\right)_{1\leqslant i,j\leqslant q}.$$

In particular, taking $\mu=0$:

(2.9') $$\det(h_{\lambda_i-i+j})=\det(e_{\lambda'_i-i+j}).$$

Power sums

For each $r\geqslant 1$ the rth *power sum* is

$$p_r=\sum x_i^r=m_{(r)}.$$

The generating function for the p_r is

$$\begin{aligned}P(t)&=\sum_{r\geqslant 1}p_rt^{r-1}=\sum_{i\geqslant 1}\sum_{r\geqslant 1}x_i^rt^{r-1}\\&=\sum_{i\geqslant 1}\frac{x_i}{1-x_it}\\&=\sum_{i\geqslant 1}\frac{d}{dt}\log\frac{1}{1-x_it}\end{aligned}$$

so that

(2.10) $$P(t)=\frac{d}{dt}\log\prod_{i\geqslant 1}(1-x_it)^{-1}=\frac{d}{dt}\log H(t)=H'(t)/H(t).$$

Likewise we have

(2.10') $$P(-t)=\frac{d}{dt}\log E(t)=E'(t)/E(t).$$

From (2.10) and (2.10') we obtain

(2.11) $$nh_n=\sum_{r=1}^{n}p_rh_{n-r},$$

(2.11') $$ne_n=\sum_{r=1}^{n}(-1)^{r-1}p_re_{n-r}$$

for $n \geqslant 1$, and these equations enable us to express the h's and the e's in terms of the p's, and vice versa. The equations (2.11′) are due to Isaac Newton, and are known as Newton's formulas. From (2.11) it is clear that $h_n \in \mathbf{Q}[p_1, \dots, p_n]$ and $p_n \in \mathbf{Z}[h_1, \dots, h_n]$, and hence that

$$\mathbf{Q}[p_1, \dots, p_n] = \mathbf{Q}[h_1, \dots, h_n].$$

Since the h_r are algebraically independent over $\mathbf{Z}$, and hence also over $\mathbf{Q}$, it follows that

(2.12) *We have*

$$\Lambda_{\mathbf{Q}} = \Lambda \otimes_{\mathbf{Z}} \mathbf{Q} = \mathbf{Q}[p_1, p_2, \dots]$$

and the p_r are algebraically independent over $\mathbf{Q}$. |

Hence, if we define

$$p_\lambda = p_{\lambda_1} p_{\lambda_2} \cdots$$

for each partition $\lambda = (\lambda_1, \lambda_2, \dots)$, then the p_λ form a $\mathbf{Q}$-basis of $\Lambda_{\mathbf{Q}}$. But they do *not* form a $\mathbf{Z}$-basis of Λ: for example, $h_2 = \frac{1}{2}(p_1^2 + p_2)$ does not have integral coefficients when expressed in terms of the p_λ.

Since the involution ω interchanges $E(t)$ and $H(t)$ it follows from (2.10) and (2.10′) that

$$\omega(p_n) = (-1)^{n-1} p_n$$

for all $n \geqslant 1$, and hence that for any partition λ we have

$$\omega(p_\lambda) = \varepsilon_\lambda p_\lambda \tag{2.13}$$

where $\varepsilon_\lambda = (-1)^{|\lambda| - l(\lambda)}$.

Finally, we shall express h_n and e_n as linear combinations of the p_λ. For any partition λ, define

$$z_\lambda = \prod_{i \geqslant 1} i^{m_i} . m_i!$$

where $m_i = m_i(\lambda)$ is the number of parts of λ equal to i. Then we have

$$\begin{aligned} H(t) &= \sum_\lambda z_\lambda^{-1} p_\lambda t^{|\lambda|}, \\ E(t) &= \sum_\lambda \varepsilon_\lambda z_\lambda^{-1} p_\lambda t^{|\lambda|}, \end{aligned} \tag{2.14}$$

or equivalently

$$h_n = \sum_{|\lambda|=n} z_\lambda^{-1} p_\lambda,$$

(2.14′)

$$e_n = \sum_{|\lambda|=n} \varepsilon_\lambda z_\lambda^{-1} p_\lambda.$$

Proof. It is enough to prove the first of the identities (2.14), since the second then follows by applying the involution ω and using (2.13). From (2.10) we have

$$\begin{aligned} H(t) &= \exp \sum_{r\geq 1} p_r t^r/r \\ &= \prod_{r\geq 1} \exp(p_r t^r/r) \\ &= \prod_{r\geq 1} \sum_{m_r=0}^{\infty} (p_r t^r)^{m_r}/r^{m_r}.m_r! \\ &= \sum_\lambda z_\lambda^{-1} p_\lambda t^{|\lambda|}. \quad | \end{aligned}$$

(2.15) *Remark.* In the language of λ-rings ([B5], [K13]) the ring Λ is the 'free λ-ring in one variable' (or, more precisely, is the underlying ring). Consequently *all* the formulas and identities in this Chapter can be translated into this language. It is not our intention to write a text on the theory of λ-rings: we shall merely provide a brief dictionary.

If R is any λ-ring and x any element of R, there exists a unique λ-homomorphism $\Lambda \to R$ under which $e_1(=h_1=p_1)$ is mapped to x. Under this homomorphism

e_r	is mapped to	$\lambda^r(x)$	(rth exterior power)
h_r		$\sigma^r(x)=(-1)^r\lambda^r(-x)$	(rth symmetric power)
$E(t)$		$\lambda_t(x)$	
$H(t)$		$\sigma_t(x)=\lambda_{-t}(-x)$	
p_r		$\psi^r(x)$	(Adams operations)

and the involution ω corresponds in R to $x \mapsto -x$. So, for example, (2.14′) becomes

$$\sigma^n(x) = \sum_{|\lambda|=n} z_\lambda^{-1} \psi^\lambda(x)$$

valid for any element x of any λ-ring (where of course $\psi^\lambda(x) = \psi^{\lambda_1}(x)\psi^{\lambda_2}(x)\dots$).

Examples

1. (a) Let $x_1 = \cdots = x_n = 1$, $x_{n+1} = x_{n+2} = \ldots = 0$. Then $E(t) = (1+t)^n$ and $H(t) = (1-t)^{-n}$, so that

$$e_r = \binom{n}{r}, \qquad h_r = \binom{n+r-1}{r}$$

and $p_r = n$ for all $n \geqslant 1$. Also

$$m_\lambda = u_\lambda \binom{n}{l(\lambda)},$$

where

$$u_\lambda = \frac{l(\lambda)!}{\prod_{i \geqslant 1} m_i(\lambda)!}$$

(b) More generally, let X be an indeterminate, and define a homomorphism $\varepsilon_X : \Lambda_{\mathbf{Q}} \to \mathbf{Q}[X]$ by $\varepsilon_X(p_r) = X$ for all $r \geqslant 1$. Then we have

$$\varepsilon_X(e_r) = \binom{X}{r}, \qquad \varepsilon_X(h_r) = \binom{X+r-1}{r} = (-1)^r \binom{-X}{r},$$

for all $r \geqslant 1$, and

$$\varepsilon_X(m_\lambda) = u_\lambda \binom{X}{l(\lambda)}.$$

For these formulas are correct when X is replaced by any positive integer n, by (a) above. Hence they are true identically.

2. Let $x_i = 1/n$ for $1 \leqslant i \leqslant n$, $x_i = 0$ for $i > n$, and then let $n \to \infty$. From Example 1 we have

$$e_r = \lim_{n \to \infty} \frac{1}{n^r} \binom{n}{r} = \frac{1}{r!}$$

and likewise $h_r = 1/r!$, so that $E(t) = H(t) = e^t$. We have $p_1 = 1$ and $p_r = 0$ for $r > 1$; more generally, $m_\lambda = 0$ for all partitions λ except $\lambda = (1^r)(r \geqslant 0)$.

3. Let $x_i = q^{i-1}$ for $1 \leqslant i \leqslant n$, and $x_i = 0$ for $i > n$, where q is an indeterminate. Then

$$E(t) = \prod_{i=0}^{n-1} (1 + q^i t) = \sum_{r=0}^{n} q^{r(r-1)/2} \begin{bmatrix} n \\ r \end{bmatrix} t^r$$

where $\begin{bmatrix} n \\ r \end{bmatrix}$ denotes the 'q-binomial coefficient' or Gaussian polynomial

$$\begin{bmatrix} n \\ r \end{bmatrix} = \frac{(1-q^n)(1-q^{n-1})\ldots(1-q^{n-r+1})}{(1-q)(1-q^2)\ldots(1-q^r)},$$

and

$$H(t) = \prod_{i=0}^{n-1} (1 - q^i t)^{-1} = \sum_{r=0}^{\infty} \begin{bmatrix} n+r-1 \\ r \end{bmatrix} t^r.$$

These identities are easily proved by induction on n. It follows that

$$e_r = q^{r(r-1)/2}\begin{bmatrix} n \\ r \end{bmatrix}, \qquad h_r = \begin{bmatrix} n+r-1 \\ r \end{bmatrix}.$$

h_r is the generating function for partitions λ such that $l(\lambda) \leqslant r$ and $l(\lambda') \leqslant n-1$, and e_r is the generating function for such partitions with all parts distinct.

4. Let $n \to \infty$ in Example 3, i.e. let $x_i = q^{i-1}$ for all $i \geqslant 1$. Then

$$E(t) = \prod_{i=0}^{\infty}(1+q^i t) = \sum_{r=0}^{\infty} q^{r(r-1)/2} t^r / \varphi_r(q),$$

$$H(t) = \prod_{i=0}^{\infty}(1-q^i t)^{-1} = \sum_{r=0}^{\infty} t^r / \varphi_r(q),$$

where

$$\varphi_r(q) = (1-q)(1-q^2)\dots(1-q^r).$$

Hence in this case

$$e_r = q^{r(r-1)/2}/\varphi_r(q), \qquad h_r = 1/\varphi_r(q)$$

and $p_r = (1-q^r)^{-1}$.

5. Since the h_r are algebraically independent we may specialize them in any way, and forget about the original variables x_i: in other words, we may take $H(t)$ (or $E(t)$) to be any power series in t with constant term 1. (We have already done this in Example 2 above, where $H(t) = e^t$.) Let a, b, q be variables and take

$$H(t) = \prod_{i=0}^{\infty} \frac{1-bq^i t}{1-aq^i t}.$$

Then we have

$$h_r = \prod_{i=1}^{r} \frac{a-bq^{i-1}}{1-q^i}, \qquad e_r = \prod_{i=1}^{r} \frac{aq^{i-1}-b}{1-q^i}$$

(see e.g. Andrews [A3], Chapter II.) Also $p_r = (a^r - b^r)/(1-q^r)$.

6. Take $H(t) = \prod_{n=1}^{\infty}(1-t^n)^{-1}$, so that $h_n = p(n)$, the number of partitions of n. Then $E(-t) = \prod_{n=1}^{\infty}(1-t^n)$, and so by Euler's pentagonal number theorem $e_n = 0$ unless n is a pentagonal number, i.e. of the form $\frac{1}{2}m(3m+1)$ for some $m \in \mathbf{Z}$; and $e_n = (-1)^{m(m+1)/2}$ if $n = \frac{1}{2}m(3m+1)$.

From (2.10) we obtain $p_r = \sigma(r)$, the sum of the divisors of r. Hence (2.11) gives in this case

$$p(n) = \frac{1}{n}\sum_{r=1}^{n} \sigma(r) p(n-r). \tag{1}$$

7. Take $H(t) = \prod_{n=1}^{\infty}(1-t^n)^{-n}$, so that $h_n = p_2(n)$, the number of plane partitions of n (§5, Example 13). From (2.10) we obtain $p_r = \sigma_2(r)$, the sum of the squares of the divisors of r. Hence by (2.11)

$$p_2(n) = \frac{1}{n}\sum_{r=1}^{n}\sigma_2(r)p_2(n-r). \tag{2}$$

It is perhaps only fair to warn the reader that the obvious generalization of (1) and (2) to m-dimensional partitions ($m > 2$) is false.

8. By solving the equations (2.6') for e_n we obtain

$$e_n = \det(h_{1-i+j})_{1 \leqslant i,j \leqslant n}$$

and dually

$$h_n = \det(e_{1-i+j})_{1 \leqslant i,j \leqslant n}.$$

Likewise from (2.11) we obtain the determinant formulas

$$p_n = \begin{vmatrix} e_1 & 1 & 0 & \dots & 0 \\ 2e_2 & e_1 & 1 & \dots & 0 \\ \vdots & \vdots & \vdots & & \vdots \\ ne_n & e_{n-1} & e_{n-2} & \dots & e_1 \end{vmatrix}$$

$$n!e_n = \begin{vmatrix} p_1 & 1 & 0 & \dots & 0 \\ p_2 & p_1 & 2 & \dots & 0 \\ \vdots & \vdots & \vdots & & \vdots \\ p_{n-1} & p_{n-2} & \cdot & \dots & n-1 \\ p_n & p_{n-1} & \cdot & \dots & p_1 \end{vmatrix}$$

and dually

$$(-1)^{n-1}p_n = \begin{vmatrix} h_1 & 1 & 0 & \dots & 0 \\ 2h_2 & h_1 & 1 & \dots & 0 \\ \vdots & \vdots & \vdots & & \vdots \\ nh_n & h_{n-1} & h_{n-2} & \dots & h_1 \end{vmatrix}$$

$$n!h_n = \begin{vmatrix} p_1 & -1 & 0 & \dots & 0 \\ p_2 & p_1 & -2 & \dots & 0 \\ \vdots & \vdots & \vdots & & \vdots \\ p_{n-1} & p_{n-2} & \cdot & \dots & -n+1 \\ p_n & p_{n-1} & \cdot & \dots & p_1 \end{vmatrix}.$$

9. (a) Let G be any subgroup of the symmetric group S_n. The *cycle indicator* of G is the symmetric function

$$c(G) = \frac{1}{|G|} \sum_{\rho} n_G(\rho) p_\rho$$

where $n_G(\rho)$ is the number of elements in G of cycle-type ρ, and the sum is over all partitions ρ of n. In particular,

$$c(S_n) = \sum_{|\rho|=n} z_\rho^{-1} p_\rho = h_n$$

by (2.14′), and for the alternating group A_n we have

$$c(A_n) = h_n + e_n.$$

(b) If G is a subgroup of S_n and H a subgroup of S_m, then $G \times H$ is a subgroup of $S_n \times S_m \subset S_{n+m}$, and we have

$$c(G \times H) = c(G)c(H).$$

(c) Let G be a subgroup of S_n and let Σ be the set of all sequences $\alpha = (\alpha_1, \ldots, \alpha_n)$ of n positive integers. For each such sequence α, define $x_\alpha = x_{\alpha_1} \ldots x_{\alpha_n}$. The group G acts on Σ by permuting the terms of these sequences, and the function $\alpha \mapsto x_\alpha$ is constant on each G-orbit. Show that

$$c(G) = \sum_{\alpha} x_\alpha \tag{1}$$

where α runs through a set of representatives of the orbits of G in Σ (Polya's theorem). (Let

$$X = |G|^{-1} \sum_{(g,\alpha)} x_\alpha$$

summed over all $(g, \alpha) \in G \times \Sigma$ such that $g\alpha = \alpha$, and show that X is equal to either side of (1).)

10. From Examples 8 and 9 it follows that the number of elements of cycle-type ρ in S_n is equal to the coefficient of p_ρ in the determinant

$$d_n = n!\,h_n = \begin{vmatrix} p_1 & -1 & 0 & \ldots & 0 \\ p_2 & p_1 & -2 & \ldots & 0 \\ \vdots & \vdots & \vdots & & \vdots \\ p_{n-1} & p_{n-2} & \cdot & \ldots & -n+1 \\ p_n & p_{n-1} & \cdot & \ldots & p_1 \end{vmatrix}.$$

Let l be a prime number. We may use this formula to count the number of conjugacy classes in S_n in which the number of elements is prime to l, by reducing the determinant d_n modulo l. Suppose that $n = a_0 + n_1 l$, where $0 \leqslant a_0 \leqslant l-1$.

Then since the multiples of l above the diagonal in d_n become zero on reduction, it follows that

$$d_n \equiv d_l^{n_1} d_{a_0} \quad (\text{mod. } l)$$

Now it is clear from the original definition of $d_n = n!c(S_n)$ that

$$d_l \equiv p_1^l - p_l \quad (\text{mod. } l)$$

and therefore we have

$$d_n \equiv (p_1^l - p_l)^{n_1} . d_{a_0} \quad (\text{mod. } l). \tag{1}$$

Hence if $n = a_0 + a_1 l + a_2 l^2 + \ldots$, with $0 \leqslant a_i \leqslant l-1$ for all $i \geqslant 0$, it follows from (1) that

$$d_n \equiv d_{a_0} \prod_{i \geqslant 1} \left(p_1^{l^i} - p_l^{l^{i-1}}\right)^{a_i} \quad (\text{mod. } l).$$

Consequently, if $\mu_l(S_n)$ denotes the number of conjugacy classes in S_n of order prime to l, we have

$$\begin{aligned} \mu_l(S_n) &= \mu_l(S_{a_0}) \prod_{i \geqslant 1} (a_i + 1) \\ &= p(a_0) \prod_{i \geqslant 1} (a_i + 1) \end{aligned}$$

where $p(a_0)$ is the number of partitions of a_0. In particular, if $l = 2$, we see that $\mu_2(S_n)$ is always a power of 2, because each a_i is then either 0 or 1: namely $\mu_2(S_n) = 2^r$ if $[n/2]$ is a sum of r distinct powers of 2.

11. Let

$$f(t) = \sum_{n=0}^{\infty} \frac{f_n t^n}{n!}, \qquad g(t) = \sum_{n=0}^{\infty} \frac{g_n t^n}{n!}$$

be formal power series (with coefficients in a commutative Q-algebra) such that $g(0) = 0$. We may substitute $g(t)$ for t in $f(t)$, and obtain say

$$H(t) = f(g(t)) = \sum_0^{\infty} \frac{H_n t^n}{n!}.$$

Clearly each coefficient H_n is of the form

$$H_n = \sum_{k=1}^{n} f_k B_{n,k}(g)$$

where the $B_{n,k}$ are polynomials in the coefficients of g, called the *partial Bell polynomials*. Since each H_n is linear in the coefficients of f, in order to compute the polynomials $B_{n,k}$ we may take $f_k = a^k$, so that $f(t) = e^{at}$. Writing

$$H(t) = \sum_{n=0}^{\infty} h_n t^n$$

as usual, we have $H(t) = \exp(ag(t))$ and therefore by (2.10)

$$P(t) = \frac{\mathrm{d}}{\mathrm{d}t} \log H(t) = ag'(t) = \sum_1^\infty \frac{ag_n t^{n-1}}{(n-1)!},$$

so that $p_n = ag_n/(n-1)!$ for all $n \geqslant 1$. Hence by (2.14′)

$$H_n = n!h_n = \sum_{|\lambda|=n} \frac{n!}{z_\lambda} p_\lambda$$

and consequently

$$B_{n,k} = \sum_\lambda \frac{n!}{z_\lambda} p_\lambda = \sum_\lambda c_\lambda g_\lambda$$

where the sum is over partitions λ of n such that $l(\lambda) = k$, and

$$g_\lambda = g_{\lambda_1} g_{\lambda_2} \dots, \qquad c_\lambda = n! \Big/ \prod_{i \geqslant 1} r_i!(i!)^{r_i}$$

if $\lambda = (1^{r_1} 2^{r_2} \dots)$. These coefficients c_λ are *integers*, because c_λ is the number of decompositions of a set of n elements into disjoint subsets containing $\lambda_1, \lambda_2, \dots$ elements. Hence each $B_{n,k}$ is a polynomial in the g_n with integer coefficients.

Particular cases:

(a) if $g(t) = \log(1+t)$, then $B_{n,k} = s(n,k)$ are the *Stirling numbers of the first kind*; $(-1)^{n-k} s(n,k)$ is the number of elements of S_n which are products of k disjoint cycles. We have

$$\sum_{n,k \geqslant 0} s(n,k) \frac{t^n}{n!} a^k = (1+t)^a = \sum_{n \geqslant 0} \binom{a}{n} t^n,$$

from which it follows that

$$\sum_{k=0}^n s(n,k) a^k = a(a-1)\dots(a-n+1)$$

and hence that $s(n,k)$ is the $(n-k)$th elementary symmetric function of $-1, -2, \dots, -n+1$.

(b) if $g(t) = e^t - 1$, so that $g_n = 1$ for all $n \geqslant 1$, then $B_{n,k} = S(n,k)$ are the *Stirling numbers of the second kind*; $S(n,k)$ is the number of decompositions of a set of n elements into k disjoint subsets, and is also the $(n-k)$th complete symmetric function of $1, 2, \dots, k$.

12. Deduce from Example 11 that if f and g are n times differentiable functions of a real variable, and if f_k, g_k, $(f \circ g)_k$ denote the kth derivatives of f, g, and $f \circ g$, then

$$(f \circ g)_n = \sum_{k=1}^n B_{n,k}(g_1, g_2, \dots)(f_k \circ g).$$

13. If $H(t)=(1-t^r)/(1-t)^r$, we have

$$h_n=\binom{n+r-1}{r-1}-\binom{n-1}{r-1}$$

and by (2.10) we find that $p_n=r$ if $n\not\equiv 0 \pmod r$, whereas $p_n=0$ if $n\equiv 0 \pmod r$. Hence from (2.14')

$$\sum_\lambda z_\lambda^{-1} r^{l(\lambda)}=\binom{n+r-1}{r-1}-\binom{n-1}{r-1} \qquad (r\geqslant 2)$$

where the sum on the left is over partitions λ of n none of whose parts is divisible by r.

In particular ($r=2$)

$$\sum_\lambda z_\lambda^{-1} 2^{l(\lambda)}=2$$

summed over all partitions of n into *odd* parts.

14. Suppose that $p_n=an^n/n!$ for $n\geqslant 1$. Then

$$h_n=a(a+n)^{n-1}/n!, \qquad e_n=a(a-n)^{n-1}/n!.$$

(Let $t=xe^{-x}$ and use Lagrange's reversion formula to show that $P(t)=ae^x/(1-x)$.)

15. Show that

$$\sum_\rho z_\rho^{-1}=\sum_\sigma z_\sigma^{-1}=\frac{1.3.5\ldots(2n-1)}{2.4.6\ldots 2n},$$

where the first sum is over all partitions ρ of $2n$ with all parts even, and the second sum is over all partitions σ of $2n$ with all parts odd.

16. Suppose that $e_n=p_n$ for each $n\geqslant 1$. Show that

$$h_n=\frac{a^n}{(n+1)!}, \qquad e_n=\frac{(-1)^n a^n B_n}{n!}$$

for some a, where B_n is the nth Bernoulli number.

17. If $h_n=n$ for each $n\geqslant 1$, the sequence $(e_n)_{n\geqslant 1}$ is periodic with period 3, and the sequence $(p_n)_{n\geqslant 1}$ is periodic with period 6.

18. (Muirhead's inequalities.) For each partition λ of n, the λ-*mean* of $x=(x_1,x_1,\ldots,x_n)$ is defined to be

$$M_\lambda(x)=\frac{1}{n!}\sum_{w\in S_n} w(x^\lambda).$$

In particular, $M_{(n)}(x)$ is the arithmetic mean of $x_1^n,\ldots,x_n^n$, and $M_{(1^n)}(x)$ is their geometric mean.

Let λ, μ be partitions of n. Then the following statements are equivalent;

(i) $\lambda \geqslant \mu$; (ii) $M_\lambda(x) \geqslant M_\mu(x)$ for all $x = (x_1, \dots, x_n) \in \mathbf{R}_+^n$.

(To show that (i) implies (ii), we may assume by (1.16) that $\lambda = R_{ij}\mu$, in which case it is enough to show that

$$x_i^{\lambda_i} x_j^{\lambda_j} + x_j^{\lambda_i} x_i^{\lambda_j} \geqslant x_i^{\mu_i} x_j^{\mu_j} + x_j^{\mu_i} x_i^{\mu_j}.$$

To show that (ii) implies (i), set $x_1 = \dots = x_r = X$ and $x_{r+1} = \dots = x_n = 1$, where X is large, and deduce that $\lambda_1 + \dots + \lambda_r \geqslant \mu_1 + \dots + \mu_r$.)

19. Let $p_n^{(r)} = \Sigma m_\lambda$, summed over all partitions λ of n of length r (so that $p_n^{(1)} = p_n$). Show that

$$\sum_{n,r} p_n^{(r)} t^{n-r} u^r = E(u-t)H(t)$$

and that

$$\sum_{n \geqslant r} p_n^{(r)} t^{n-r} = \frac{(-1)^r}{r!} E^{(r)}(-t)H(t)$$

where $E^{(r)}(t)$ is the rth derivative of $E(t)$ with respect to t. Deduce that

$$p_n^{(r)} = \sum_{a+b=n} (-1)^{a-r} \binom{a}{r} e_a h_b$$

and that (if $n \geqslant r$) $p_n^{(r)}$ is equal to the determinant of the matrix $(a_{ij})_{0 \leqslant i,j \leqslant n-r}$, where $a_{i0} = \binom{r+i}{r} e_{r+i}$ and $a_{ij} = e_{i-j+1}$ if $j \geqslant 1$.

20. For any partition λ, let $u_\lambda = l(\lambda)!/\prod_{i \geqslant 1} m_i(\lambda)!$, as in Example 1(a). Show that

$$p_n = n \sum_{|\lambda|=n} \frac{(-1)^{l(\lambda)-1}}{l(\lambda)} u_\lambda h_\lambda$$

$$= -n \sum_{|\lambda|=n} \frac{\varepsilon_\lambda}{l(\lambda)} u_\lambda e_\lambda$$

and that

$$e_n = \sum_{|\lambda|=n} \varepsilon_\lambda u_\lambda h_\lambda.$$

(Let $H^+(t) = \Sigma_{n \geqslant 1} h_n t^n$, and pick out the coefficients of t^n in $\log(1 + H^+(t))$ and $(1 + H^+(t))^{-1}$, expanded in powers of $H^+(t)$.)

21. Let $x_n = 1/n^2$ for each $n \geqslant 1$. Then

$$E(-t^2) = \prod_{n \geqslant 1} \left(1 - \frac{t^2}{n^2}\right) = \frac{\sin \pi t}{\pi t} \tag{1}$$

so that $e_n = \pi^{2n}/(2n+1)!$ for each $n \geqslant 0$. We have

$$p_r = \sum_{n \geqslant 1} \frac{1}{n^{2r}} = \zeta(2r)$$

where ζ is Riemann's zeta function. Deduce from (1) that

$$-2t^2P(t^2) = t\frac{d}{dt}\log E(-t^2) = \pi t \cot \pi t - 1,$$

and hence that

$$\zeta(2r) = (-1)^{r-1}2^{2r-1}\pi^{2r}B_{2r}/(2r)!,$$

where B_{2r} is the $2r$th Bernoulli number.

22. (a) From Jacobi's triple product identity

$$\sum_{n \in \mathbf{Z}} t^{n^2}u^n = \prod_{n \geqslant 1}(1-t^{2n})(1+t^{2n-1}u)(1+t^{2n-1}u^{-1})$$

by setting $u = -1$ we obtain

$$\sum_{n \in \mathbf{Z}} (-1)^n t^{n^2} = \prod_{n \geqslant 1} \frac{1-t^n}{1+t^n}.$$

Deduce that if

$$E(t) = \prod_{n \geqslant 1} \frac{1+t^n}{1-t^n} \tag{1}$$

then $h_n = 2$ or 0 according as $n \geqslant 1$ is a square or not.

(b) Deduce from (1) and (2.10′) that

$$p_n = 2(-1)^{n-1}\sigma'(n),$$

where $\sigma'(n)$ is the sum of the divisors $d \geqslant 1$ of n such that n/d is odd.

(c) Let $N_r(n)$ denote the number of representations of n as a sum of r squares, that is to say the number of integer vectors $(x_1, \dots, x_r) \in \mathbf{Z}^r$ such that $x_1^2 + \dots + x_r^2 = n$. Deduce from (b) above and Example 8 that

$$N_r(n) = \frac{(2r)^n}{n!}\begin{vmatrix} \sigma'(1) & 1/2r & 0 & \dots & 0 \\ \sigma'(2) & \sigma'(1) & 2/2r & \dots & 0 \\ \vdots & \vdots & & & \vdots \\ \sigma'(n-1) & \sigma'(n-2) & \dots & \dots & (n-1)/2r \\ \sigma'(n) & \sigma'(n-1) & \dots & \dots & \sigma'(1) \end{vmatrix}.$$

23. If G is a finite group and d is a positive integer, let $w_d(G)$ denote the number of solutions of $x^d = 1$ in G. Show that

$$\sum_{n \geqslant 0} \frac{1}{n!} w_d(S_n) t^n = \exp\left(\sum_{r \mid d} \frac{1}{r} t^r\right). \tag{1}$$

(If $x \in S_n$ has cycle-type λ, then $x^d = 1$ if and only if all the parts of the partition λ divide d. Hence

(2)
$$w_d(S_n)/n! = \sum z_\lambda^{-1}$$

summed over such partitions λ of n. Let $\varphi_d: \Lambda_{\mathbf{Q}} \to \mathbf{Z}$ be the homomorphism defined by $\varphi_d(p_r) = 1$ or 0 according as r does or does not divide d. Then the right-hand side of (2) is by (2.14′) equal to $\varphi_d(h_n)$ and hence the generating function (1) is $\varphi_d(H(t))$.)

24. Another involution on the ring Λ may be defined as follows. Let

(1)
$$u = tH(t) = t + h_1 t^2 + h_2 t^3 + \dots .$$

Then t can be expressed as a power series in u, say

(2)
$$t = u + h_1^* u^2 + h_2^* u^3 + \dots ,$$

with coefficients $h_r^* \in \Lambda'$ for each $r \geqslant 1$. The formulas (1) and (2) show that the ring homomorphism $\psi: \Lambda \to \Lambda$ defined by $\psi(h_r) = h_r^*$ for each $r \geqslant 1$ is an *involution* on Λ.

For each $f \in \Lambda$, let $f^* = \psi(f)$. Thus for example $h_\lambda^* = h_{\lambda_1}^* h_{\lambda_2}^* \dots$ for each partition $\lambda = (\lambda_1, \lambda_2, \dots)$, and the h_λ^* form a $\mathbf{Z}$-basis of Λ.

(a) To calculate h_n^* explicitly, we may argue as follows. From (2) we have

$$dt = \sum_{n \geqslant 0} (n+1) h_n^* u^n \, du$$

and therefore $(n+1)h_n^*$ is the residue of the differential

$$dt/u^{n+1} = dt/t^{n+1} H(t)^{n+1},$$

hence is equal to the coefficient of t^n in the expansion of $H(t)^{-n-1}$ in powers of t. Writing $H(t) = 1 + H^+(t)$ as in Example 20, it follows that

$$(n+1)h_n^* = \sum_{|\lambda| = n} (-1)^{l(\lambda)} \binom{n + l(\lambda)}{n} u_\lambda h_\lambda$$

with u_λ as in Example 20.

(b) Show likewise that

$$p_n^* = \sum_{|\lambda| = n} z_\lambda^{-1} (-n)^{l(\lambda)} p_\lambda$$

and that

$$(n-1)e_n^* = - \sum_{|\lambda| = n} \binom{n-1}{l(\lambda)} u_\lambda e_\lambda .$$

25. Let $f(x), g(x) \in \mathbf{Z}[[x]]$ be formal power series with constant term 1. Let $t = xf(x)$ and express $g(x)$ as a power series in t:

$$g(x) = H(t) = \sum_{n \geqslant 0} h_n t^n .$$

Let $\varphi_n(f,g)$ denote the coefficient of x^n in $(f+xf')g/f^{n+1}$, where f' is the derivative of f. Then we have

$$h_n = \varphi_n(f,g), \qquad e_n = (-1)^n \varphi_n(f,g^{-1})$$

and (with the notation of Example 24)

$$h_n^* = \varphi_n(fg,g^{-1}), \qquad e_n^* = (-1)^n \varphi_n(fg,g).$$

Moreover, if $\psi_n(f,g)$ is the coefficient of x^{n-1} in $g'/f^n g$, we have

$$p_n = \psi_n(f,g), \qquad p_n^* = \psi_n(fg,g^{-1}).$$

(a) Take $f(x) = (1+x)^{-\alpha}$ and $g(x) = (1+x)^{\beta}$. Then

$$\varphi_n(f,g) = \frac{\beta}{n\alpha+\beta}\binom{n\alpha+\beta}{n} = u_n(\alpha,\beta),$$

say, and therefore

$$h_n = u_n(\alpha,\beta), \qquad e_n = u_n(1-\alpha,\beta),$$

$$h_n^* = u_n(\alpha-\beta,-\beta), \qquad e_n^* = u_n(1-\alpha+\beta,-\beta),$$

and

$$p_n = \frac{\beta}{\alpha}\binom{n\alpha}{n}, \qquad p_n^* = \frac{\beta}{\beta-\alpha}\binom{n\alpha-n\beta}{n}.$$

In particular, $u_n(2,1)$ is the nth Catalan number $C_n = (n+1)^{-1}\binom{2n}{n}$, and $u_n(2,2) = C_{n+1}$, $u_n(2,-1) = -C_{n-1}$. Hence when $p_n = a\binom{2n}{n}$ with $a = \frac{1}{2}, 1, -\frac{1}{2}$ we have respectively $h_n = C_n, C_{n+1}, -C_{n-1}$, and $h_n^* = -\delta_{1n}, (-1)^n(n+1), n^{-1}\binom{3n}{n-1}$.

(b) Take $f(x) = e^{-\alpha x}$, $g(x) = e^{\beta x}$. Then

$$\varphi_n(f,g) = \beta(n\alpha+\beta)^{n-1}/n! = v_n(\alpha,\beta),$$

say, and hence

$$h_n = v_n(\alpha,\beta), \qquad e_n = v_n(-\alpha,\beta),$$

$$h_n^* = v_n(\alpha-\beta,-\beta), \qquad e_n^* = v_n(\beta-\alpha,-\beta),$$

and

$$p_n = \frac{\beta(n\alpha)^{n-1}}{(n-1)!}, \qquad p_n^* = \frac{-\beta(n\alpha-n\beta)^{n-1}}{(n-1)!}.$$

26. Let k be a finite field with q elements, and let V be a k-vector space of dimension n. Let $\mathbf{S} = \mathbf{S}(V)$ be the symmetric algebra of V over k (so that if $x_1, \ldots, x_n$ is any k-basis of V, then $\mathbf{S} = k[x_1, \ldots, x_n]$). Let

$$f_V(t) = \prod_{v \in V} (t + v),$$

a polynomial in $\mathbf{S}[t]$.

(a) Show that

$$f(at + bu) = af(t) + bf(u)$$

for all $a, b \in k$ and indeterminates t, u. (If $a \in k$, $a \neq 0$ we have

$$f(at) = \prod_{v \in V} (at + v) = a^{q^n} \prod_{v \in V} (t + a^{-1}v)$$
$$= af(t).$$

Next, let

$$f(t + u) - f(t) - f(u) = \sum_{r > 0} t^r g_r(u)$$

with $g_r(u) \in \mathbf{S}[u]$; since $f(t + v) = f(t)$ and $f(v) = 0$ for all $v \in V$, it follows that $g_r(v) = 0$ for all $v \in V$. Since g_r has degree $< q^n$, we conclude that $g_r = 0$ for each r, and hence that $f(t + u) = f(t) + f(u)$.)

(b) Deduce from (a) that $f_V(t)$ is of the form

$$f_V(t) = t^{q^n} + a_1(V)t^{q^{n-1}} + \ldots + a_n(V)t,$$

where each $a_r(V) \in \mathbf{S}$, and in particular $a_n(V)$ is the product of the non-zero vectors in V.

Show that

$$\sum_{L < V} a_1(L) = 0 \tag{1}$$

where the sum is over all lines (i.e. one-dimensional subspaces) L in V. (Since each $v \neq 0$ in V lies in a unique line $L = kv$, it follows that

$$f_V(t) = t \prod_{L < V} t^{-1} f_L(t) = t \prod_{L < V} (t^{q-1} + a_1(L))$$

and therefore the sum (1) is equal to the coefficient of $t^{q^n - 1}$ in $f_V(t)$, which is clearly zero.)

(c) Let U be a vector subspace of V. The mapping $v \mapsto f_U(v)$ of V into $\mathbf{S}$ is k-linear, by (a) above, and its kernel is U. Hence its image $f_U(V)$ is isomorphic to the quotient of V by U, and we shall denote it by V/U. Each element of V/U is a product of the form $\prod_{u \in U}(v + u)$ for some $v \in V$, i.e. it is the product in $\mathbf{S}$ of the elements of a coset of U in V.

Show that

$$f_V(t) = f_{V/U}(f_U(t))$$

and that if T is a vector subspace of U then

$$V/U = (V/T)/(U/T).$$

(d) Show that for $1 \leqslant r \leqslant n-1$ we have

$$a_r(V) = \sum_{L<V} a_r(V/L) \tag{2}$$

summed as before over all lines L in V. (From (c) above we have $f_V(t) = f_{V/L}(f_L(t))$, from which it follows that

$$a_r(V) = a_r(V/L) + a_{r-1}(V/L)a_1(L)^{q^{n-r}}.$$

Since the number of lines L in V is $1 + q + \ldots + q^{n-1} \equiv 1$ (mod. q), it follows that

$$a_r(V) = \sum_L a_r(V/L) + \sum_L a_{r-1}(V/L)a_1(L)^{q^{n-r}} \tag{3}$$

By induction on $n = \dim V$ we may assume that

$$a_{r-1}(V/L) = \sum_M a_{r-1}(V/M) \tag{4}$$

summed over all two-dimensional subspaces M in V that contain L. The second term on the right-hand side of (3) is therefore equal to

$$\sum_M a_{r-1}(V/M)\Bigl(\sum_{L<M} a_1(L)\Bigr)^{q^{n-r}}$$

which is zero by (1) above.)

(e) Deduce from (d) that

$$a_r(V) = \sum_U a_r(V/U)$$

summed over all subspaces U of V of dimension $n-r$, where $a_r(V/U)$ is the product of the vectors $v \in V$ such that $v \notin U$.

27. As in Example 26, let k be a finite field with q elements, let $x_1, \ldots, x_n$ be independent indeterminates over k, and let $k[x] = k[x_1, \ldots, x_n]$. Let V be the k-vector space spanned by $x_1, \ldots, x_n$, and let

$$f_V(t) = t^{q^n} + a_1 t^{q^{n-1}} + \ldots + a_n t \tag{1}$$

be the monic polynomial whose roots are the elements of V. Let $k[a] = k[a_1, \ldots, a_n] \subset k[x]$.

(a) The coefficients $a_1,\dots,a_n$ in (1) are algebraically independent over k. (Let $b_1,\dots,b_n$ be independent indeterminates over k, and let W be the set of roots of the polynomial

$$g(t)=t^{q^n}+b_1t^{q^{n-1}}+\dots+b_nt$$

in a splitting field. The roots are all distinct, since $g'(t)=b_n\neq 0$, and since $g(\alpha t+\beta u)=\alpha g(t)+\beta g(u)$ for $\alpha,\beta\in k$ it follows that W is a k-vector space of dimension n. Choose a basis $(y_1,\dots,y_n)$ of W and let $\theta: k[x]\to k[y]$ be the k-algebra homomorphism that sends x_i to y_i $(1\leqslant i\leqslant n)$. Then $\theta(a_i)=b_i$ $(1\leqslant i\leqslant n)$; since the b_i are algebraically independent over k, so are the a_i.)

(b) Let $G=GL(V)\cong GL_n(k)$ be the group of automorphisms of the vector space V. Then G acts on the algebra $k[x]$; let $k[x]^G$ denote the subalgebra of G-invariants. Since G permutes the roots of the polynomial $f_V(t)$, it fixes each of the coefficients a_i, so that $k[a]\subset k[x]^G$. In fact (see (d) below) $k[a]=k[x]^G$.

(c) $k[x]$ is a free $k[a]$-module with basis $(x^\alpha)_{\alpha\in E}$, where

$$E=\{\alpha=(\alpha_1,\dots,\alpha_n):0\leqslant\alpha_i\leqslant q^n-q^{i-1}-1\}.$$

(Let V_r denote the subspace of V spanned by $x_1,\dots,x_r$, for $0\leqslant r\leqslant n-1$ (so that $V_0=0$). The polynomial $g_r(t)=f_V(t)/f_{V_r}(t)$ is monic of degree q^n-q^r, has coefficients in the ring $k[a,x_1,\dots,x_r]$, and has x_{r+1} as a root.

Now let $h\in k[x]$. Use the polynomial g_{n-1} to reduce the degree of h in x_n below q^n-q^{n-1}. Then use g_{n-2} to reduce the degree of h in x_{n-1} below q^n-q^{n-2}, and so on. In the end we shall obtain say

$$h=\sum_{\alpha\in E}h_\alpha x^\alpha \tag{2}$$

with coefficients $h_\alpha\in k[a]\subset k[x]^G$.

Hence the x^α, $\alpha\in E$, span $k[x]$ as $k[x]^G$-module. They therefore also span the field $k(x)=k(x_1,\dots,x_n)$ as vector space over $k(x)^G$, since every element of $k(x)$ can be written in the form u/v with $u\in k[x]$ and $v\in k[x]^G$. But by Galois theory the dimension of $k(x)$ over $k(x)^G$ is

$$|G|=\prod_{i=1}^n(q^n-q^{i-1})=|E|;$$

hence $(x^\alpha)_{\alpha\in E}$ is a *basis* of $k(x)$ over $k(x)^G$, i.e. the expression (2) for $h\in k[x]$ is unique.)

(d) Suppose now that $h\in k[x]^G$. Then in (2) we must have $h_\alpha=0$ if $\alpha\neq(0,\dots,0)$, and $h=h_{0,\dots,0}\in k[a]$. Hence $k[x]^G=k[a]$ (Dickson's theorem).

Notes and references

Example 11. For more information on Bell polymomials, Stirling numbers etc., see for example L. Comtet's book [C3].

Example 13. This example is due to A. O. Morris [M15].

Example 27. The proof of Dickson's theorem given here I learnt from R. Steinberg.

3. Schur functions

Suppose to begin with that the number of variables is finite, say $x_1, \dots, x_n$. Let $x^\alpha = x_1^{\alpha_1} \dots x_n^{\alpha_n}$ be a monomial, and consider the polynomial a_α obtained by antisymmetrizing x^α: that is to say,

$$a_\alpha = a_\alpha(x_1, \dots, x_n) = \sum_{w \in S_n} \varepsilon(w) . w(x^\alpha)$$

where $\varepsilon(w)$ is the *sign* (± 1) of the permutation w. This polynomial a_α is skew-symmetric, i.e. we have

$$w(a_\alpha) = \varepsilon(w) a_\alpha$$

for any $w \in S_n$; in particular, therefore, a_α vanishes unless $\alpha_1, \dots, \alpha_n$ are all distinct. Hence we may as well assume that $\alpha_1 > \alpha_2 > \dots > \alpha_n \geqslant 0$, and therefore we may write $\alpha = \lambda + \delta$, where λ is a partition of length $\leqslant n$, and $\delta = (n-1, n-2, \dots, 1, 0)$. Then

$$a_\alpha = a_{\lambda+\delta} = \sum_w \varepsilon(w) . w(x^{\lambda+\delta})$$

which can be written as a determinant:

$$a_{\lambda+\delta} = \det(x_i^{\lambda_j + n - j})_{1 \leqslant i,j \leqslant n}.$$

This determinant is divisible in $\mathbf{Z}[x_1, \dots, x_n]$ by each of the differences $x_i - x_j$ $(1 \leqslant i < j \leqslant n)$, and hence by their product, which is the *Vandermonde determinant*

$$\prod_{1 \leqslant i < j \leqslant n} (x_i - x_j) = \det(x_i^{n-j}) = a_\delta .$$

So $a_{\lambda+\delta}$ is divisible by a_δ in $\mathbf{Z}[x_1, \dots, x_n]$, and the quotient

(3.1) $$s_\lambda = s_\lambda(x_1, \dots, x_n) = a_{\lambda+\delta}/a_\delta$$

is *symmetric*, i.e. is in Λ_n. It is called the *Schur function* in the variables $x_1, \dots, x_n$, corresponding to the partition λ (where $l(\lambda) \leqslant n$), and is homogeneous of degree $|\lambda|$.

Notice that the definition (3.1) makes sense for any integer vector $\lambda \in \mathbf{Z}^n$ such that $\lambda + \delta$ has no negative parts. If the numbers $\lambda_i + n - i$ $(1 \leqslant i \leqslant n)$ are not all distinct, then $s_\lambda = 0$. If they are all distinct, then we have $\lambda + \delta = w(\mu + \delta)$ for some $w \in S_n$ and some partition μ, and $s_\lambda = \varepsilon(w) s_\mu$.

The polynomials $a_{\lambda+\delta}$, where λ runs through all partitions of length $\leqslant n$, form a basis of the $\mathbf{Z}$-module A_n of skew-symmetric polynomials in $x_1, \dots, x_n$. Multiplication by a_δ is an isomorphism of Λ_n onto A_n (i.e. A_n is the free Λ_n-module generated by a_δ), and therefore

(3.2) *The Schur functions* $s_\lambda(x_1,\ldots,x_n)$, *where* $l(\lambda)\leqslant n$, *form a* **Z**-*basis of* Λ_n. |

Now let us consider the effect of increasing the number of variables. If $l(\alpha)\leqslant n$, it is clear that $a_\alpha(x_1,\ldots,x_n,0)=a_\alpha(x_1,\ldots,x_n)$. Hence

$$\rho_{n+1,n}(s_\lambda(x_1,\ldots,x_{n+1}))=s_\lambda(x_1,\ldots,x_n)$$

in the notation of §2. It follows that for each partition λ the polynomials $s_\lambda(x_1,\ldots,x_n)$, as $n\to\infty$, define a unique element $s_\lambda\in\Lambda$, homogeneous of degree $|\lambda|$. From (3.2) we have immediately:

(3.3) *The* s_λ *form a* **Z**-*basis of* Λ, *and for each* $k\geqslant 0$ *the* s_λ *such that* $|\lambda|=k$ *form a* **Z**-*basis of* Λ^k. |

From (2.4) and (2.8) it follows that each Schur function s_λ can be expressed as a polynomial in the elementary symmetric functions e_r, and as a polynomial in the complete symmetric functions h_r. The formulas are:

(3.4)
$$s_\lambda=\det(h_{\lambda_i-i+j})_{1\leqslant i,j\leqslant n}$$

where $n\geqslant l(\lambda)$, and

(3.5)
$$s_\lambda=\det(e_{\lambda'_i-i+j})_{1\leqslant i,j\leqslant m}$$

where $m\geqslant l(\lambda')$.

By (2.9′), it is enough to prove one of these formulas, say (3.4). We shall work with n variables $x_1,\ldots,x_n$. For $1\leqslant k\leqslant n$ let $e_r^{(k)}$ denote the elementary symmetric functions of $x_1,\ldots,x_{k-1},x_{k+1},\ldots,x_n$ (omitting x_k), and let M denote the $n\times n$ matrix

$$M=\left((-1)^{n-i}e_{n-i}^{(k)}\right)_{1\leqslant i,k\leqslant n}.$$

The formula (3.4) will be a consequence of

(3.6) *For any* $\alpha=(\alpha_1,\ldots,\alpha_n)\in\mathbf{N}^n$, *let*

$$A_\alpha=(x_j^{\alpha_i}),\qquad H_\alpha=(h_{\alpha_i-n+j})$$

($n\times n$ matrices). Then $A_\alpha=H_\alpha M$.

Proof. Let

$$E^{(k)}(t)=\sum_{r=0}^{n-1}e_r^{(k)}t^r=\prod_{i\neq k}(1+x_it).$$

Then

$$H(t)E^{(k)}(-t)=(1-x_kt)^{-1}.$$

By picking out the coefficient of t^{α_i} on either side, we obtain

$$\sum_{j=1}^{n} h_{\alpha_i-n+j}\cdot(-1)^{n-j}e^{(k)}_{n-j}=x_k^{\alpha_i}$$

and hence $H_\alpha M=A_\alpha$. |

Now take determinants in (3.6): we obtain

$$a_\alpha=\det(A_\alpha)=\det(H_\alpha)\det(M)$$

for any $\alpha\in\mathbf{N}^n$, and in particular $\det M=a_\delta$, since $\det(H_\delta)=1$. Hence

(3.7) $$a_\alpha=a_\delta\det(H_\alpha)$$

or equivalently

(3.7′) $$a_\alpha=a_\delta\sum_{w\in S_n}\varepsilon(w)h_{\alpha-w\delta}$$

for any $\alpha\in\mathbf{N}^n$. Taking $\alpha=\lambda+\delta$ in (3.7), we obtain (3.4), or equivalently from (3.7′)

(3.4′) $$s_\lambda=\sum_{w\in S_n}\varepsilon(w)h_{\lambda+\delta-w\delta}\,.$$

From (3.4) and (3.5) it follows that

(3.8) $$\omega(s_\lambda)=s_{\lambda'}$$

for all partitions λ.

Also from (3.4) and (3.5) we obtain, in particular,

(3.9) $$s_{(n)}=h_n,\qquad s_{(1^n)}=e_n\,.$$

Finally, the formula (3.4) or (3.4′) which expresses s_λ as a polynomial in the h's can also be expressed in terms of raising operators (§1):

(3.4″) $$s_\lambda=\prod_{i<j}(1-R_{ij})h_\lambda$$

where, for any raising operator R, Rh_λ means $h_{R\lambda}$.†

† It should be remarked that if R, R' are raising operators, $RR'h_\lambda=h_{RR'\lambda}$ is not necessarily equal to $R(R'h_\lambda)$. For it may well happen that $R'\lambda$ has a negative component, but $RR'\lambda$ does not, in which case $R'h_\lambda=0$ but $RR'h_\lambda\neq 0$. See [G3] for a discussion of this point.

Proof. In the ring $\mathbf{Z}[x_1^{\pm 1},\ldots,x_n^{\pm 1}]$ we have

$$\sum_{w\in S_n}\varepsilon(w)x^{\lambda+\delta-w\delta}=x^{\lambda+\delta}a_{-\delta}=x^{\lambda+\delta}\prod_{i<j}\left(x_i^{-1}-x_j^{-1}\right)$$

$$=\prod_{i<j}\left(1-x_ix_j^{-1}\right).x^{\lambda}$$

$$=\prod_{i<j}(1-R_{ij})x^{\lambda}$$

where $R(x^{\lambda})=x^{R\lambda}$ for any raising operator R. If we now apply the $\mathbf{Z}$-linear map $\varphi:\mathbf{Z}[x_1^{\pm 1},\ldots,x_n^{\pm 1}]\to\Lambda_n$ defined by $\varphi(x^{\alpha})=h_{\alpha}$ for all $\alpha\in\mathbf{Z}^n$, we see that

$$\sum_{w\in S_n}\varepsilon(w)h_{\lambda+\delta-w\delta}=\prod_{i<j}(1-R_{ij})h_{\lambda}$$

and therefore (3.4″) follows from (3.4′). |

(3.10) *Remark.* In view of (2.15) we may use (3.4) or (3.5) to define 'Schur operations' in any λ-ring R. If μ is any partition and x is any element of R, we define

$$S^{\mu}(x)=\det(\sigma^{\mu_i-i+j}(x))_{1\leqslant i,j\leqslant n}$$

$$=\det(\lambda^{\mu'_i-i+j}(x))_{1\leqslant i,j\leqslant m}$$

where $n\geqslant l(\mu)$ and $m\geqslant l(\mu')$. We have

$$S^{\mu}(-x)=(-1)^{|\mu|}S^{\mu'}(x)$$

and in particular

$$S^{(n)}(x)=\sigma^n(x),\qquad S^{(1^n)}(x)=\lambda^n(x).$$

For example, the results of Examples 1–3 below evaluate $S^{\lambda}(1+q+\ldots+q^{n-1})$, $S^{\lambda}((1-q)^{-1})$ and $S^{\lambda}((a-b)/(1-q))$, where a,b,q are elements of rank 1 in a λ-ring R such that $1-q$ is a unit in R.

Since each $f\in\Lambda$ is an integral linear combination of the s_{μ}, say

$$f=\sum a_{\mu}s_{\mu},$$

it follows that f determines a 'natural operation'

$$F=\sum a_{\mu}S^{\mu}$$

on the category of λ-rings. F is *natural* in the sense that it commutes with all λ-homomorphisms (because it is a polynomial in the λ^r). Conversely, any natural operation F arises in this way, from $f=F(e_1)$.

Examples

1. Take $x_i = q^{i-1}$ ($1 \leqslant i \leqslant n$) as in §2, Example 3. If λ is any partition of length $\leqslant n$, we have

$$a_{\lambda+\delta} = \det(q^{(i-1)(\lambda_j+n-j)})_{1 \leqslant i,j \leqslant n}$$

which is a Vandermonde determinant in the variables q^{λ_j+n-j} ($1 \leqslant j \leqslant n$), so that

$$a_{\lambda+\delta} = \prod_{i<j} (q^{\lambda_j+n-j} - q^{\lambda_i+n-i})$$

$$= q^{n(\lambda)+n(n-1)(n-2)/6} \prod_{i<j} (1 - q^{\lambda_i-\lambda_j-i+j})$$

which by use of §1, Example 1 is equal to

$$\frac{q^{n(\lambda)+n(n-1)(n-2)/6} \prod_{i \geqslant 1} \varphi_{\lambda_i+n-i}(q)}{\prod_{x \in \lambda} (1 - q^{h(x)})}$$

where $h(x)$ is the hook-length at $x \in \lambda$, and $\varphi_r(q) = (1-q)\dots(1-q^r)$. Hence (§1, Example 3)

$$s_\lambda = a_{\lambda+\delta}/a_\delta = q^{n(\lambda)} \prod_{x \in \lambda} \frac{1 - q^{n+c(x)}}{1 - q^{h(x)}}$$

where $c(x)$ is the content (§1, Example 3) of $x \in \lambda$.

For any partition λ define

$$\begin{bmatrix} n \\ \lambda \end{bmatrix} = \prod_{x \in \lambda} \frac{1 - q^{n-c(x)}}{1 - q^{h(x)}}$$

(which when $\lambda = (r)$ agrees with the notation $\begin{bmatrix} n \\ r \end{bmatrix}$ for the q-binomial coefficients introduced in §2, Example 3). Then we have

$$s_\lambda(1, q, \dots, q^{n-1}) = q^{n(\lambda)} \begin{bmatrix} n \\ \lambda' \end{bmatrix}.$$

$\begin{bmatrix} n \\ \lambda \end{bmatrix}$ is a polynomial in q, of degree

$$d = \sum_{x \in \lambda} (n - c(x) - h(x)) = \sum_{i=1}^{n} (n+1-2i)\lambda'_i$$

by using §1, Examples 2 and 3. If a_i is the coefficient of q^i in $\begin{bmatrix} n \\ \lambda \end{bmatrix}$ for $0 \leqslant i \leqslant d$, then clearly $a_i = a_{d-i}$. We shall show in §8, Example 4 that $\begin{bmatrix} n \\ \lambda \end{bmatrix}$ is *unimodal* (or 'spindle-shaped'), i.e. that $a_0 \leqslant a_1 \leqslant \dots \leqslant a_{[d/2]}$.

Finally, we can express $\begin{bmatrix} n \\ \lambda \end{bmatrix}$ as a determinant in the q-binomial coefficients $\begin{bmatrix} n \\ r \end{bmatrix}$, by using (3.5).

2. Let $n \to \infty$ in Example 1, so that $H(t) = \prod_{i=0}^{\infty}(1-q^i t)^{-1}$. From Example 1 we have

$$s_\lambda = q^{n(\lambda)} \prod_{x \in \lambda} (1 - q^{h(x)})^{-1} = q^{n(\lambda)} H_\lambda(q)^{-1}$$

where $H_\lambda(q)$ is the 'hook polynomial' $\prod_{x\in\lambda}(1-q^{h(x)})$.

3. More generally, let

$$H(t) = \prod_{i=0}^{\infty} \frac{1-bq^i t}{1-aq^i t}$$

as in §2, Example 5. Then

$$(*) \qquad s_\lambda = q^{n(\lambda)} \prod_{x \in \lambda} \frac{a - bq^{c(x)}}{1-q^{h(x)}}.$$

For if we replace t by $a^{-1}t$, the effect is to replace s_λ by $a^{-|\lambda|}s_\lambda$. Hence we may assume that $a=1$. Both sides of $(*)$ are then polynomials in b, hence it is enough to show that they are equal for infinitely many values of b. But when $b = q^n$ and $a = 1$ we are back in the situation of Example 1, and $(*)$ is therefore true for $b = q^n$.

4. Suppose $x_i = 1$ $(1 \leqslant i \leqslant n)$, $x_i = 0$ for $i > n$. Then $E(t) = (1+t)^n$, and

$$s_\lambda = \prod_{x\in\lambda} \frac{n + c(x)}{h(x)}$$

by setting $q = 1$ in Example 1.

More generally, if $E(t) = (1+t)^X$, where X need not be a positive integer, then

$$s_\lambda = \prod_{x\in\lambda} \frac{X + c(x)}{h(x)}$$

for the same reason as in Example 3: both sides are polynomials in X which take the same values at all positive integers.

These polynomials may be regarded as generalized binomial coefficients, and they take integer values whenever X is an integer. For any partition λ define

$$\binom{X}{\lambda} = \prod_{x\in\lambda} \frac{X - c(x)}{h(x)}$$

(which is consistent with the usual notation for binomial coefficients). Then

$$\binom{X}{\lambda} = \det\left(\binom{X}{\lambda_i - i + j}\right)$$

by (3.5). Also

$$\binom{-X}{\lambda} = (-1)^{|\lambda|}\binom{X}{\lambda'}.$$

5. As in §2, Example 2, take $x_i = 1/n$ for $1 \leqslant i \leqslant n$, $x_i = 0$ for $i > n$, and let $n \to \infty$. Then $E(t) = H(t) = e^t$, and from Example 4 we have

$$s_\lambda = \lim_{n\to\infty} \frac{1}{n^{|\lambda|}} \prod_{x\in\lambda} \frac{n+c(x)}{h(x)}$$

$$= \prod_{x\in\lambda} h(x)^{-1}.$$

6. Let $p(n)$ denote the number of partitions of n. Then

$$\det(p(i-j+1))_{1\leqslant i,j\leqslant n}$$

is equal to ± 1 or 0 according as n is or is not a pentagonal number. (Use §2, Example 6 together with (3.4).)

7. Let m be a positive integer. Then

$$\prod_{1\leqslant i<j\leqslant n} \frac{x_i^m - x_j^m}{x_i - x_j} = \frac{a_{m\delta}}{a_\delta} = s_{(m-1)\delta}$$

$$= \det(h_{(m-1)(n-i)-i+j})_{1\leqslant i,j\leqslant n}$$

$$= \det(h_{mi-j})_{1\leqslant i,j\leqslant n-1}.$$

In particular,

$$\prod_{i<j}(x_i + x_j) = \det(h_{2i-j}).$$

8. Consider the ring $Q_n = \mathbf{Q}[x_1^{\pm 1},\ldots,x_n^{\pm 1}]$ of polynomials in $x_1,\ldots,x_n$ and their inverses. For each $\alpha \in \mathbf{Z}^n$ the monomial $x^\alpha = x_1^{\alpha_1}\ldots x_n^{\alpha_n}$ generates a symmetric function

$$\tilde{m}_\alpha = \sum_{w\in S_n} x^{w\alpha}$$

and the $\tilde{m}_\alpha$ such that $\alpha_1 \geqslant \alpha_2 \geqslant \ldots \geqslant \alpha_n$ form a basis of $Q_n^{S_n}$.

Define a linear mapping $\varphi: Q_n^{S_n} \to \Lambda_n \otimes \mathbf{Q}$ by $\varphi(\tilde{m}_\alpha) = h_\alpha$ (with the usual convention that $h_\alpha = 0$ if any α_i is negative).

(a) For all $\alpha, \beta \in \mathbf{Z}^n$ we have

$$\varphi(a_\alpha a_\beta) = \det(h_{\alpha_i+\beta_j})_{1\leqslant i,j\leqslant n}.$$

For

$$a_\alpha a_\beta = \sum_{w_1,w_2\in S_n} \varepsilon(w_1 w_2) x^{w_1\alpha + w_2\beta}$$

$$= \sum_{w\in S_n} \varepsilon(w) \sum_{w_1\in S_n} x^{w_1(\alpha+w\beta)}$$

$$= \sum_{w\in S_n} \varepsilon(w) \tilde{m}_{\alpha+w\beta}$$

so that

$$\varphi(a_\alpha a_\beta) = \sum_{w \in S_n} \varepsilon(w) h_{\alpha + w\beta} = \det(h_{\alpha_i + \beta_j}).$$

(b) In particular, if λ is any partition of length $\leqslant n$, we have

$$\varphi(s_\lambda a_\delta a_{-\delta}) = \varphi(a_{\lambda+\delta} a_{-\delta}) = \det(h_{\lambda_i - i + j}) = s_\lambda$$

by (3.4). Since the s_λ form a $\mathbf{Z}$-basis of Λ_n, it follows that $\varphi(f a_\delta a_{-\delta}) = f$ for all $f \in \Lambda_n$.

(c) Let $\alpha, \beta \in \mathbf{N}^n$ and let $\bar{\beta} = (\beta_n, \ldots, \beta_1)$ be the reverse of β. Then $s_{\bar{\beta}} = a_{\beta - \delta}/a_{-\delta}$, and hence from (a), (b) we have

$$s_\alpha s_{\bar{\beta}} = \varphi(a_{\alpha+\delta} a_{\beta - \delta}) = \det(h_{\alpha_i + \beta_j - i + j})_{1 \leqslant i, j \leqslant n},$$

a formula which expresses the product of two Schur functions (in a finite number of variables) as a determinant in the h_r.

9. Let $a, b \geqslant 0$, then $(a \mid b)$ is the Frobenius notation (§1) for the partition $(a+1, 1^b)$. From the determinant formula (3.4) we have

$$s_{(a \mid b)} = h_{a+1} e_b - h_{a+2} e_{b-1} + \ldots + (-1)^b h_{a+b+1}$$

If a or b is negative, we *define* $s_{(a \mid b)}$ by this formula. It follows that (when a or b is negative) $s_{(a \mid b)} = 0$ except when $a + b = -1$, in which case $s_{(a \mid b)} = (-1)^b$.

Now let λ be any partition of length $\leqslant n$. By multiplying the matrix $(h_{\lambda_i - i + j})_{1 \leqslant i, j \leqslant n}$ on the right by the matrix $((-1)^{j-1} e_{n+1-j-k})_{1 \leqslant j, k \leqslant n}$, we obtain the matrix $(s_{(\lambda_i - i \mid n-k)})_{1 \leqslant i, k \leqslant n}$. By taking determinants and using §1, Example 4 we arrive at the formula

$$s_{(\alpha \mid \beta)} = \det(s_{(\alpha_i \mid \beta_j)})_{1 \leqslant i, j \leqslant r}$$

where $(\alpha \mid \beta) = (\alpha_1, \ldots, \alpha_r \mid \beta_1, \ldots, \beta_r)$.

10. $$s_\lambda(1 + x_1, 1 + x_2, \ldots, 1 + x_n) = \sum_\mu d_{\lambda\mu} s_\mu(x_1, \ldots, x_n)$$

summed over all partitions $\mu \subset \lambda$, where

$$d_{\lambda\mu} = \det\left(\binom{\lambda_i + n - i}{\mu_j + n - j}\right)_{1 \leqslant i, j \leqslant n}.$$

(Calculate $a_{\lambda+\delta}(1 + x_1, \ldots, 1 + x_n)$ and observe that $a_\delta(1 + x, \ldots, 1 + x_n) = a_\delta(x_1, \ldots, x_n)$.)

This formula can be used to calculate the Chern classes of the exterior square $\Lambda^2 E$ and symmetric square $\mathbf{S}^2 E$ of a vector bundle E. If $c(E) = \prod_{i=1}^m (1 + x_i)$ is the total Chern class of E, then

$$\begin{aligned} c(\Lambda^2 E) &= \prod_{i<j} (1 + x_i + x_j) \\ &= 2^{-m(m-1)/2} \prod_{i<j} (1 + 2x_i + 1 + 2x_j) \\ &= 2^{-m(m-1)/2} s_\delta (1 + 2x_1, \dots, 1 + 2x_m) \end{aligned}$$

by Example 7, where $\delta = (m-1, m-2, \dots, 0)$, and therefore

$$c(\Lambda^2 E) = 2^{-m(m-1)/2} \sum_{\mu \subset \delta} d_{\delta\mu} 2^{|\mu|} s_\mu(x_1, \dots, x_m).$$

Likewise

$$\begin{aligned} c(\mathbf{S}^2 E) &= \prod_{i \leqslant j} (1 + x_i + x_j) \\ &= 2^{-m(m-1)/2} \sum_{\nu \subset \varepsilon} d_{\varepsilon\nu} 2^{|\nu|} s_\nu(x_1, \dots, x_m) \end{aligned}$$

where $\varepsilon = (m, m-1, \dots, 1)$.

11. Let $\mu = (\mu_1, \dots, \mu_n)$ be a partition of length $\leqslant n$, and r a positive integer. Then, the variables being $x_1, \dots, x_n$, we have

$$a_{\mu+\delta} p_r = \sum_{q=1}^{n} a_{\mu+\delta+r\varepsilon_q} \tag{1}$$

where ε_q is the sequence with 1 in the qth place and 0 elsewhere. We shall rearrange the sequence $\mu + \delta + r\varepsilon_q$ in descending order. If it has two terms equal, it will contribute nothing to (1). We may therefore assume that for some $p \leqslant q$ we have

$$\mu_{p-1} + n - p + 1 > \mu_q + n - q + r > \mu_p + n - p,$$

in which case $a_{\mu+\delta+r\varepsilon_q} = (-1)^{q-p} a_{\lambda+\delta}$, where λ is the partition

$$\lambda = (\mu_1, \dots, \mu_{p-1}, \mu_q + p - q + r, \mu_p + 1, \dots, \mu_{q-1} + 1, \mu_{q+1}, \dots, \mu_n),$$

and therefore $\theta = \lambda - \mu$ is a border strip of length r. Recall (§1) that the *height* $\mathrm{ht}(\theta)$ of a border strip θ is one less than the number of rows it occupies. With this terminology, the preceding discussion shows that

$$s_\mu p_r = \sum_\lambda (-1)^{\mathrm{ht}(\lambda-\mu)} s_\lambda \tag{2}$$

summed over all partitions $\lambda \supset \mu$ such that $\lambda - \mu$ is a border strip of length r.

From (2) it follows that, for any partitions λ, μ, ρ such that $|\lambda| = |\mu| + |\rho|$, the coefficient of s_λ in $s_\mu p_\rho$ is

$$\sum_S (-1)^{\mathrm{ht}(S)}$$

summed over all sequences of partitions $S = (\lambda^{(0)}, \lambda^{(1)}, \ldots, \lambda^{(m)})$ such that $\mu = \lambda^{(0)} \subset \lambda^{(1)} \subset \ldots \subset \lambda^{(m)} = \lambda$, with each $\lambda^{(i)} - \lambda^{(i-1)}$ a border strip of length ρ_i, and

$$\mathrm{ht}(S) = \sum_i \mathrm{ht}(\lambda^{(i)} - \lambda^{(i-1)}).$$

12. Let $\sigma: \mathbf{Z}[x_1, \ldots, x_n] \to \Lambda_n$ be the $\mathbf{Z}$-linear mapping defined by $\sigma(x^\alpha) = s_\alpha$ for all $\alpha \in \mathbf{N}^n$. Then σ is Λ_n-linear, i.e. $\sigma(fg) = f\sigma(g)$ for $f \in \Lambda_n$ and $g \in \mathbf{Z}[x_1, \ldots, x_n]$. For $\sigma(x^\alpha) = a_\delta^{-1} a(x^{\alpha+\delta})$, where

$$a = \sum_{w \in S_n} \varepsilon(w) w$$

is the antisymmetrization operator. By linearity it follows that $\sigma(g) = a_\delta^{-1} a(gx^\delta)$ for all $g \in \mathbf{Z}[x_1, \ldots, x_n]$, and the result follows from the fact that a is Λ_n-linear.

13. If $a, b \geqslant 0$ we have

$$(a+b+1)a!\,b!\,s_{(a|b)} = \begin{vmatrix} p_1 & c_1 & 0 & \cdots & 0 \\ p_2 & p_1 & c_2 & \cdots & 0 \\ \vdots & & & & \vdots \\ p_{a+b} & & & & c_{a+b} \\ p_{a+b+1} & & & \cdots & p_1 \end{vmatrix}$$

where $(c_1, \ldots, c_{a+b}) = (-1, -2, \ldots, -a, b, b-1, \ldots, 1)$.

(Use the relation $s_{(a|b)} + s_{(a+1|b-1)} = h_{a+1} e_b$ which follows from the first formula in Example 9, together with the determinant formulas of §2, Example 8, and induction on b.)

14.
$$\prod \frac{1 + ux_i}{1 - tx_i} = E(u)H(t) = 1 + (t+u) \sum_{a,b \geqslant 0} s_{(a|b)} t^a u^b.$$

15. Let M be an $n \times n$ matrix with eigenvalues $x_1, \ldots, x_n$. Then for each integer $r \geqslant 0$ we have

$$M^{n+r} = \sum_{p=0}^{n-1} (-1)^p s_{(r|p)}(x_1, \ldots, x_n) M^{n-p-1}.$$

(If $M^{n+r} = \sum a_p M^{n-p-1}$, we have $x_i^{n+r} = \sum a_p x_i^{n-p-1}$ for $1 \leqslant i \leqslant n$; now solve these equations for $a_0, \ldots, a_{p-1}$.)

16. Let λ, μ be partitions of length $\leqslant n$, and let

$$P_n(\lambda, \mu) = \det(p_{\lambda_i + \mu_j + 2n - i - j})_{1 \leqslant i,j \leqslant n},$$

with the understanding that $p_0 = n$. Then in Λ_n we have

$$s_\lambda = P_n(\lambda, \mu)/P_n(0, \mu),$$

$$s_\lambda s_\mu = P_n(\lambda, \mu)/P_n(0,0).$$

(Observe that $P_n(\lambda, \mu) = a_{\lambda+\delta} a_{\mu+\delta}$, by multiplication of determinants.)

17. (a) Let p be an integer $\geqslant 2$ and let $\omega = e^{2\pi i/p}$. If λ is any partition of length $\leqslant p$, we have

$$(*) \qquad s_\lambda(1, \omega, \dots, \omega^{p-1}) = \prod_{1 \leqslant j < k \leqslant p} \frac{\omega^{\lambda_j + p - j} - \omega^{\lambda_k + p - k}}{\omega^{p-j} - \omega^{p-k}}$$

from which it follows that $s_\lambda(1, \omega, \dots, \omega^{p-1}) = \pm 1$ if $\lambda \sim_p 0$ (§1, Example 8(f)) and is zero otherwise. More precisely, if $\lambda \sim_p 0$ we have $s_\lambda(1, \omega, \dots, \omega^{p-1}) = \sigma_p(\lambda)$, where $\sigma_p(\lambda)$ is the sign $\varepsilon(w)$ of the unique permutation $w \in S_p$ such that $\lambda + \delta_p \equiv w\delta_p (\text{mod.}\, p)$, where $\delta_p = (p-1, p-2, \dots, 1, 0)$.

(b) Assume from now on that p is an odd prime. Then

$$E(t) = \prod_{r=1}^{p} (1 + \omega^{r-1} t) = 1 + t^p$$

$$\equiv (1+t)^p \quad (\text{mod.}\, p)$$

and therefore

$$(1) \qquad s_\lambda(1, \omega, \dots, \omega^{p-1}) \equiv s_\lambda(1, \dots, 1) \quad (\text{mod.}\, p)$$
$$= \binom{p}{\lambda'}$$

(Example 4). Hence $\binom{p}{\lambda} \equiv 0 \ (\text{mod.}\, p)$ unless $\lambda \sim {}_p 0$

(c) Let $q \neq p$ be another odd prime and let $\lambda = (q-1)\delta_p$. Then

$$(2) \qquad s_\lambda(x_1, \dots, x_p) = \prod_{i<j} (x_i^q - x_j^q)/(x_i - x_j)$$

from which it follows that

$$(3) \qquad s_\lambda(1, \dots, 1) = q^{p(p-1)/2} \equiv \left(\frac{q}{p}\right) \quad (\text{mod.}\, p)$$

where $\left(\frac{q}{p}\right)$ is the Legendre symbol, equal to $+1$ or -1 according as q is or is not a square modulo p. From (1) and (3) we deduce that

$$(4) \qquad s_\lambda(1, \omega, \dots, \omega^{p-1}) = \left(\frac{q}{p}\right).$$

(d) Let G_p (resp. G_q) denote the set of complex pth (resp. qth) roots of unity, and choose $S \subset G_p$ and $T \subset G_q$ such that

$$(x^p - 1)/(x-1) = \prod_{\sigma \in S} (x-\sigma)(x-\sigma^{-1}),$$

$$(x^q - 1)/(x-1) = \prod_{\tau \in T} (x-\tau)(x-\tau^{-1}).$$

From (2) and (4) we have

$$\left(\frac{q}{p}\right) = \prod (\alpha^q - \beta^q)/(\alpha - \beta) \tag{5}$$

the product being taken over all two-element subsets $\{\alpha, \beta\}$ of G_p, where we may assume that $\alpha^{-1}\beta \in S$. We have

$$\begin{aligned}(\alpha^q - \beta^q)/(\alpha-\beta) &= \prod_{\tau \in T} (\alpha - \beta\tau)(\alpha - \beta\tau^{-1}) \\ &= \prod_{\tau \in T} \alpha\beta(1-\sigma\tau)(\sigma^{-1} - \tau^{-1})\end{aligned} \tag{6}$$

where $\sigma = \alpha^{-1}\beta \in S$. Deduce from (5) and (6) that

$$\left(\frac{q}{p}\right) = \prod_{\sigma,\tau} (1-\sigma\tau)(\sigma^{-1} - \tau^{-1}) \tag{7}$$

(product over all $\sigma \in S, \tau \in T$). (Observe that $\prod \alpha\beta = 1$, that each $\sigma \in S$ arises as $\alpha^{-1}\beta$ from p subsets $\{\alpha, \beta\}$, and that $(1-\sigma\tau)(\sigma^{-1} - \tau^{-1})$ is a real number.)

By interchanging q and p we have likewise

$$\left(\frac{p}{q}\right) = \prod_{\sigma,\tau} (1-\sigma\tau)(\tau^{-1} - \sigma^{-1}) \tag{7'}$$

and therefore

$$\left(\frac{p}{q}\right) = (-1)^{(p-1)(q-1)/4} \left(\frac{q}{p}\right)$$

(the law of quadratic reciprocity).

18. Let k be a finite field with q elements, let $x_1, \dots, x_n$ be independent indeterminates over k, and let $k[x] = k[x_1, \dots, x_n]$. Let $\alpha = (\alpha_1, \dots, \alpha_n) \in \mathbf{N}^n$, let $q^\alpha = (q^{\alpha_1}, \dots, q^{\alpha_n})$, and let $A_\alpha \in k[x]$ be the polynomial obtained by antisymmetrizing the monomial x^{q^α}, so that

$$A_\alpha = A_\alpha(x_1, \dots, x_n) = \det(x_i^{q^{\alpha_j}})_{1 \leqslant i,j \leqslant n}. \tag{1}$$

As in the text, we may assume that $\alpha_1 > \alpha_2 > \dots > \alpha_n \geqslant 0$, so that $\alpha = \lambda + \delta$, where λ is a partition of length $\leqslant n$.

(a) As in §2, Example 26, let V denote the k-vector space spanned by $x_1,\dots,x_n$ in $k[x]$, and let $G = GL(V) \cong GL_n(k)$, acting on $k[x]$. Show that

$$gA_\alpha = (\det g)A_\alpha \tag{2}$$

for $g \in G$, and deduce that A_α is divisible in $k[x]$ by each non-zero $v \in V$. (The definition (1) shows that A_α is divisible by each x_i, and G acts transitively on $V - \{0\}$.)

(b) Consider in particular the case $\lambda = 0$, i.e. $\alpha = \delta = (n-1, n-2, \dots, 1, 0)$. The polynomial A_δ is homogeneous of degree

$$q^{n-1} + q^{n-2} + \dots + 1 = (q^n - 1)/(q-1),$$

and has leading term q^δ. Let V_0 denote the set of all non-zero vectors $v = \Sigma\, a_i x_i$ in V for which the first nonvanishing coefficient a_i is equal to 1. Then we have

$$A_\delta = \prod_{v \in V_0} v$$

and each A_α is divisible in $k[x]$ by A_δ.

(c) Now define, for any partition λ of length $\leqslant n$,

$$S_\lambda = S_\lambda(x_1,\dots,x_n) = A_{\lambda+\delta}/A_\delta. \tag{3}$$

From (2) above it follows that S_λ is G-invariant, and hence depends only on λ and V, not on the particular basis $(x_1,\dots,x_n)$ of V. Accordingly we shall write $S_\lambda(V)$ in place of $S_\lambda(x_1,\dots,x_n)$. It is a homogeneous polynomial of degree $\Sigma_{i=1}^n (q^{\lambda_i} - 1)q^{n-i}$.

(d) If t is another indeterminate, we have from (b) above

$$\begin{aligned} A_{\delta_{n+1}}(t, x_1,\dots,x_n) &= A_{\delta_n}(x_1,\dots,x_n) \prod_{v \in V} (t+v) \\ &= A_{\delta_n}(x_1,\dots,x_n) f_V(t) \end{aligned}$$

in the notation of §2, Example 25. By expanding the determinant $A_{\delta_{n+1}}(t, x_1,\dots,x_n)$ along the top row, show that

$$f_V(t) = t^{q^n} - E_1(V)t^{q^{n-1}} + \dots + (-1)^n E_n(V)t,$$

where

$$E_r(V) = S_{(1^r)}(V) \qquad (1 \leqslant r \leqslant n).$$

The $E_r(V)$ are the analogues of the elementary symmetric functions, and in the notation of §2, Example 26 we have

$$\begin{aligned} E_r(V) &= (-1)^r a_r(V) \\ &= (-1)^r \sum_U a_r(V/U) \end{aligned}$$

summed over subspaces U of V of codimension r.

19. In continuation of Example 18, let

$$H_r(V) = S_{(r)}(V) \qquad (r \geqslant 0)$$

with the usual convention that $H_r(V) = 0$ if $r < 0$. (Likewise we define $E_r(V)$ to be zero if $r < 0$ or $r > n$.) Let $\mathbf{S}(V)$ $(= k[x])$ be the symmetric algebra of V over k, and let $\varphi: \mathbf{S}(V) \to \mathbf{S}(V)$ denote the Frobenius map $u \mapsto u^q$, which is a k-algebra endomorphism of $\mathbf{S}(V)$, its image being $k[x_1^q, \dots, x_n^q]$. Since we shall later encounter negative powers of φ, it is convenient to introduce

$$\hat{\mathbf{S}}(V) = \bigcup_{r \geqslant 0} \mathbf{S}(V)^{q^{-r}},$$

where $\mathbf{S}(V)^{q^{-r}} = k[x_1^{q^{-r}}, \dots, x_n^{q^{-r}}]$. On $\hat{\mathbf{S}}(V)$, φ is an automorphism.

(a) Let $E(V), H(V)$ denote the (infinite) matrices

$$H(V) = \left(\varphi^{i+1} H_{j-i}(V)\right)_{i,j \in \mathbf{Z}},$$

$$E(V) = \left((-1)^{j-i} \varphi^j E_{j-i}(V)\right)_{i,j \in \mathbf{Z}}.$$

Both are upper triangular, with 1's on the diagonal. Show that

$$E(V) = H(V)^{-1}.$$

(We have to show that

$$\sum_j (-1)^{k-j} \varphi^k(E_{k-j}) \varphi^{i+1}(H_{j-i}) = \delta_{ik}$$

for all i, k. This is clear if $i \geqslant k$. If $i < k$, we may argue as follows: since $f_V(x_i) = 0$ it follows from Example 18 that

$$\varphi^n(x_i) - E_1 \varphi^{n-1}(x_i) + \dots + (-1)^n E_n x_i = 0$$

and hence that

(1) $$\varphi^{n+r-1}(x_i) - \varphi^{r-1}(E_1)\varphi^{n+r-2}(x_i) + \dots + (-1)^n \varphi^{r-1}(E_n)\varphi^{r-1}(x_i) = 0$$

for all $r \geqslant 0$ and $1 \leqslant i \leqslant n$. On the other hand, by expanding the determinant $A_{(r)+\delta}$ down the first column, it is clear that $H_r = H_r(V)$ is of the form

(2) $$H_r = \sum_{i=1}^n u_i \varphi^{n+r-1}(x_i)$$

with coefficients $u_i \in k(x)$ independent of r. From (1) and (2) it follows that

(3) $$H_r - \varphi^{r-1}(E_1) H_{r-1} + \dots + (-1)^n \varphi^{r-1}(E_n) H_{r-n} = 0$$

for each $r \geqslant 0$. Putting $r = k - i$ and operating on (3) with φ^{i+1}, we obtain the desired relation.)

(b) Let λ be a partition of length $\leqslant n$. Then

$$S_\lambda(V) = \det\left(\varphi^{1-j}H_{\lambda_i-i+j}(V)\right)$$
$$= \det\left(\varphi^{j-1}E_{\lambda'_i-i+j}(V)\right)$$

in strict analogy with (3.4) and (3.5). (Let $\alpha = (\alpha_1, \dots, \alpha_n) \in \mathbf{N}^n$. From equation (2) above we have

$$\varphi^{1-j}(H_{\alpha_i-n+j}) = \sum_{i=1}^{n} \varphi^{\alpha_i}(x_k)\varphi^{1-j}(u_k)$$

which shows that the matrix $(\varphi^{1-j}H_{\alpha_i-n+j})_{i,j}$ is the product of the matrices $(\varphi^{\alpha_i}x_k)_{i,k}$ and $(\varphi^{1-j}u_k)_{k,j}$—all three matrices having n rows and n columns. On taking determinants it follows that

(4) $$\det\left(\varphi^{1-j}H_{\alpha_i-n+j}\right) = A_\alpha B$$

where $B = \det(\varphi^{1-j}u_k)$. In particular, taking $\alpha = \delta$ (so that $\alpha_i - n + j = j - i$) the left-hand side of (4) becomes equal to 1, so that $A_\delta B = 1$ and therefore

$$\det\left(\varphi^{1-j}H_{\alpha_i-n+j}\right) = A_\alpha / A_\delta$$

for all $\alpha \in \mathbf{N}^n$. Taking $\alpha = \lambda + \delta$, we obtain the first formula. The second (involving the E's) is then deduced from it and the result of (a) above, exactly as in the text.)

20. Let R be any commutative ring and let $a = (a_n)_{n \in \mathbf{Z}}$ be any (doubly infinite) sequence of elements of R. For each $r \in \mathbf{Z}$ we define $\tau^r a$ to be the sequence whose nth term is a_{n+r}. Let

$$(x \mid a)^r = (x + a_1) \dots (x + a_r)$$

for each $r \geqslant 0$.

Now let $x = (x_1, \dots, x_n)$ be a sequence of independent indeterminates over R, and for each $\alpha = (\alpha_1, \dots, \alpha_n) \in \mathbf{N}^n$ define

(1) $$A_\alpha(x \mid a) = \det((x_i \mid a)^{\alpha_j})_{1 \leqslant i,j \leqslant n}.$$

In particular, when $\alpha = \delta = (n-1, n-2, \dots, 1, 0)$, since $(x_i \mid a)^{n-j}$ is a monic polynomial in x_i of degree $n - j$, it follows that

(2) $$A_\delta(x \mid a) = \det(x_i^{n-j}) = a_\delta(x)$$

is the Vandermonde determinant, independent of the sequence a. Since $A_\alpha(x \mid a)$ is a skew-symmetric polynomial in $x_1, \dots, x_n$, it is therefore divisible by $A_\delta(x \mid a)$ in $R[x_1, \dots, x_n]$. As in the text, we may assume that $\alpha_1 > \alpha_2 > \dots > \alpha_n \geqslant 0$, i.e. that $\alpha = \lambda + \delta$ where λ is a partition of length $\leqslant n$. It follows therefore that

(3) $$s_\lambda(x \mid a) = A_{\lambda+\delta}(x \mid a) / A_\delta(x \mid a)$$

is a symmetric (but not homogeneous) polynomial in $x_1, \dots, x_n$ with coefficients in R. Moreover it is clear from the definitions that

$$A_{\lambda+\delta}(x \mid a) = a_{\lambda+\delta}(x) + \text{lower terms},$$

and hence that

$$s_\lambda(x \mid a) = s_\lambda(x) + \text{lower terms}.$$

Hence the $s_\lambda(x \mid a)$ form an R-basis of the ring $\Lambda_{n,R}$.

When $\lambda = (r)$ we shall write

$$h_r(x \mid a) = s_{(r)}(x \mid a) \qquad (r \geqslant 0)$$

with the usual convention that $h_r(x \mid a) = 0$ if $r < 0$; and when $\lambda = (1^r)$ $(0 \leqslant r \leqslant n)$ we shall write

$$e_r(x \mid a) = s_{(1^r)}(x \mid a)$$

with the convention that $e_r(x \mid a) = 0$ if $r < 0$ or $r > n$.

(a) Let t be another indeterminate and let

$$f(t) = \prod_{i=1}^{n} (t - x_i).$$

Show that

$$f(t) = \sum_{r=0}^{n} (-1)^r e_r(x \mid a)(t \mid a)^{n-r}. \tag{4}$$

(From (2) above it follows that

$$f(t) = A_{\delta_{n+1}}(t, x_1, \dots, x_n \mid a) / A_{\delta_n}(x_1, \dots, x_n \mid a);$$

now expand the determinant $A_{\delta_{n+1}}$ along the top row.)

(b) Let $E(x \mid a), H(x \mid a)$ be the (infinite) matrices

$$H(x \mid a) = \left(h_{j-i}(x \mid \tau^{i+1}a)\right)_{i,j \in \mathbf{Z}},$$

$$E(x \mid a) = \left((-1)^{j-i} e_{j-i}(x \mid \tau^j a)\right)_{i,j \in \mathbf{Z}}.$$

Both are upper triangular, with 1's on the diagonal. Show that

$$E(x \mid a) = H(x \mid a)^{-1}.$$

(We have to show that

$$\sum_j (-1)^{k-j} e_{k-j}(x \mid \tau^k a) h_{j-i}(x \mid \tau^{i+1} a) = \delta_{ik}$$

for all i, k. This is clear if $i > k$, so we may assume $i < k$. Since $f(x_i) = 0$ it follows from (4) above that

$$\sum_{r=0}^{n} (-1)^r e_r(x \mid a)(x_i \mid a)^{n-r} = 0$$

and hence, replacing a by $\tau^{s-1}a$ and multiplying by $(x_i \mid a)^{s-1}$, that

$$\sum_{r=0}^{n} (-1)^r e_r(x \mid \tau^{s-1}a)(x_i \mid a)^{n-r+s-1} = 0 \tag{5}$$

for all $s > 0$ and $1 \leqslant i \leqslant n$. Now it is clear, by expanding the determinant $A_{(m)+\delta}(x \mid a)$ down the first column, that $h_m(x \mid a)$ is of the form

$$h_m(x \mid a) = \sum_{i=1}^{n} (x_i \mid a)^{m+n-1} u_i(x) \tag{6}$$

with coefficients $u_i(x)$ rational functions of $x_1, \ldots, x_n$ independent of m. (In fact, $u_i(x) = 1/f'(x_i)$.)

From (5) and (6) it follows that

$$\sum_{r=0}^{n} (-1)^r e_r(x \mid \tau^{s-1}a) h_{s-r}(x \mid a) = 0$$

for each $s > 0$. Putting $s = k - i$ and replacing a by $\tau^{i+1}a$, we obtain the desired relation.)

(c) Let λ be a partition of length $\leqslant n$. Then

$$s_\lambda(x \mid a) = \det\left(h_{\lambda_i - i + j}(x \mid \tau^{1-j}a)\right),$$

$$= \det\left(e_{\lambda'_i - i + j}(x \mid \tau^{j-1}a)\right),$$

again in strict analogy with (3.4) and (3.5). (Let $\alpha = (\alpha_1, \ldots, \alpha_n) \in \mathbf{N}^n$. From (6) above we have

$$h_{\alpha_i - n + j}(x \mid \tau^{1-j}a) = \sum_{k=1}^{n} (x_k \mid \tau^{1-j}a)^{\alpha_i + j - 1} u_k(x)$$

$$= \sum_{k=1}^{n} (x_k \mid a)^{\alpha_i} (x_k \mid \tau^{1-j}a)^{j-1} u_k(x)$$

which shows that the matrix $H_\alpha = (h_{\alpha_i - n + j}(x \mid \tau^{1-j}a))_{i,j}$ is the product of the matrices $((x_k \mid a)^{\alpha_i})_{i,k}$ and $B = ((x_k \mid \tau^{1-j}a)^{j-1} u_k(x))_{k,j}$. On taking determinants it follows that

$$\det(H_\alpha) = A_\alpha \det(B).$$

In particular, when $\alpha = \delta$ the matrix H_δ is unitriangular and hence has determinant equal to 1. It follows that $A_\delta \det(B) = 1$ and hence that $\det(H_\alpha) = A_\alpha / A_\delta$ for all $\alpha \in \mathbf{N}^n$. Taking $\alpha = \lambda + \delta$, we obtain the first formula. The second formula, involving the e's, is then deduced from it and the result of (b) above, exactly as in the text.)

The results of this Example, and their proofs, should be compared with those of Example 19. It should also be remarked that when a is the zero sequence ($a_n = 0$ for all n) then $s_\lambda(x \mid a)$ is the Schur function $s_\lambda(x)$.

21. Let R be a commutative ring and let $S = R[h_{rs} : r \geqslant 1, s \in \mathbf{Z}]$, where the h_{rs} are independent indeterminates over R. Also, for convenience, define $h_{0s} = 1$ and $h_{rs} = 0$ for $r < 0$ and all $s \in \mathbf{Z}$. Let $\varphi: S \to S$ be the automorphism defined by $\varphi(h_{rs}) = h_{r,s+1}$. Thus $h_{rs} = \varphi^s(h_r)$, where $h_r = h_{r0}$, and we shall use this notation henceforth.

For any partition λ we define

$$\tilde{s}_\lambda = \det\left(\varphi^{-j+1} h_{\lambda_i - i + j}\right)_{1 \leqslant i,j \leqslant n} \tag{1}$$

where $n \geqslant l(\lambda)$. These 'Schur functions' include as special cases the symmetric functions of the last two Examples: in Example 19 we take $R = k$ and specialize h_{rs} to $\varphi^s H_r(V)$, and in Example 20 we specialize h_{rs} to $h_r(x \mid \tau^s a)$.

From (1) it follows that $h_r = \tilde{s}_{(r)}$. We define

$$e_r = \tilde{s}_{(1^r)}$$

for all $r \geqslant 0$, and set $e_r = 0$ for $r < 0$.

(a) Let E, H be the (infinite) matrices

$$H = \left(\varphi^{i+1} h_{j-i}\right)_{i,j \in \mathbf{Z}},$$

$$E = \left((-1)^{j-i} \varphi^j e_{j-i}\right)_{i,j \in \mathbf{Z}}.$$

Both are upper triangular with 1's on the diagonal. Show that

$$E = H^{-1}.$$

(We have to show that

$$\sum_j (-1)^{j-i} \varphi^j(e_{j-i}) \varphi^{j+1}(h_{k-j}) = \delta_{ik} \tag{2}$$

for all i, k. This is obvious if $i \geqslant k$, so assume $i < k$ and let $r = k - i$. Then (2) is equivalent to

$$\sum_{s=0}^{r} (-1)^s \varphi^{-s}(e_{r-s}) \varphi^{-s+1}(h_s) = 0$$

which follows from $e_r = \det(\varphi^{1-j} h_{1-i+j})_{1 \leqslant i,j \leqslant r}$ by expanding the determinant along the top row.)

(b) Deduce from (a) that

$$\tilde{s}_\lambda = \det\left(\varphi^{j-1} e_{\lambda'_i - i + j}\right).$$

(c) Let $\omega: S \to S$ be the R-algebra homomorphism that maps $\varphi^s h_r$ to $\varphi^{-s} e_r$ for all r, s. Then ω is an involution and $\omega \tilde{s}_\lambda = \tilde{s}_{\lambda'}$ for all partitions λ.

(d) Let $\lambda = (\alpha_1, \dots, \alpha_r \mid \beta_1, \dots, \beta_r)$ in Frobenius notation (§1). Show that with $\tilde{s}_\lambda$ as defined in (1) we have

$$\tilde{s}_\lambda = \det\left(\tilde{s}_{(\alpha_i \mid \beta_j)}\right)_{1 \le i, j \le r}.$$

(Copy the proof in Example 9 above.)

22. Let $x_1, \dots, x_n, u_1, \dots, u_n$ be independent indeterminates over $\mathbf{Z}$, and let $f(t) = (t - x_1) \dots (t - x_n)$. For each $r \in \mathbf{Z}$ let

$$H_r = \sum_{i=1}^{n} u_i x_i^{n+r-1} / f'(x_i)$$

and for each sequence $\alpha = (\alpha_1, \dots, \alpha_r) \in \mathbf{Z}^r$, where $r \le n$, let M_α denote the $n \times n$ matrix

$$M_\alpha = \left(u_{ij} x_i^{\alpha_j + n - j}\right)$$

where $u_{ij} = u_i$ if $1 \le j \le r$, $u_{ij} = 1$ if $j > r$, and $\alpha_j = 0$ if $j > r$. Then let

$$S_\alpha = a_\delta(x)^{-1} \det(M_\alpha).$$

Show that

$$S_\alpha = \det(H_{\alpha_i - i + j})_{1 \le i, j \le r}.$$

(Multiply the matrix M_α on the left by the matrix whose (i, j) element is $x_j^{i-1} / f'(x_j)$ if $i \le r$, and is δ_{ij} if $i > r$.)

23. The ring Λ_n of symmetric polynomials in n variables $x_1, \dots, x_n$ is the image of Λ under the homomorphism ρ_n of §2, which maps the formal power series $E(t) = \sum_{r \geq 0} e_r t^r$ to the polynomial $\prod_{i=1}^{n} (1 + x_i t)$ of degree n. More generally, we may specialize $E(t)$ to a rational function of t, say

$$(*) \qquad E_{x/y}(t) = \prod_{i=1}^{m} (1 + x_i t) \Big/ \prod_{j=1}^{n} (1 + y_j t).$$

(In the language of λ-rings, this amounts to considering the difference $x - y$, where x has rank m and y has rank n.)

Let $x^{(m)} = (x_1, \dots, x_m)$, $y^{(n)} = (y_1, \dots, y_n)$ and let $s_\lambda(x^{(m)} / y^{(n)})$ (or just $s_\lambda(x/y)$) denote the image of the Schur function s_λ under this specialization. From $(*)$ we have

$$e_r(x^{(m)} / y^{(n)}) = \sum_{i+j=r} (-1)^j e_i(x) h_j(y)$$

and the formula (3.5) shows that $s_\lambda(x^{(m)} / y^{(n)})$ is a polynomial in the x's and the y's, symmetric in each set of variables separately.

These polynomials have the following properties:

(1) (homogeneity) $s_\lambda(x^{(m)}/y^{(n)})$ is homogeneous of degree $|\lambda|$.
(2) (restriction) The result of setting $x_m = 0$ (resp. $y_n = 0$) in $s_\lambda(x^{(m)}/y^{(n)})$ is $s_\lambda(x^{(m-1)}/y^{(n)})$ (resp. $s_\lambda(x^{(m)}/y^{(n-1)})$).
(3) (cancellation) The result of setting $x_m = y_n$ in $s_\lambda(x^{(m)}/y^{(n)})$ is $s_\lambda(x^{(m-1)}/y^{(n-1)})$ (if $m, n \geqslant 1$).
(4) (factorization) If the partition λ satisfies $\lambda_m \geqslant n \geqslant \lambda_{m+1}$, so that λ can be written in the form $\lambda = ((n^m) + \alpha) \cup \beta'$, where α (resp. β) is a partition of length $\leqslant m$ (resp. $\leqslant n$), then

$$s_\lambda(x^{(m)}/y^{(n)}) = (-1)^{|\beta|} R(x^{(m)}, y^{(n)}) s_\alpha(x^{(m)}) s_\beta(y^{(n)})$$

where $R(x^{(m)}, y^{(n)})$ is the product of the mn factors $x_i - y_j$ $(1 \leqslant i \leqslant m, 1 \leqslant j \leqslant n)$.

It is clear from the definition that the s_λ satisfy (1), (2), and (3), and it may be shown directly that they satisfy (4) (see the notes and references at the end of this section). We shall not stop to give a direct proof here, because this property (4) will be a consequence of the results of this Example and the next.

Moreover, these four properties *characterize* the $s_\lambda(x/y)$. More precisely, let $\Lambda_{m,n} (\cong \Lambda_m \otimes \Lambda_n)$ denote the ring $\mathbf{Z}[x_1, \dots, x_m, y_1, \dots, y_n]^{S_m \times S_n}$ of polynomials in the x's and y's that are symmetric in each set of variables separately, and suppose that we are given polynomials $s_\lambda^*(x^{(m)}/y^{(n)}) \in \Lambda_{m,n}$ for each partition λ and each pair of non-negative integers m, n, satisfying the conditions (1*)–(4*) obtained from (1)–(4) by replacing s_λ by s_λ^* throughout. Then $s_\lambda^*(x^{(m)}/y^{(n)}) = s_\lambda(x^{(m)}/y^{(n)})$ for all λ, m, n.

(a) First of all, when $n = 0$ it follows from (4*) that

(5) $$s_\lambda^*(x^{(m)}/\varnothing) = s_\lambda(x^{(m)})$$

for partitions λ of length $\leqslant m$, and likewise when $n = 0$ that

(5′) $$s_{\lambda'}^*(\varnothing/y^{(n)}) = (-1)^{|\lambda|} s_\lambda(y^{(n)})$$

for partitions λ of length $\leqslant n$.

(b) Next, let $\Lambda_{m/n}$ denote the subring of $\Lambda_{m,n}$ consisting of the polynomials f in which the result of setting $x_m = y_n = t$ is independent of t. It follows from (3*) that $s_\lambda^*(x^{(m)}/y^{(n)}) \in \Lambda_{m/n}$.

Let $\Gamma_{m,n}$ be the set of lattice points $(i, j) \in \mathbf{Z}^2$ such that $i \geqslant 1$, $j \geqslant 1$ and either $i \leqslant m$ or $j \leqslant n$. We shall show that

(6) *the* $s_\lambda^*(x^{(m)}/y^{(n)})$ *such that* $\lambda \subset \Gamma_{m,n}$ *span* $\Lambda_{m/n}$.

This is true when $m = 0$ or $n = 0$, by (5) and (5′) above. Assume then that $m, n \geqslant 1$ and that (6) is true for $m-1, n-1$. Let $f \in \Lambda_{m/n}$ and let $f_0 = f|_{x_m = y_n = 0}$, so that $f_0 \in \Lambda_{m-1/n-1}$ and therefore is of the form

$$f_0 = \sum_\lambda a_\lambda s_\lambda^*(x^{(m-1)}/y^{(m-1)})$$

summed over partitions $\lambda \subset \Gamma_{m-1,n-1}$, with coefficients $a_\lambda \in \mathbf{Z}$. Let

(7) $$g = f - \sum_\lambda a_\lambda s^*_\lambda(x^{(m)}/y^{(n)}).$$

Then $g \in \Lambda_{m/n}$, so that $g|_{x_m = y_n} = g|_{x_m = y_n = 0} = 0$. Consequently g is divisible in $\Lambda_{m,n}$ by $x_m - y_n$ and hence (by symmetry) by $R(x^{(m)}, y^{(n)})$: say $g = Rh$ where $h \in \Lambda_{m,n}$. By writing h in the form

$$h = \sum_{\alpha,\beta} b_{\alpha,\beta} s_\alpha(x^{(m)}) s_\beta(y^{(n)})$$

summed over partitions α, β such that $l(\alpha) \leqslant m$ and $l(\beta) \leqslant n$, it follows from (4*) that $g = Rh$ is a linear combination of the $s^*_\mu(x^{(m)}, y^{(n)})$ such that $(n^m) \subset \mu \subset \Gamma_{m,n}$. In view of (7) above, this establishes (6).

(c) Now let λ be any partition. In order to show that $s^*_\lambda(x^{(m)}/y^{(n)}) = s_\lambda(x^{(m)}/y^{(n)})$ we may, by virtue of (2) and (2*), assume that m and n are large, and in particular that $m \geqslant |\lambda|$. Since $s_\lambda(x^{(m)}/y^{(n)}) \in \Lambda_{m/n}$ by (3), it follows from (6) above that we may write

(8) $$s_\lambda(x^{(m)}/y^{(n)}) = \sum_\mu c_\mu s^*_\mu(x^{(m)}/y^{(n)})$$

where (by (1) and (1*)) the sum is over partitions μ such that $|\mu| = |\lambda|$ and hence $l(\mu) \leqslant m$ (since $|\lambda| \leqslant m$). If we now set $y_1 = \ldots = y_n = 0$ in (8), we obtain

$$s_\lambda(x^{(m)}) = \sum_\mu c_\mu s_\mu(x^{(m)})$$

by virtue of (2), (2*), and (5). Hence $c_\mu = \delta_{\lambda\mu}$ and finally $s_\lambda = s^*_\lambda$.

24. (a) Let $x = (x_1, \ldots, x_m)$, $y = (y_1, \ldots, y_n)$. If λ is any partition let

$$f_\lambda(x, y) = \prod_{(i,j) \in \lambda} (x_i - y_j)$$

with the understanding that $x_i = 0$ if $i > m$, and $y_j = 0$ if $j > n$. Also let

$$\Delta(x) = \prod_{1 \leqslant i < j \leqslant m} \left(1 - x_i^{-1} x_j\right), \qquad \Delta(y) = \prod_{1 \leqslant i < j \leqslant n} \left(1 - y_i^{-1} y_j\right).$$

Then we have

(1) $$s_\lambda(x/y) = \sum_{w \in S_m \times S_n} w(f_\lambda(x, y)/\Delta(x)\Delta(y)).$$

In this formula, S_m permutes the x's and S_n permutes the y's.

(Let $s^*_\lambda(x/y)$ denote the right-hand side of (1). We have

(2) $$s^*_\lambda(x/y) = a_{\delta_m}(x)^{-1} a_{\delta_n}(y)^{-1} \sum_{w \in S_m \times S_n} \varepsilon(w) . w(x^{\delta_m} y^{\delta_n} f_\lambda(x, y))$$

from which it is clear that $s^*_\lambda \in \Lambda_{m,n}$, in the notation of Example 23, and it is enough to verify that s^*_λ has the properties (1*)–(4*), *loc. cit*. Of these, all but the cancellation property (3*) are obviously satisfied. As to (3*), let $\varphi(u) = s_\lambda^*(x/y)|_{x_m=y_n=u}$, which is a polynomial in u of degree say d. It will be enough to show that $\varphi(u) = \varphi(0)$, i.e. that $d = 0$. If $w \in S_m \times S_n$, let $w^{-1}(x_m) = x_i$ and $w^{-1}(y_n) = y_j$. If $(i, j) \in \lambda$, then $x_m - y_n = w(x_i - y_j)$ is a factor of $wf_\lambda(x, y)$, which therefore vanishes when $x_m = y_n = u$. Hence $\varphi(u)$ is a sum over those $w \in S_m \times S_n$ such that $(i, j) \notin \lambda$, that is to say $i > \lambda'_j$ and $j > \lambda_i$. For such a permutation w, the degree in u of $w(x^{\delta_m} y^{\delta_n} f_\lambda(x, y))|_{x_m=y_n=u}$ is

$$m - i + n - j + \lambda_i + \lambda'_j \leqslant m + n - 2;$$

on the other hand, the degree in u of $a_{\delta_m}(x) a_{\delta_n}(y)|_{x_m=y_n=u}$ is $(m-1) + (n-1) = m + n - 2$. It now follows from (2) that $d = 0$, as required.)

(b) When $n = 0$, we have $f_\lambda(x, y) = x^\lambda$, and the formula (1) reduces to the definition (3.1) of $s_\lambda(x)$. Next, when $m = 0$, it follows from the definition (Example 23) that $s_\lambda(x/y)$ becomes $(-1)^{|\lambda|} \omega s_\lambda(y)$. On the other hand, $f_\lambda(x, y)$ becomes $(-1)^{|\lambda|} y^{\lambda'}$, so that the formula (1) in the case $m = 0$ reduces to (3.8). Finally, as we have already remarked, the factorization property (4) of Example 23 is an immediate consequence of (1), and so is the fact that $s_\lambda(x/y) = 0$ unless $\lambda \subset \Gamma_{m,n}$.

(c) The $s_\lambda(x/y)$ such that $\lambda \subset \Gamma_{m,n}$ form a $\mathbf{Z}$-basis of the ring $\Lambda_{m/n}$ defined in Example 23(b). (It follows from Example 23 that the $s_\lambda(x/y)$ such that $\lambda \subset \Gamma_{m,n}$ span $\Lambda_{m/n}$, and it remains to be shown that they are linearly independent. This is clearly true if $m = 0$ or $n = 0$, by (3.2) and (3.8). Hence we may assume $m \geqslant 1$ and $n \geqslant 1$ and the result true when either m is replaced by $m - 1$, or n by $n - 1$. Suppose then that

$$\sum_\lambda a_\lambda s_\lambda(x/y) = 0$$

where the sum is over partitions $\lambda \subset \Gamma_{m,n}$. By setting $x_m = 0$ (resp. $y_n = 0$) we see that $a_\lambda = 0$ for $\lambda \subset \Gamma_{m-1,n}$ (resp. $\lambda \subset \Gamma_{m,n-1}$). Since $\Gamma_{m,n}$ is the union of $\Gamma_{m-1,n}$ and $\Gamma_{m,n-1}$, it follows that $a_\lambda = 0$ for all $\lambda \in \Gamma_{m,n}$, as required.)

Notes and references

Schur functions, despite their name, were first considered by Jacobi [J3], as quotients of skew-symmetric polynomials by the polynomial a_δ, just as we have introduced them. Their relevance to the representation theory of the symmetric groups and the general linear groups, which we shall describe later, was discovered by Schur [S4] much later. The identity (3.4) which expresses s_λ in terms of the h's is due originally to Jacobi (*loc. cit.*), and is often called the Jacobi–Trudi identity.

The results of Examples 1–4 may be found in Littlewood [L9], Chapter VII, which gives other results of the same sort. The formula in Example 8 for the product of two Schur functions as a determinant in the h's is essentially due to Jacobi (*loc. cit.*), though rediscovered since. The result of Example 9 is due to Giambelli [G8]. Example 10 is due to A. Lascoux [L1]. The proof of the law of quadratic reciprocity sketched in Example 17 is essentially Eisenstein's proof (see Serre [S13], Chapter I). The presentation here is due to V. G. Kac [K1].

For Examples 18–21 see [M8]. For Examples 23 and 24, and the history of the Sergeev–Pragacz formula (i.e., formula (1) of Example 24), see [P3], also [P2] and [S27]. Other proofs of this formula, in the context of Schubert polynomials, are due to A. Lascoux (see for example [M7]). The factorization property (4) in Example 23 (which is a special case of the Sergeev–Pragacz formula) is due to Berele and Regev [B2].

4. Orthogonality

Let $x = (x_1, x_2, \dots)$ and $y = (y_1, y_2, \dots)$ be two finite or infinite sequences of independent variables. We shall denote the symmetric functions of the x's by $s_\lambda(x), p_\lambda(x)$, etc., and the symmetric functions of the y's by $s_\lambda(y), p_\lambda(y)$, etc.

We shall give three series expansions for the product

$$\prod_{i,j}(1 - x_i y_i)^{-1}.$$

The first of these is

$$\prod_{i,j}(1 - x_i y_j)^{-1} = \sum_\lambda z_\lambda^{-1} p_\lambda(x) p_\lambda(y) \tag{4.1}$$

summed over all partitions λ.

This follows from (2.14), applied to the set of variables $x_i y_j$.

Next we have

$$\prod_{i,j}(1 - x_i y_j)^{-1} = \sum_\lambda h_\lambda(x) m_\lambda(y) = \sum_\lambda m_\lambda(x) h_\lambda(y) \tag{4.2}$$

summed over all partitions λ.

Proof. We have

$$\begin{aligned}\prod_{i,j}(1 - x_i y_j)^{-1} &= \prod_j H(y_j) \\ &= \prod_j \sum_{r=0}^{\infty} h_r(x) y_j^r \\ &= \sum_\alpha h_\alpha(x) y^\alpha \\ &= \sum_\lambda h_\lambda(x) m_\lambda(y)\end{aligned}$$

where α runs through all sequences $(\alpha_1, \alpha_2, \ldots)$ of non-negative integers such that $\sum \alpha_i < \infty$, and λ runs through all partitions. |

The third identity is

(4.3) $$\prod_{i,j}(1-x_iy_j)^{-1} = \sum_{\lambda} s_\lambda(x)s_\lambda(y)$$

summed over all partitions λ.

Proof. This is a consequence of (4.2) and (3.7′). Let $x=(x_1,\ldots,x_n)$, $y=(y_1,\ldots,y_n)$ be two finite sets of variables, and as usual let $\delta=(n-1, n-2,\ldots,0)$. Then from (4.2) we have

(4.4) $$a_\delta(x)a_\delta(y)\prod_{i,j=1}^{n}(1-x_iy_j)^{-1} = a_\delta(x)\sum_{\alpha,w} h_\alpha(x)\varepsilon(w)y^{\alpha+w\delta}$$

summed over $\alpha \in \mathbf{N}^n$ and $w \in S_n$,

$$= a_\delta(x)\sum_{\beta,w}\varepsilon(w)h_{\beta-w\delta}(x)y^\beta$$

$$= \sum_{\beta} a_\beta(x)y^\beta$$

by (3.7′). Since $a_{w\beta}=\varepsilon(w)a_\beta$, it follows that this last sum is equal to $\sum a_\gamma(x)a_\gamma(y)$ summed over $\gamma_1 > \gamma_2 > \ldots > \gamma_n \geqslant 0$, i.e. to

$$\sum_{\lambda} a_{\lambda+\delta}(x)a_{\lambda+\delta}(y),$$

summed over partitions λ of length $\leqslant n$. This proves (4.3) in the case of n variables x_i and n variables y_i; now let $n\to\infty$ as usual. |

We now define a scalar product on Λ, i.e. a $\mathbf{Z}$-valued bilinear form $\langle u,v\rangle$, by requiring that the bases (h_λ) and (m_λ) should be dual to each other:

(4.5) $$\langle h_\lambda, m_\mu\rangle = \delta_{\lambda\mu}$$

for all partitions λ, μ, where $\delta_{\lambda\mu}$ is the Kronecker delta.

(4.6) *For each* $n \geqslant 0$, *let* (u_λ), (v_λ) *be* $\mathbf{Q}$*-bases of* $\Lambda^n_{\mathbf{Q}}$, *indexed by the partitions of* n. *Then the following conditions are equivalent:*

(a) $\langle u_\lambda, v_\mu\rangle = \delta_{\lambda\mu}$ *for all* λ, μ;

(b) $\sum_\lambda u_\lambda(x)v_\lambda(y) = \prod_{i,j}(1-x_iy_j)^{-1}$.

Proof. Let

$$u_\lambda = \sum_\rho a_{\lambda\rho} h_\rho, \qquad v_\mu = \sum_\sigma b_{\mu\sigma} m_\sigma .$$

Then

$$\langle u_\lambda, v_\mu \rangle = \sum_\rho a_{\lambda\rho} b_{\mu\rho}$$

so that (a) is equivalent to

(a′) $$\sum_\rho a_{\lambda\rho} b_{\mu\rho} = \delta_{\lambda\mu} .$$

Also (b) is equivalent to the identity

$$\sum_\lambda u_\lambda(x) v_\lambda(y) = \sum_\rho h_\rho(x) m_\rho(y)$$

by (4.2), hence is equivalent to

(b′) $$\sum_\lambda a_{\lambda\rho} b_{\lambda\sigma} = \delta_{\rho\sigma} .$$

Since (a′) and (b′) are equivalent, so are (a) and (b). |

From (4.6) and (4.1) it follows that

(4.7) $$\langle p_\lambda, p_\mu \rangle = \delta_{\lambda\mu} z_\lambda$$

so that the p_λ form an *orthogonal* basis of $\Lambda_{\mathbf{Q}}$. Likewise from (4.6) and (4.3) we have

(4.8) $$\langle s_\lambda, s_\mu \rangle = \delta_{\lambda\mu}$$

so that the s_λ form an *orthonormal* basis of Λ, and the s_λ such that $|\lambda| = n$ form an orthonormal basis of Λ^n. Any other orthonormal basis of Λ^n must therefore be obtained from the basis (s_λ) by transformation by an orthogonal integer matrix. The only such matrices are signed permutation matrices, and therefore (4.8) characterizes the s_λ, up to order and sign.

Also from (4.7) or (4.8) we see that

(4.9) *The bilinear form $\langle u, v \rangle$ is symmetric and positive definite.* |

(4.10) *The involution ω is an isometry, i.e.* $\langle \omega u, \omega v \rangle = \langle u, v \rangle$.

Proof. From (2.13) we have $\omega(p_\lambda) = \pm p_\lambda$, hence by (4.7)

$$\langle \omega(p_\lambda), \omega(p_\mu) \rangle = \langle p_\lambda, p_\mu \rangle$$

which proves (4.10), since the p_λ form a $\mathbf{Q}$-basis of $\Lambda_{\mathbf{Q}}$ (2.12). |

Finally, from (4.10) and (4.5) we have

$$\langle e_\lambda, f_\mu \rangle = \delta_{\lambda\mu}$$

where $f_\mu = \omega(m_\mu)$, i.e. (e_λ) and (f_λ) are dual bases of Λ.

Remarks. 1. By applying the involution ω to the symmetric functions of the x variables we obtain from (4.1), (4.2), and (4.3) three series expansions for the product $\prod_{i,j}(1+x_i y_i)$, namely

$$(4.1') \qquad \prod_{i,j}(1+x_i y_j) = \sum_\lambda \varepsilon_\lambda z_\lambda^{-1} p_\lambda(x) p_\lambda(y),$$

$$(4.2') \qquad \prod_{i,j}(1+x_i y_j) = \sum_\lambda m_\lambda(x) e_\lambda(y) = \sum_\lambda e_\lambda(x) m_\lambda(y),$$

$$(4.3') \qquad \prod_{i,j}(1+x_i y_j) = \sum_\lambda s_\lambda(x) s_{\lambda'}(y),$$

the last by virtue of (3.8).

2. If x, y are elements of a λ-ring R, we have

$$\begin{aligned}\sigma_t(xy) &= \sum_\lambda z_\lambda^{-1} \psi^\lambda(x)\psi^\lambda(y) t^{|\lambda|} \\ &= \sum_\lambda S^\lambda(x) S^\lambda(y) t^{|\lambda|}\end{aligned}$$

from (4.1) and (4.3), and

$$\begin{aligned}\lambda_t(xy) &= \sum_\lambda \varepsilon_\lambda z_\lambda^{-1} \psi^\lambda(x)\psi^\lambda(y) t^{|\lambda|} \\ &= \sum_\lambda S^\lambda(x) S^{\lambda'}(y) t^{|\lambda|}\end{aligned}$$

from (4.1′) and (4.3′).

Examples

1. If we take $y_1 = \dots = y_n = t$, $y_{n+1} = y_{n+2} = \dots = 0$ in (4.3′), we obtain

$$\begin{aligned}E(t)^n &= \sum_\lambda s_\lambda(x) s_{\lambda'}(y) \\ &= \sum_\lambda \binom{n}{\lambda} s_\lambda(x) t^{|\lambda|}\end{aligned}$$

in the notation of §3, Example 4.

The coefficients of the powers of t on each side are polynomials in n (with coefficients in Λ) which are equal for all positive integral values of n, and hence identically equal. Consequently we have

$$E(t)^X = \sum_\lambda \binom{X}{\lambda} s_\lambda t^{|\lambda|}$$

for all X. By replacing X, t by $-X$, $-t$ we obtain

$$H(t)^X = \sum_\lambda \binom{X}{\lambda'} s_\lambda t^{|\lambda|}.$$

These identities generalize the binomial theorem.

2. Let $y_i = q^{i-1}$ for $1 \leqslant i \leqslant n$, and $y_i = 0$ for $i > n$. From (4.3') we obtain

$$\prod_{i=1}^n E(q^{i-1}) = \sum_\lambda q^{n(\lambda')} \begin{bmatrix} n \\ \lambda \end{bmatrix} s_\lambda$$

in the notation of §3, Example 1. Likewise, from (4.3),

$$\prod_{i=1}^n H(q^{i-1}) = \sum_\lambda q^{n(\lambda)} \begin{bmatrix} n \\ \lambda' \end{bmatrix} s_\lambda.$$

In these formulas we may let $n \to \infty$ and obtain

$$\prod_{i,j \geqslant 1} \left(1 + x_j q^{i-1}\right) = \sum_\lambda \frac{q^{n(\lambda')}}{H_\lambda(q)} s_\lambda(x),$$

$$\prod_{i,j \geqslant 1} \left(1 - x_j q^{i-1}\right)^{-1} = \sum_\lambda \frac{q^{n(\lambda)}}{H_\lambda(q)} s_\lambda(x),$$

where $H_\lambda(q) = \prod_{x \in \lambda}(1 - q^{h(x)})$ is the hook-length polynomial corresponding to the partition λ.

3. Let $y_1 = \ldots = y_n = t/n$, $y_{n+1} = y_{n+2} = \ldots = 0$, and then let $n \to \infty$. We have

$$\prod_i \left(1 + \frac{x_i t}{n}\right)^n \to \prod_i \exp(x_i t) = \exp(e_1 t)$$

and

$$\frac{1}{n^{|\lambda|}} \binom{n}{\lambda} \to \prod_{x \in \lambda} h(x)^{-1} = h(\lambda)^{-1}$$

where $h(\lambda)$ is the product of the hook-lengths of λ. Hence from (4.3') we obtain

$$\exp(e_1 t) = \sum_\lambda \frac{s_\lambda}{h(\lambda)} t^{|\lambda|}$$

and therefore

$$e_1^n = \sum_{|\lambda| = n} \frac{n!}{h(\lambda)} s_\lambda$$

or equivalently

$$\langle e_1^n, s_\lambda \rangle = n!/h(\lambda).$$

4. From (2.14′) and (4.7) we have

$$\langle h_n, p_\lambda \rangle = 1$$

for all partitions λ of n. Dually,

$$\langle e_n, p_\lambda \rangle = \varepsilon_\lambda.$$

5.
$$\prod_{i=1}^{m} \prod_{j=1}^{n} (x_i + y_j) = \sum_\lambda s_\lambda(x) s_{\tilde{\lambda}'}(y)$$

summed over all partitions $\lambda = (\lambda_1, \dots, \lambda_m)$ such that $\lambda_1 \leqslant n$ (i.e. $\lambda \subset (n^m)$), where $\tilde{\lambda}' = (m - \lambda'_n, \dots, m - \lambda'_1)$. (Replace y_i by y_i^{-1} in (4.3′), and clear of fractions.)

Hence from §3, Example 10 we have

$$\prod_{i=1}^{m} \prod_{j=1}^{n} (1 + x_i + y_j) = \sum_{\lambda, \mu} d_{\lambda\mu} s_\mu(x) s_{\tilde{\lambda}'}(y)$$

summed over pairs of partitions λ, μ such that $\mu \subset \lambda \subset (n^m)$. (This formula gives the Chern classes of a tensor product $E \otimes F$ of vector bundles, since if $c(E) = \prod(1 + x_i)$ and $c(F) = \prod(1 + y_j)$ are the total Chern classes of E and F, we have $c(E \otimes F) = \prod(1 + x_i + y_j)$.)

6. Let $\Delta = \det((1 - x_i y_j)^{-1})_{1 \leqslant i,j \leqslant n}$ (Cauchy's determinant). Then

$$\Delta = a_\delta(x) a_\delta(y) \prod_{i,j=1}^{n} (1 - x_i y_j)^{-1}.$$

For if we multiply each element of the ith row of the matrix $((1 - x_i y_j)^{-1})$ by $\prod_{j=1}^{n}(1 - x_i y_j)$, we shall obtain a matrix D whose (i, j) element is

$$\prod_{r \neq j} (1 - x_i y_r) = \sum_{k=1}^{n} x_i^{n-k} . (-1)^{n-k} e_{n-k}^{(j)}(y)$$

in the notation of (3.6). This shows that $D = A_\delta(x) M(y)$, so that $\det(D) = a_\delta(x) a_\delta(y)$. On the other hand, it is clear from the definition of D that $\det(D) = \Delta . \prod(1 - x_i y_j)$.

Since also

$$\begin{aligned}\Delta &= \det\left(1 + x_i y_j + x_i^2 y_j^2 + \dots\right) \\ &= \sum_\alpha \det(x_i^{\alpha_j} y_j^{\alpha_j}) \\ &= \sum_\alpha a_\alpha(x) y^\alpha\end{aligned}$$

the summation being over all $\alpha = (\alpha_1, \ldots, \alpha_n) \in \mathbf{N}^n$, it follows that

$$\Delta = \sum_\lambda a_{\lambda+\delta}(x) a_{\lambda+\delta}(y)$$

summed over all partitions λ of length $\leqslant n$. Hence we have another proof of (4.3).

7. Likewise the identity (4.3′) can be proved directly, without recourse to duality. Consider the Vandermonde determinant $a_\delta(x, y)$ in $2n$ variables $x_1, \ldots, x_n, y_1, \ldots, y_n$; on the one hand, this is equal to $a_\delta(x) a_\delta(y) \prod (x_i - y_j)$; on the other hand, expanding the determinant by Laplace's rule, we see that it is equal to

$$\sum_\mu (-1)^{e(\mu)} a_\mu(x) a_{\bar{\mu}}(y), \tag{1}$$

summed over $\mu \in \mathbf{N}^n$ such that $2n-1 \geqslant \mu_1 > \mu_2 > \ldots > \mu_n \geqslant 0$, where $\bar{\mu}$ is the strictly decreasing sequence consisting of the integers in $[0, 2n-1]$ not equal to any of the μ_i, and $e(\mu) = \sum (2n - i - \mu_i)$. By writing $\mu = \lambda + \delta$ and using (1.7), we see that (1) is equal to

$$(-1)^n (y_1 \ldots y_n)^{2n-1} \sum_\lambda a_{\lambda+\delta}(x) a_{\lambda'+\delta}(-y^{-1})$$

summed over all partitions λ such that $l(\lambda) \leqslant n$ and $l(\lambda') \leqslant n$. If we now replace each y_i by y_i^{-1}, we obtain (4.3′).

8. Let M be a module over a commutative ring A, and let $\varphi: M \times M \to A$ be an A-bilinear form on M. The *standard extension* of φ to the symmetric algebra $\mathbf{S}(M)$ is the bilinear form defined on each $\mathbf{S}^n(M)$ by

$$\Phi(u, v) = \sum_{w \in S_n} \prod_{i=1}^n \varphi(u_i, v_{w(i)})$$

where $u = u_1 \ldots u_n$, $v = v_1 \ldots v_n$ and the u_i, v_j are elements of M. In other words, $\Phi(u, v)$ is the *permanent* of the $n \times n$ matrix $(\varphi(u_i, v_j))$.

In particular, let $A = \mathbf{Q}$ and let M be the $\mathbf{Q}$-vector space with basis $p_1, p_2, \ldots$, so that $\mathbf{S}(M) = \mathbf{Q}[p_1, p_2, \ldots] = \Lambda_{\mathbf{Q}}$. Define $\varphi(p_r, p_s) = r\delta_{rs}$ for all $r, s \geqslant 1$. Then the scalar product (4.5) on $\Lambda_{\mathbf{Q}}$ is the standard extension of φ.

9. Let $C(x, y) = \prod_{i,j} (1 - x_i y_j)^{-1}$. Then for all $f \in \Lambda$ we have

$$\langle C(x, y), f(x) \rangle = f(y)$$

where the scalar product is taken in the x variables. (By linearity, it is enough to prove this when $f = s_\lambda$, and then it follows from (4.3) and (4.8).)

In other words, $C(x, y)$ is a 'reproducing kernel' for the scalar product.

10. Let $p_n^{(r)} = \sum m_\lambda$ summed over partitions λ of n of length r, as in §2, Example 19. Show that

$$p_n^{(r)} = \sum_{a+b=n} (-1)^{b+r-1} \binom{b}{r-1} s_{(a, 1^b)}.$$

(Under the specialization $h_r \mapsto X$ for all $r \geqslant 1$, the Jacobi–Trudi formula (3.4) shows that $s_\lambda \mapsto 0$ if $\lambda_2 > 1$, that is to say if λ is not a hook; and if $\lambda = (a, 1^b)$ it is easily shown that $s_\lambda \mapsto X(X-1)^b$. Hence from the identity ((4.2), (4.3))

$$\sum m_\lambda(x) h_\lambda(y) = \sum s_\lambda(x) s_\lambda(y)$$

we obtain

$$\sum m_\lambda(x) X^{l(\lambda)} = \sum_{a,b \geqslant 0} s_{(a,1^b)}(x) X(X-1)^b$$

from which the result follows.)

11. (a) Let λ be a partition of n. Show that when m_λ is expressed as a polynomial in $p_1, \dots, p_n$, the coefficient of p_n is non-zero. (The coefficient in question is $n^{-1}\langle p_n, m_\lambda \rangle$, which is also the coefficient of h_λ in $n^{-1}p_n$ expressed as a polynomial in the h's, and is given explicitly in §2, Example 20.)

(b) For each integer $r \geqslant 1$ let u_r be a monomial symmetric function of degree r (i.e. $u_r = m_\lambda$ for some partition λ of r). Show that the u_r are algebraically independent over $\mathbf{Q}$ and that $\Lambda_\mathbf{Q} = \mathbf{Q}[u_1, u_2, \dots]$. (From (a) above we have $u_r = c_r p_r +$ a polynomial in $p_1, \dots, p_{r-1}$, where $c_r \neq 0$. This shows by induction on r that $\mathbf{Q}[u_1, \dots, u_r] = \mathbf{Q}[p_1, \dots, p_r]$ for each $r \geqslant 1$. Hence for each $m \geqslant 1$ the monomials of degree m in the u variables span $\Lambda^m_\mathbf{Q}$, and are therefore linearly independent over $\mathbf{Q}$.)

Notes and references

The scalar product on Λ was apparently first introduced by Redfield [R1] and later popularized by P. Hall [H3]. Example 5 is due to A. Lascoux [L1], and Example 11 to D. G. Mead [M10].

5. Skew Schur functions

Any symmetric function $f \in \Lambda$ is uniquely determined by its scalar products with the s_λ: namely

$$f = \sum_\lambda \langle f, s_\lambda \rangle s_\lambda$$

since the s_λ form an orthonormal basis of Λ (4.8).

Let λ, μ be partitions, and define a symmetric function $s_{\lambda/\mu}$ by the relations

$$\langle s_{\lambda/\mu}, s_\nu \rangle = \langle s_\lambda, s_\mu s_\nu \rangle \tag{5.1}$$

for all partitions ν. The $s_{\lambda/\mu}$ are called *skew Schur functions*. Equivalently, if $c^{\lambda}_{\mu\nu}$ are the integers defined by

(5.2)
$$s_\mu s_\nu = \sum_\lambda c^{\lambda}_{\mu\nu} s_\lambda,$$

then we have

(5.3)
$$s_{\lambda/\mu} = \sum_\nu c^{\lambda}_{\mu\nu} s_\nu .$$

In particular, it is clear that $s_{\lambda/0} = s_\lambda$, where 0 denotes the zero partition. Also $c^{\lambda}_{\mu\nu} = 0$ unless $|\lambda| = |\mu| + |\nu|$, so that $s_{\lambda/\mu}$ is homogeneous of degree $|\lambda| - |\mu|$, and is zero if $|\lambda| < |\mu|$. (We shall see shortly that $s_{\lambda/\mu} = 0$ unless $\lambda \supset \mu$.)

Now let $x = (x_1, x_2, \dots)$ and $y = (y_1, y_2, \dots)$ be two sets of variables. Then

$$\sum_\lambda s_{\lambda/\mu}(x) s_\lambda(y) = \sum_{\lambda,\nu} c^{\lambda}_{\mu\nu} s_\nu(x) s_\lambda(y)$$
$$= \sum_\nu s_\nu(x) s_\mu(y) s_\nu(y)$$

by (5.2) and (5.3), and therefore

$$\sum_\lambda s_{\lambda/\mu}(x) s_\lambda(y) = s_\mu(y) \sum_\nu h_\nu(x) m_\nu(y)$$

by (4.2) and (4.3). Now suppose that $y = (y_1, \dots, y_n)$, so that the sums above are restricted to partitions λ and ν of length $\leqslant n$. Then the previous equation can be rewritten in the form

$$\sum_\lambda s_{\lambda/\mu}(x) a_{\lambda+\delta}(y) = \sum_\nu h_\nu(x) m_\nu(y) a_{\mu+\delta}(y)$$
$$= \sum_\alpha h_\alpha(x) \sum_{w \in S_n} \varepsilon(w) y^{\alpha + w(\mu+\delta)}$$

summed over $\alpha \in \mathbf{N}^n$. Hence $s_{\lambda/\mu}(x)$ is equal to the coefficient of $y^{\lambda+\delta}$ in this sum, i.e. we have

$$s_{\lambda/\mu} = \sum_{w \in S_n} \varepsilon(w) h_{\lambda+\delta-w(\mu+\delta)}$$

with the usual convention that $h_\alpha = 0$ if any component α_i of α is negative. This formula can also be written as a determinant

(5.4)
$$s_{\lambda/\mu} = \det\left(h_{\lambda_i - \mu_j - i + j}\right)_{1 \leqslant i,j \leqslant n}$$

where $n \geqslant l(\lambda)$.

When $\mu = 0$, (5.4) becomes (3.4).

From (5.4) and (2.9) we have also

(5.5) $$s_{\lambda/\mu} = \det(e_{\lambda'_i - \mu'_j - i + j})_{1 \leqslant i,j \leqslant m}$$

where $m \geqslant l(\lambda')$, and therefore

(5.6) $$\omega(s_{\lambda/\mu}) = s_{\lambda'/\mu'}.$$

From (5.4) it follows that $s_{\lambda/\mu} = 0$ unless $\lambda_i \geqslant \mu_i$ for all i, i.e. unless $\lambda \supset \mu$. For if $\lambda_r < \mu_r$ for some r, we have $\lambda_i \leqslant \lambda_r < \mu_r \leqslant \mu_j$ for $1 \leqslant j \leqslant r \leqslant i \leqslant n$, and therefore $\lambda_i - \mu_j - i + j < 0$ for this range of values of (i, j). Consequently the matrix $(h_{\lambda_i - \mu_j - i + j})$ has an $(n - r + 1) \times r$ block of zeros in the bottom left-hand corner, and therefore its determinant vanishes.

The same considerations show that if $\lambda \supset \mu$ and $\mu_r \geqslant \lambda_{r+1}$ for some $r < n$, the matrix $(h_{\lambda_i - \mu_j - i + j})$ is of the form $\begin{pmatrix} A & C \\ 0 & B \end{pmatrix}$, where A has r rows and columns, and B has $n - r$ rows and columns, so that its determinant is equal to $\det(A)\det(B)$. Hence if the skew diagram $\lambda - \mu$ consists of two disjoint pieces θ, φ (each of which is a skew diagram), then we have $s_{\lambda/\mu} = s_\theta \, . \, s_\varphi$. To summarize:

(5.7) *The skew Schur function* $s_{\lambda/\mu}$ *is zero unless* $\lambda \supset \mu$, *in which case it depends only on the skew diagram* $\lambda - \mu$. *If* θ_i *are the components* (§1) *of* $\lambda - \mu$, *we have* $s_{\lambda/\mu} = \prod s_{\theta_i}$. |

If the number of variables x_i is finite, we can say more:

(5.8) *We have* $s_{\lambda/\mu}(x_1, \ldots, x_n) = 0$ *unless* $0 \leqslant \lambda'_i - \mu'_i \leqslant n$ *for all* $i \geqslant 1$.

Proof. Suppose that $\lambda'_r - \mu'_r > n$ for some $r \geqslant 1$. Since $e_{n+1} = e_{n+2} = \ldots = 0$, it follows as above that the matrix $(e_{\lambda'_i - \mu'_j - i + j})$ has a rectangular block of zeros in the top right-hand corner, with one vertex of the rectangle on the main diagonal, hence its determinant vanishes. |

Now let $x = (x_1, x_2, \ldots)$, $y = (y_1, y_2, \ldots)$, $z = (z_1, z_2, \ldots)$ be three sets of independent variables. Then by (5.2) we have

(a) $$\sum_{\lambda, \mu} s_{\lambda/\mu}(x) s_\lambda(z) s_\mu(y) = \sum_\mu s_\mu(y) s_\mu(z) \, . \, \prod_{i,k} (1 - x_i z_k)^{-1}$$

which by (4.3) is equal to

$$\prod_{i,k} (1 - x_i z_k)^{-1} \, . \, \prod_{j,k} (1 - y_j z_k)^{-1}$$

and therefore also equal to

(b) $$\sum_\lambda s_\lambda(x, y) s_\lambda(z)$$

where $s_\lambda(x,y)$ denotes the Schur function corresponding to λ in the set of variables $(x_1, x_2, \dots, y_1, y_2, \dots)$. From the equality of (a) and (b) we conclude that

(5.9)
$$\begin{aligned} s_\lambda(x,y) &= \sum_\mu s_{\lambda/\mu}(x)s_\mu(y) \\ &= \sum_{\mu,\nu} c^\lambda_{\mu\nu} s_\mu(y)s_\nu(x). \end{aligned}$$

More generally, we have

(5.10)
$$s_{\lambda/\mu}(x,y) = \sum_\nu s_{\lambda/\nu}(x)s_{\nu/\mu}(y)$$

summed over partitions ν such that $\lambda \supset \nu \supset \mu$.

Proof. From (5.9) we have

$$\begin{aligned} \sum_\mu s_{\lambda/\mu}(x,y)s_\mu(z) &= s_\lambda(x,y,z) \\ &= \sum_\nu s_{\lambda/\nu}(x)s_\nu(y,z) \\ &= \sum_{\mu,\nu} s_{\lambda/\nu}(x)s_{\nu/\mu}(y)s_\mu(z) \end{aligned}$$

by (5.9) again; now equate the coefficients of $s_\mu(z)$ at either end of this chain of equalities. |

The formula (5.10) may clearly be generalized, as follows. Let $x^{(1)}, \dots, x^{(n)}$ be n sets of variables, and let λ, μ be partitions. Then

(5.11)
$$s_{\lambda/\mu}(x^{(1)}, \dots, x^{(n)}) = \sum_{(\nu)} \prod_{i=1}^n s_{\nu^{(i)}/\nu^{(i-1)}}(x^{(i)})$$

summed over all sequences $(\nu) = (\nu^{(0)}, \nu^{(1)}, \dots, \nu^{(n)})$ of partitions, such that $\nu^{(0)} = \mu$, $\nu^{(n)} = \lambda$, and $\nu^{(0)} \subset \nu^{(1)} \subset \dots \subset \nu^{(n)}$. |

We shall apply (5.11) in the case where each set of variables $x^{(i)}$ consists of a single variable x_i. For a single x, it follows from (5.8) that $s_{\lambda/\mu}(x) = 0$ unless $\lambda - \mu$ is a *horizontal strip* (§1), in which case $s_{\lambda/\mu}(x) = x^{|\lambda - \mu|}$. Hence each of the products in the sum on the right-hand side of (5.11) is a monomial $x_1^{\alpha_1} \dots x_n^{\alpha_n}$, where $\alpha_i = |\nu^{(i)} - \nu^{(i-1)}|$, and hence we have $s_{\lambda/\mu}(x_1, \dots, x_n)$ expressed as a sum of monomials x^α, one for each *tableau*

(§1) T of shape $\lambda - \mu$. If the weight of T is $\alpha = (\alpha_1, \ldots, \alpha_n)$, we shall write x^T for x^α. Then:

(5.12) $$s_{\lambda/\mu} = \sum_T x^T$$

summed over all tableaux T of shape $\lambda - \mu$. |

For each partition ν such that $|\nu| = |\lambda - \mu|$, let $K_{\lambda-\mu,\nu}$ denote the number of tableaux of shape $\lambda - \mu$ and weight ν. From (5.12) we have

(5.13) $$s_{\lambda/\mu} = \sum_\nu K_{\lambda-\mu,\nu} m_\nu$$

and therefore

(5.14) $$K_{\lambda-\mu,\nu} = \langle s_{\lambda/\mu}, h_\nu \rangle = \langle s_\lambda, s_\mu h_\nu \rangle$$

so that

(5.15) $$s_\mu h_\nu = \sum_\lambda K_{\lambda-\mu,\nu} s_\lambda.$$

In particular, suppose that $\nu = (r)$, a partition with only one non-zero part. Then $K_{\lambda-\mu,(r)}$ is 1 or 0 according as $\lambda - \mu$ is or is not a horizontal r-strip, and therefore from (5.15) we have

(5.16) (Pieri's formula) $$s_\mu h_r = \sum_\lambda s_\lambda$$

summed over all partitions λ such that $\lambda - \mu$ is a horizontal r-strip. |

By applying the involution ω to (5.16), we obtain

(5.17) $$s_\mu e_r = \sum_\lambda s_\lambda$$

summed over all partitions λ such that $\lambda - \mu$ is a vertical r-strip. |

Remarks. 1. It is easy to give a direct proof of (5.17). Consider (for a finite set of variables $x_1, \ldots, x_n$) the product

$$a_{\mu+\delta} e_r = \sum_{w \in S_n} \varepsilon(w) x^{w(\mu+\delta)} \sum_\alpha x^\alpha$$

$$= \sum_\alpha a_{\mu+\alpha+\delta}$$

where the sum is over all $\alpha \in \mathbf{N}^n$ such that each α_i is 0 or 1, and $\sum \alpha_i = r$. For each such α, the sequence

$$\mu + \alpha + \delta = (\mu_1 + \alpha_1 + n - 1, \mu_2 + \alpha_2 + n - 2, \ldots, \mu_n + \alpha_n)$$

is in descending order, so that we have only to reject those α for which two consecutive terms are equal. We are then left with those α for which $\lambda = \mu + \alpha$ is a *partition*, i.e. such that $\lambda - \mu$ is a vertical r-strip. This proves (5.17), hence also (5.16) by duality. We can now play back the rest of the argument: (5.16) implies (5.15) by induction on the length of ν, hence (5.14), which in turn is merely a restatement of (5.13).

2. Proposition (5.12) is the origin of the application of Schur functions to enumeration of plane partitions (see the examples at the end of this section). For this reason, combinatorialists often prefer to take (5.12) as the definition of Schur functions (see e.g. Stanley [S23]). This approach has the advantage of starting directly with a simple explicit definition, but it is not clear a priori why one should be led to make such a definition in the first place.

3. In any λ-ring we can define operations $S^{\lambda/\mu}$ by the formula (5.3):

$$S^{\lambda/\mu} = \sum_{\nu} c^{\lambda}_{\mu\nu} S^{\nu}$$

Then (5.9), for example, takes the form of an addition theorem:

$$S^{\lambda}(x+y) = \sum_{\mu} S^{\lambda/\mu}(x) S^{\mu}(y)$$

for any two elements x, y of a λ-ring. Similarly for the other formulas in this section.

4. The formula (5.4) shows that the skew Schur functions $s_{\lambda/\mu}(x)$, where λ and μ are partitions of length $\leqslant p$, are the $p \times p$ minors of the matrix $H_x = (h_{i-j}(x))$, i.e. they are the entries in the pth exterior power $\Lambda^p(H_x)$. The relation (5.10) is therefore equivalent to

$$\Lambda^p(H_{x,y}) = \Lambda^p(H_x)\Lambda^p(H_y).$$

Thus it is a consequence of the functoriality of Λ^p, since $H_{x,y} = H_x H_y$.

Examples

1. Let $\lambda - \mu$ be a horizontal strip. Then $s_{\lambda/\mu} = h_\nu = h_{\nu_1} h_{\nu_2} \ldots$ where the integers ν_i are the lengths of the components of the strip. (Use (5.7).) Likewise, if $\lambda - \mu$ is a vertical strip, we have $s_{\lambda/\mu} = e_{\nu_1} e_{\nu_2} \ldots$, where again the ν_i are the lengths of the components of the strip.

2. (a) Let λ be a partition of n. Then the number of standard tableaux of shape λ is

$$K_{\lambda,(1^n)} = \langle s_\lambda, h_1^n \rangle$$

by (5.14). By §4, Example 3 it follows that the number of standard tableaux of shape λ is equal to $n!/h(\lambda)$, where $h(\lambda)$ is the product of the hook-lengths of λ.

This result is true more generally if λ is a skew diagram all of whose connected components are right diagrams (i.e. diagrams of partitions).

(b) Let p be a positive integer and let λ be a partition, $\tilde{\lambda}$ its p-core (§1, Example 8). A *p-tableau* of shape $\lambda - \tilde{\lambda}$ is a sequence of partitions

$$\tilde{\lambda} = \mu^{(0)} \subset \mu^{(1)} \subset \ldots \subset \mu^{(m)} = \lambda$$

such that $\mu^{(i)} - \mu^{(i-1)}$ is a border strip of length p, for $1 \leqslant i \leqslant m$. (Thus when $p = 1$, a p-tableau is just a standard tableau (and $\tilde{\lambda} = 0$); when $p = 2$, it is also called a domino tableau.)

Let λ^* be the p-quotient of λ, thought of as a skew diagram with components $\lambda^{(i)}$ $(0 \leqslant i \leqslant p-1)$, and let $h(\lambda^*) = \prod h(\lambda^{(i)})$ be the product of the hook-lengths of λ^*. From §1, Example 8 it follows that the p-tableaux of shape $\lambda - \tilde{\lambda}$ are in one–one correspondence with the *standard* tableaux of shape λ^*, and hence by (a) above the number of p-tableaux of shape $\lambda - \tilde{\lambda}$ is equal to $m!/h(\lambda^*)$, where $m = |\lambda^*|$. This number is also equal to $p^m m!/h_p(\lambda)$, where $h_p(\lambda)$ is the product of the hook-lengths of λ that are divisible by p (§1, Example 8(d)).

3. For each symmetric function $f \in \Lambda$, let $f^\perp : \Lambda \to \Lambda$ be the adjoint of multiplication by f, i.e.

$$\langle f^\perp u, v\rangle = \langle u, fv\rangle$$

for all $u, v \in \Lambda$. Then $f \mapsto f^\perp : \Lambda \to \mathrm{End}(\Lambda)$ is a ring homomorphism.

(a) Since $\langle s_\mu^\perp s_\lambda, s_\nu\rangle = \langle s_\lambda, s_\mu s_\nu\rangle = \langle s_{\lambda/\mu}, s_\nu\rangle$ for all partitions λ, μ, ν, it follows that $s_\mu^\perp s_\lambda = s_{\lambda/\mu}$. Hence from (5.9) we have

$$s_\lambda(x, y) = \sum_\mu s_\mu^\perp s_\lambda(x) \, . \, s_\mu(y)$$

and therefore, for any $f \in \Lambda$,

$$f(x, y) = \sum_\mu s_\mu^\perp f(x) \, . \, s_\mu(y).$$

(b) We have $h_\lambda^\perp m_\mu = 0$ unless $\mu = \lambda \cup \nu$ for some partition ν, and in that case $h_\lambda^\perp m_\mu = m_\nu$. For

$$\langle h_\lambda^\perp m_\mu, h_\nu\rangle = \langle m_\mu, h_\lambda h_\nu\rangle = \langle m_\mu, h_{\lambda \cup \nu}\rangle$$

which is zero unless $\mu = \lambda \cup \nu$.

In particular, $h_n^\perp m_\mu = 0$ if n is not a part of μ, and $h_n^\perp m_\mu = m_\nu$ if n is a part of μ, where ν is the partition obtained by removing one part n from μ. It follows that for every $f(x_0, x_1, x_2, \ldots) \in \Lambda$, $(h_n^\perp f)(x_1, \ldots, x_n)$ is the coefficient of x_0^n in f.

(c) Next consider $p_n^\perp$. If $N \geqslant n$ we have

$$\langle p_n^\perp h_N, p_\lambda\rangle = \langle h_N, p_n p_\lambda\rangle = \langle h_{N-n}, p_\lambda\rangle$$

for all partitions λ of $N-n$, by §4, Example 4. Hence

$$p_n^{\perp} h_N = h_{N-n}$$

and therefore

$$p_n^{\perp} = \sum_{r \geq 0} h_r \, \partial/\partial h_{n+r}$$

acting on symmetric functions expressed as polynomials in the h's.

Dually

$$p_n^{\perp} = (-1)^{n-1} \sum_{r \geq 0} e_r \, \partial/\partial e_{n+r}$$

acting on symmetric functions expressed as polynomials in the e's.

Further, we have $\langle p_n^{\perp} p_\lambda, p_\mu \rangle = \langle p_\lambda, p_n p_\mu \rangle$, which is zero if $\lambda \neq \mu \cup (n)$, and is equal to z_λ if $\lambda = \mu \cup (n)$. It follows that $p_n^{\perp} p_\lambda = z_\lambda z_\mu^{-1} p_\mu$ if n is a part of λ, and μ is the partition obtained by removing one part n from λ. From the definition of z_λ it follows that $z_\lambda z_\mu^{-1} = n \, . \, m_n(\lambda)$, where $m_n(\lambda)$ is the multiplicity of n as a part of λ, and therefore

$$p_n^{\perp} = n \, \partial/\partial p_n$$

acting on symmetric functions expressed as polynomials in the p's. In particular, each $p_n^{\perp}$ is a derivation of Λ.

Since each $f \in \Lambda$ can be expressed as a polynomial $\varphi(p_1, p_2, \ldots)$ with rational coefficients, it follows that

$$f^{\perp} = \varphi(\partial/\partial p_1, 2\partial/\partial p_2, \ldots)$$

is a linear differential operator with constant coefficients.

(d) For each $n \in \mathbf{Z}$, let $\pi_n : \Lambda \to \Lambda$ be the operator defined as follows: if $n \geq 1$, π_n is multiplication by p_n; if $n \leq -1$, then $\pi_n = p_{-n}^{\perp}$; and π_0 is the identity. Then we have

$$[\pi_m, \pi_n] = n \delta_{m+n,0} \pi_0$$

for all $m, n \in \mathbf{Z}$, so that the linear span of the π_n is a Heisenberg Lie algebra.

4. We have

$$\sum_\lambda s_\lambda = \prod_i (1-x_i)^{-1} \prod_{i<j} (1-x_i x_j)^{-1},$$

where the sum on the left is over all partitions λ.

It is enough to prove this for a finite set of variables $x_1, \ldots, x_n$. Let $\Phi(x_1, \ldots, x_n)$ denote $\Sigma_\lambda s_\lambda(x_1, \ldots, x_n)$, which is now a sum over partitions λ of length $\leq n$. By induction on n, it is enough to show that

$$\Phi(x_1, \ldots, x_n, y) = \Phi(x_1, \ldots, x_n)(1-y)^{-1} \prod_{i=1}^{n} (1-x_i y)^{-1}.$$

From (5.9) it follows that

$$\Phi(x_1,\dots,x_n,y)=\sum_{\lambda,\mu} y^{|\lambda-\mu|}s_\mu(x_1,\dots,x_n)$$

where the sum on the right is over all pairs of partitions $\lambda\supset\mu$ such that $l(\mu)\leqslant n$ and $\lambda-\mu$ is a horizontal strip. For each such pair λ, μ, define $\nu\subset\mu$ by $\mu_i-\nu_i=\lambda_{i+1}-\mu_{i+1}$ $(i\geqslant 1)$, so that $|\lambda-\mu|=\lambda_1-\mu_1+|\mu-\nu|$. Then λ can be reconstructed from μ, ν, and the integer $\lambda_1-\mu_1$, and hence

$$(*)\qquad \sum_{\lambda,\mu} y^{|\lambda-\mu|}s_\mu(x_1,\dots,x_n)=\sum_{\mu,\nu} y^{|\mu-\nu|}(1-y)^{-1}s_\mu(x_1,\dots,x_n),$$

the sum on the right being over pairs of partitions $\mu\supset\nu$ such that $l(\mu)\leqslant n$ and $\mu-\nu$ is a horizontal strip. By (5.16), the right-hand side of $(*)$ is equal to

$$\sum_{\nu,r} y^r(1-y)^{-1}h_r(x_1,\dots,x_n)s_\nu(x_1,\dots,x_n)$$

summed over all partitions ν of length $\leqslant n$, and all integers $r\geqslant 0$; and this last sum is equal to $(1-y)^{-1}\prod_{i=1}^n(1-x_iy)^{-1}\Phi(x_1,\dots,x_n)$, as required.

5. (a) We have

$$\sum_{\mu\text{ even}} s_\mu=\prod_i(1-x_i^2)^{-1}\prod_{i<j}(1-x_ix_j)^{-1},$$

where the sum on the left is over all *even* partitions μ (i.e. with all parts μ_i even).

Each partition λ can be reduced to an even partition μ by removing a vertical strip, in exactly one way: we take $\mu_i=\lambda_i$ if λ_i is even, and $\mu_i=\lambda_i-1$ if λ_i is odd. From this observation and (5.17) it follows that

$$\Big(\sum_{\mu\text{ even}} s_\mu\Big)\Big(\sum_{r\geqslant 0} e_r\Big)=\sum_\lambda s_\lambda,$$

the sum on the right being over *all* partitions λ. Since $\sum e_r=\prod(1+x_i)$, the result now follows from Example 4.

(b) We have

$$\sum_{\nu'\text{ even}} s_\nu=\prod_{i<j}(1-x_ix_j)^{-1}.$$

The proof is dual to that of (a): each partition λ can be reduced to one with even columns by removing a horizontal strip, in exactly one way. From this observation and (5.16) it follows that

$$\Big(\sum_{\nu'\text{ even}} s_\nu\Big)\Big(\sum_{r\geqslant 0} h_r\Big)=\sum_\lambda s_\lambda,$$

and since $\sum_{r\geqslant 0}h_r=\prod(1-x_i)^{-1}$, the result again follows from Example 4.

The involution ω interchanges the identities in (a) and (b).

6. We have

$$\sum_\lambda (-1)^{n(\lambda)} s_\lambda = \prod_i (1-x_i)^{-1} \prod_{i<j} (1+x_i x_j)^{-1}.$$

For if we replace each variable x_i by $\sqrt{-1}\,.\,x_i$ in Example 5(b), we shall obtain

$$\prod_{i<j} (1+x_i x_j)^{-1} = \sum_\nu (-1)^{|\nu|/2} s_\nu$$

summed over partitions ν with all columns of even length. Each partition λ is obtained uniquely from such a partition ν by adjoining a horizontal strip, and therefore

$$\begin{aligned}\prod_i (1-x_i)^{-1} \prod_{i<j} (1+x_i x_j)^{-1} &= \sum_{r\geq 0} \sum_\nu (-1)^{|\nu|/2} s_\nu h_r \\ &= \sum_\lambda (-1)^{f(\lambda)} s_\lambda\end{aligned}$$

where $f(\lambda) = \sum_{i\geq 1} [\frac{1}{2}\lambda_i']$ (and $[x]$ is the greatest integer $\leqslant x$). Since $\left[\frac{r}{2}\right] \equiv \binom{r}{2}$ (mod 2), it follows from (1.6) that $f(\lambda) \equiv n(\lambda)$ (mod 2).

7. The same argument as in Example 5(b) shows that

$$\sum_\lambda t^{c(\lambda)} s_\lambda = \prod_i (1-tx_i)^{-1} \prod_{i<j} (1-x_i x_j)^{-1}$$

where the sum is over all partitions λ, and $c(\lambda)$ is the number of columns of odd length in λ. This includes the identities of Example 4 (when $t=1$) and Example 5(b) (when $t=0$).

8. By applying the involution ω to Example 7 we obtain

$$\sum_\lambda t^{r(\lambda)} s_\lambda = \prod_i \frac{1+tx_i}{1-x_i^2} \prod_{i<j} \frac{1}{1-x_i x_j}$$

where the sum is over all partitions λ, and $r(\lambda)$ is the number of rows of odd length in λ. When $t=1$ this reduces to Example 4, and when $t=0$ it reduces to Example 5(a).

9. The products

$$\prod_{i<j} (1-x_i x_j), \quad \prod_i (1-x_i) \prod_{i<j} (1-x_i x_j), \quad \prod_i (1-x_i^2) \prod_{i<j} (1-x_i x_j).$$

(i.e. the reciprocals of those of Examples 4, 5(a) and 5(b)) can also be expanded as series of Schur functions. The expansions may be derived from Weyl's identity for the root-systems of types D_n, B_n, C_n respectively. (If R is a root system with Weyl group W, R^+ a system of positive roots, ρ half the sum of the positive roots, then Weyl's identity ([B8], p. 185) is

$$(*) \qquad \sum_{w\in W} \varepsilon(w) e^{w\rho} = \prod_{\alpha\in R^+} (e^{\alpha/2} - e^{-\alpha/2})$$

where $\varepsilon(w)$ is the sign of $w \in W$, and the e's are formal exponentials.)

(a) When R is of type D_n, the identity $(*)$ leads to

$$\sum_{\pi}(-1)^{|\pi|/2}s_\pi(x_1,\ldots,x_n) = \prod_{i<j}(1-x_ix_j)$$

summed over all partitions $\pi = (\alpha_1-1,\ldots,\alpha_p-1\,|\,\alpha_1,\ldots,\alpha_p)$ in Frobenius notation, where $\alpha_1 \leqslant n-1$.

(b) When R is of type C_n, we obtain from $(*)$

$$\sum_{\rho}(-1)^{|\rho|/2}s_\rho(x_1,\ldots,x_n) = \prod_{i}(1-x_i^2)\prod_{i<j}(1-x_ix_j)$$

summed over all partitions $\rho = (\alpha_1+1,\ldots,\alpha_p+1\,|\,\alpha_1,\ldots,\alpha_p)$, where $\alpha_1 \leqslant n-1$.

(c) When R is of type B_n, we obtain from $(*)$

$$\sum_{\sigma}(-1)^{(|\sigma|+p(\sigma))/2}s_\sigma(x_1,\ldots,x_n) = \prod_{i}(1-x_i)\prod_{i<j}(1-x_ix_j)$$

summed over all self-conjugate partitions $\sigma = (\alpha_1,\ldots,\alpha_p\,|\,\alpha_1,\ldots,\alpha_p)$ such that $\alpha_1 \leqslant n-1$, where $p(\sigma) = p$.

10. In the language of λ-rings, the identities of Examples 5(a), 5(b), and 9 give series expansions (in terms of Schur operations) for $\sigma_t(\sigma^2(x))$, $\sigma_t(\lambda^2(x))$, $\lambda_t(\sigma^2(x))$, and $\lambda_t(\lambda^2(x))$, namely

$$\sigma_t(\sigma^2(x)) = \sum_{\mu\ \text{even}} S^\mu(x)t^{|\mu|/2},$$

$$\sigma_t(\lambda^2(x)) = \sum_{\nu'\ \text{even}} S^\nu(x)t^{|\nu|/2},$$

$$\lambda_t(\sigma^2(x)) = \sum_{\rho} S^\pi(x)t^{|\rho|/2},$$

$$\lambda_t(\lambda^2(x)) = \sum_{\pi} S^\pi(x)t^{|\pi|/2},$$

the last two summations being over partitions $\rho = (\alpha_1+1,\ldots,\alpha_p+1\,|\,\alpha_1,\ldots,\alpha_p)$ and $\pi = (\alpha_1-1,\ldots,\alpha_p-1\,|\,\alpha_1,\ldots,\alpha_p)$.

11. Let $x_1 = \ldots = x_N = t$, $x_{N+1} = x_{N+2} = \ldots = 0$ in the formula of Example 4. Then $s_\lambda = \binom{N}{\lambda'}t^{|\lambda|}$ (§3, Example 4) and hence, for each $n \geqslant 0$,

$$\sum_{|\lambda|=n}\binom{N}{\lambda} = \text{coefficient of } t^n \text{ in } (1-t)^{-N}(1-t^2)^{-N(N-1)/2}$$

$$= \text{coefficient of } t^n \text{ in } (1-t)^{-N(N+1)/2}(1+t)^{-N(N-1)/2}.$$

Since this is true for all positive integers N, it is a polynomial identity, i.e.

$$\sum_{|\lambda|=n}\binom{X}{\lambda} = \text{coefficient of } t^n \text{ in } (1-t)^{-X(X+1)/2}(1+t)^{-X(X-1)/2}.$$

12. Let $x_1 = \dots = x_N = t/N$, $x_{N+1} = x_{N+2} = \dots = 0$ in the identity of Example 4, and let $N \to \infty$. Then from Example 11 we obtain

$$\sum_{|\lambda| = n} h(\lambda)^{-1} = \text{coefficient of } t^n \text{ in } \exp(t + \tfrac{1}{2}t^2)$$

where (§4, Example 3) $h(\lambda)$ is the product of the hook lengths of λ. From Example 2 it follows that the total number of standard tableaux of weight (1^n) is equal to $n!$ multiplied by the coefficient of t^n in $\exp(t + \frac{1}{2}t^2)$. This number is also the number of permutations $w \in S_n$ such that $w^2 = 1$.

13. Let λ be a partition. A *plane partition of shape* λ is a mapping π from (the diagram of) λ to the positive integers such that $\pi(x_1) \geqslant \pi(x_2)$ whenever x_2 lies below or to the right of x_1 in λ. The numbers $\pi(x)$ are the *parts* of π, and

$$|\pi| = \sum_{x \in \lambda} \pi(x)$$

is called the *weight* of π. Any plane partition π determines a sequence $\lambda = \lambda^{(0)} \supset \lambda^{(1)} \supset \dots$ of (linear) partitions such that $\pi^{-1}(i) = \lambda^{(i-1)} - \lambda^{(i)}$ for each $i \geqslant 1$.

If $\pi(x_1) > \pi(x_2)$ whenever x_2 lies directly below x_1 (i.e. if the parts of π decrease strictly down each column) then π is said to be *column-strict*. Clearly π is column-strict if and only if each skew diagram $\pi^{-1}(i) = \lambda^{(i-1)} - \lambda^{(i)}$ is a horizontal strip.

A plane partition π has a three-dimensional *diagram*, consisting of the points (i, j, k) with integer coordinates such that $(i, j) \in \lambda$ and $1 \leqslant k \leqslant \pi(i, j)$. Alternatively, we may think of the diagram of π as a set of unit cubes, such that $\pi(x)$ cubes are stacked vertically on each square $x \in \lambda$. As in the case of ordinary (linear) partitions, we shall use the same symbol π to denote a plane partition and its diagram.

If S is any set of plane partitions, the *generating function* of S is the polynomial or formal power series

$$\sum_{\pi \in S} q^{|\pi|}$$

in which the coefficient of q^n is the number of plane partitions of weight n which belong to S.

(a) Consider column-strict plane partitions of shape λ, with all parts $\leqslant n$. By (5.12) the generating function for these is $s_\lambda(q^n, q^{n-1}, \dots, q)$, which by §3, Example 1 is

$$q^{|\lambda| + n(\lambda)} \prod_{x \in \lambda} \frac{1 - q^{n + c(x)}}{1 - q^{h(x)}}.$$

(b) Let l, m, n be three positive integers, and consider the set of plane partitions π with all parts $\leqslant n$ and shape λ such that $l(\lambda) \leqslant l$ and $l(\lambda') \leqslant m$: that is, the set of three-dimensional diagrams π which fit inside a box B with side-lengths l, m, n. By adding $l + 1 - i$ to each part in the ith row of π, for $1 \leqslant i \leqslant l$, we convert π

into a column-strict plane partition of shape $(m,\ldots,m)=(m^l)$ and largest part $\leqslant l+n$. From (a) above, the generating function for the plane partitions $\pi\subset B$ is therefore

(1) $$\prod_{x\in(m^l)}\frac{1-q^{l+n+c(x)}}{1-q^{h(x)}}.$$

In this form the result does not display the symmetry which it must have as a function of l, m, and n. It may be rewritten as follows: for each $y=(i,j,k)\in B$, define the *height* of y to be $\mathrm{ht}(y)=i+j+k-2$ (so that the point $(1,1,1)$ has height 1). Then the generating function (1) may be written in the form

(2) $$\sum_{\pi\subset B}q^{|\pi|}=\prod_{y\in B}\frac{1-q^{1+\mathrm{ht}(y)}}{1-q^{\mathrm{ht}(y)}}.$$

(c) We may now let any or all of l, m, n become infinite. The most striking result is obtained by letting all of l, m, n tend to ∞: the box B is then replaced by the positive octant, and for each $n\geqslant 1$ the number of lattice points (i,j,k) with $i+j+k-2=n$ and $i,j,k\geqslant 1$ is equal to the coefficient of t^{n-1} in $(1-t)^{-3}$, hence to $\frac{1}{2}n(n+1)$. It follows that the generating function for *all* plane partitions is

(3) $$\prod_{n=1}^{\infty}\left(\frac{1-q^{n+1}}{1-q^n}\right)^{n(n+1)/2}=\prod_{n=1}^{\infty}(1-q^n)^{-n}.$$

(d) Likewise, the generating function for all plane partitions with largest part $\leqslant m$ is

(4) $$\prod_{n=1}^{\infty}(1-q^n)^{-\min(m,n)}.$$

14. From Example 13(a), by letting $n\to\infty$, the generating function for all column-strict plane partitions of shape λ is

(1) $$q^{|\lambda|+n(\lambda)}H_\lambda(q)^{-1}$$

where $H_\lambda(q)=\prod_{x\in\lambda}(1-q^{h(x)})$.

Another way of obtaining this generating function is as follows. Let π be a column-strict plane partition of shape λ, and let S be the set of pairs $(\pi(i,j),j)$ where $(i,j)\in\lambda$. The elements of S are all distinct, because π is column-strict. We order S as follows: (r,j) precedes (r',j') if either $r>r'$, or $r=r'$ and $j<j'$. This is a linear ordering of S. Define a standard tableau $T(\pi)$ of shape λ as follows: $T(i,j)=k\Leftrightarrow(\pi(i,j),j)$ is the kth element of S in the linear ordering defined above. For example, if π is

33211
22
1

then S is the ordered set

$$(3,1),(3,2),(2,1),(2,2),(2,3),(1,1),(1,4),(1,5),$$

and $T(\pi)$ is the standard tableau

$$\begin{matrix}12578\\34\\6\end{matrix}$$

Conversely, let T be a standard tableau of shape λ, and let π be a column-strict plane partition such that $T(\pi)=T$. Let $|\lambda|=n$, and for $1\leqslant k\leqslant n$ let a_k be the part of π in the square occupied by k in T. Then $a_1\geqslant\ldots\geqslant a_n\geqslant 1$ and $a_k>a_{k+1}$ whenever $k\in R(T)$, where $R(T)$ is the set of integers $k\in[1,n-1]$ such that $k+1$ lies in a lower row than k in the tableau T. Now let

$$b_k=\begin{cases}a_k-a_{k+1} & \text{if} \quad k\notin R(T) \quad \text{and } k\neq n\\ a_k-a_{k+1}-1 & \text{if} \quad k\in R(T)\\ a_n-1 & \text{if} \quad k=n\end{cases}$$

so that $b_k\geqslant 0$ for $k=1,2,\ldots,n$. Then we have

$$\sum_{k=1}^{n} a_k=n+r(T)+\sum_{k=1}^{n} kb_k$$

where

$$r(T)=\sum\{k: k+1 \text{ lies in a lower row than } k \text{ in } T\}$$

and therefore the generating function for the column-strict plane partitions π such that $T(\pi)=T$ is

$$q^{n+r(T)}\varphi_n(q)^{-1}$$

where as usual $\varphi_n(q)=(1-q)\ldots(1-q^n)$.

Hence the generating function for column-strict plane partitions of shape λ is

$$q^n\left(\sum_T q^{r(T)}\right)\Big/\varphi_n(q) \tag{2}$$

summed over all standard tableaux T of shape λ.

From (1) and (2) it follows that

$$\sum_T q^{r(T)}=q^{n(\lambda)}\varphi_n(q)/H_\lambda(q). \tag{3}$$

15. Let S be any set of positive integers. From (5.12) and Example 4 it follows that the generating function for column-strict plane partitions all of whose parts belong to S is

$$\prod_{i\in S}(1-q^i)^{-1}\prod_{\substack{i,j\in S\\ i<j}}(1-q^{i+j})^{-1}.$$

(a) Take S to consist of all the positive integers. Then the generating function for *all* column-strict plane partitions, of arbitrary shape, is

(1) $$\prod_{n=1}^{\infty}(1-q^n)^{-[(n+1)/2]}.$$

(b) Take S to consist of all the *odd* positive integers. We obtain the generating function

(2) $$\prod_{n=1}^{\infty}(1-q^{2n-1})^{-1}(1-q^{2n})^{-[n/2]}.$$

Now the column-strict plane partitions with all parts odd are in one-to-one correspondence with the *symmetrical* plane partitions π (i.e. such that $\pi(i,j)=\pi(j,i)$). For the diagram of a symmetrical plane partition may be thought of as a sequence of diagrams of symmetrical (linear) partitions $\pi^{(1)}\supset\pi^{(2)}\supset\dots$, piled one on top of the other; each $\pi^{(i)}$ is of the form $(\alpha_1,\dots,\alpha_p\,|\,\alpha_1,\dots,\alpha_p)$ in Frobenius notation, and hence determines a linear partition $\sigma^{(i)}=(2\alpha_1+1,\dots,2\alpha_p+1)$ with odd parts, all distinct; and the $\sigma^{(i)}$ can be taken as the columns of a column-strict plane partition with odd parts. It follows that (2) is the generating function for the set of all symmetrical plane partitions.

16. Let $\Phi(x_1,\dots,x_n)=\prod_i(1-x_i)^{-1}\prod_{i<j}(1-x_ix_j)^{-1}$ as in Example 4. By setting $t=0$ in the identity of Chapter III, §5, Example 5 we obtain

(1) $$\sum_{m,\lambda}u^m s_\lambda(x_1,\dots,x_n)=\sum_{\varepsilon}\Phi(x_1^{\varepsilon_1},\dots,x_n^{\varepsilon_n})/(1-u\prod x_i^{(1-\varepsilon_i)/2})$$

where the sum on the left is over all partitions $\lambda=(\lambda_1,\dots,\lambda_n)$ of length $\leqslant n$, and integers $m\geqslant\lambda_1$; and the sum on the right is over all $\varepsilon=(\varepsilon_1,\dots,\varepsilon_n)$ with each $\varepsilon_i=\pm1$.

We shall rewrite (1) in the notation of root-systems. Let $v_1,\dots,v_n$ be the standard basis of $\mathbf{R}^n$. Then the set of vectors

$$R=\{\pm v_i(1\leqslant i\leqslant n),\ \pm v_i\pm v_j(1\leqslant i<j\leqslant n)\}$$

is a root-system of type B_n, for which

$$R^+=\{v_i(1\leqslant i\leqslant n), v_i\pm v_j(1\leqslant i<j\leqslant n)\}$$

is a system of positive roots, so that

$$\rho=\tfrac{1}{2}((2n-1)v_1+(2n-3)v_2+\dots+v_n)$$

is half the sum of the positive roots. The subset R_0 of R defined by

$$R_0=\{v_i-v_j: i\neq j\}$$

is a subsystem of R of type A_{n-1}, and $R_0^+=R^+\cap R_0$ is a system of positive roots for R_0. The Weyl group W_0 of R_0 is the symmetric group S_n, acting by permutations of $v_1,\dots,v_n$, and the Weyl group W of R is the semidirect product of W_0

with the group (of order 2^n) of transformations $w_\varepsilon: v_i \mapsto \varepsilon_i v_i (1 \leqslant i \leqslant n)$, where as before $\varepsilon = (\varepsilon_1, \ldots, \varepsilon_n)$ and each ε_i is ± 1. In this notation,

$$\Phi(e^{-v_1}, \ldots, e^{-v_n}) = \prod_{\alpha \in R_0^+} (1 - e^{-\alpha}) \Big/ \prod_{\alpha \in R^+} (1 - e^{-\alpha})$$

$$= \left(\sum_{w \in W_0} \varepsilon(w) e^{w\rho} \right) \Big/ \left(\sum_{w \in W} \varepsilon(w) e^{w\rho} \right),$$

by virtue of Weyl's identity (Example 9). It follows that the right-hand side of (1) (with x_i replaced by e^{-v_i}) may be written as a sum over W, and by equating the coefficients of u^m on either side of (1) we arrive at the identity

$$\sum_\lambda s_\lambda(e^{-v_1}, \ldots, e^{-v_n}) = e^{-m\theta} J(m\theta + \rho)/J(\rho) \tag{2}$$

where $\theta = \frac{1}{2}(v_1 + \ldots + v_n)$, and for any vector v

$$J(v) = \sum_{w \in W} \varepsilon(w) e^{wv},$$

and the sum on the left is over all partitions λ such that $l(\lambda) \leqslant n$ and $l(\lambda') \leqslant m$ (i.e. such that $\lambda \subset (m^n)$).

(If preferred, the right-hand side of (2) can be written as a quotient of determinants:

$$\sum_\lambda s_\lambda(x_1, \ldots, x_n) = D_m / D_0 \tag{2'}$$

where $D_m = \det(x_j^{m+2n-i} - x_j^{i-1})_{1 \leqslant i,j \leqslant n}$, and the summation is as before over partitions $\lambda \subset (m^n)$.)

This identity (2) is a polynomial identity in n independent variables e^{-v_i}. We may therefore specialize it to obtain identities in one variable q, by replacing each e^{-v_i} by q^{f_i}, where the f_i are arbitrary integers. This means that each exponential e^{-v} is replaced by $q^{\langle v, f \rangle}$, where $f = \sum f_i v_i$ and $\langle v, f \rangle$ is the standard scalar product on $\mathbf{R}^n$. In this way we obtain

$$\sum_\lambda s_\lambda(q^{f_1}, \ldots, q^{f_n}) = q^{m\langle \theta, f \rangle} \frac{\sum \varepsilon(w) q^{-\langle m\theta + \rho, wf \rangle}}{\sum \varepsilon(w) q^{-\langle \rho, wf \rangle}}, \tag{3}$$

the sum on the left being over all partitions $\lambda \subset (m^n)$.

17. In formula (3) of Example 16 let us take $f = 2\rho$, the sum of the positive roots of R, so that $f_i = 2n - 2i + 1$. On the right-hand side, the alternating sum

$$\sum \varepsilon(w) q^{-\langle m\theta + \rho, 2w\rho \rangle}$$

is by Weyl's identity (Example 9) equal to the product

$$\prod_{\alpha \in R^+} (q^{-\langle m\theta + \rho, \alpha \rangle} - q^{\langle m\theta + \rho, \alpha \rangle})$$

and therefore the right-hand side of (3) is equal to

$$\prod_{\alpha \in R^+} \frac{q^{\langle 2m\theta + 2\rho, \alpha \rangle} - 1}{q^{\langle 2\rho, \alpha \rangle} - 1}.$$

In this product the positive roots $v_i - v_j (i < j)$ make no contribution, because they are orthogonal to $\theta = \frac{1}{2} \sum v_i$. Hence we obtain the identity

(4)

$$\sum_{\lambda \subset (m^n)} s_\lambda(q^{2n-1}, q^{2n-3}, \ldots, q) = \prod_{i=1}^{n} \frac{q^{m+2i-1} - 1}{q^{2i-1} - 1} \prod_{1 \leqslant i < j \leqslant n} \frac{q^{2(m+i+j-1)} - 1}{q^{2(i+j-1)} - 1}.$$

The left-hand side of (4) is the generating function for column-strict plane partitions with odd parts $\leqslant 2n - 1$, and with at most m columns and at most n rows; or equivalently (Example 15) it is the generating function for *symmetrical* plane partitions π whose diagrams are contained in the box $B = B_{(n,n,m)} = \{(i,j,k): 1 \leqslant i, j \leqslant n, 1 \leqslant k \leqslant m\}$.

The right-hand side of (4) can be rewritten in a form analogous to that of Example 13, formula (2), as follows. Let G_2 be the group of two elements consisting of the identity and the mapping $(i,j,k) \mapsto (j,i,k)$, so that the box B is stable under G_2. For each orbit η of G_2 in B let $|\eta|$ ($= 1$ or 2) be the number of elements of η, and let

$$\mathrm{ht}(\eta) = \sum_{y \in \eta} \mathrm{ht}(y)$$

where $\mathrm{ht}(i,j,k) = i + j + k - 2$ as in Example 13. Then the generating function for symmetrical plane partitions $\pi \subset B$ is

(5)
$$\prod_{\eta \in B/G_2} \frac{1 - q^{\mathrm{ht}(\eta) + |\eta|}}{1 - q^{\mathrm{ht}(\eta)}}.$$

18. Let G_3 be the group of three elements generated by $(i,j,k) \mapsto (j,k,i)$ and let C_n be the cube $\{(i,j,k): 1 \leqslant i,j,k \leqslant n\}$. The formula (5) of Example 17 suggests the following conjecture: the generating function for *cyclically symmetric* plane partitions π (i.e. those whose diagrams are stable under G_3) contained in the cube C_n should be

(6)
$$\prod_{\eta \in C_n/G_3} \frac{1 - q^{\mathrm{ht}(\eta) + |\eta|}}{1 - q^{\mathrm{ht}(\eta)}}.$$

This conjecture has since been proved by Mills, Robbins, and Rumsey [M11].

Next, let G_6 be the group of all permutations of (i,j,k), and call a plane partition *completely symmetric* if its diagram is stable under G_6. The obvious analogue of (5) and (6) for G_6 is trivially *false*, because the rational function

$$\prod_{\eta \in C_n/G_6} \frac{1 - q^{\mathrm{ht}(\eta) + |\eta|}}{1 - q^{\mathrm{ht}(\eta)}}$$

is not a polynomial if $n \geqslant 3$. However, it seems likely† that the *number* of completely symmetric plane partitions with all parts $\leqslant n$ is correctly given by setting $q = 1$ in this expression, i.e. it is equal to

$$\prod_{\eta \in C_n/G_6} \frac{\mathrm{ht}(\eta) + |\eta|}{\mathrm{ht}(\eta)}.$$

19. With the notation of Example 16, the set of vectors

$$R_1 = \{\pm 2v_i \, (1 \leqslant i \leqslant n), \ \pm v_i \pm v_j \, (1 \leqslant i < j \leqslant n)\}$$

is a root system of type C_n, for which

$$R_1^+ = \{2v_i \, (1 \leqslant i \leqslant n), v_i \pm v_j \, (1 \leqslant i < j \leqslant n)\}$$

is a system of positive roots, so that

$$\rho_1 = nv_1 + (n-1)v_2 + \ldots + v_n$$

is half the sum of the positive roots. The Weyl group is the same group W as in Example 16.

We shall take $f = \rho_1$ in formula (3) of Example 16, so that e^{-v_i} is replaced by q^{n-i+1}. As in Example 17, by virtue of Weyl's identity we have

$$\sum \varepsilon(w) q^{-\langle m\theta + \rho, w\rho_1 \rangle} = \prod_{\alpha \in R_1^+} (q^{-\langle m\theta + \rho, \alpha/2 \rangle} - q^{\langle m\theta + \rho, \alpha/2 \rangle})$$

and therefore the right-hand side of (3) is equal to

$$\prod_{\alpha \in R_1^+} \frac{q^{\langle m\theta + \rho, \alpha \rangle} - 1}{q^{\langle \rho, \alpha \rangle} - 1}.$$

Again the roots $v_i - v_j$ $(i < j)$ make no contribution to this product, and hence we obtain

$$\sum_{\lambda \subset (m^n)} s_\lambda(q^n, \ldots, q) = \prod_{1 \leqslant i \leqslant j \leqslant n} \frac{q^{m+i+j-1} - 1}{q^{i+j-1} - 1}. \tag{7}$$

The left-hand side of (7) is the generating function for column-strict plane partitions with largest part $\leqslant n$ and at most m columns, and the right-hand side can be written in terms of the height function introduced in Example 13, namely as

$$\prod_{y \in D} \frac{1 - q^{\mathrm{ht}(y)+1}}{1 - q^{\mathrm{ht}(y)}} \tag{8}$$

where D is the prism $\{(i, j, k) : 1 \leqslant i \leqslant j \leqslant n, 1 \leqslant k \leqslant m\}$.

† This conjecture has recently been proved by J. Stembridge [S30].

20. (a) Let $A=(a_{ij})$ and $B=(b_{ij})$ be $n\times n$ matrices such that $\det A=1$. Let c_{ij} be the determinant of the matrix obtained from A by replacing its ith column by the jth column of B, so that

$$c_{ij}=\sum_{k=1}^{n} a^{ik}b_{kj},$$

where a^{ik} is the cofactor of a_{ki} in A, which since $\det A=1$ is equal to the (i,k) element of A^{-1}. It follows that c_{ij} is the (i,j) element of $A^{-1}B$, and hence that $\det(c_{ij})=\det B$.

Now let $r<n$ and suppose that for each $j>r$ the jth column of A is equal to the jth column of B. Then $c_{ij}=\delta_{ij}$ whenever $j>r$, and hence we have

$$\det(c_{ij})_{1\leqslant i,j\leqslant r}=\det B. \tag{1}$$

(b) Let $\lambda=(\alpha_1,\ldots,\alpha_r\mid\beta_1,\ldots,\beta_r)$ be a partition of length $\leqslant n$, in Frobenius notation, and let

$$\lambda^{(i,j)}=\left(\alpha_1,\ldots,\hat{\alpha}_i,\ldots,\alpha_r\mid\beta_1,\ldots,\hat{\beta}_j,\ldots,\beta_r\right)$$

for $i,j=1,2,\ldots,r$, where the circumflexes indicate that the symbols they cover are to be omitted. Then the skew diagram $[\alpha_i\mid\beta_j]=\lambda-\lambda^{(i,j)}$ (which of course depends on λ as well as α_i, β_j) is a border strip, and more precisely is that part of the border of λ consisting of the squares (h,k) such that $h\geqslant i$ and $k\geqslant j$. With this notation established, we have

$$s_\lambda=\det(s_{[\alpha_i\mid\beta_j]})_{1\leqslant i,j\leqslant r}. \tag{2}$$

(Let $\xi_k=k-\lambda_k$ $(1\leqslant k\leqslant n)$. The sequence $\xi=(\xi_k)$ is obtained from the sequence $(-\alpha_1,\ldots,-\alpha_r,1,2,\ldots,n)$ by deletion of $\beta_r+1,\ldots,\beta_1+1$. Hence the corresponding sequence for the partition $\lambda^{(i,j)}$ is obtained from ξ by deleting $-\alpha_i$ and inserting β_j+1. It follows that, up to sign, $s_{[\alpha_i,\beta_j]}$ is equal to the determinant c_{ij} of the $n\times n$ matrix obtained from $A=(h_{-\xi_i+\xi_j})$ by replacing its ith column by the (β_j+1)th column of the matrix $B=(h_{\lambda_i-i+j})=(h_{-\xi_i+j})$. Moreover, the sign involved is $(-1)^{\beta_i-i+j}$, so that we have

$$\det(s_{[\alpha_i\mid\beta_j]})=(-1)^{|\beta|}\det(c_{ij}). \tag{3}$$

On the other hand, by (1) above, $\det(c_{ij})$ is equal to the determinant of the matrix obtained from B by rearranging its columns in the sequence $(\beta_1+1,\ldots,\beta_r+1,1,2,\ldots,n)$, and therefore

$$\det(c_{ij})=(-1)^{|\beta|}\det B=(-1)^{|\beta|}s_\lambda. \tag{4}$$

The formula (2) now follows from (3) and (4).)

21. Let θ, φ be two skew diagrams. Let a be the rightmost square in the top row of θ, and let b be the leftmost square in the bottom row of φ. Let φ^v (resp. φ^h)

denote the diagram obtained from φ by a shift sending b to the square immediately above a (resp. immediately to the right of a) and let $\theta \underset{v}{*} \varphi$ (resp. $\theta \underset{h}{*} \varphi$) denote the skew diagram $\theta \cup \varphi^v$ (resp. $\theta \cup \varphi^h$).

(a) Show that

(1) $$s_\theta s_\varphi = s_{\theta \underset{v}{*} \varphi} + s_{\theta \underset{h}{*} \varphi}.$$

(From (5.12) it follows that

$$s_\theta s_\varphi = \sum_{T,U} x^T x^U$$

summed over all pairs of tableaux T, U of respective shapes θ, φ. Split up the set of these pairs (T,U) into two subsets according as $T(a) \leqslant U(b)$ or $T(a) > U(b)$, where $T(a)$ is the symbol occupying the square a in T, and likewise for $U(b)$.)

(b) In view of (5.7), the repeated use of (1) enables us to express any skew Schur function as a sum of skew Schur functions corresponding to connected diagrams. In particular, we have

(2) $$h_1^n = \sum_\theta s_\theta$$

summed over the 2^{n-1} border strips (or ribbons) of length n. Taking the coefficient of $x_1 \dots x_n$ in both sides of (2) (or, equivalently, the scalar product with h_1^n) we see that this decomposition describes the partition of the symmetric group S_n into the subsets of permutations having a given set of descents.

(c) Let $\lambda = (\alpha \mid \beta) = (\alpha_1, \dots, \alpha_r \mid \beta_1, \dots, \beta_r)$ be a partition in Frobenius notation, and let $S_{(\alpha \mid \beta)}$, $S_{[\alpha \mid \beta]}$ denote the $r \times r$ matrices

$$S_{(\alpha \mid \beta)} = (s_{(\alpha_i \mid \beta_j)}), \qquad S_{[\alpha \mid \beta]} = \left((-1)^{i+j} s_{[\alpha_i \mid \beta_j]}\right)$$

in the notation introduced in Example 20. Show that

(3) $$S_{(\alpha \mid \beta)} = H_\alpha S_{[\alpha \mid \beta]} E_\beta$$

where $H_\alpha = (h_{\alpha_i - \alpha_j})$, $E_\beta = (e_{\beta_j - \beta_i})$. (Use (1) above.)

By taking determinants in (3) and using §3, Example 9, we obtain another proof of the formula (2) of Example 20.

22. (a) Let $\lambda = (\alpha_1, \dots, \alpha_r \mid \beta_1, \dots, \beta_r)$, $\mu = (\gamma_1, \dots, \gamma_s \mid \varepsilon_1, \dots, \varepsilon_s)$ be two partitions in Frobenius notation. Define matrices

$$S_{(\alpha \mid \beta)} = (s_{(\alpha_i \mid \beta_j)})_{1 \leqslant i,j \leqslant r},$$

$$H_{\alpha,\gamma} = (h_{\alpha_i - \gamma_j})_{1 \leqslant i \leqslant r, 1 \leqslant j \leqslant s},$$

$$E_{\beta,\varepsilon} = (e_{\beta_j - \varepsilon_i})_{1 \leqslant i \leqslant s, 1 \leqslant j \leqslant r}.$$

Then we have

(1) $$s_{\lambda/\mu}=(-1)^s\det\begin{pmatrix} S_{(\alpha\mid\beta)} & H_{\alpha,\gamma}\\ E_{\beta,\varepsilon} & 0\end{pmatrix}.$$

When $\mu=0$ (so that $s=0$) this formula reduces to that of §3, Example 9.

(Choose $n\geqslant \max(l(\lambda),l(\mu))$ and let $I=(-\alpha_1,\dots,-\alpha_r,1,2,\dots,n)$, $J=(-\gamma_1,\dots,-\gamma_s,1,2,\dots,n)$. Then the sequence $(i-\lambda_i)_{1\leqslant i\leqslant n}$ is obtained from the sequence I by deleting the terms β_i+1 $(1\leqslant i\leqslant r)$, and likewise the sequence $(j-\mu_j)_{1\leqslant j\leqslant n}$ is obtained from J by deleting the terms ε_j+1 $(1\leqslant j\leqslant s)$. Hence the matrix $(h_{\lambda_i-\mu_j-i+j})$ is obtained from the matrix $M=(h_{j-i})_{(i,j)\in I\times J}$ by deleting the rows with indices $r+\beta_i+1$ $(1\leqslant i\leqslant r)$ and the columns with indices $s+\varepsilon_j+1$ $(1\leqslant j\leqslant s)$. Hence if U is the matrix of $n+r$ rows and r columns which has 1's in the positions $(r+\beta_i+1,i)$ $(1\leqslant i\leqslant r)$ and zeros elsewhere, and if V is the matrix of s rows and $n+s$ columns that has 1's in the positions $(j,s+\varepsilon_j+1)$ $(1\leqslant j\leqslant s)$ and zeros elsewhere, we shall have

$$s_{\lambda/\mu}=\det(h_{\lambda_i-\mu_j-i+j})=(-1)^k\det B$$

where B is the matrix (of $N=n+r+s$ rows and columns) $\begin{pmatrix}0 & V\\ U & M\end{pmatrix}$, and $k=r^2+rs+s^2+|\beta|+|\varepsilon|$.

Now let A be the $N\times N$ matrix whose first $r+s$ rows are those of the unit matrix 1_N, and whose last n rows coincide with the last n rows of the matrix B. From the definition of M it follows that $\det A=1$. We now apply the method of Example 20(a) (but with rows replacing columns): if c_{ij} denotes the determinant of the matrix obtained by replacing the jth row of A by the ith row of B, for $1\leqslant i,j\leqslant r+s$, then $\det B=\det(c_{ij})$. By calculating the c_{ij} and paying careful attention to the signs involved, we obtain the desired formula (1).)

(b) Let $H_\alpha=(h_{\alpha_i-\alpha_j})$, $E_\beta=(e_{\beta_j-\beta_i})$ as in Example 21(c). From (1) it follows that

(2) $$s_{\lambda/\mu}=(-1)^s\det\begin{pmatrix} H_\alpha^{-1}S_{(\alpha\mid\beta)}E_\beta^{-1} & H_\alpha^{-1}H_{\alpha,\gamma}\\ E_{\beta,\varepsilon}E_\beta^{-1} & 0\end{pmatrix}.$$

From Example 21(c) we have

$$H_\alpha^{-1}S_{(\alpha\mid\beta)}E_\beta^{-1}=\left((-1)^{i+j}s_{(\alpha_i\mid\beta_j)}\right)_{1\leqslant i,j\leqslant r}.$$

Consider the matrix $H_\alpha^{-1}H_{\alpha,\gamma}$. Its (i,j) element, say c_{ij}, is equal to the determinant of the matrix obtained from H_α by replacing its ith column by the jth column of $H_{\alpha,\gamma}$. There are three possibilities:

(i) if $\gamma_j=\alpha_k$ for some k, then $c_{ij}=\delta_{ik}$;
(ii) if γ_j is not equal to any α_k, and $\alpha_i<\gamma_j$, then $c_{ij}=0$;
(iii) if γ_j is not equal to any α_k, and $\alpha_i\geqslant\gamma_j$, then c_{ij} is (up to sign) the Schur function corresponding to the border strip consisting of those squares (a,b) in the border of λ such that $\gamma_j<b-a\leqslant\alpha_i$.

Similar considerations apply to the matrix $E_{\beta,\varepsilon}E_\beta^{-1}$. Hence (2) leads to an expression for the skew Schur function $s_{\lambda/\mu}$ as a determinant of Schur functions of border strips. We shall leave the precise formulation as an exercise to the reader.

23. As in §3, Example 21, let $s_\lambda(x/y)$ denote the Schur functions associated with the rational function

$$E_{x/y}(t)=\sum_{r\geq 0} e_r(x/y)t^r=\prod_{i=1}^{m}(1+x_it)\prod_{j=1}^{n}(1+y_jt)^{-1}.$$

If ω_y denotes the involution ω acting on symmetric functions of the y variables, then $s_\lambda(x/y)$ is obtained from $\omega_y s_\lambda(x,y)$ by replacing each y_j by $-y_j$. Hence by (5.9) and (5.6) we have

$$s_\lambda(x/y)=\sum_\mu(-1)^{|\lambda-\mu|}s_\mu(x)s_{\lambda'/\mu'}(y) \tag{1}$$

and

$$s_\lambda(y/x)=(-1)^{|\lambda|}s_{\lambda'}(x/y). \tag{2}$$

(a) Let $\Gamma_{m,n}$ denote the set of lattice points $(i,j)\in\mathbf{Z}^2$ such that $i,j\geq 1$ and either $i\leq m$ or $j\leq n$. By (5.8) and (1) above, $s_\lambda(x/y)\neq 0$ if and only if there is a partition $\mu\subset\lambda$ such that $l(\mu)\leq m$ and $\lambda_i-\mu_i\leq n$ for all $i\geq 1$, and this condition is equivalent to $\lambda\subset\Gamma_{m,n}$.

(b) A *bitableau* T of type (m,n) and shape $\lambda-\mu$ (where $\lambda\supset\mu$) is a sequence of partitions

$$\mu=\lambda^{(0)}\subset\lambda^{(1)}\subset\ldots\subset\lambda^{(m+n)}=\lambda \tag{3}$$

such that the skew diagram $\theta^{(i)}=\lambda^{(i)}-\lambda^{(i-1)}$ is a horizontal strip for $1\leq i\leq m$ and a vertical strip for $m+1\leq i\leq m+n$. Graphically, T may be described by filling each square of $\theta^{(i)}$ with the symbol i, for $1\leq i\leq m$, and each square of $\theta^{(m+j)}$ with the symbol j', for $1\leq j\leq n$. Thus the symbols follow the order $1<2<\ldots<m<1'<2'<\ldots<n'$, and the conditions on T are

(i) in each row (resp. column) of T the symbols increase in the weak sense from left to right (resp. from top to bottom);
(ii) there is at most one marked symbol j' in each row, and at most one unmarked symbol i in each column.

With each such bitableau T we associate a monomial $(x/y)^T$ obtained by replacing each symbol i (resp. j') by x_i (resp. $-y_j$) and then forming the product of all these x's and $-y$'s. It follows then from (2) above and (5.12) that

$$s_{\lambda/\mu}(x/y)=\sum_T(x/y)^T \tag{4}$$

summed over all bitableaux T of type (m,n) and shape $\lambda-\mu$.

(c) In (b) above the symbols followed the order $1<2<\ldots<m<1'<2'<\ldots<n'$. Show that (4) will remain true if this order is replaced by any other total ordering

of the set $\{1,2,\ldots,m,1',2',\ldots,n'\}$. (In λ-ring terms, this is simply the observation that the summands in $x_1+\ldots+x_m-y_1-\ldots-y_n$ may be permuted in any way.)

24. (a) Let p be an integer $\geqslant 2$ and let $\varphi_p:\Lambda\to\Lambda$ be the ring homomorphism defined as follows: $\varphi_p(h_n)=h_{n/p}$ if p divides n, and $\varphi_p(h_n)=0$ otherwise. (The effect of φ_p is to replace each variable x_i by its set of pth roots.)

If λ is a partition of length $\leqslant n$, let $\lambda^*=(\lambda^{(r)})_{0\leqslant r\leqslant p-1}$ denote its p-quotient, as in §1, Example 8. If $\mu\subset\lambda$ let

$$s_{\lambda^*/\mu^*}=\prod_{r=0}^{p-1}s_{\lambda^{(r)}/\mu^{(r)}}.$$

Show that $\varphi_p(s_{\lambda/\mu})=\pm s_{\lambda^*/\mu^*}$ if λ, μ have the same p-core, and that $\varphi_p(s_{\lambda/\mu})=0$ otherwise. (Even when λ and μ have the same p-core, we do not necessarily have $\mu^{(r)}\subset\lambda^{(r)}$, so that s_{λ^*/μ^*} may be zero also in this case.)

(Let $\xi=\lambda+\delta$, $\eta=\mu+\delta$, where $\delta=(n-1,n-2,\ldots,1,0)$, so that $s_{\lambda/\mu}=\det(h_{\xi_i-\eta_j})$ by (5.4). For each $r=0,1,\ldots,p-1$ let A_r (resp. B_r) be the set of indices i between 1 and n such that $\xi_i\equiv r(\operatorname{mod} p)$ (resp. $\eta_i\equiv r(\operatorname{mod} p)$). Then $\varphi_p(s_{\lambda/\mu})$ is equal to $\pm\det M$, where M is the matrix which is the diagonal sum of the (not necessarily square) matrices $M_r=(h_{(\xi_i-\eta_j)/p})$, $(i,j)\in A_r\times B_r$. It follows that $\varphi_p(s_{\lambda/\mu})=0$ unless $|A_r|=|B_r|$ for each r, i.e. unless λ and μ have the same p-core. If this condition is satisfied, then $\varphi_p(s_{\lambda/\mu})=\pm\prod\det M_r=\pm s_{\lambda^*/\mu^*}$.)

(b) Let $\omega=e^{2\pi i/p}$. Deduce from (a) that $s_{\lambda/\mu}(1,\omega,\ldots,\omega^{p-1})$ is zero unless

(i) λ and μ have the same p-core;
(ii) $\lambda^{(r)}\supset\mu^{(r)}$ for $r=0,1,\ldots,p-1$;
(iii) each $\lambda^{(r)}-\mu^{(r)}$ is a horizontal strip.

Conditions (i) and (ii) together mean that $\lambda-\mu$ is a union of border strips of length p. We shall write $\lambda\approx_p\mu$ to mean that conditions (i)–(iii) are satisfied.

If $\lambda\approx_p\mu$ then $s_{\lambda/\mu}(1,\omega,\ldots,\omega^{p-1})=\pm 1$. The sign, $\sigma_p(\lambda/\mu)$ say, may be determined as follows. Let $r_i\in[0,p-1]$ be the remainder when $\xi_i=\lambda_i+n-i$ is divided by p, and rearrange the sequence $\xi=(\xi_1,\ldots,\xi_n)$ so that ξ_i precedes ξ_j if and only if either $r_i<r_j$, or $r_i=r_j$ and $\xi_i>\xi_j$. Let $\varepsilon_p(\lambda)$ be the sign of the corresponding permutation, and define $\varepsilon_p(\mu)$ likewise. Then $\sigma_p(\lambda/\mu)=\varepsilon_p(\lambda)\varepsilon_p(\mu)$.

25. We shall identify the elements of the ring $\Lambda\otimes_{\mathbf{Z}}\Lambda$ with the functions of two sets of variables (x) and (y), symmetric in each set separately: thus $f\otimes g$ corresponds to $f(x)g(y)$. Clearly, for any $f\in\Lambda$ the function $f(x,y)$ lies in $\Lambda\otimes\Lambda$, and we define the *diagonal map* (or *comultiplication*) $\Delta:\Lambda\to\Lambda\otimes\Lambda$ by

$$(\Delta f)(x,y)=f(x,y).$$

Also the *counit* $\varepsilon:\Lambda\to\mathbf{Z}$ is the linear mapping which vanishes on Λ^n for each $n\geqslant 1$ and is such that $\varepsilon(1)=1$ (so that $\varepsilon(f)$ is the constant term of f).

It is easy to verify that this comultiplication and counit endows Λ with the structure of a cocommutative Hopf algebra over $\mathbf{Z}$ (for the definition of such an object see, for example, [Z2]; the most important axiom is that Δ is a ring homomorphism).

(a) The definitions at once imply that

$$\Delta h_n = \sum_{p+q=n} h_p \otimes h_q, \qquad \Delta e_n = \sum_{p+q=n} e_p \otimes e_q,$$

$$\Delta p_n = p_n \otimes 1 + 1 \otimes p_n \qquad (n \geqslant 1).$$

This last relation signifies that the p_n are *primitive* elements of Λ.

(b) From (5.9) we have

$$\Delta s_\lambda = \sum_\mu s_{\lambda/\mu} \otimes s_\mu$$

for any partition λ. Hence, with the notation of Example 3,

$$\Delta f = \sum_\mu s_\mu^\perp f \otimes s_\mu$$

for any $f \in \Lambda$, and more generally

$$\Delta f = \sum_\mu u_\mu^\perp f \otimes v_\mu$$

whenever (u_μ), (v_μ) are dual bases of Λ.

(c) The ring $\Lambda \otimes \Lambda$ carries a scalar product, defined by

$$\langle f_1 \otimes g_1, f_2 \otimes g_2 \rangle = \langle f_1, f_2 \rangle \langle g_1, g_2 \rangle$$

for $f_1, f_2, g_1, g_2 \in \Lambda$. With respect to this scalar product, the diagonal map $\Delta : \Lambda \to \Lambda \otimes \Lambda$ is the adjoint of the multiplication map $\Lambda \otimes \Lambda \to \Lambda$ (i.e. $f \otimes g \mapsto fg$). In other words,

$$\langle \Delta f, g \otimes h \rangle = \langle f, gh \rangle$$

for all $f, g, h \in \Lambda$. (By linearity it is enough to check that $\langle \Delta s_\lambda, s_\mu \otimes s_\nu \rangle = \langle s_\lambda, s_\mu s_\nu \rangle$, which follows from (b) above together with (5.1) and (4.8).) Also the counit $\varepsilon : \Lambda \to \mathbf{Z}$ is the adjoint of the natural embedding $e : \mathbf{Z} \to \Lambda$ with respect to the scalar product $\langle m, n \rangle = mn$ on $\mathbf{Z}$.

These properties signify that the Hopf algebra Λ is self-dual.

(d) It follows easily from (c) that if $f \in \Lambda$ and $\Delta f = \sum a_i \otimes b_i$ then

$$f^\perp(gh) = \sum_i a_i^\perp(g) b_i^\perp(h)$$

for all $g, h \in \Lambda$. In particular,

$$s_\lambda^\perp(gh) = \sum_{\mu,\nu} c_{\mu\nu}^\lambda s_\mu^\perp(g) s_\nu^\perp(h).$$

(e) From (c) it also follows that an element $p \in \Lambda^n$ $(n \geqslant 1)$ is primitive (i.e. $\Delta p = p \otimes 1 + 1 \otimes p$) if and only if p is orthogonal to all products fg, where $f, g \in \Lambda$ are homogeneous of positive degree. In particular, $\langle p, h_\lambda \rangle = 0$ for all partitions λ

of n except $\lambda=(n)$. Since (h_λ) and (m_λ) are dual bases of Λ, it follows that p is a scalar multiple of p_n. Hence the p_n $(n \geqslant 1)$ are a $\mathbf{Z}$-basis of the space of primitive elements of Λ.

26. The identities (4.3) and (4.3′) can be generalized as follows: for any two partitions λ, μ we have

$$\sum_\rho s_{\rho/\lambda}(x)s_{\rho/\mu}(y)=\prod_{i,j}(1-x_iy_j)^{-1}\sum_\tau s_{\mu/\tau}(x)s_{\lambda/\tau}(y); \tag{1}$$

$$\sum_\rho s_{\rho/\lambda'}(x)s_{\rho'/\mu}(y)=\prod_{i,j}(1+x_iy_j)\sum_\tau s_{\mu'/\tau}(x)s_{\lambda/\tau'}(y). \tag{2}$$

(By applying the involution ω to the symmetric functions of the x variables we see that (2) follows from (1), so it is enough to prove (1). Let $P(x,y)=\prod_{i,j}(1-x_iy_j)^{-1}$ and let (x), (y), (z), and (u) be four sets of independent variables. We shall decompose the product

$$P=P(x,y)P(x,u)P(z,y)P(z,u)$$

in two different ways.

Firstly, from (4.3) and (5.9) we have

$$P=\sum_\rho s_\rho(x,z)s_\rho(y,u)=\sum_{\rho,\lambda,\mu} s_{\rho/\lambda}(x)s_{\rho/\mu}(y)s_\lambda(z)s_\mu(u). \tag{a}$$

On the other hand, using (4.3) and (5.9) we see that

$$\begin{aligned} P&=P(x,y)\sum_{\sigma,\nu,\tau} s_\sigma(x)s_\sigma(u)s_\nu(y)s_\nu(z)s_\tau(z)s_\tau(u) \\ &=P(x,y)\sum_{\sigma,\nu,\tau,\lambda,\mu} c^\mu_{\sigma\tau}s_\sigma(x)c^\lambda_{\nu\tau}s_\nu(y)s_\lambda(z)s_\mu(u) \\ &=P(x,y)\sum_{\tau,\lambda,\mu} s_{\mu/\tau}(x)s_{\lambda/\tau}(y)s_\lambda(z)s_\mu(u). \end{aligned} \tag{b}$$

Now compare the coefficients of $s_\lambda(z)s_\mu(u)$ in (a) and (b).)

In view of Example 25 we may say that (1) is obtained from (4.3) by applying the diagonal map Δ.

27. (a) By applying the same arguments as in Example 26 to the identities of Examples 4 and 5 we obtain

$$\sum_\rho s_{\rho/\lambda}=\prod_i(1-x_i)^{-1}\prod_{i<j}(1-x_ix_j)^{-1}\sum_\tau s_{\lambda/\tau}, \tag{3}$$

$$\sum_{\rho\text{ even}} s_{\rho/\lambda}=\prod_{i<j}(1-x_ix_j)^{-1}\sum_{\tau\text{ even}} s_{\lambda/\tau}, \tag{4}$$

$$\sum_{\rho'\text{ even}} s_{\rho/\lambda}=\prod_{i<j}(1-x_ix_j)^{-1}\sum_{\tau'\text{ even}} s_{\lambda/\tau}. \tag{5}$$

(b) Likewise, from the identities of Example 9 we obtain

$$\sum_{\pi} s_{\pi/\lambda} = \prod_{i<j}(1+x_ix_j)\sum_{\rho} s_{\lambda'/\rho}, \tag{6}$$

$$\sum_{\rho} s_{\rho/\lambda} = \prod_{i<j}(1+x_ix_j)\sum_{\pi} s_{\lambda'/\pi}, \tag{7}$$

$$\sum_{\sigma} a_\sigma s_{\sigma/\lambda}(x) = \prod_{i<j}(1-x_i)\prod_{i<j}(1-x_ix_j)\sum_{\sigma} a_\sigma s_{\lambda'/\sigma}(-x). \tag{8}$$

Here, as in Example 9, π runs through partitions of the form $(\alpha_1-1,\dots,\alpha_p-1|\alpha_1,\dots,\alpha_p)$ in Frobenius notation; ρ through partitions of the form $(\alpha_1,\dots,\alpha_p|\alpha_1-1,\dots,\alpha_p-1)$; σ through self-conjugate partitions $(\alpha_1,\dots,\alpha_p|\alpha_1,\dots,\alpha_p)$; and finally $a_\sigma=(-1)^{(|\sigma|+p)/2}$.

28. Show that

(a) $$\sum_{\rho,\lambda} q^{|\rho|}s_{\rho/\lambda}(x)s_{\rho/\lambda}(y) = \prod_{i\geq 1}\left((1-q^i)^{-1}\prod_{j,k}\left(1-q^ix_jy_k\right)^{-1}\right),$$

(b) $$\sum_{\rho,\lambda} q^{|\rho|}s_{\rho'/\lambda}(x)s_{\rho/\lambda'}(y) = \prod_{i\geq 1}\left((1-q^i)^{-1}\prod_{j,k}\left(1+q^ix_jy_k\right)\right),$$

(c) $$\sum_{\rho,\lambda} q^{|\rho|}s_{\rho/\lambda}(x)s_{\rho'/\lambda}(y) = \prod_{i\geq 1}\left((1+q^{2i-1})\prod_{j,k}\frac{1+q^{2i-1}x_jy_k}{1-q^{2i}x_jy_k}\right).$$

(Let $F(x,y)$ denote the left-hand side of (a). Then it follows from equation (1) in Example 26 that

$$F(x,y) = \prod_{j,k}(1-qx_jy_k)^{-1}.F(qx,y)$$

and hence that

$$F(x,y) = \prod_{i\geq 1}\prod_{j,k}\left(1-q^ix_jy_k\right)^{-1}.F(0,y).$$

But

$$F(0,y) = \sum_{\rho,\lambda} q^{|\rho|}s_{\rho/\lambda}(0)s_{\rho/\lambda}(y) = \sum_{\rho} q^{|\rho|}$$

$$= \prod_{i\geq 1}(1-q^i)^{-1}$$

and (a) is proved.

The identity (b) follows from (a) by applying the involution ω to the symmetric functions of the x's.

Finally, let $G(x,y)$ denote the left-hand side of (c). Again from Example 26 we obtain

$$G(x,y) = \prod_{j,k}(1+qx_jy_k).\,\omega_yG(qx,y)$$

where ω_y is the involution ω acting on symmetric functions in the y variables, and hence

$$G(x,y) = \prod_{j,k} \frac{1+qx_jy_k}{1-q^2x_jy_k} . G(q^2x,y).$$

The proof may now be completed as in (a) above.)

29. With the notation of Example 3, let

$$E^{\perp}(t) = \sum_{n \geqslant 0} e_n^{\perp} t^n,$$

$$H^{\perp}(t) = \sum_{n \geqslant 0} h_n^{\perp} t^n.$$

Both $E^{\perp}(t)$ and $H^{\perp}(t)$ map Λ into $\Lambda[t]$, and $H^{\perp}(t)\omega = \omega E^{\perp}(t)$.

(a) Show that $E^{\perp}(t)$ and $H^{\perp}(t)$ are ring homomorphisms. (Use Example 25(a), (d).)

(b) Since $h_m^{\perp}(h_n) = h_{n-m}$ (with the usual convention that $h_r = 0$ if $r < 0$) it follows that

$$\begin{aligned} H^{\perp}(t)(H(u)) &= \sum_{m,n \geqslant 0} h_{n-m} t^m u^n \\ &= (1-tu)^{-1} H(u) \end{aligned} \tag{1}$$

and hence, using (a) above, that

$$H^{\perp}(t)(H(u)f) = (1-tu)^{-1} H(u) H^{\perp}(t) f$$

for all $f \in \Lambda$, so that

$$H^{\perp}(t) \circ H(u) = (1-tu)^{-1} H(u) \circ H^{\perp}(t) \tag{2}$$

where $H(u)$ is regarded as a multiplication operator.

Show likewise that

$$H^{\perp}(t) \circ E(u) = (1+tu) E(u) \circ H^{\perp}(t), \tag{3}$$

$$E^{\perp}(t) \circ E(u) = (1-tu)^{-1} E(u) \circ E^{\perp}(t), \tag{4}$$

$$E^{\perp}(t) \circ H(u) = (1+tu) H(u) \circ E^{\perp}(t). \tag{5}$$

(c) For all $f \in \Lambda$ we have

$$H^{\perp}(t) f(x_1, x_2, \ldots) = f(t, x_1, x_2, \ldots).$$

(By (a) above, it is enough to prove this when $f = h_n$, in which case it follows from (1).)

(d) Now define

$$B(t) = \sum_{n \in \mathbf{Z}} B_n t^n = H(t) \circ E^{\perp}(-t^{-1}),$$

$$B^{\perp}(t) = \sum_{n \in \mathbf{Z}} B_n^{\perp} t^n = E(-t^{-1}) \circ H^{\perp}(t),$$

so that

$$B_n = \sum_{r \geqslant 0} (-1)^r h_{n+r} \circ e_r^{\perp},$$

$$B_n^{\perp} = \sum_{r \geqslant 0} (-1)^r e_r \circ h_{n+r}^{\perp}.$$

Each B_n is a linear endomorphism of Λ, and $B_n^{\perp}$ is the adjoint of B_n with respect to the usual scalar product on Λ. Also $B_n^{\perp} = (-1)^n \omega B_n \omega$ for all $n \in \mathbf{Z}$.

Since

$$H(t) = \exp\left(\sum_{n \geqslant 1} \frac{1}{n} p_n t^n \right),$$

$$E(t) = \exp\left(\sum_{n \geqslant 1} \frac{(-1)^{n-1}}{n} p_n t^n \right),$$

we have

$$B(t) = \exp\left(\sum_{n \geqslant 1} \frac{1}{n} p_n t^n \right) \circ \exp\left(\sum_{n \geqslant 1} \frac{-1}{n} p_n^{\perp} t^{-n} \right). \tag{6}$$

More generally, if $t_1, \ldots, t_n$ are independent variables, let

$$B(t_1, \ldots, t_n) = H(t_1) \ldots H(t_n) E^{\perp}(-t_1^{-1}) \ldots E^{\perp}(-t_n^{-1}).$$

Deduce from (5) that

$$B(t_1) \ldots B(t_n) = \prod_{i<j} \left(1 - t_i^{-1} t_j\right) . B(t_1, \ldots, t_n)$$

and hence that

$$B(t_1) \ldots B(t_n) 1 = \prod_{i<j} \left(1 - t_i^{-1} t_j\right) H(t_1) \ldots H(t_n). \tag{7}$$

Let λ be a partition of length $\leqslant n$. By equating the coefficients of t^{λ} on either side of (7), show that

$$B_{\lambda_1} \ldots B_{\lambda_n}(1) = \prod_{i<j} (1 - R_{ij}) h_{\lambda}$$

and hence by (3.4″) that

$$s_{\lambda} = B_{\lambda_1} \ldots B_{\lambda_n}(1).$$

(e) Let $\varphi(tu) = \sum_{n \in \mathbf{Z}} t^n u^n$. Show that

$$H(t)E(-u^{-1})\varphi(tu) = \varphi(tu) \tag{8}$$

and deduce from (2), (4), and (8) that

$$B(t)B^{\perp}(u) + tuB^{\perp}(u)B(t) = \varphi(tu),$$

or equivalently

$$B_r B_s^{\perp} + B_{s-1}^{\perp} B_{r-1} = \delta_{rs}$$

for all $r, s \in \mathbf{Z}$.

30. With the notation of §3, Example 21 define, for any two partitions λ and μ

$$\tilde{s}_{\lambda/\mu} = \det\left(\varphi^{\mu_j - j + 1}(h_{\lambda_i - \mu_j - i + j})\right)_{1 \leqslant i, j \leqslant m}$$

where $m \geqslant \max(l(\lambda), l(\mu))$. (Thus $\tilde{s}_{\lambda/0} = \tilde{s}_\lambda$ as defined *loc. cit.*). Show that $\tilde{s}_{\lambda/\mu} = 0$ unless $\lambda \supset \mu$.

(a) Assume then that $\lambda \supset \mu$, and let $\theta = \lambda - \mu$. The function $\tilde{s}_\theta = \tilde{s}_{\lambda/\mu}$ depends not only on the skew shape θ, but also on its location in the lattice plane. For each $(p, q) \in \mathbf{Z}^2$ let $\tau_{p,q}$ denote the translation $(i, j) \mapsto (i + p, j + q)$. Show that

$$\tilde{s}_{\tau_{0,1}(\theta)} = \varphi \tilde{s}_\theta, \qquad \tilde{s}_{\tau_{1,0}(\theta)} = \varphi^{-1} \tilde{s}_\theta$$

and hence that

$$\tilde{s}_{\tau_{p,q}(\theta)} = \varphi^{q-p} \tilde{s}_\theta .$$

In particular, $\tilde{s}_\theta$ is invariant under diagonal translation ($p = q$).

(b) Let $\hat{\theta}$ be the result of rotating θ through 180° about a point on the main diagonal. Show that $\tilde{s}_{\hat{\theta}} = \varepsilon \tilde{s}_\theta$, where ε is the involution defined by $\varphi^s h_r \mapsto \varphi^{1-r-s} h_r$.

(c) Let $\theta' = \lambda' - \mu'$ be the reflection of θ in the main diagonal. Show that $\tilde{s}_{\theta'} = \omega \tilde{s}_\theta$, where ω is the involution (§3, Example 21(c)) defined by $\omega(\varphi^s h_r) = \varphi^{-s} e_r$ for all r, s. Equivalently,

$$s_{\lambda/\mu} = \det\left(\varphi^{-\mu'_j + j - 1}(e_{\lambda'_i - \mu'_j - i + j})\right).$$

(The proof is the same as that of (2.9), using the relation $E = H^{-1}$ of §3, Example 21(a).)

(d) Extend the results of Examples 20, 21, and 22 to the present situation. (Example 20 goes through unchanged; so does Example 21(a), provided that the contents (§1) of the squares a, b are related by $c(b) = c(a) + 1$. In Example 21(c), the matrices H_α and E_β should now read

$$H_\alpha = \left(\varphi^{\alpha_j + 1} h_{\alpha_i - \alpha_j}\right), \qquad E_\beta = \left(\varphi^{-\beta_i - 1} e_{\beta_j - \beta_i}\right).$$

In Example 22(a), the matrices $H_{\alpha,\gamma}$ and $E_{\beta,\varepsilon}$ should now read

$$H_{\alpha,\gamma} = \left(\varphi^{\gamma_j+1} h_{\alpha_i-\gamma_j}\right), \qquad E_{\beta,\varepsilon} = \left(\varphi^{-\varepsilon_i-1} e_{\beta_j-\varepsilon_i}\right).)$$

31. As usual let $H(t) = \sum_{n \geq 0} h_n t^n$, and assume that the h_n are real numbers (with $h_0 = 1$). The power series $H(t)$ is called a *P-series* if $s_\lambda = \det(h_{\lambda_i - i + j}) \geq 0$ for all partitions λ. In particular, the coefficients h_n are ≥ 0.

(a) If $H(t)$ is a P-series, so also is $H(-t)^{-1}$.

(b) The product of P-series is a P-series. (Use (5.9) and the fact that the coefficients $c_{\mu\nu}^\lambda$ are ≥ 0.)

(c) If $H(t)$ is a P-series and $h_n = 0$ for some n, then $H(t)$ is a polynomial of degree $< n$. (For $h_{n+1} \leq h_1 h_n$ since $s_{(n,1)} \geq 0$, and hence $h_{n+1} = 0$.)

(d) Every P-series has a positive radius of convergence. (We may assume that $H(t)$ is not a polynomial, hence $h_n > 0$ for all $n \geq 1$, by (c) above. Since $s_{(n^2)} \geq 0$ we have $h_{n+1}/h_n \leq h_n/h_{n-1}$, and hence the sequence of positive real numbers h_{n+1}/h_n converges.)

(e) Let $\sum_{n \geq 1} x_n$ be a convergent series of positive real numbers. Then $H(t) = \prod(1 - x_n t)^{-1}$ is a P-series (by (5.12)).

(f) If $a > 0$, e^{at} is a P-series. (For $s_\lambda = a^{|\lambda|}/h(\lambda) > 0$, by §3, Example 5.)

From (a), (b), (e), and (f) it follows that any $H(t)$ of the form

$$H(t) = e^{at} \prod_{n=1}^{\infty} \frac{1 + x_n t}{1 - y_n t},$$

where $a > 0$ and $\sum x_n$, $\sum y_n$ are convergent series of positive terms, is a P-series. Conversely, every P-series is of this form (but this is harder to prove: see [T3]).

Notes and references

Example 2. The fact that the number of standard tableaux of shape λ is equal to $n!/h(\lambda)$ is due to Frame, Robinson, and Thrall [F8]. For a purely combinatorial proof, see [F9], and for a probabilistic proof see [G15] or [S2].

Example 3. The operators $e_n^\perp$, $h_n^\perp$ were introduced by Hammond [H4], and the $s_\lambda^\perp$ by Foulkes [F5], in both cases defined as differential operators. See also Foulkes [F7].

Example 4. This identity is usually ascribed to Littlewood [L9], p. 238; however, in an equivalent form it was stated by Schur in 1918 (*Ges. Abhandlungen*, vol. 3, p. 456). Bender and Knuth found an elegant combinatorial proof (reproduced in [S23], p. 177), using the properties of Knuth's correspondence.

Examples 5 and 9. These identities are all due to Littlewood [L9]. The observation that the identities of Example 9 follow naturally from Weyl's identity for the classical root systems is, I believe, new.

Examples 13, 14, and 15. Plane partitions were first investigated by MacMahon [M9], and the generating functions (1), (3), and (4) of Example 13 are due originally to him, but proved differently. The application of Schur functions to these problems is due to Stanley [S23], who gives more details and references.

Examples 16 and 17. MacMahon [M9] conjectured the generating function (Example 17, (4)) for symmetrical plane partitions, but was unable to prove it. It remained a conjecture until proved by Andrews [A4].

Example 18. For an account of recent developments in the study of generating functions of plane partitions satisfying prescribed symmetry conditions, see Stembridge [S31] and the references given there.

Example 19. The generating function (7) was established by Gordon (see Stanley [S23], p. 265) who did not publish his proof. It too was proved by Andrews [A4].

Examples 20, 21, and 22. These results are due to A. Lascoux and P. Pragacz [L2], [L3]. Examples 21(a) and (b) were contributed by A. Zelevinsky to the Russian translation of the first edition of this book.

Examples 25, 26, and 27. These examples were also contributed by A. Zelevinsky. The Hopf algebra structure on Λ (Example 25) is discussed in [G6], and it is shown in [Z2] that the whole theory of symmetric functions can be systematically derived from this structure. The results of Examples 26 and 27 appear to have been discovered independently by various people (Lascoux, Towber, Stanley, Zelevinsky).

Example 29. The operators B_n were introduced by J. N. Bernstein (see [Z2], p. 69).

6. Transition matrices

In this section we shall be dealing with matrices whose rows and columns are indexed by the partitions of a positive integer n. We shall regard the partitions of n as arranged in reverse lexicographical order (§1), so that (n) comes first and (1^n) comes last. It follows from (1.10) that λ precedes μ if $\lambda \geqslant \mu$ (but not conversely). A matrix $(M_{\lambda\mu})$ indexed by the partitions of n will be said to be *strictly upper triangular* if $M_{\lambda\mu} = 0$ unless $\lambda \geqslant \mu$, and *strictly upper unitriangular* if also $M_{\lambda\lambda} = 1$ for all λ. Likewise we define *strictly lower triangular* and *strictly lower unitriangular*.

Let U_n (resp. U_n') denote the set of strictly upper (resp. lower) unitriangular matrices with integer entries, indexed by the partitions of n.

(6.1) U_n, U_n' *are groups (with respect to matrix multiplication).*

Proof. Suppose $M, N \in U_n$. Then $(MN)_{\lambda\mu} = \Sigma_\nu M_{\lambda\nu} N_{\nu\mu}$ is zero unless there exists a partition ν such that $\lambda \geqslant \nu \geqslant \mu$, i.e. unless $\lambda \geqslant \mu$. For the same reason, $(MN)_{\lambda\lambda} = M_{\lambda\lambda} N_{\lambda\lambda} = 1$. Hence $MN \in U_n$.

Now let $M \in U_n$. The set of equations

(1) $$\sum_\mu M_{\lambda\mu} x_\mu = y_\lambda$$

is equivalent to

(2) $$\sum (M^{-1})_{\lambda\mu} y_\mu = x_\lambda.$$

For a fixed λ, the equations (1) for y_ν where $\nu \leqslant \lambda$, involve only the x_μ for $\mu \leqslant \nu$, hence for $\mu \leqslant \lambda$. Hence the same is true of the equations (2), and therefore $(M^{-1})_{\lambda\mu} = 0$ unless $\mu \leqslant \lambda$. It follows that $M^{-1} \in U_n$. |

Let J denote the transposition matrix:

$$J_{\lambda\mu} = \begin{cases} 1 & \text{if } \lambda' = \mu, \\ 0 & \text{otherwise.} \end{cases}$$

(6.2) *M is strictly upper triangular* (resp. *unitriangular*) *if and only if JMJ is strictly lower triangular* (resp. *unitriangular*).

Proof. If $N = JMJ$, we have $N_{\lambda\mu} = M_{\lambda'\mu'}$. By (1.11), $\lambda' \geqslant \mu'$ if and only if $\mu \geqslant \lambda$, whence the result. |

If (u_λ), (v_λ) are any two **Q**-bases of Λ^n, each indexed by the partitions of n, we denote by $M(u, v)$ the matrix $(M_{\lambda\mu})$ of coefficients in the equations

$$u_\lambda = \sum_\mu M_{\lambda\mu} v_\mu ;$$

$M(u, v)$ is called the *transition matrix* from the basis (u_λ) to the basis (v_λ). It is a non-singular matrix of rational numbers.

(6.3) *Let (u_λ), (v_λ), (w_λ) be* **Q**-*bases of Λ^n. Then*

(1) $$M(u, v) M(v, w) = M(u, w),$$

(2) $$M(v, u) = M(u, v)^{-1}.$$

Let (u'_λ), (v'_λ) be the bases dual to (u_λ), (v_λ) respectively (with respect to the scalar product of §4). *Then*

(3) $$M(u', v') = M(v, u)' = M(u, v)^*$$

(where M' denotes the transpose and M^* the transposed inverse of a matrix M).

(4) $$M(\omega u, \omega v) = M(u, v)$$

where $\omega: \Lambda \to \Lambda$ is the involution defined in §2.

All of these assertions are obvious.

Consider now the five **Z**-bases of Λ^n defined in §2 and §3: (e_λ), (f_λ),

(h_λ), (m_λ), (s_λ). We shall show that all the transition matrices relating pairs of these bases can be expressed in terms of the matrix

$$K = M(s, m)$$

and the transposition matrix J.

Since (m_λ) and (h_λ) are dual bases, and the basis (s_λ) is self-dual (4.8), we have

$$M(s, h) = K^*$$

by (6.3)(3). If we now apply the involution ω and observe that

$$M(\omega s, s) = J,$$

by virtue of (3.8), we have

$$M(s, e) = M(\omega s, h) = M(\omega s, s) M(s, h) = JK^*$$

by (6.3)(1) and (4). Finally, by (6.3)(3) again,

$$M(s, f) = M(s, e)^* = (JK^*)^* = JK.$$

We can now use (6.3)(1) and (2) to complete the following table of transition matrices, in which the entry in row u and column v is $M(u, v)$:

Table 1

	e	h	m	f	s
e	1	$K'JK^*$	$K'JK$	$K'K$	$K'J$
h	$K'JK^*$	1	$K'K$	$K'JK$	K'
m	$K^{-1}JK^*$	$K^{-1}K^*$	1	$K^{-1}JK$	K^{-1}
f	$K^{-1}K^*$	$K^{-1}JK^*$	$K^{-1}JK$	1	$K^{-1}J$
s	JK^*	K^*	K	JK	1

Some of the transition matrices in Table 1 have combinatorial interpretations. From (5.13) it follows that

(6.4) *$K_{\lambda\mu}$ is the number of tableaux of shape λ and weight μ.* |

The numbers $K_{\lambda\mu}$ are sometimes called *Kostka numbers*. By (6.4) they are non-negative. Moreover,

(6.5) *The matrix $(K_{\lambda\mu})$ is strictly upper unitriangular.*

Proof. If T is a tableau of shape λ and weight μ, then for each $r \geqslant 1$ there are altogether $\mu_1 + \ldots + \mu_r$ symbols $\leqslant r$ in T, which must all be located in the top r rows of T (because of the condition of strict increase down the columns of a tableau). Hence $\mu_1 + \ldots + \mu_r \leqslant \lambda_1 + \ldots + \lambda_r$ for all $r \geqslant 1$, i.e. $\mu \leqslant \lambda$. Hence $K_{\lambda\mu} = 0$ unless $\lambda \geqslant \mu$, and for the same reason $K_{\lambda\lambda} = 1$. |

From Table 1 and (6.5) we can read off:

(6.6) (i) *$M(s,h)$ and $M(h,s)$ are strictly lower unitriangular.*
(ii) *$M(s,m)$ and $M(m,s)$ are strictly upper unitriangular.*
(iii) *$M(e,m) = M(h,f)$ and is symmetric.*
(iv) *$M(e,f) = M(h,m)$ and is symmetric.*
(v) *$M(e,h) = M(h,e) = M(m,f)' = M(f,m)'$.*
(vi) *$M(h,s) = M(s,m)'$.*
(vii) *$M(e,s) = M(s,f)'$.*

In fact (Example 5 below) $M(e,h)$ is strictly upper triangular (and therefore by (6.6)(v) $M(m,f)$ is strictly lower triangular).

(6.7) (i) *$M(e,m)_{\lambda\mu} = \sum_\nu K_{\nu\lambda}K_{\nu'\mu}$ is the number of matrices of zeros and ones with row sums λ_i and column sums μ_j.*
(ii) *$M(h,m)_{\lambda\mu} = \sum_\nu K_{\nu\lambda}K_{\nu\mu}$ is the number of matrices of non-negative integers with row sums λ_i and column sums μ_j.*

Proof. (i) Consider the coefficient of a monomial x^μ (where μ is a partition of n) in $e_\lambda = e_{\lambda_1}e_{\lambda_2}\dots$. Each monomial in e_{λ_i} is of the form $\prod_j x_j^{a_{ij}}$, where each a_{ij} is 0 or 1, and $\sum_j a_{ij} = \lambda_i$; hence we must have

$$\prod_{i,j} x_j^{a_{ij}} = \prod_j x_j^{\mu_j}$$

so that $\sum_i a_{ij} = \mu_j$. Hence the matrix (a_{ij}) has row sums λ_i and column sums μ_j.

For (ii) the proof is similar: the only difference is that e_λ is replaced by h_λ, and consequently the exponents a_{ij} can now be arbitrary integers $\geqslant 0$. ∎

Remark. From (6.4) and (6.7)(i) it follows that the number of (0, 1)-matrices with row sums λ_i and column sums μ_j is equal to the number of pairs of tableaux of conjugate shapes and weights λ, μ. In fact one can set up an explicit one–one correspondence between these two sets of objects (Knuth's dual correspondence [K12], [S23]).

Likewise, there is an explicit one–one correspondence, also due to Knuth, between the set of matrices of non-negative integers with row sums λ_i and column sums μ_j, and pairs of tableaux of the same shape and weights λ, μ.

We shall now consider transition matrices involving the **Q**-basis (p_λ). For this purpose we introduce the following notation: if λ is a partition of length r, and f is any mapping of the interval $[1,r]$ of **Z** into the set $\mathbf{N}^+$ of positive integers, let $f(\lambda)$ denote the vector whose ith component is

$$f(\lambda)_i = \sum_{f(j)=i} \lambda_j,$$

for each $i \geqslant 1$.

Let L denote the transition matrix $M(p,m)$:

$$p_\lambda = \sum_\mu L_{\lambda\mu} m_\mu. \tag{6.8}$$

With the notation introduced above we have.

(6.9) *$L_{\lambda\mu}$ is equal to the number of f such that $f(\lambda) = \mu$.*

Proof. On multiplying out $p_\lambda = p_{\lambda_1} p_{\lambda_2} \dots$, we shall obtain a sum of monomials $x_{f(1)}^{\lambda_1} x_{f(2)}^{\lambda_2} \dots$, where f is any mapping $[1, l(\lambda)] \to \mathbf{N}^+$. Hence

$$p_\lambda = \sum_f x^{f(\lambda)}$$

from which the result follows, since $L_{\lambda\mu}$ is equal to the coefficient of x^μ in p_λ. |

If $\mu = f(\lambda)$ as above, the parts μ_i of μ are partial sums of the λ_j; equivalently, λ is of the form $\bigcup_{i \geqslant 1} \lambda^{(i)}$, where each $\lambda^{(i)}$ is a partition of μ_i. We say that λ is a *refinement* of μ, and write $\lambda \leqslant_R \mu$ to express this relationship between λ and μ. Clearly, $\leqslant_R$ is a partial order on the set $\mathscr{P}_n$ of partitions of n, for each integer $n \geqslant 0$.

$$\lambda \leqslant_R \mu \Rightarrow \lambda \leqslant \mu. \tag{6.10}$$

Proof. Let $I_k = f([1,k])$ for $1 \leqslant k \leqslant l(\lambda)$. Since $\mu_i = \sum_{f(j)=i} \lambda_j$, it follows that

$$\lambda_1 + \dots + \lambda_k \leqslant \sum_{i \in I_k} \mu_i \leqslant \mu_1 + \dots + \mu_k,$$

the last inequality because I_k has at most k elements. |

Remark. The converse of (6.10) is false, since for example two distinct partitions λ and μ of n such that $l(\lambda) = l(\mu)$ are always incomparable for the relation $\leqslant_R$, but may well be comparable for $\leqslant$.

From (6.9) and (6.10) it follows that $L_{\lambda\mu} = 0$ unless $\lambda \leqslant_R \mu$, hence unless $\lambda \leqslant \mu$. The matrix L is therefore *strictly lower triangular.*

The transition matrices $M(p,e)$, $M(p,f)$, and $M(p,h)$ may all be expressed in terms of L:

(6.11) (i) $M(p,e) = \varepsilon z L^*$,
(ii) $M(p,f) = \varepsilon L$,
(iii) $M(p,h) = z L^*$,

where L^* is the transposed inverse of L, and ε (resp. z) denotes the diagonal matrix (ε_λ) (resp. (z_λ)).

Proof. Since the bases dual to (h_λ) and (p_λ) are respectively (m_λ) and $(z_\lambda^{-1}p_\lambda)$, it follows from (6.3) that

$$M(p,h) = M(z^{-1}p,m)^* = (z^{-1}L)^* = zL^*.$$

Hence

$$M(p,e) = M(\omega p, \omega e) = M(\varepsilon p, h) = \varepsilon z L^*,$$

and

$$M(p,f) = M(\omega p, \omega f) = M(\varepsilon p, m) = \varepsilon L. \quad |$$

Finally, we have

$$\text{(6.12)} \qquad M(p,s) = M(p,m)M(s,m)^{-1} = LK^{-1}.$$

We shall see in the next section that $M(p,s)$, restricted to partitions of n, is the character table of the symmetric group S_n.

Finally, the relations between the six bases e, f, h, m, s, and p are summarized in the graph below, in which the symbol attached to a directed edge uv is the transition matrix $M(u,v)$. (In the cases where $M(u,v) = M(v,u)$, the edge uv carries no arrow.) For the sake of clarity, the diagonals of the hexagon have been omitted.

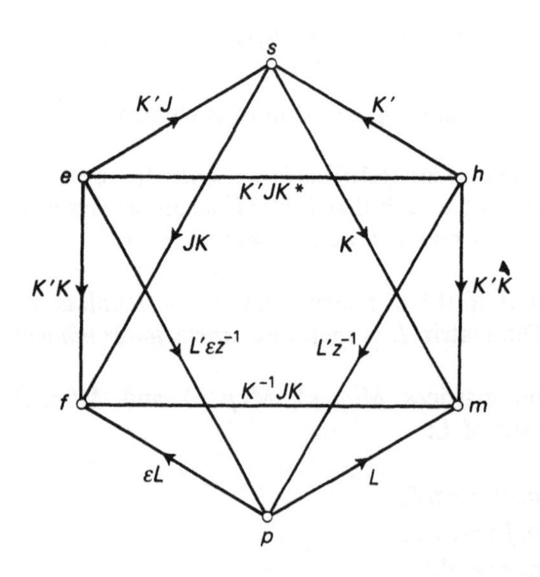

Examples

1. $M(h,m)_{\lambda\mu}$ is equal to the number of double cosets $S_\lambda w S_\mu$ in S_n, where $S_\lambda = S_{\lambda_1} \times S_{\lambda_2} \times \dots$, $S_\mu = S_{\mu_1} \times S_{\mu_2} \times \dots$.

2. For certain choices of λ or μ there are closed formulas for the Kostka number $K_{\lambda\mu}$.

(a) For any partition λ of n, $K_{\lambda,(1^n)}$ is by (6.4) the number of standard tableaux of shape λ, so that by §5, Example 2 we have $K_{\lambda,(1^n)} = n!/h(\lambda)$, where $h(\lambda)$ is the product of the hook-lengths of λ.

(b) Let $\lambda = (a, 1^b)$ be a hook. Then for any partition μ of $n = a+b$ we have $K_{\lambda\mu} = \binom{l(\mu)-1}{b}$. For the tableaux of shape λ and weight μ are determined by the entries in the first column, which must all be distinct, and the top entry (i.e. the number in the corner square of λ) must be 1.

(c) Suppose that λ and μ are partitions of n such that each part of μ is $\geqslant \lambda_2$. In any tableau T of shape λ and weight μ, the 1's must occupy the first μ_1 squares in the top row, and hence T is determined by its entries in the remaining rows, which constitute a tableau of shape $\nu = (\lambda_2, \lambda_3, \dots)$ and arbitrary weight. Hence it follows from (5.12) and §3, Example 4 that $K_{\lambda\mu} = \binom{l(\mu)-1}{\nu'}$.

3. The (infinite) matrix K is the diagonal sum of matrices K_n $(n \geqslant 0)$, where $K_n = (K_{\lambda\mu})_{\lambda,\mu \in \mathscr{P}_n}$. We have $K_0 = K_1 = (1)$, and for $n = 2,3,\dots,6$ the matrices K_n are shown on p. 111.

The matrices K_n and their inverses have the following stability property, when their rows and columns are ordered in reverse lexicographical order, as in the examples on p. 111. Let $m = \sum_{r \leqslant n/2} p(r)$, where $p(r)$ is the number of partitions of r; then the principal $m \times m$ submatrix in the top left-hand corner of K_n (resp. K_n^{-1}) is equal to the corresponding principal submatrix of $K_{n'}$ (resp. $K_{n'}^{-1}$), for all $n' > n$.

This is a consequence of the following fact: if $\mu_1 \geqslant \lambda_2$, then $K_{\lambda\mu} = K_{\lambda+(r),\mu+(r)}$ for all $r \geqslant 1$. (Each tableau of shape $\lambda + (r) = (\lambda_1 + r, \lambda_2, \lambda_3, \dots)$ and weight $\mu + (r) = (\mu_1 + r, \mu_2, \mu_3, \dots)$ can be obtained from one of shape λ and weight μ by moving the top row r squares to the right and inserting r 1's in the squares vacated.)

4. Let $K_{\lambda\mu}^{(-1)}$ denote the (λ, μ) entry of the inverse Kostka matrix K^{-1}. Thus $K_{\lambda\mu}^{(-1)}$ is the coefficient of h_λ in $s_\mu = \det(h_{\mu_i - i + j})$, and is also the coefficient of s_μ in m_λ expressed as a sum of Schur functions. We have $K_{\lambda\mu}^{(-1)} = 0$ unless $\lambda \geqslant \mu$, by (6.1) and (6.5), and $K_{\lambda\lambda}^{(-1)} = 1$.

(a) Let r, n be positive integers. Then

$$\sum_{\substack{l(\lambda)=r \\ |\lambda|=n}} K_{\lambda\mu}^{(-1)} = (-1)^{b+r-1}\binom{b}{r-1}$$

if μ is of the form $(a, 1^b)$, where $a + b = n$; and is zero otherwise. (This follows from §4, Example 10.)

(b) Suppose $\mu=(a,1^m)$ is a hook. Then

$$K_{\lambda\mu}^{(-1)}=(-1)^{l(\lambda)+l(\mu)}\frac{(l(\lambda)-1)!}{\prod_{i\geqslant 1}m_i(\lambda)!}\lambda'_a.$$

(By expanding the determinant $\det(h_{\mu_i-i+j})$ along the top row, we obtain

$$s_\mu=h_ae_m-h_{a+1}e_{m-1}+\ldots+(-1)^m h_{a+m}$$

from which it follows that s_μ is the coefficient of t^n (where $n=a+m=|\mu|$) in

(1) $$(-1)^m E(-t)\sum_{r\geqslant a}h_rt^r=(-1)^m H(t)^{-1}\sum_{r\geqslant a}h_rt^r.$$

Since

$$\begin{aligned}H(t)^{-1}&=\left(1+\sum_{r\geqslant 1}h_rt^r\right)^{-1}\\&=\sum_{n\geqslant 0}(-1)^n\left(\sum_{r\geqslant 1}h_rt^r\right)^n\end{aligned}$$

it follows that (1) is equal to

$$(-1)^m\left(\sum_{r\geqslant a}h_rt^r\right)\sum_{\nu}(-1)^{l(\nu)}\frac{l(\nu)!}{\prod_{i\geqslant 1}m_i(\nu)!}h_\nu t^{|\nu|},$$

the right-hand sum being over all partitions ν. Now $K_{\lambda\mu}^{(-1)}$ is equal to the coefficient of h_λ in this expression.)

(c) Suppose μ is a partition of the form $(a_1,\ldots,a_r,1^m)$, where $r\geqslant 1$, $a_i-a_{i+1}\geqslant r-2$ for $1\leqslant i\leqslant r-1$, and $a_r\geqslant r$. Then for any partition λ of weight $m+a_1+\ldots+a_r$ we have

$$K_{\lambda\mu}^{(-1)}=(-1)^{l(\lambda)+l(\mu)}\frac{(l(\lambda)-r)!}{\prod_{i\geqslant 1}m_i(\lambda)!}\det(\lambda'_{a_i-i+j}-i+1)_{i\leqslant i,j\leqslant r}$$

(By expanding the determinant of the matrix (h_{μ_i-i+j}) down the first column, we shall obtain

$$s_\mu=\sum_{i=1}^{r}(-1)^{i-1}h_{a_i-i+1}s_{\mu^{(i)}},$$

where $\mu^{(i)} = (a_1+1,\dots,a_{i-1}+1,a_{i+1},\dots,a_r,1^m)$. From this it follows that

$$K_{\lambda\mu}^{(-1)} = \sum (-1)^{i-1} K_{\lambda-(a_i-i+1),\mu^{(i)}}^{(-1)}$$

where the sum is over integers $i = 1,2,\dots,r$ such that $a_i - i + 1$ is a part of λ, and $\lambda - (a_i - i + 1)$ denotes the partition obtained by deleting this part. The formula may now be proved by induction on r; the starting point $r = 1$ of the induction is the formula of (b) above.)

(d) Let μ be a partition of length l. By expanding the determinant $\det(h_{\mu_i - i + j})$ down the *last* column, we shall obtain

$$s_\mu = \sum_{i=1}^{l} (-1)^{l-i} h_{\mu_i + l - i} s_{\nu^{(i)}}$$

where $\nu^{(i)} = (\mu_1,\dots,\mu_{i-1},\mu_{i+1}-1,\dots,\mu_l - 1)$. These partitions $\nu^{(i)}$ are all contained in μ, and the skew diagram $\mu - \nu^{(i)}$ is a border strip (or ribbon) of length $\mu_i + l - i$ and height $l - i$ which starts at the square $(r,1)$, i.e. intersects the first column. Hence if we define a *special* border strip to be one that intersects the first column, we have

$$K_{\lambda\mu}^{(-1)} = \sum_S (-1)^{\mathrm{ht}(S)} \tag{2}$$

summed over all sequences $S = (\mu^{(0)},\mu^{(1)},\dots,\mu^{(r)})$ of partitions such that $r = l(\lambda)$, $0 = \mu^{(0)} \subset \mu^{(1)} \subset \dots \subset \mu^{(r)} = \mu$, with each $\theta_i = \mu^{(i)} - \mu^{(i-1)}$ a special border strip, and such that the lengths of the θ_i are the parts of λ in some order; and finally $\mathrm{ht}(S) = \sum_{i=1}^{r} \mathrm{ht}(\theta_i)$.

(e) The combinatorial formula (2) above may be used to derive closed formulas for $K_{\lambda\mu}^{(-1)}$ in certain cases, of which the following is a sample:

(i) if $\mu = (a,1^m)$ is a hook (as in (b) above), θ_1 must be a hook of length $\lambda_i \geqslant a$, and $\theta_2,\dots,\theta_r$ are vertical strips. Hence the formula (2) leads to the result of (b) above.

(ii) $K_{(n)\mu}^{(-1)} = (-1)^b$ if $\mu = (a,1^b)$, and is zero otherwise.

(iii) If $\lambda = (a,1^b)$ with $a > 1$, then $K_{\lambda\mu}^{(-1)}$ is equal to $(-1)^{a+1}(b+1)$ if $\mu = (1^{a+b})$; to $(-1)^{a-\alpha}$ if $\mu = (\alpha,2^\beta,1^\gamma)$ with $\alpha+\beta \leqslant a$; and is zero otherwise.

(iv) If λ is of the form (r^s), then $K_{\lambda\mu}^{(-1)} = 0$ or ± 1, because there is at most one choice for S.

5. A *domino* is a connected horizontal strip, i.e. a set of consecutive squares in the same row. If λ and μ are partitions of n, a *domino tabloid* of shape λ and type μ is a filling of the diagram of λ with non-overlapping dominos of lengths $\mu_1,\mu_2,\dots$, dominos of the same length being regarded as indistinguishable. Let $d_{\lambda\mu}$ denote the number of domino tabloids of shape λ and type μ. Then we have

$$M(e,h)_{\lambda\mu} = M(h,e)_{\lambda\mu} = \varepsilon_\mu d_{\lambda\mu}$$

where as usual $\varepsilon_\mu = (-1)^{|\mu| - l(\mu)}$.

(Since

$$E(t) = H(-t)^{-1} = \left(1 - \sum_{i \geqslant 1} (-1)^{i-1} h_i t^i\right)^{-1}$$

$$= \sum_{r \geqslant 0} \left(\sum_{i \geqslant 1} (-1)^{i-1} h_i t^i\right)^r$$

it follows that, for each integer $n \geqslant 0$,

$$e_n = \sum_{\alpha} \varepsilon_\alpha h_\alpha$$

summed over all finite sequences $\alpha = (\alpha_1, \dots, \alpha_r)$ of positive integers such that $\sum \alpha_i = n$, where $\varepsilon_\alpha = (-1)^{\Sigma(\alpha_i - 1)}$. Hence for any partition λ we have

$$e_\lambda = \sum_{\beta} \varepsilon_\beta h_\beta$$

summed over all sequences $\beta = (\beta_1, \dots, \beta_S)$ of positive integers that can be partitioned into consecutive blocks with sums $\lambda_1, \lambda_2, \dots$. Such sequences are in one–one correspondence with domino tabloids of shape λ.)

It follows that $M(e,h)_{\lambda\mu} = 0$ unless $\mu \leqslant_R \lambda$, and that $M(e,h)_{\lambda\lambda} = \varepsilon_\lambda$. Thus the matrix $M(e,h)\varepsilon$ is a strictly upper unitriangular matrix of non-negative integers.

6. Since by (6.7)(v) the transition matrix $M(f,m)$ is the transpose of $M(e,h)$, it follows from Example 5 that the 'forgotten' symmetric functions f_λ are given by

$$f_\lambda = \varepsilon_\lambda \sum_{\mu} d_{\mu\lambda} m_\mu$$

so that $\varepsilon_\lambda f_\lambda$ is the generating function for domino tabloids of type λ. In particular:

(a) $f_{(1^n)} = \sum m_\mu = h_n$, since for each partition μ of n there is just one domino tabloid of shape μ and type (1^n).

(b) If $\lambda = (r1^{n-r})$ where $r \geqslant 2$ we have

$$f_\lambda = (-1)^{r-1} \sum_{\mu} c_\mu m_\mu$$

where $c_\mu = \mu'_r + \mu'_{r+1} + \dots$. For a domino tabloid of shape μ and type $(r1^{n-r})$ is determined by the position of the single domino of length r, which can lie in any row of μ of length $\mu_i \geqslant r$, and has $\mu_i - r + 1$ possible positions in that row. Hence $c_\mu = \sum_{\mu_i \geqslant r} (\mu_i - r + 1) = \sum_{i \geqslant r} \mu'_i$.

7. A *domino tableau* of shape μ and type λ is a numbering of the squares of the diagram of μ with positive integers, increasing along each row, and such that for each $i \geqslant 1$ the squares numbered i form a domino (Example 5) of length λ_i. In this

terminology, (6.9) may be restated as follows: the number of domino tableaux of shape μ and type λ is equal to $L_{\lambda\mu}$. (For each such domino tableau determines a mapping $f:[1,l(\lambda)]\to[1,l(\mu)]$ such that $f(\lambda)=\mu$ by the rule that $f(i)=j$ if the squares numbered i lie in the jth row of μ; and conversely each such mapping determines a domino tableau of shape μ and type λ, in which the λ_i squares numbered i lie in row $f(i)$.)

8. The *weight* of a domino tabloid (Example 5) is defined to be the product of the lengths of the leftmost dominos in the successive rows. Let $w_{\lambda\mu}$ denote the sum of the weights of all domino tabloids of shape λ and type μ. Then we have

(1) $$M(p,h)_{\lambda\mu}=\varepsilon_\lambda\varepsilon_\mu w_{\lambda\mu}.$$

(Since

$$\begin{aligned}P(t)&=H'(t)H(t)^{-1}\\&=\sum_{i\geqslant 0}(-1)^i\Big(\sum_{r\geqslant 1}h_rt^r\Big)^i\Big(\sum_{r\geqslant 1}rh_rt^{r-1}\Big)\end{aligned}$$

it follows that, for each integer $n\geqslant 1$,

$$p_n=\sum_\alpha(-1)^{l(\alpha)-1}\alpha_1h_\alpha$$

summed over all finite sequences $\alpha=(\alpha_1,\dots,\alpha_r)$ of positive integers such that $\sum\alpha_i=n$, where $l(\alpha)$ is the length r of the sequence. Hence for any partition λ we have

$$p_\lambda=\sum_\beta(-1)^{l(\beta)-l(\lambda)}u_\beta h_\beta$$

summed over all sequences $\beta=(\beta_1,\dots,\beta_S)$ of positive integers that can be partitioned into consecutive blocks $(\beta_1,\dots,\beta_{i_1}),(\beta_{i_1+1},\dots,\beta_{i_2}),\dots$ with sums $\lambda_1,\lambda_2,\dots$; where u_β is the product $\beta_1\,\beta_{i_1+1}\,\beta_{i_2+1}\dots$ of the first terms in each block. As in Example 5, such sequences are in one-one correspondence with domino tabloids of shape λ, and u_β is the weight of the tabloid.)

From (1) and (6.11) it follows that $M(p,e)_{\lambda\mu}=\varepsilon_\mu w_{\lambda\mu}$, $M(f,p)_{\lambda\mu}=\varepsilon_\lambda z_\mu^{-1}w_{\lambda\mu}$ and $M(m,p)_{\lambda\mu}=\varepsilon_\lambda\varepsilon_\mu z_\mu^{-1}w_{\mu\lambda}$. Hence

$$\varepsilon_\lambda f_\lambda=\sum_\mu z_\mu^{-1}w_{\lambda\mu}p_\mu$$

is a polynomial in the power-sums with positive rational coefficients. Also $\varepsilon L^{-1}\varepsilon z$ is a matrix of non-negative integers, with (λ,μ) entry $w_{\mu\lambda}$.

9. From (6.11) we have $M(h,m)=L'z^{-1}L$ and $M(m,f)=L^{-1}\varepsilon L$. Comparison with Table 1 shows that the matrices K and L are related by

$$K'K=L'z^{-1}L,\qquad K^{-1}JK=L^{-1}\varepsilon L.$$

Hence the matrix $X = LK^{-1} = M(p, s)$ satisfies

$$XX' = z, \qquad X^{-1}\varepsilon X = J.$$

10. For each partition μ let

$$u_\mu = \prod_{i \geq 1} m_i(\mu)!$$

and let U denote the diagonal matrix (u_μ). Then the matrix LU^{-1} is a strictly lower unitriangular matrix of non-negative integers.

(Let λ, μ be partitions of n and let $E_{\lambda\mu}$ denote the set of mappings $f: [1, l(\lambda)] \to [1, l(\mu)]$ such that $f(\lambda) = \mu$, so that by (6.9) we have $L_{\lambda\mu} = |E_{\lambda\mu}|$. Also let $S^\mu = \prod S_{m_i(\mu)}$ be the subgroup of $S_{l(\mu)}$ that fixes μ. If $f \in E_{\lambda\mu}$ and $w \in S^\mu$ it is clear that $wf \in E_{\lambda\mu}$, so that S^μ acts faithfully on $E_{\lambda\mu}$. Hence $L_{\lambda\mu} = |E_{\lambda\mu}|$ is divisible by $|S^\mu| = u_\mu$. Finally if $\lambda = \mu$ we have $E_{\mu\mu} = S^\mu$, whence $L_{\mu\mu} = u_\mu$.)

The matrix LU^{-1} is the transition matrix $M(p, \tilde{m})$ between the power-sum products and the 'augmented' monomial symmetric functions $\tilde{m}_\mu = u_\mu m_\mu$. We have

$$\tilde{m}_\mu = \sum x_{i_1}^{\mu_1} \dots x_{i_l}^{\mu_l}$$

where $l = l(\mu)$ and the sum is over all sequences $(i_1, \dots, i_l)$ of pairwise distinct positive integers. Since LU^{-1} is unitriangular, $(\tilde{m}_\mu)$ is a $\mathbf{Z}$-basis of the subring $\mathbf{Z}[p_1, p_2, \dots]$ of Λ.

11. Let $\lambda = (\lambda_1, \dots, \lambda_n)$ be a partition of n. Show that

$$m_\lambda = \sum_\alpha s_\alpha$$

summed over all derangements α of λ. (Multiply $m_\lambda(x_1, \dots, x_n) = \sum x^\alpha$ by $a_\delta(x_1, \dots, x_n)$.) Hence $K_{\lambda\mu}^{(-1)} = r - s$, where r (resp. s) is the number of derangements α of λ such that $s_\alpha = s_\mu$ (resp. $-s_\mu$).

Notes and references

The relations between the various transition matrices contained in Table 1 and (6.6) were known to Kostka [K16], who also computed the matrices K_n and K_n^{-1} up to $n = 11$ [K17]. See also Foulkes [F7]. The formulas in Example 4(a), (b), and (c) are taken from [K17], and those in (d) and (e) from [E1]. The results in Examples 5–8 are due to Eğecioğlu and Remmel [E2]. See also [D6].

	2	1^2
2	1	1
1^2		1

	3	21	1^3
3	1	1	1
21		1	2
1^3			1

	4	31	2^2	21^2	1^4
4	1	1	1	1	1
31		1	1	2	3
2^2			1	1	2
21^2				1	3
1^4					1

	5	41	32	31^2	2^21	21^3	1^5
5	1	1	1	1	1	1	1
41		1	1	2	2	3	4
32			1	1	2	3	5
31^2				1	1	3	6
2^21					1	2	5
21^3						1	4
1^5							1

	6	51	42	41^2	3^2	321	31^3	2^3	2^21^2	21^4	1^6
6	1	1	1	1	1	1	1	1	1	1	1
51		1	1	2	1	2	3	2	3	4	5
42			1	1	1	2	3	3	4	6	9
41^2				1	0	1	3	1	3	6	10
3^2					1	1	1	1	2	3	5
321						1	2	2	4	8	16
31^3							1	0	1	4	10
2^3								1	1	2	5
2^21^2									1	3	9
21^4										1	5
1^6											1

7. The characters of the symmetric groups

In this section we shall take for granted the elementary facts about representations and characters of finite groups.

If G is a finite group and f, g are functions on G with values in a commutative $\mathbf{Q}$-algebra, the *scalar product* of f and g is defined by

$$\langle f, g\rangle_G = \frac{1}{|G|}\sum_{x\in G} f(x)g(x^{-1}).$$

If H is a subgroup of G and f is a character of H, the induced character of G will be denoted by $\mathrm{ind}_H^G(f)$. If g is a character of G, its restriction to H will be denoted by $\mathrm{res}_G^H(g)$.

Each permutation $w \in S_n$ factorizes uniquely as a product of disjoint cycles. If the orders of these cycles are $\rho_1, \rho_2, \dots$, where $\rho_1 \geqslant \rho_2 \geqslant \dots$, then $\rho(w) = (\rho_1, \rho_2, \dots)$ is a partition of n called the *cycle-type* of w. It determines w up to conjugacy in S_n, and the conjugacy classes of S_n are indexed in this way by the partitions of n.

We define a mapping $\psi: S_n \to \Lambda^n$ as follows:

$$\psi(w) = p_{\rho(w)}.$$

If m, n are positive integers, we may embed $S_m \times S_n$ in S_{m+n} by making S_m and S_n act on complementary subsets of $\{1, 2, \dots, m+n\}$. Of course there are many different ways of doing this, but the resulting subgroups of S_{m+n} are all conjugate. Hence if $v \in S_m$ and $w \in S_n$, $v \times w \in S_{m+n}$ is well-defined up to conjugacy in S_{m+n}, with cycle-type $\rho(v \times w) = \rho(v) \cup \rho(w)$, so that

$$\psi(v \times w) = \psi(v)\psi(w). \tag{7.1}$$

Let R^n denote the $\mathbf{Z}$-module generated by the irreducible characters of S_n, and let

$$R = \bigoplus_{n \geqslant 0} R^n,$$

with the understanding that $S_0 = \{1\}$, so that $R^0 = \mathbf{Z}$. The $\mathbf{Z}$-module R has a ring structure, defined as follows. Let $f \in R^m$, $g \in R^n$, and embed $S_m \times S_n$ in S_{m+n}. Then $f \times g$ is a character of $S_m \times S_n$, and we define

$$f.g = \mathrm{ind}_{S_m \times S_n}^{S_{m+n}}(f \times g),$$

which is a character of S_{m+n}, i.e. an element of R^{m+n}. Thus we have defined a bilinear multiplication $R^m \times R^n \to R^{m+n}$, and it is not difficult to verify that with this multiplication R is a commutative, associative, graded ring with identity element.

Moreover, R carries a scalar product: if $f, g \in R$, say $f = \Sigma f_n$, $g = \Sigma g_n$ with $f_n, g_n \in R^n$, we define

$$\langle f, g \rangle = \sum_{n \geq 0} \langle f_n, g_n \rangle_{S_n}.$$

Next we define a **Z**-linear mapping

$$\mathrm{ch}\colon R \to \Lambda_{\mathbf{C}} = \Lambda \otimes_{\mathbf{Z}} \mathbf{C}$$

as follows: if $f \in R^n$, then

$$\mathrm{ch}(f) = \langle f, \psi \rangle_{S_n} = \frac{1}{n!} \sum_{w \in S_n} f(w) \psi(w)$$

(since $\psi(w) = \psi(w^{-1})$). If f_ρ is the value of f at elements of cycle type ρ, we have

(7.2) $$\mathrm{ch}(f) = \sum_{|\rho| = n} z_\rho^{-1} f_\rho p_\rho.$$

$\mathrm{ch}(f)$ is called the *characteristic* of f, and ch is the *characteristic map*. From (7.2) and (4.7) it follows that, for f and g in R^n,

$$\langle \mathrm{ch}(f), \mathrm{ch}(g) \rangle = \sum_{|\rho| = n} z_\rho^{-1} f_\rho g_\rho = \langle f, g \rangle_{S_n}$$

and hence that ch is an *isometry*.

The basic fact is now

(7.3) *The characteristic map is an isometric isomorphism of R onto Λ.*

Proof. Let us first verify that ch is a ring homomorphism. If $f \in R^m$ and $g \in R^n$, we have

$$\begin{aligned} \mathrm{ch}(f.g) &= \langle \mathrm{ind}_{S_m \times S_n}^{S_{m+n}}(f \times g), \psi \rangle_{S_{m+n}} \\ &= \langle f \times g, \mathrm{res}_{S_{m+n}}^{S_m \times S_n}(\psi) \rangle_{S_m \times S_n} \end{aligned}$$

by Frobenius reciprocity,

$$= \langle f, \psi \rangle_{S_m} \langle g, \psi \rangle_{S_n} = \mathrm{ch}(f).\mathrm{ch}(g)$$

by (7.1).

Next, let η_n be the identity character of S_n. Then

$$\mathrm{ch}(\eta_n) = \sum_{|\rho| = n} z_\rho^{-1} p_\rho = h_n$$

by (7.2) and (2.14′). If now $\lambda = (\lambda_1, \lambda_2, \ldots)$ is any partition of n, let η_λ denote $\eta_{\lambda_1} \cdot \eta_{\lambda_2} \cdot \ldots$. Then η_λ is a character of S_n, namely the character induced by the identity character of $S_\lambda = S_{\lambda_1} \times S_{\lambda_2} \times \ldots$, and we have $\mathrm{ch}(\eta_\lambda) = h_\lambda$.

Now define, for each partition λ of n,

$$\chi^\lambda = \det(\eta_{\lambda_i - i + j})_{1 \leqslant i,j \leqslant n} \in R^n, \tag{7.4}$$

i.e. χ^λ is a (possibly virtual) character of S_n, and by (3.4) we have

$$\mathrm{ch}(\chi^\lambda) = s_\lambda. \tag{7.5}$$

Since ch is an isometry, it follows from (4.8) that $\langle \chi^\lambda, \chi^\mu \rangle = \delta_{\lambda\mu}$ for any two partitions λ, μ, and hence in particular that the χ^λ are, up to sign, irreducible characters of S_n. Since the number of conjugacy classes in S_n is equal to the number of partitions of n, these characters exhaust all the irreducible characters of S_n; hence the χ^λ for $|\lambda| = n$ form a basis of R^n, and hence ch is an isomorphism of R^n onto Λ^n for each n, hence of R onto Λ. |

(7.6) (i) *The irreducible characters of* S_n *are* $\chi^\lambda(|\lambda| = n)$ *defined by* (7.4).
(ii) *The degree of* χ^λ *is* $K_{\lambda,(1^n)}$, *the number of standard tableaux of shape* λ.

Proof. From the proof of (7.3), we have only to show that χ^λ and not $-\chi^\lambda$ is an irreducible character; for this purpose it will suffice to show that $\chi^\lambda(1) > 0$. Now we have from (7.5) and (7.2)

$$s_\lambda = \mathrm{ch}(\chi^\lambda) = \sum_\rho z_\rho^{-1} \chi_\rho^\lambda p_\rho$$

where χ_ρ^λ is the value of χ^λ at elements of cycle-type ρ. Hence

$$\chi_\rho^\lambda = \langle s_\lambda, p_\rho \rangle \tag{7.7}$$

by (4.7), and in particular

$$\chi^\lambda(1) = \chi_{(1^n)}^\lambda = \langle s_\lambda, p_1^n \rangle$$

so that

$$h_1^n = p_1^n = \sum_{|\lambda| = n} \chi^\lambda(1) s_\lambda$$

and therefore $\chi^\lambda(1) = M(h, s)_{(1^n),\lambda} = K_{\lambda,(1^n)}$ from Table 1. |

(7.8) *The transition matrix* $M(p, s)$ *is the character table of* S_n, i.e.

$$p_\rho = \sum_\lambda \chi_\rho^\lambda s_\lambda.$$

Hence χ_ρ^λ *is equal to the coefficient of* $x^{\lambda+\delta}$ *in* $a_\delta p_\rho$.

This is a restatement of (7.7).

Remark. From Table 1 in §6 we have

$$h_\lambda = s_\lambda + \sum_{\mu > \lambda} K_{\mu\lambda} s_\mu,$$

$$e_{\lambda'} = s_\lambda + \sum_{\mu < \lambda} K_{\mu'\lambda'} s_\mu.$$

Hence

$$\eta_\lambda = \chi^\lambda + \sum_{\mu > \lambda} K_{\mu\lambda} \chi^\mu,$$

$$\varepsilon_{\lambda'} = \chi^\lambda + \sum_{\mu < \lambda} K_{\mu'\lambda'} \chi^\mu.$$

These relations give the decomposition of the induced modules $H_\lambda = \mathrm{ind}_{S_\lambda}^{S_n}(1)$ and $E_{\lambda'} = \mathrm{ind}_{S_{\lambda'}}^{S_n}(\varepsilon)$ respectively. It follows that H_λ and $E_{\lambda'}$ have a unique common irreducible component, namely the irreducible S_n-module with character χ^λ. This observation leads to a simple construction of the irreducible S_n-modules: see Example 15 below.

Let λ, μ, ν be partitions of n, and let

$$\gamma_{\mu\nu}^\lambda = \langle \chi^\lambda, \chi^\mu \chi^\nu \rangle_{S_n} = \frac{1}{n!} \sum_{w \in S_n} \chi^\lambda(w) \chi^\mu(w) \chi^\nu(w)$$

which is symmetrical in λ, μ, ν. Then we have, for two sets of variables $x = (x_1, x_2, \dots)$ and $y = (y_1, y_2, \dots)$

(7.9)
$$s_\lambda(xy) = \sum_{\mu, \nu} \gamma_{\mu\nu}^\lambda s_\mu(x) s_\nu(y),$$

where (xy) means the set of variables $x_i y_j$. (Compare (5.9).)

Proof. For all partitions ρ we have $p_\rho(xy) = p_\rho(x) p_\rho(y)$ and hence from (7.8)

$$\sum_\lambda \chi^\lambda s_\lambda(xy) = \sum_{\mu, \nu} \chi^\mu \chi^\nu s_\mu(x) s_\nu(y)$$

so that $s_\lambda(xy)$ is the coefficient of χ^λ in the right-hand side. |

Let $f, g \in \Lambda^n$, say $f = \mathrm{ch}(u)$, $g = \mathrm{ch}(v)$ where u, v are class-functions on S_n. The *internal product* of f and g is defined to be

$$f * g = \mathrm{ch}(uv)$$

where uv is the function $w \mapsto u(w)v(w)$ on S_n. With respect to this product, Λ^n becomes a commutative and associative ring, with identity element h_n.

It is convenient to extend this product by linearity to the whole of Λ, and indeed to its completion $\hat{\Lambda}$ (§2); if

$$f = \sum_{n \geqslant 0} f^{(n)}, \qquad g = \sum_{n \geqslant 0} g^{(n)}$$

with $f^{(n)}, g^{(n)} \in \Lambda^n$, we define

$$f * g = \sum_{n \geqslant 0} f^{(n)} * g^{(n)}.$$

For this product, $\hat{\Lambda}$ is a commutative ring with identity element $\hat{1} = \Sigma h_n$, and Λ is a subring (but does not contain this identity element).

If λ, μ, ν are partitions of n, we have

(7.10) $$s_\lambda * s_\mu = \sum_\nu \gamma_{\lambda\mu}^\nu s_\nu$$

so that by (7.9) and the symmetry of the coefficients $\gamma_{\lambda\mu}^\nu$

(7.11) $$s_\lambda(xy) = \sum_\mu s_\mu(x)(s_\lambda * s_\mu)(y).$$

Also we have

(7.12) $$p_\lambda * p_\mu = \delta_{\lambda\mu} z_\lambda p_\lambda$$

so that the elements $z_\lambda^{-1} p_\lambda \in \Lambda_{\mathbf{Q}}^n$ are pairwise orthogonal idempotents, and their sum over all partitions λ of n is by (2.14) the identity element h_n of Λ^n.

Finally, for all $f, g \in \Lambda$ we have

(7.13) $$\langle f, g \rangle = (f * g)(1)$$

where $(f * g)(1)$ means $f * g$ evaluated at $(x_1, x_2, \ldots) = (1, 0, 0, \ldots)$. (By linearity it is enough to verify (7.13) when $f = p_\lambda$ and $g = p_\mu$, and in that case it follows from (7.12) and (4.7), since $p_\lambda(1) = 1$ for all partitions λ.)

Examples

1. $\chi^{(n)} = \eta_n$ is the trivial character of S_n, and $\chi^{(1^n)} = \varepsilon_n = \varepsilon$ is the sign character. (Compare (7.10) with (2.14').)

2. For any partition λ of n, $\chi^{\lambda'} = \varepsilon_n \chi^\lambda$. For

$$\chi_\rho^{\lambda'} = \langle s_{\lambda'}, p_\rho \rangle = \langle s_\lambda, \varepsilon_\rho p_\rho \rangle = \varepsilon_\rho \chi_\rho^\lambda$$

since $\omega(s_{\lambda'}) = s_\lambda$ and $\omega(p_\rho) = \varepsilon_\rho p_\rho$. Hence the involution ω on Λ corresponds to multiplication by ε_n in R^n. Equivalently, $e_n * f = \omega(f)$ for all $f \in \Lambda^n$.

3. Corresponding to each skew diagram $\lambda - \mu$ of weight n, there is a character $\chi^{\lambda/\mu}$ of S_n defined by $\mathrm{ch}(\chi^{\lambda/\mu}) = s_{\lambda/\mu}$. If $|\mu| = m$ we have from (5.1) and (7.3)

$$\begin{aligned}\langle \chi^{\lambda/\mu}, \chi^\nu \rangle_{S_n} &= \langle \chi^\lambda, \chi^\mu . \chi^\nu \rangle_{S_{m+n}} \\ &= \langle \mathrm{res}^{S_m \times S_n}_{S_{m+n}} \chi^\lambda, \chi^\mu \times \chi^\nu \rangle_{S_m \times S_n}\end{aligned}$$

by Frobenius reciprocity, and therefore the restriction of χ^λ to $S_m \times S_n$ is $\sum_{|\mu| = m} \chi^\mu \times \chi^{\lambda/\mu}$.

The *degree* of $\chi^{\lambda/\mu}$ is equal to $\langle s_{\lambda/\mu}, e_1^n \rangle = K_{\lambda/\mu,(1^n)}$, i.e. to the number of standard tableaux of shape $\lambda - \mu$.

4. Let G be a subgroup of S_n, and let $c(G)$ be the cycle indicator of G (§2, Example 9). Then $c(G) = \mathrm{ch}(\chi_G)$, where χ_G is the character of S_n induced by the trivial character 1_G of G. For $\mathrm{ch}(\chi_G) = \langle \chi_G, \psi \rangle_{S_n} = \langle 1_G, \psi | G \rangle_G$ (by Frobenius reciprocity) $= c(G)$.

If G, H are subgroups of S_n, $\langle c(G), c(H) \rangle$ is the number of (G, H) double cosets in S_n.

5. From §3, Example 11 and (7.8) we obtain the following combinatorial rule for computing χ^λ_ρ:

$$\chi^\lambda_\rho = \sum_S (-1)^{\mathrm{ht}(S)}$$

summed over all sequences of partitions $S = (\lambda^{(0)}, \lambda^{(1)}, \dots, \lambda^{(m)})$ such that $m = l(\rho)$, $0 = \lambda^{(0)} \subset \lambda^{(1)} \subset \dots \subset \lambda^{(m)} = \lambda$, and such that each $\lambda^{(i)} - \lambda^{(i-1)}$ is a border strip (§3, Example 11) of length ρ_i, and $\mathrm{ht}(S) = \sum_i \mathrm{ht}(\lambda^{(i)} - \lambda^{(i-1)})$.

6. The degree $f^\lambda = \chi^\lambda(1)$ of χ^λ may also be computed as follows. By (7.8), it is the coefficient of $x^{\lambda+\delta}$ in $(\sum x_i)^n \sum_{w \in S_n} \varepsilon(w) x^{w\delta}$. If we put $\mu = \lambda + \delta$ (so that $\mu_i = \lambda_i + n - i$, $1 \leqslant i \leqslant n$), this coefficient is

$$\sum_{w \in S_n} \varepsilon(w) n! \Big/ \prod_{i=1}^n (\mu_i - n + w(i))!$$

which is the determinant $n! \det(1/(\mu_i - n + j)!)$, hence equal to

$$\begin{aligned}&\frac{n!}{\mu!} \det(\mu_i(\mu_i - 1) \dots (\mu_i - n + j + 1)) \\ &= \frac{n!}{\mu!} \det(\mu_i^{n-j}) = \frac{n!}{\mu!} \Delta(\mu_1, \dots, \mu_n)\end{aligned}$$

where $\mu! = \prod_i \mu_i!$ and $\Delta(\mu_1, \dots, \mu_n) = \prod_{i<j} (\mu_i - \mu_j)$.

7. Let $\rho = (r, 1^{n-r})$, so that χ^λ_ρ is the value of the character χ^λ of S_n at an r-cycle $(1 \leqslant r \leqslant n)$. By (7.8), χ^λ_ρ is the coefficient of $x^\mu = x^{\lambda+\delta}$ in $(\sum x_i^r)(\sum x_i)^{n-r} \sum_{w \in S_n} \varepsilon(w) x^{w\delta}$. From the result of Example 6, this coefficient is

$$\sum_i \frac{(n-r)! \Delta(\mu_1, \dots, \mu_i - r, \dots, \mu_n)}{\mu_1! \dots (\mu_i - r)! \dots \mu_n!}$$

and therefore

$$\chi_\rho^\lambda / f^\lambda = \frac{(n-r)!}{n!} \sum_{i=1}^{n} \frac{\mu_i!}{(\mu_i - r)!} \prod_{j \neq i} \frac{\mu_i - \mu_j - r}{\mu_i - \mu_j}.$$

If we put $\varphi(x) = \prod(x - \mu_i)$ and $h_\rho = n!/z_\rho = n!/(n-r)!r$, this formula becomes

$$-r^2 h_\rho \chi_\rho^\lambda / f^\lambda = \sum_{i=1}^{n} \mu_i(\mu_i - 1)\dots(\mu_i - r + 1)\varphi(\mu_i - r)/\varphi'(\mu_i)$$

which is equal to the coefficient of x^{-1} in the expansion of

$$x(x-1)\dots(x-r+1)\varphi(x-r)/\varphi(x)$$

in *descending* powers of x.

In particular, when $r = 2$ we obtain

$$h_\rho \chi_\rho^\lambda / f^\lambda = n(\lambda') - n(\lambda).$$

8. If λ is a partition of n, let

$$f_\lambda(q) = q^{n(\lambda)} \varphi_n(q) \Big/ \prod_{x \in \lambda} (1 - q^{h(x)}).$$

Let $w \in S_n$ be an n-cycle and let ζ be a primitive complex nth root of unity. Then for all $r \geqslant 0$ we have

$$\chi^\lambda(w^r) = f_\lambda(\zeta^r). \tag{1}$$

(Since $f_\lambda(q) = \varphi_n(q) s_\lambda(1, q, q^2, \dots)$ (§3, Example 2) it follows that

$$\sum_{|\lambda| = n} f_\lambda(q) s_\lambda(x)$$

is equal to the coefficient of t^n in $\varphi_n(q) \prod_{i,j} (1 - x_i q^{j-1} t)^{-1}$, and hence by the q-binomial theorem (§2, Example 4) is equal to

$$\sum_{|\alpha| = n} g_\alpha(q) x^\alpha$$

where $g_\alpha(q) = \varphi_n(q) / \prod_{i \geqslant 1} \varphi_{\alpha_i}(q)$, and the sum is over all sequences $\alpha = (\alpha_1, \alpha_2, \dots)$ of non-negative integers such that $|\alpha| = \sum \alpha_i = n$. Now ζ^r is a primitive sth root of unity, where n/s is the highest common factor of r and n. Show that $g_\alpha(\zeta^r) = 0$ unless each α_i is divisible by s, and that if $\alpha = s\beta$ then $g_\alpha(\zeta^r)$ is equal to the multinomial coefficient $(n/s)!/\prod \beta_i!$ Deduce that

$$\sum_{|\lambda| = n} f_\lambda(\zeta^r) s_\lambda(x) = \left(\sum x_i^s\right)^{n/s} = p_s^{n/s}. \tag{2}$$

On the other hand we have by (7.8)

$$\sum_{|\lambda|=n} \chi^\lambda(w^r)s_\lambda(x) = p_s^{n/s}, \tag{3}$$

and (1) now follows from comparison of (2) and (3).)

9. (a) Let $\Lambda_+^n \subset \Lambda^n$ denote the semigroup consisting of the non-negative integral linear combinations of the Schur functions $s_\lambda, |\lambda| = n$, and let $\Lambda_+ = \bigoplus_{n \geqslant 0} \Lambda_+^n$. If $f \in \Lambda$, then $f \in \Lambda_+$ if and only if $\langle f, g\rangle \geqslant 0$ for all $g \in \Lambda_+$.

By (7.5) and (7.6), Λ_+^n consists of the characteristics of ordinary (not virtual) characters of S_n. From (7.3) it follows that $\Lambda_+^m . \Lambda_+^n \subset \Lambda_+^{m+n}$ for all $m, n \geqslant 0$, so that Λ_+ is closed under multiplication as well as addition. (Equivalently, the coefficients $c_{\mu\nu}^\lambda = \langle s_\lambda, s_\mu s_\nu\rangle$ are non-negative.) In particular, h_λ and e_λ lie in Λ_+ for all partitions λ. From Examples 2 and 3 it follows that Λ_+ is stable under the involution ω, and that $s_{\lambda/\mu} \in \Lambda_+$ for all partitions λ, μ.

(b) Define a partial order on Λ by letting $f \geqslant g$ if and only if $f - g \in \Lambda_+$. Show that the following conditions on partitions λ, μ of n are equivalent:

(a) $h_\lambda \leqslant h_\mu$; (a′) $e_\lambda \leqslant e_\mu$; (b) $s_\lambda \leqslant h_\mu$; (b′) $s_{\lambda'} \leqslant e_\mu$; (c) $M(e,m)_{\lambda'\mu} > 0$; (d) $\lambda \geqslant \mu$.

(Since Λ_+ is stable under ω, we see that (a) $\Leftrightarrow$ (a′) and (b) $\Leftrightarrow$ (b′). Next, the relation $h_\lambda = \sum K_{\mu\lambda} s_\mu$ (6.7) shows that $s_\lambda \leqslant h_\lambda$ and hence that (a) $\Rightarrow$ (b). Since $e_{\lambda'} \geqslant s_\lambda$ we have $M(e,m)_{\lambda'\mu} = \langle e_{\lambda'}, h_\mu\rangle \geqslant \langle s_\lambda, h_\mu\rangle$, whence (b) $\Rightarrow$ (c). The next implication (c) $\Rightarrow$ (d) follows from (6.6) (i). Finally, to show that (d) $\Rightarrow$ (a) we may by (1.16) assume that $\lambda = R_{ij}\mu$, and then we have

$$h_\mu - h_\lambda = h_\nu(h_{\mu_i}h_{\mu_j} - h_{\mu_i+1}h_{\mu_j-1}) = h_\nu s_{(\mu_i,\mu_j)} \geqslant 0,$$

where ν is the partition obtained from μ by deleting μ_i and μ_j.)

The equivalence (c) $\Leftrightarrow$ (d) is known as the Gale–Ryser theorem: there exists a matrix of zeros and ones with row sums λ_i and column sums μ_j if and only if $\lambda' \geqslant \mu$ (use (6.6)).

Another combinatorial corollary is the following: if $\lambda \geqslant \mu$ then $K_{\theta\lambda} \leqslant K_{\theta\mu}$ for any skew diagram θ. (Take the scalar product of both sides of (a) with s_θ, and use (5.13).) In particular, we have $K_{\lambda\mu} > 0$ whenever $\lambda \geqslant \mu$ (because $K_{\lambda\lambda} = 1$).

10. (a) We have

$$\sum_\lambda s_\lambda * s_\lambda = \prod_{k \geqslant 1} (1 - p_k)^{-1}$$

where the sum on the left is over all partitions λ. For

$$s_\lambda * s_\lambda = \sum_\rho z_\rho^{-1} (\chi_\rho^\lambda)^2 p_\rho$$

and

$$\sum_\lambda (\chi_\rho^\lambda)^2 = z_\rho$$

by orthogonality of characters, so that

$$\sum_{|\lambda|=n} s_\lambda * s_\lambda = \sum_{|\rho|=n} p_\rho$$

which is equivalent to the result stated.

(b) We have

$$\prod_{i,j,k}(1-x_iy_jz_k)^{-1} = \sum_{\lambda,\mu} s_\lambda(x)s_\mu(y)(s_\lambda * s_\mu)(z),$$

$$\prod_{i,j,k}(1+x_iy_jz_k) = \sum_{\lambda,\mu} s_\lambda(x)s_{\mu'}(y)(s_\lambda * s_\mu)(z).$$

11. Let $\varphi = \Sigma_{|\lambda|=n}\,\chi^\lambda$. If $w \in S_n$ has cycle-type ρ, then

$$\varphi(w) = \sum_\lambda \chi_\rho^\lambda = \sum_\lambda \langle s_\lambda, p_\rho\rangle = \langle s, p_\rho\rangle$$

where (§5, Example 4)

$$s = \prod_i (1-x_i)^{-1}\prod_{i<j}(1-x_ix_j)^{-1}.$$

By calculating $\log s$, show that

$$s = \prod_{n \text{ odd}} \exp\left(\frac{p_n}{n} + \frac{p_n^2}{2n}\right)\prod_{n \text{ even}} \exp\frac{p_n^2}{2n}$$

and hence that $\varphi(w) = \Pi_{i \geqslant 1}\, a_i^{(m_i(\rho))}$, where $a_i^{(m)}$ is the coefficient of t^m in $\exp(t + \frac{1}{2}it^2)$ or $\exp(\frac{1}{2}it^2)$ according as i is odd or even. In particular, $\varphi(w) = 0$ if w contains an odd number of $2r$-cycles, for any $r \geqslant 1$.

12. Let C_n be the cyclic subgroup of S_n generated by an n-cycle, and let θ be a faithful character of C_n. Show that the induced character $\varphi_n = \operatorname{ind}_{C_n}^{S_n}(\theta)$ is independent of the choice of θ, and that

$$\operatorname{ch}(\varphi_n) = \frac{1}{n}\sum_{d|n} \mu(d) p_d^{n/d}$$

where μ is the Möbius function.

(Let V be a finite-dimensional vector space over a field of characteristic 0, and let $L(V) = \oplus_{n>0} L^n(V)$ be the free Lie algebra generated by V. Then for each $n \geqslant 0$, L^n is a homogeneous polynomial functor of degree n, and $\alpha(L_n) = \varphi_n$ in the notation of Appendix A, (5.4).)

13. For each permutation w and each integer $r \geqslant 1$, let $a_r(w)$ denote the number of cycles of length r in the cycle decomposition of w. The a_r are functions on the disjoint union of the symmetric groups S_n. As such they are algebraically independent over $\mathbf{Q}$, for if f is a polynomial in r variables such that $f(a_1(w),\dots,a_r(w)) = 0$

for all permutations w, then $f(m_1,\dots,m_r)=0$ for all choices of non-negative integers m_i, and hence f is the zero polynomial. Hence

$$A=\mathbf{Q}[a_1,a_2,\dots]$$

is a polynomial ring. (The multiplication in A is pointwise multiplication of functions: $(fg)(w)=f(w)g(w)$, not the induction product defined in the text.)

(a) For each partition $\rho=(1^{m_1}2^{m_2}\dots)$ let

$$\binom{a}{\rho}=\prod_{r\geqslant 1}\binom{a_r}{m_r}=\prod_{r\geqslant 1}\frac{a_r^{(m_r)}}{m_r!},$$

where $a_r^{(m_r)}$ is the 'falling factorial'

$$a_r^{(m_r)}=a_r(a_r-1)\dots(a_r-m_r+1).$$

Since the a_r are algebraically independent, the monomials $a_1^{m_1}a_2^{m_2}\dots$ form a $\mathbf{Q}$-basis of A, and hence the polynomials $\binom{a}{\rho}$ form another $\mathbf{Q}$-basis.

Define a linear mapping $\theta\colon A\to\hat{\Lambda}_{\mathbf{Q}}$ by

$$\theta(f)=\sum_{n\geqslant 0}\mathrm{ch}(f\mid S_n)$$

for $f\in A$. If $f=\binom{a}{\rho}$ then

$$\mathrm{ch}(f\mid S_n)=\frac{1}{n!}\sum_{w\in S_n}\binom{a}{\rho}(w)\psi(w).$$

If w has cycle-type $\tau=(1^{n_1}2^{n_2}\dots)$ we have $a_r(w)=n_r$ and hence $\binom{a}{\rho}(w)=\prod_{r\geqslant 1}\binom{n_r}{m_r}$, which is equal to $z_\tau/z_\rho z_\sigma$ if $\tau=\rho\cup\sigma$ for some partition σ (i.e. if ρ is a subsequence of τ) and is zero otherwise. It follows that

$$\theta\binom{a}{\rho}=\sum_\sigma z_\rho^{-1}z_\sigma^{-1}p_\rho p_\sigma$$

$$=z_\rho^{-1}p_\rho H,$$

where

$$H=\sum_\sigma z_\sigma^{-1}p_\sigma=\sum_{n\geqslant 0}h_n.$$

(b) Define a linear mapping $\varphi\colon\Lambda_{\mathbf{Q}}\to A$ by

$$\varphi(p_\rho)=z_\rho\binom{a}{\rho}=\prod_{r\geqslant 1}r^{m_r}a_r^{(m_r)}$$

for each partition $\rho = (1^{m_1} 2^{m_2} \dots)$. From (a) above it follows that $\theta\varphi: \Lambda_{\mathbf{Q}} \to \hat{\Lambda}_{\mathbf{Q}}$ is multiplication by H. The mapping φ is bijective and hence defines a new product $f * g$ on A by the rule

$$f * g = \varphi(\varphi^{-1}(f)\varphi^{-1}(g)).$$

We have

$$\binom{a}{\rho} * \binom{a}{\sigma} = \frac{z_{\rho \cup \sigma}}{z_\rho z_\sigma} \binom{a}{\rho \cup \sigma}$$

or equivalently

$$a^{(m)} * a^{(n)} = a^{(m+n)}$$

for any two finite vectors $m = (m_1, m_2, \dots)$, $n = (n_1, n_2, \dots)$ of non-negative integers, where $a^{(m)} = a_1^{(m_1)} a_2^{(m_2)} \dots$, and likewise for $a^{(n)}$.

(c) Let

$$P = \prod_{r \geq 1} (1 + p_r)^{a_r}.$$

Then $\varphi(f) = \langle P, f \rangle$ for all $f \in \Lambda$.

(Introducing variables $x = (x_1, x_2, \dots)$, we have

$$P(x) = \prod_{r \geq 1} (1 + p_r(x))^{a_r} \tag{1}$$

$$= \prod_{r \geq 1} \sum_{m_r \geq 0} \binom{a_r}{m_r} p_r(x)^{m_r}$$

$$= \sum_\rho \binom{a}{\rho} p_\rho(x).$$

Let $C(x, y) = \prod_{i,j}(1 - x_i y_j)^{-1} = \sum_\rho z_\rho^{-1} p_\rho(x) p_\rho(y)$, as in §4, Example 9. Then it follows from (1) and the definition of φ that $P(x) = \varphi_y C(x, y)$, where φ_y acts on symmetric functions in the y's. Hence (*loc. cit.*)

$$\langle P, f \rangle = \varphi_y \langle C(x, y), f(x) \rangle = \varphi_y f(y) = \varphi(f).)$$

14. Let $\lambda = (\lambda_1, \lambda_2, \dots)$ be a partition of n and let $(N, \lambda) = (N) \cup \lambda = (N, \lambda_1, \lambda_2, \dots)$ where N is any integer $\geq \lambda_1$. From (7.8) it follows that, for each $w \in S_{N+n}$, $\chi^{(N,\lambda)}(w)$ is equal to the coefficient of $x_0^{N+n} x_1^{\lambda_1 + n - 1} \dots x_n^{\lambda_n}$ in the product

$$a_\delta(x_0, x_1, \dots, x_n) \prod_{r \geq 1} p_r(x_0, x_1, \dots, x_n)^{a_r(w)} \tag{1}$$

where, as in Example 13, $a_r(w)$ is the number of r-cycles in the cycle decomposition of w. Since the polynomial (1) is homogeneous in $x_0, x_1, \dots, x_n$, nothing is lost by setting $x_0 = 1$, which shows that $\chi^{(N,\lambda)}(w)$ is equal to the coefficient of $x_1^{\lambda_1+n-1} \dots x_n^{\lambda_n}$ in

$$a_\delta(x_1,\dots,x_n)\prod_{i=1}^{n}(1-x_i)\prod_{r\geqslant 1}(1+p_r(x_1,\dots,x_n))^{a_r(w)}$$

and therefore is equal to the coefficient of s_λ in

$$\prod_{i\geqslant 1}(1-x_i)\prod_{r\geqslant 1}(1+p_r)^{a_r(w)}.$$

It follows that (as functions on S_{N+n}) we have

(2) $$\chi^{(N,\lambda)} = \langle EP, s_\lambda\rangle$$

where

$$E = \prod_{i\geqslant 1}(1-x_i) = \sum_{r\geqslant 0}(-1)^r e_r$$

and

(3) $$P = \prod_{r\geqslant 1}(1+p_r)^{a_r} = \sum_\rho \binom{a}{\rho} p_\rho$$

as in Example 13.

The scalar product $\langle EP, s_\lambda\rangle$ on the right-hand side of (2) is a polynomial $X^\lambda \in A = \mathbf{Q}[a_1, a_2, \dots]$ called the *character polynomial* corresponding to the partition λ. It has the property that $X^\lambda \mid S_{N+n} = \chi^{(N,\lambda)}$ for all $N \geqslant \lambda_1$.

There are various explicit formulas for the polynomials X^λ:

(a) Since by (2.14)

$$E = \sum_\sigma (-1)^{l(\sigma)} z_\sigma^{-1} p_\sigma$$

it follows from (3) that

(4) $$X^\lambda = \sum_{\rho,\sigma}(-1)^{l(\sigma)} z_\sigma^{-1}\chi^\lambda_{\rho\cup\sigma}\binom{a}{\rho}$$

summed over partitions ρ, σ such that $|\rho| + |\sigma| = |\lambda|$.

(b) Alternatively, we have

$$\begin{aligned} X^\lambda &= \sum_{r\geqslant 0}(-1)^r\langle e_r P, s_\lambda\rangle \\ &= \sum_{r\geqslant 0}(-1)^r\langle P, s_{\lambda/(1^r)}\rangle \\ &= \sum_\mu (-1)^{|\lambda-\mu|}\langle P, s_\mu\rangle \end{aligned}$$

summed over partitions $\mu \subset \lambda$ such that $\lambda - \mu$ is a vertical strip. Hence

(5) $$X^\lambda = \sum_\mu (-1)^{|\lambda-\mu|} \sum_\rho \chi^\mu_\rho \binom{a}{\rho}$$

summed over μ as above, and ρ such that $|\rho| = |\mu|$.

(c) Let $u_\lambda = E^\perp s_\lambda = \Sigma_{r \geq 0}(-1)^r s_{\lambda/(1^r)}$. From (5.4) it follows that

(6) $$u_\lambda = \begin{vmatrix} 1 & 1 & \dots & 1 \\ h_{\lambda_1 - 1} & h_{\lambda_1} & \dots & h_{\lambda_1+n-1} \\ h_{\lambda_2-2} & h_{\lambda_2-1} & \dots & h_{\lambda_2+n-2} \\ \vdots & \vdots & & \vdots \\ h_{\lambda_n-n} & h_{\lambda_n-n+1} & \dots & h_{\lambda_n} \end{vmatrix}.$$

From Example 13(c) we have

$$X^\lambda = \langle EP, s_\lambda \rangle = \langle P, u_\lambda \rangle = \varphi(u_\lambda)$$

where $\varphi: \Lambda_{\mathbf{Q}} \to A$ is the mapping defined in Example 13(b). Hence if we define

$$\pi_r = \varphi(h_r) = \sum_{|\rho|=r} \binom{a}{\rho} \in A$$

for all r (so that $\pi_r = 0$ if $r < 0$), it follows from (6) that

(7) $$X^\lambda = \begin{vmatrix} 1 & 1 & \dots & 1 \\ \pi_{\lambda_1 - 1} & \pi_{\lambda_1} & \dots & \pi_{\lambda_1+n-1} \\ \pi_{\lambda_2-2} & \pi_{\lambda_2-1} & \dots & \pi_{\lambda_2+n-2} \\ \vdots & \vdots & & \vdots \\ \pi_{\lambda_n-n} & \pi_{\lambda_n-n+1} & \dots & \pi_{\lambda_n} \end{vmatrix}^*$$

where the asterisk indicates that the multiplication in A is that defined in Example 13(b).

(d) By subtracting column from column in the determinant (7) we obtain

(8) $$X^\lambda = \det{}^*(\pi'_{\lambda_i - i + j})_{1 \leq i,j \leq n}$$

where $\pi'_r = \pi_r - \pi_{r-1}$, and again the asterisk indicates that the determinant is to be expanded using the $*$-product.

15. In general, if A is a ring and $x, y \in A$, then Axy is a submodule of Ay and is the image of Ax under the homomorphism $a \mapsto ay$ $(a \in A)$, hence is isomorphic to a quotient of Ax.

(a) Let λ be a partition of n and let T be any numbering of the diagram of λ with the numbers $1, 2, \dots, n$. Let R (resp. C) denote the subgroup of S_n that stabilizes each row (resp. column) of T, so that $R \cong S_\lambda$ and $C \cong S_{\lambda'}$. Let $A = \mathbf{Q}[S_n]$ be the group algebra of S_n and let

$$a = \sum_{u \in C} \varepsilon(u)u, \qquad s = \sum_{v \in R} v.$$

Then As is the induced module $\text{ind}_R^{S_n}(1)$, isomorphic to H_λ, and likewise $Aa \cong E_{\lambda'}$.

Let $e = as \in A$. Since $R \cap C = \{1\}$, the products uv $(u \in C,\ v \in R)$ are all distinct, and hence

(1) $$e = 1 + \ldots \neq 0.$$

From the observation above, $M_\lambda = Ae$ is a submodule of As and is isomorphic to a quotient of Aa. From the remark following (7.8) it follows that M_λ is the irreducible A-module (or S_n-module) with character χ^λ.

(b) Let $\varphi: A \to A$ be right multiplication by e. Then $\varphi(M_\lambda) = M_\lambda e = Ae^2 \subset Ae = M_\lambda$. Since M_λ is irreducible, it follows from Schur's lemma that $\varphi | M_\lambda$ is a scalar $c \in \mathbf{Q}$, and therefore $e^2 = \varphi(e) = ce$. Hence $\varphi^2 = c\varphi$, so the only eigenvalues of φ are 0 and c, and the eigenvalue c has multiplicity equal to the dimension of M_λ. Hence

(2) $$\text{trace}\ \varphi = c \dim M_\lambda = cn!/h(\lambda)$$

by (7.6) and §5, Example 2. On the other hand, it follows from (1) above that for each $w \in S_n$ the coefficient of w in $\varphi(w) = we$ is equal to 1; hence relative to the basis S_n of A the matrix of φ has all its diagonal elements equal to 1, and therefore

(3) $$\text{trace}\ \varphi = n!$$

From (2) and (3) it follows that $c = h(\lambda)$ and hence that $\hat{e} = h(\lambda)^{-1}as$ is a primitive idempotent of A affording the character χ^λ.

(c) With the notation of (a) above, let $m_T \in \mathbf{Q}[x_1, \ldots, x_n]$ denote the monomial $x_1^{d(1)} \ldots x_n^{d(n)}$, where $d(i) = r - 1$ if i lies in the rth row of T, and let f_T denote the product $\prod(x_i - x_j)$ taken over all pairs (i, j) such that j lies due north of i in T. Thus f_T is the product of the Vandermonde determinants corresponding to the columns of T, and m_T is its leading term, so that $f_T = am_T$.

Let $\theta: A \to \mathbf{Q}[x_1, \ldots, x_n]$ be the mapping $u \mapsto um_T$. Since $d(i) = d(j)$ if and only if i and j lie in the same row of T, it follows that the subgroup of S_n that fixes m_T is the row-stabilizer R, and hence that $\theta(A) \cong As$. Consequently $\theta | M_\lambda$ is an isomorphism, and we may therefore identify M_λ with its image $\theta(M_\lambda) = Aasm_T = Aam_T = Af_T$. In this incarnation M_λ is the *Specht module* corresponding to the partition λ: it is the $\mathbf{Q}$-vector space spanned by all $n!$ polynomials f_T, for all numberings T of the diagram of λ, and the symmetric group acts by permuting the x's.

(d) The dimension of M_λ is equal to the number of standard tableaux of shape λ, by (7.6). In fact the polynomials f_T, where T is a standard tableau, are linearly independent over any field, and hence form a *basis* of M_λ.

(Order the monomials x^α, $\alpha \in \mathbf{N}^n$ as follows: $x^\alpha < x^\beta$ if and only if α precedes β in the lexicographical order on $\mathbf{N}^n$.

(i) Suppose that $i<j$ and $d(i)<d(j)$ in T, and let w be the transposition (ij). Then $m_T < m_{wT}$.
(ii) Deduce from (i) that if T is standard then $f_T = m_T +$ later monomials.
(iii) Let $T_1, \dots, T_r$ be the standard tableaux of shape λ. The monomials $m_{T_1}, \dots, m_{T_r}$ are all distinct, and we may assume that $m_{T_1} < \dots < m_{T_r}$. Use (ii) to show that the f_{T_i} are linearly independent: if $\sum c_i f_{T_i} = 0$, the coefficient of m_{T_1} in the left-hand side is equal to c_1, hence $c_1 = 0$; repeating the argument gives $c_2 = 0$, and so on.)

(e) From (d) it follows that each f_T is a linear combination of the f_{T_i}, say $f_T = \sum c_i f_{T_i}$ with coefficients $c_i \in \mathbf{Q}$. Show that each c_i is an integer. (Let m be the common denominator of the rational numbers c_i, and let $c_i = m_i/m$ where the m_i are integers. Then we have

$$mf_T = \sum m_i f_{T_i}. \tag{4}$$

If $m>1$, let p be a prime dividing m, and reduce (4) mod. p. Since not all the m_i are divisible by p, we conclude that the f_{T_i} are linearly dependent over the field of p elements, contrary to (d) above.)

Hence for each permutation $w \in S_n$ the entries in the matrix representing w, relative to the basis (f_{T_i}) of M_λ, are all integers.

16. For each partition λ of n, let R_λ be an irreducible matrix representation of S_n with character χ^λ, such that $R_\lambda(w)$ is a matrix of integers for each $w \in S_n$. (Example 15(c) provides an example.) For each partition ρ of n, let $\bar{c}_\rho$ denote the sum (in the group ring $\mathbf{Z}[S_n]$) of all elements of cycle-type ρ in S_n. Then $\bar{c}_\rho$ commutes with each $w \in S_n$, and hence by Schur's lemma $R_\lambda(\bar{c}_\rho)$ is a scalar multiple of the unit matrix, say

$$R_\lambda(\bar{c}_\rho) = \omega_\rho^\lambda 1_d, \tag{1}$$

where ω_ρ^λ is an integer and $d = n!/h(\lambda)$ is the degree of χ^λ. By taking traces in (1) we obtain

$$\frac{n!}{z_\rho}\chi_\rho^\lambda = \frac{n!}{h(\lambda)}\omega_\rho^\lambda$$

and therefore

$$\omega_\rho^\lambda = \frac{h(\lambda)}{z_\rho}\chi_\rho^\lambda \tag{2}$$

is an *integer* for all λ, ρ.

Let C_n denote the centre of the group ring $\mathbf{Z}[S_n]$: it is a commutative ring with $\mathbf{Z}$-basis $(\bar{c}_\rho)_{|\rho|=n}$. For each partition λ of n, the linear mapping $\omega^\lambda : C_n \to \mathbf{Z}$ defined by $\omega^\lambda(\bar{c}_\rho) = \omega_\rho^\lambda$ is a ring homomorphism, since $R_\lambda(\bar{c}_\rho \bar{c}_\sigma) = R_\lambda(\bar{c}_\rho)R_\lambda(\bar{c}_\sigma)$. Moreover the ω^λ, $|\lambda| = n$, are a $\mathbf{Z}$-basis of $\mathrm{Hom}(C_n, \mathbf{Z})$.

17. (a) For each partition λ, the 'augmented Schur function' $\tilde{s}_\lambda$ is defined by

$$\tilde{s}_\lambda = h(\lambda)s_\lambda = \sum_\rho \frac{h(\lambda)}{z_\rho}\chi_\rho^\lambda p_\rho = \sum_\rho \omega_\rho^\lambda p_\rho$$

in the notation of Example 16. Thus $\tilde{s}_\lambda$ is a polynomial in the power sums p_r with integer coefficients, i.e. $\tilde{s}_\lambda \in \Psi$, where $\Psi = \mathbf{Z}[p_1, p_2, \dots]$ is the subring of Λ generated by the power sums. Moreover, $\tilde{s}_\lambda$ is the smallest integer multiple of s_λ that lies in Ψ, since the coefficient of p_1^n in $\tilde{s}_\lambda$ is 1.

(b) Let l be a prime number (the letter p being preempted) and let Ψ_l' denote the subring of Ψ generated by the p_r with r prime to l. Show that $\tilde{s}_\lambda \in \Psi_l'$ if and only if λ is an l-core (§1, Example 8).

(If λ is an l-core, all the hook-lengths in λ are prime to l, and hence all the border strips of λ have lengths prime to l. From this observation and Example 5 it follows that $\chi_\rho^\lambda = 0$ if ρ has any parts divisible by l, and hence that $\tilde{s}_\lambda \in \Psi_l'$. Conversely, if λ is not an l-core, let $\tilde{\lambda}$ and λ^* denote the l-core and l-quotient of λ (§1, Example 8). Let $|\lambda^*| = b \geqslant 1$ and let $|\tilde{\lambda}| = a$. Then for the partition $\rho = (l^b 1^a)$ we have $\chi_\rho^\lambda = \pm a!b!/h(\tilde{\lambda})h(\lambda^*) \neq 0$, and hence $\tilde{s}_\lambda \notin \Psi_l'$.)

(c) Let X be an indeterminate and $\varepsilon_X : \Psi \to \mathbf{Z}[X]$ the homomorphism defined by $\varepsilon_X(p_r) = X$ for all $r \geqslant 1$. Then

$$\varepsilon_X(\tilde{s}_\lambda) = c_\lambda(X)$$

where $c_\lambda(X)$ is the content polynomial of λ (§1, Example 11).

18. Let λ be a partition of n, let l be a prime number, and let κ be the l-core of λ. Then

$$\tilde{s}_\lambda \equiv \tilde{s}_\kappa (p_1^l - p_l)^r \pmod{l} \tag{1}$$

in $\Psi = \mathbf{Z}[p_1, p_2, \dots]$, where $r = (n - |\kappa|)/l$ and $\tilde{s}_\lambda$ is as defined in Example 17.

The proof of (1) uses some concepts from modular representation theory, for which we refer to [P4]: namely the notion of the defect group of a block (or equivalently of a central character) and the Brauer homomorphism.

Let F denote the field of l elements. For any finite group G, let $C(G)$ denote the centre of the group algebra $F[G]$. For each partition λ of n we have a character $\varpi^\lambda : C(S_n) \to F$ of the F-algebra $C(S_n)$, obtained from ω^λ (Example 16) by reduction modulo l. Each defect group D^λ of ϖ^λ is a Sylow l-subgroup of the centralizer of an element of cycle-type (l^m) in $S_{lm} \subset S_n$, where $0 \leqslant m \leqslant n/l$ ([J9], 6.2.39). It follows that if $D^\lambda \neq \{1\}$, then D^λ contains a subgroup Q of order l, generated by an l-cycle. On the other hand, if $D^\lambda = \{1\}$, there is a partition ρ of n for which both z_ρ and ω_ρ^λ are prime to l, and hence (Example 16) $h(\lambda)\chi_\rho^\lambda = z_\rho \omega_\rho^\lambda$ is prime to l. Consequently $h(\lambda)$ is prime to l and therefore (§1, Example 10) λ is an l-core.

Assume now that λ is not an l-core, so that D^λ contains Q as above. Let $H = Q \times S_{n-l}$ be the centralizer of Q in S_n. Then the mapping $\varphi : C(S_n) \to C(H) = C(Q) \otimes_F C(S_{n-l})$ defined by restriction to H is an F-algebra homomorphism (the Brauer homomorphism), and ϖ^λ factors through φ: say $\varpi^\lambda = (\varepsilon \otimes \varpi^\mu) \circ \varphi$, where μ is some partition of $n - l$, and ε is the unique (trivial) character of $C(Q)$.

If now ρ is a partition of n, the conjugacy class $\bar{c}_\rho$ in S_n meets H only if ρ is of the form $(1^l) \cup \sigma$ or $(l) \cup \sigma$ for some partition σ of $n - l$. If $\rho = (1^l) \cup \sigma$ we have $\varpi^\lambda(\bar{c}_\rho) = \varpi^\mu(\bar{c}_\sigma)$, and if $\rho = (l) \cup \sigma$ we have $\varpi^\lambda(\bar{c}_\rho) = (l-1)\varpi^\mu(\bar{c}_\sigma) = -\varpi^\mu(\bar{c}_\sigma)$. Since $\varpi^\lambda(\bar{c}_\rho)$ is the reduction modulo l of ω_ρ^λ (Example 16) it follows that modulo l we have $\omega_\rho^\lambda \equiv \omega_\sigma^\mu$ if $\rho = (1^l) \cup \sigma$, $\omega_\rho^\lambda \equiv -\omega_\sigma^\mu$ if $\rho = (l) \cup \sigma$, and $\omega_\rho^\lambda \equiv 0$ for all other ρ. Hence

$$\tilde{s}_\lambda = \sum_\rho \omega_\rho^\lambda p_\rho \equiv \left(\sum_\sigma \omega_\sigma^\mu p_\sigma\right)(p_1^l - p_l)$$

so that

$$\tilde{s}_\lambda \equiv \tilde{s}_\mu (p_1^l - p_l) \quad (\text{mod. } l).$$

If μ is not an l-core we may repeat the argument. In this way we shall obtain

$$(2) \qquad \tilde{s}_\lambda \equiv \tilde{s}_\nu (p_1^l - p_l)^q$$

for some integer $q \geqslant 1$ and some l-core ν. Now apply the specialization $\varepsilon_X : p_r \mapsto X$ ($r \geqslant 1$) (Example 17(c)): we obtain

$$c_\lambda(X) \equiv c_\nu(X)(X^l - X)^q \quad (\text{mod. } l)$$
$$\equiv c_\pi(X)$$

where $\pi = \nu + (lq)$ has l-core ν. From §1, Example 11(c) we now conclude that the partitions λ and π have the same l-core, so that $\nu = \kappa$ and $q = r$, completing the proof.

19. Let λ, μ be partitions of n, and let l be a prime number. Then with the notation of Example 17, $\tilde{s}_\lambda \equiv \tilde{s}_\mu$ (mod. l) if and only if λ, μ have the same l-core ('Nakayama's conjecture'). (If $\tilde{s}_\lambda \equiv \tilde{s}_\mu$, then $c_\lambda(X) \equiv c_\mu(X)$ by Example 17(c), and hence λ, μ have the same l-core by §1, Example 11(c). Conversely, if λ, μ have the same l-core, then it follows from Example 18 that $\tilde{s}_\lambda \equiv \tilde{s}_\mu$.)

20. As in §5, Example 25 we shall identify each $f \otimes g \in \Lambda \otimes \Lambda$ with $f(x)g(y)$, where (x) and (y) are two sets of independent variables. Define a comultiplication $\Delta^* : \Lambda \to \Lambda \otimes \Lambda$ and a counit $\varepsilon^* : \Lambda \to \mathbf{Z}$ by

$$\Delta^* f = f(xy)$$

where (xy) is the set of all products $x_i y_j$, and

$$\varepsilon^* f = f(1, 0, 0, \ldots)$$

for all $f \in \Lambda$.

With respect to Δ^* and ε^*, Λ is a cocommutative Hopf algebra over $\mathbf{Z}$; both Δ^* and ε^* are ring homomorphisms, and $(1 \otimes \varepsilon^*) \circ \Delta^*$ is the identity mapping.

(a) Show that, for all $n \geqslant 1$,

$$(1) \qquad \Delta^* h_n = \sum_{|\lambda| = n} s_\lambda \otimes s_\lambda, \qquad \varepsilon^* h_n = 1;$$

$$(2) \qquad \Delta^* e_n = \sum_{|\lambda| = n} s_\lambda \otimes s_{\lambda'}, \qquad \varepsilon^* e_n = \delta_{1n};$$

$$(3) \qquad \Delta^* p_n = p_n \otimes p_n, \qquad \varepsilon^* p_n = 1.$$

Also we have (7.9)

$$\Delta^* s_\lambda = \sum_{\mu,\nu} \gamma^\lambda_{\mu\nu} s_\mu \otimes s_\nu. \tag{4}$$

(b) As in §5, Example 25 define a scalar product on $\Lambda \otimes \Lambda$ by

$$\langle f_1 \otimes g_1, f_2 \otimes g_2 \rangle = \langle f_1, f_2 \rangle \langle g_1, g_2 \rangle$$

for $f_1, f_2, g_1, g_2 \in \Lambda$. With respect to this scalar product, Δ^* is the adjoint of the internal product: in other words, we have

$$\langle \Delta^* f, g \otimes h \rangle = \langle f, g * h \rangle \tag{5}$$

for all $f, g, h \in \Lambda$. (By linearity we may take $f = s_\lambda$, $g = s_\mu$, $h = s_\nu$, and then both sides of (5) are equal to $\gamma^\lambda_{\mu\nu}$ by (4) above.)

21. For any commutative ring A, let $\mathrm{G}(A)$ denote the set of all unital ring homomorphisms $\alpha: \Lambda \to A$. Each such homomorphism is determined by the formal power series

$$\alpha H(t) = \sum_{i \geq 0} \alpha(h_i) t^i \in A[[t]]$$

with constant term $\alpha(h_0) = 1$, and we may therefore identify $\mathrm{G}(A)$ with the set of formal power series in $A[[t]]$ with constant term equal to 1.

(a) The comultiplication $\Delta: \Lambda \to \Lambda \otimes \Lambda$ defined in §5, Example 25 induces an abelian group structure on $\mathrm{G}(A)$ as follows. If $\alpha, \beta \in \mathrm{G}(A)$, we define

$$\alpha + \beta = m_A \circ (\alpha \otimes \beta) \circ \Delta$$

where $m_A: A \otimes A \to A$ is the multiplication in A. We have then (*loc. cit.*)

$$(\alpha + \beta) h_k = \sum_{i+j=k} \alpha(h_i) \beta(h_j)$$

so that

$$(\alpha + \beta) H(t) = (\alpha H(t))(\beta H(t)),$$

the *product* of the power series $\alpha H(t)$ and $\beta H(t)$ in $A[[t]]$.

Next let $\varpi: \Lambda \to \Lambda$ be the involution defined by $\varpi(h_i) = (-1)^i e_i$ $(i \geq 1)$, so that on Λ^n, ϖ is $(-1)^n \omega$. Then define

$$-\alpha = \alpha \circ \varpi;$$

we have $(-\alpha) H(t) = \alpha(E(-t)) = (\alpha H(t))^{-1}$, so that $(-\alpha)(H(t))$ is the *inverse* in $A[[t]]$ of the power series $\alpha H(t)$.

Finally, the zero element $\mathbf{0}$ of $\mathrm{G}(A)$ is induced by the counit ε: namely $\mathbf{0} = e_A \circ \varepsilon$, where $e_A: \mathbf{Z} \to A$ is the unique homomorphism of $\mathbf{Z}$ into A. Since $\varepsilon(h_i) = 0$ for each $i \geq 1$, it follows that $\mathbf{0} H(t) = 1$.

(b) The comultiplication Δ^* of Example 20 induces a multiplication in $\mathrm{G}(A)$ by the rule

$$\alpha\beta = m_A \circ (\alpha \otimes \beta) \circ \Delta^*.$$

This product may be described as follows: if we formally factorize the power series $\alpha H(t)$ and $\beta H(t)$, say

$$\alpha H(t) = \prod (1 + \xi_i t), \qquad \beta H(t) = \prod (1 + \eta_i t),$$

then

$$(\alpha\beta)H(t) = \prod_{i,j} (1 + \xi_i \eta_j t).$$

The element $1 \in G(A)$ defined by

$$1 = e_A \circ \varepsilon^*$$

with ε^* the counit defined in Example 20, is the identity element for this multiplication, and $1(H(t)) = (1-t)^{-1}$.

(c) With addition and multiplication as defined in (a) and (b), $G(A)$ is a commutative ring, with zero element 0 and identity element 1. If $\varphi: A \to B$ is a homomorphism of A into a commutative ring B, then $G(\varphi): G(A) \to G(B)$ defined by $G(\varphi)\alpha = \varphi \circ \alpha$ is a ring homomorphism. Thus G is a covariant functor on the category of commutative rings.

22. Define an internal product on $\Lambda \otimes \Lambda$ by

$$(f_1 \otimes f_2) * (g_1 \otimes g_2) = (f_1 * g_1) \otimes (f_2 * g_2)$$

for $f_1, f_2, g_1, g_2 \in \Lambda$. Show that $\Delta(f * g) = (\Delta f) * (\Delta g)$ for all $f, g \in \Lambda$, but that in general $\Delta^*(f * g) \neq (\Delta^* f) * (\Delta^* g)$.

23. (a) Let $f, g, h \in \Lambda$. Then the scalar product $\langle f * g, h \rangle$ is symmetrical in f, g, and h.

(b) Let $(u_\lambda), (v_\lambda)$ be dual bases of Λ, and let $f \in \Lambda$. Then

$$\Delta^* f = \sum_\lambda (u_\lambda * f) \otimes v_\lambda.$$

(For $\langle \Delta^* f, g \otimes u_\lambda \rangle = \langle f, g * u_\lambda \rangle = \langle u_\lambda * f, g \rangle$ by (a) above and Example 20(b).)

(c) Let $f, g, h \in \Lambda$ and let $\Delta h = \sum a_i \otimes b_i$. Then

$$(fg) * h = \sum (f * a_i)(g * b_i).$$

In particular,

$$(fg) * s_\lambda = \sum_\mu (f * s_{\lambda/\mu})(g * s_\mu).$$

(d) Let λ, μ be partitions. Then

$$h_\lambda * s_\mu = \sum \prod_{i \geqslant 1} s_{\mu^{(i)}/\mu^{(i-1)}}$$

summed over all sequences $(\mu^{(0)}, \mu^{(1)}, \dots)$ of partitions such that $0 = \mu^{(0)} \subset \mu^{(1)} \subset \dots \subset \mu$ and $|\mu^{(i)} - \mu^{(i-1)}| = \lambda_i$ for each $i \geqslant 1$. (Use (c).)

(e) If $M = (m_{ij})$ is a matrix of non-negative integers, let $h_M = \prod_{i,j} h_{m_{ij}}$. If λ and μ are partitions, show that

$$h_\lambda * h_\mu = \sum h_M$$

summed over all matrices M of non-negative integers with $l(\lambda)$ rows and $l(\mu)$ columns and row sums λ_i, column sums μ_j.

24. The symmetric group S_n embeds naturally in S_{n+1} as the subgroup that fixes $n+1$. The union

$$S_\infty = \bigcup_{n \geqslant 0} S_n$$

is the group of permutations of the set of positive integers that fix all but a finite subset. If $w \in S_n$ has cycle-type ρ, where $\rho = (\rho_1, \ldots, \rho_r)$ is a partition of length r, then when regarded as an element of S_{n+k} the permutation w has cycle type $(\rho_1, \ldots, \rho_r, 1, \ldots, 1) = \rho \cup (1^k)$. We are therefore led to define the *modified cycle-type* of w to be the partition $(\rho_1 - 1, \ldots, \rho_r - 1)$. This modified cycle-type is stable under the embedding of S_n in S_{n+k}.

For each partition λ, let C_λ denote the set of all $w \in S_\infty$ whose modified cycle-type is λ. As λ runs through all partitions, the C_λ are the conjugacy classes of the group S_∞. For example, $C_{(1)}$ is the class of transpositions, and C_0 consists of the identity permutation.

For each $n \geqslant 0$, let Z_n denote the centre of the group ring $\mathbb{Z}[S_n]$, and for each partition λ let $c_\lambda(n) \in Z_n$ denote the sum of all $w \in S_n$ whose modified cycle-type is λ, i.e. the sum of all $w \in S_n \cap C_\lambda$. We have $c_\lambda(n) \neq 0$ if and only if $|\lambda| + l(\lambda) \leqslant n$.

Now let λ, μ be partitions. The product $c_\lambda(n)c_\mu(n)$ in Z_n will be a linear combination of the $c_\nu(n)$, say

$$c_\lambda(n)c_\mu(n) = \sum_\nu a^\nu_{\lambda\mu}(n)c_\nu(n) \tag{1}$$

with coefficients $a^\nu_{\lambda\mu}(n) \in \mathbb{N}$, and zero unless $|\nu| \leqslant |\lambda| + |\mu|$. For example, when $\lambda = \mu = (1)$, $c_{(1)}(n)^2$ is the sum of all products $(ij)(kl)$ of two transpositions in S_n, and a simple calculation gives

$$c_{(1)}(n)^2 = 3c_{(2)}(n) + 2c_{(11)}(n) + \tfrac{1}{2}n(n-1)c_0(n). \tag{2}$$

In general (see [F2]) the coefficients $a^\nu_{\lambda\mu}(n)$ are polynomial functions of n, and are independent of n if and only if $|\nu| = |\lambda| + |\mu|$. We may therefore, following [F2], construct a ring as follows. Let R be the subring of the polynomial ring $\mathbb{Q}[t]$ consisting of polynomials that take integer values at all integers (the binomial coefficients $\binom{t}{n}$, $n \geqslant 0$, form a $\mathbb{Z}$-basis of R), and let F be the commutative R-algebra with R-basis (c_λ) indexed by partitions λ and multiplication defined by

$$c_\lambda c_\mu = \sum_\nu a^\nu_{\lambda\mu} c_\nu, \tag{3}$$

where $a^\nu_{\lambda\mu} \in R$ takes the value $a^\nu_{\lambda\mu}(n)$ at the integer n, and the sum is over partitions ν such that $|\nu| \leqslant |\lambda| + |\mu|$. If we assign each c_λ the degree $|\lambda|$, F is not a *graded* ring, because the right-hand side of (3) is not homogeneous; but it is a *filtered* ring: if F^r is the subspace of F spanned by the c_λ such that $|\lambda| \leqslant r$, then $F^r . F^s \subset F^{r+s}$. We may therefore form the associated graded ring $G_R = Gr(F)$: G_R is the direct sum $\oplus_{r \geqslant 0} G_R^r$, where $G_R^r = F^r/F^{r-1}$, and the multiplication in G_R is induced from that in F. The effect of passing from F to G_R is simply to throw out the terms of lower degree in (3): in G_R the multiplication is defined by

$$c_\lambda c_\mu = \sum_{|\nu| = |\lambda| + |\mu|} a_{\lambda\mu}^\nu c_\nu \tag{4}$$

and (as remarked above) the structure constants $a_{\lambda\mu}^\nu$ such that $|\nu| = |\lambda| + |\mu|$ are non-negative *integers*. (For example, in G_R we have $c_{(1)}^2 = 3c_{(2)} + 2c_{(11)}$, from (2) above.) It follows that $G_R = R \otimes_{\mathbf{Z}} G$, where G is the free $\mathbf{Z}$-module with basis (c_λ) and multiplicative structure given by (4).

Let us write c_r in place of $c_{(r)}$, $r \geqslant 1$ ($c_0 = 1$ is the identity element of F and G). Show that

(a) if $|\lambda| + r = m$, then

$$a_{\lambda(r)}^{(m)} = \begin{cases} (m+1)r! \Big/ \prod_{i \geqslant 0} m_i(\lambda)! & \text{if } l(\lambda) \leqslant r+1, \\ 0 & \text{otherwise,} \end{cases} \tag{5}$$

where $m_0(\lambda) = r + 1 - l(\lambda)$;

(b) if $|\lambda| + r = |\nu|$ then

$$a_{\lambda(r)}^\nu = \sum a_{\mu(r)}^{(\nu_i)} \tag{6}$$

summed over pairs (i, μ) such that $\mu \cup \nu = \lambda \cup (\nu_i)$. Deduce that $a_{\lambda(r)}^\nu = 0$ unless $\nu \geqslant \lambda \cup (r)$, and that $a_{\lambda(r)}^{\lambda \cup (r)} > 0$.

From (b) it follows that, for each partition $\lambda = (\lambda_1, \lambda_2, \dots)$, $c_{\lambda_1} c_{\lambda_2} \dots$ is of the form

$$c_{\lambda_1} c_{\lambda_2} \dots = \sum_{\mu \geqslant \lambda} d_{\lambda\mu} c_\mu$$

with $d_{\lambda\lambda} > 0$. Hence $c_1, c_2, \dots$ are algebraically independent elements of G, and generate G over $\mathbf{Q}$, i.e. $G \otimes \mathbf{Q} = \mathbf{Q}[c_1, c_2, \dots]$. Moreover, the multiplicative structure of G is uniquely determined by (a) and (b).

25. Let ψ be the involution on Λ defined in §2, Example 24. With the notation of that Example, the $h_\lambda^* = \psi(h_\lambda)$ form a $\mathbf{Z}$-basis of Λ. Let (g_λ) be the dual basis, so that $\langle g_\lambda, h_\mu^* \rangle = \delta_{\lambda\mu}$. Equivalently, $g_\lambda = \psi^\perp(m_\lambda)$, where $\psi^\perp$ is the adjoint of ψ relative to the scalar product. We shall show that, in the notation of Example 24 above, the linear mapping $\varphi: \Lambda \to G$ defined by $\varphi(g_\lambda) = c_\lambda$ for all partitions λ is a ring isomorphism.

(a) From §2, Example 24 we have

$$h_n^* = -h_n + \sum_{\mu < (n)} u_{(n)\mu} h_\mu$$

for suitable integers $u_{(n)\mu}$, and hence h_λ^* is of the form

$$h_\lambda^* = (-1)^{l(\lambda)} h_\lambda + \sum_{\mu<\lambda} u_{\lambda\mu} h_\mu .$$

This shows that the transition matrix $M(h^*, h)$ $(= M(h, h^*))$ is strictly upper triangular, with diagonal elements $(-1)^{l(\lambda)}$. By (6.3)(3) we have $M(g, m) = M(h, h^*)'$, hence

$$g_\lambda = (-1)^{l(\lambda)} m_\lambda + \sum_{\mu>\lambda} u_{\mu\lambda} m_\mu .$$

In particular, $g_{(n)} = -m_{(n)} = -p_n$.

(b) Let $b_{\lambda\mu}^\nu = \langle g_\lambda g_\mu, h_\nu^* \rangle$ be the coefficient of g_ν in $g_\lambda g_\mu$. Since the $g_{(n)} = -p_n$ generate $\Lambda_{\mathbf{Q}}$, in order to prove that φ defined above is a ring isomorphism it will be enough to show that $b_{\lambda\mu}^\nu = a_{\lambda\mu}^\nu$ whenever μ is a one-part partition (r), and for this purpose it will suffice to show that the b's satisfy the counterparts of the two relations (5) and (6) of Example 24.

Consider first

$$b_{\lambda(r)}^{(m)} = -\langle g_\lambda p_r, h_m^* \rangle = -\langle g_\lambda, p_r^\perp h_m^* \rangle \tag{1}$$

in the notation of §5, Example 3. From §2, Example 24, h_m^* is the coefficient of t^m in

$$\frac{1}{m+1} H(t)^{-m-1} = \frac{1}{m+1} \exp\left(-(m+1) \sum_{r\geq 1} \frac{p_r t^r}{r} \right).$$

Hence $-p_r^\perp h_m^* = -r\,\partial h_m^*/\partial p_r$ is equal to the coefficient of t^m in $t^r H(t)^{-m-1}$, that is to say it is the residue of the differential

$$t^r\,\mathrm{d}t/(tH(t))^{m+1} = \frac{1}{r+1}\,\mathrm{d}(t^{r+1})/u^{m+1} \tag{2}$$

where $u = tH(t)$ as in §2, Example 24. Now

$$\begin{aligned} t^{r+1} &= u^{r+1}\left(\sum h_n^* u^n\right)^{r+1} \\ &= \sum \frac{(r+1)!}{\prod_{i\geq 0} m_i(\lambda)!} h_\lambda^* u^{|\lambda|+r+1}, \end{aligned}$$

summed over partitions λ of length $l(\lambda) \leq r+1$, where $m_0(\lambda) = r+1-l(\lambda)$. Hence from (2) we obtain

$$-p_r^\perp h_m^* = \sum \frac{r!(m+1)}{\prod_{i\geq 0} m_i(\lambda)!} h_\lambda^*$$

summed over partitions λ such that $l(\lambda) \leq r+1$ and $|\lambda| = m-r$. From (1) it now follows by comparison with relation (5) of Example 24 that $b_{\lambda(r)}^{(m)} = a_{\lambda(r)}^{(m)}$ whenever $|\lambda| + r = m$.

(c) Finally consider

$$b^{\nu}_{\lambda(r)} = -\langle g_\lambda p_r, h^*_\nu \rangle = -\langle g_\lambda, p_r^\perp h^*_\nu \rangle.$$

Since $p_r^\perp = r\,\partial/\partial p_r$ is a derivation (§5, Example 3) we have

$$p_r^\perp h^*_\nu = \sum h^*_{\nu^{(i)}} p_r^\perp h^*_{\nu_i},$$

where $\nu^{(i)} = (\nu_1, \dots, \hat{\nu}_i, \dots)$, and the sum is over $i \geqslant 1$ such that $\nu_i \leqslant r$. Hence

$$b^{\nu}_{\lambda(r)} = -\sum \langle g_\lambda, h^*_{\nu^{(i)}} p_r^\perp h^*_{\nu_i} \rangle, \tag{3}$$

in which the scalar product on the right-hand side is the coefficient of h^*_λ in $h^*_{\nu^{(i)}} p_r^\perp h^*_{\nu_i}$, hence is zero unless $\lambda = \nu^{(i)} \cup \mu$ for some partition μ, i.e. $\mu \cup \nu = \lambda \cup (\nu_i)$; and then is equal to the coefficient of h^*_μ in $p_r^\perp h^*_{\nu_i}$, which is $\langle g_\mu, p_r^\perp h^*_{\nu_i} \rangle = -b^{(\nu_i)}_{\mu(r)}$ by (1). Hence (3) takes the form

$$b^{\nu}_{\lambda(r)} = \sum b^{(\nu_i)}_{\mu(r)}$$

summed over pairs (i, μ) such that $\mu \cup \nu = \lambda \cup (\nu_i)$. We have thus established the counterpart of relation (6) of Example 24, and the proof is complete.

26. Let $\delta: R \to R \otimes R$ be the comultiplication on R that corresponds to $\Delta: \Lambda \to \Lambda \otimes \Lambda$ (§5, Example 25) under the characteristic map (so that $(\mathrm{ch} \otimes \mathrm{ch}) \circ \delta = \Delta \circ \mathrm{ch}$). If $f \in R_n$, show that

$$\delta f = \bigoplus_{p+q=n} f \mid S_p \times S_q.$$

Notes and references

The representation theory of finite groups was founded by Frobenius in a series of papers published in the last years of the nineteenth century, and reproduced in Vol. 3 of his collected works; in particular, he obtained the irreducible characters of the symmetric groups in 1900 [F10], and our exposition does not differ substantially from his.

The internal product $f * g$ occurs first (as far as I am aware) in the 1927 paper of Redfield [R1], and later in [L11]. (Littlewood calls it the inner product: we have avoided this terminology, because inner product is sometimes taken as synonymous with scalar product.)

Example 5 is due to Littlewood and Richardson [L13], but is commonly known as the Murnaghan–Nakayama rule ([M18], [N1]). Examples 6 and 7 are due to Frobenius (*loc. cit.*). Example 9 was contributed by A. Zelevinsky.

Examples 13 and 14. Character polynomials occur already in Frobenius' 1904 paper [F11]. The formulas (7) and (8) are due to Specht [S19].

Examples 24 and 25. For proofs of Example 24(a), (b) see [F2]. A better proof of the result of Example 25 will be found in [G9].

8. Plethysm

In this section we shall study briefly another sort of multiplication in Λ, called *plethysm* or *composition*, and defined as follows. Let $f, g \in \Lambda$, and write g as a sum of monomials:

$$g = \sum_{\alpha} u_{\alpha} x^{\alpha}.$$

Now introduce the set of fictitious variables y_i defined by

$$\prod (1 + y_i t) = \prod_{\alpha} (1 + x^{\alpha} t)^{u_{\alpha}} \tag{8.1}$$

and define

$$f \circ g = f(y_1, y_2, \ldots). \tag{8.2}$$

If $f \in \Lambda^m$ and $g \in \Lambda^n$, then clearly $f \circ g \in \Lambda^{mn}$. Also e_1 acts as a two sided identity: $f \circ e_1 = e_1 \circ f = f$ for all $f \in \Lambda$.

From the definition (8.2) it is clear that

(8.3) *For each $g \in \Lambda$, the mapping $f \mapsto f \circ g$ is an endomorphism of the ring Λ.* |

By taking logarithms of both sides of (8.1) we obtain

$$p_n(y) = \sum_{\alpha} u_{\alpha} (x^{\alpha})^n \qquad (n \geqslant 1)$$

so that

$$p_n \circ g = g \circ p_n = g(x_1^n, x_2^n, \ldots) \tag{8.4}$$

for all $g \in \Lambda$. In particular,

$$p_n \circ p_m = p_m \circ p_n = p_{mn}. \tag{8.5}$$

From (8.4) it follows that

(8.6) *For each $n \geqslant 1$, the mapping $g \mapsto p_n \circ g$ is an endomorphism of the ring Λ.* |

Plethysm is associative: for all $f, g, h \in \Lambda$ we have

$$(f \circ g) \circ h = f \circ (g \circ h). \tag{8.7}$$

Proof. Since the p_n generate $\Lambda_{\mathbf{Q}}$ (2.12), by virtue of (8.3) and (8.6) it is enough to verify associativity when $f = p_m$ and $g = p_n$, in which case it is obvious from (8.4) and (8.5). |

For plethysm involving Schur functions, there are the following formulas: from (5.9) it follows that

$$
\begin{aligned}
s_\lambda \circ (g+h) &= \sum_{\mu,\nu} c^\lambda_{\mu\nu}(s_\mu \circ g)(s_\nu \circ h) \\
&= \sum_\mu (s_{\lambda/\mu} \circ g)(s_\mu \circ h)
\end{aligned} \tag{8.8}
$$

and from (7.9) that

$$
s_\lambda \circ (gh) = \sum_{\mu,\nu} \gamma^\lambda_{\mu\nu}(s_\mu \circ g)(s_\nu \circ h). \tag{8.9}
$$

The sum in (8.8) is over pairs of partitions $\mu, \nu \subset \lambda$, and in (8.9) over pairs of partitions μ, ν such that $|\mu| = |\nu| = |\lambda|$.

Finally, let λ, μ be partitions. Then $s_\lambda \circ s_\mu$ is an integral linear combination of Schur functions, say

$$
s_\lambda \circ s_\mu = \sum_\pi a^\pi_{\lambda\mu} s_\pi \tag{8.10}
$$

summed over partitions π such that $|\pi| = |\lambda| . |\mu|$. We shall prove in Appendix A that *the coefficients* $a^\pi_{\lambda\mu}$ *are all* $\geqslant 0$.

Remarks. 1. We have observed in (3.10) that to each $f \in \Lambda$ there corresponds a natural operation F on the category of λ-rings. In this correspondence, plethysm corresponds to composition of operations: if $f, g \in \Lambda$ correspond to the natural operations F, G, then $f \circ g$ corresponds to $F \circ G$.

2. Plethysm is defined in the ring R of §7 via the characteristic map: for $u, v \in R$, $u \circ v$ is defined to be $\mathrm{ch}^{-1}(\mathrm{ch}\, u \circ \mathrm{ch}\, v)$. If u (resp. v) is an irreducible character of S_m (resp. S_n), then $u \circ v$ is a character of S_{mn} which may be described as follows: if U (resp. V) is an S_m-module with character u (resp. an S_n-module with character v), the wreath product $S_n \sim S_m$ (which is the normalizer of $S_n^m = S_n \times \ldots \times S_n$ in S_{mn}) acts on U and on the mth tensor power $\mathbf{T}^m(V)$, hence also on $U \otimes \mathbf{T}^m(V)$; and $u \circ v$ is the character of the S_{mn}-module induced by $U \otimes \mathbf{T}^m(V)$. See Appendix A to this Chapter.

Examples

1. (a) Let $f \in \Lambda^m$, $g \in \Lambda^n$. Show that

$$
\omega(f \circ g) = \begin{cases} f \circ (\omega g) & \text{if } n \text{ is even,} \\ (\omega f) \circ (\omega g) & \text{if } n \text{ is odd,} \end{cases}
$$

and that

$$f \circ (-g) = (-1)^m (\omega f) \circ g.$$

(b) If λ, μ are partitions, let $\lambda \circ \mu (= \mu \circ \lambda)$ denote the partition whose parts are $\lambda_i \mu_j$. Then we have

$$p_\lambda \circ p_\mu = p_\mu \circ p_\lambda = p_{\lambda \circ \mu}.$$

(c) Show that $\omega(h_r \circ p_s) = (-1)^{r(s-1)} e_r \circ p_s$.

2. Since $s_{\lambda/\mu} = s_\mu^\perp (s_\lambda)$ in the notation of §5, Example 3, it follows from (8.8) that

$$f \circ (g+h) = \sum_\mu \left((s_\mu^\perp f) \circ g \right) (s_\mu \circ h)$$

for all $f, g, h \in \Lambda$. Also

$$f \circ (gh) = \sum_\mu ((s_\mu * f) \circ g)(s_\mu \circ h).$$

3. We have

$$h_n \circ (fg) = \sum_{|\lambda| = n} (s_\lambda \circ f)(s_\lambda \circ g),$$

$$e_n \circ (fg) = \sum_{|\lambda| = n} (s_\lambda \circ f)(s_{\lambda'} \circ g).$$

These formulas are particular cases of (8.9) (and are consequences of (4.3) and (4.3′)).

4. Let λ be a partition of length $\leqslant n$, and consider $(s_\lambda \circ s_{(n-1)})(x_1, x_2)$. By definition this is equal to $s_\lambda(x_1^{n-1}, x_1^{n-2} x_2, \dots, x_2^{n-1})$, i.e. to

$$x_2^{(n-1)|\lambda|} s_\lambda(q^{n-1}, q^{n-2}, \dots, 1),$$

where $q = x_1 x_2^{-1}$. On the other hand, by the positivity of the coefficients in (8.10), $(s_\lambda \circ s_{n-1})(x_1, x_2)$ is a linear combination of the $s_\pi(x_1, x_2)$ with non-negative integer coefficients, where $\pi = (\pi_1, \pi_2)$ and $\pi_1 + \pi_2 = (n-1)|\lambda| = d$ say. Now

$$\begin{aligned} s_\pi(x_1, x_2) &= x_1^{\pi_1} x_2^{\pi_2} + x_1^{\pi_1 - 1} x_2^{\pi_2 + 1} + \dots + x_1^{\pi_2} x_2^{\pi_1} \\ &= x_2^d (q^{\pi_1} + q^{\pi_1 - 1} + \dots + q^{\pi_2}). \end{aligned}$$

Hence $s_\lambda(q^{n-1}, q^{n-2}, \dots, 1)$ is a non-negative linear combination of the polynomials $q^{\pi_1} + q^{\pi_1 - 1} + \dots + q^{\pi_2}$, where $\pi_1 \geqslant \pi_2$ and $\pi_1 + \pi_2 = d$. It follows that $s_\lambda(q^{n-1}, q^{n-2}, \dots, 1)$ is a *unimodal* symmetrical polynomial in q, i.e. that if a_i is the coefficient of q^i in this polynomial, for $0 \leqslant i \leqslant d$, then $a_i + a_{d-i}$ (symmetry) and

$$a_0 \leqslant a_1 \leqslant \dots \leqslant a_{[d/2]}$$

(unimodality).

From §3, Example 1, it follows that the generalized Gaussian polynomial

$$\begin{bmatrix} n \\ \lambda \end{bmatrix} = \prod_{x \in \lambda} \frac{1 - q^{n - c(x)}}{1 - q^{h(x)}}$$

is symmetrical and unimodal for all n and λ.

5. Let G be a subgroup of S_m and H a subgroup of S_n, so that $G \sim H$ is a subgroup of the wreath product $S_m \sim S_n \subset S_{mn}$. Then the cycle-indicator (§2, Example 9) of $G \sim H$ is

$$c(G \sim H) = c(H) \circ c(G).$$

6. Closed formulas for the plethysms $h_r \circ h_2$, $h_r \circ e_2$, $e_r \circ h_2$, $e_r \circ e_2$ may be derived from the series expansions of §5, Examples 5 and 9:

(a) $h_r \circ h_2 = \sum_\mu s_\mu$, summed over all *even* partitions μ of $2r$ (i.e. partitions with all parts even).

(b) $h_r \circ e_2 = \sum_\mu s_{\mu'}$, summed over even partitions μ, as in (a).

(c) $e_r \circ e_2 = \sum_\pi s_\pi$, summed over partitions π of the form $(\alpha_1 - 1, \dots, \alpha_p - 1 | \alpha_1, \dots, \alpha_p)$, where $\alpha_1 > \dots > \alpha_p > 0$ and $\alpha_1 + \dots + \alpha_p = r$.

(d) $e_r \circ h_2 = \sum_\pi s_{\pi'}$, summed over partitions π as in (c).

7. Let $h_n^{(r)} = p_r \circ h_n = h_n \circ p_r$, so that

$$h_n^{(r)}(x_1, x_2, \dots) = h_n(x_1^r, x_2^r, \dots)$$

which is the coefficient of t^{nr} in

$$\prod_{i \geqslant 1} (1 - x_i^r t^r)^{-1} = \prod_{i \geqslant 1} \prod_{j=1}^{r} (1 - x_i \omega^j t)^{-1},$$

where $\omega = e^{2\pi i/r}$. By (4.3) this product is equal to

$$\sum_\mu s_\mu(x) s_\mu(1, \omega, \dots, \omega^{r-1}) t^{|\mu|}.$$

Now (§3, Example 17)

$$s_\mu(1, \omega, \dots, \omega^{r-1}) = \sigma_r(\mu) = \pm 1$$

if $l(\mu) \leqslant r$ and μ is an r-core, and $s_\mu(1, \omega, \dots, \omega^{r-1}) = 0$ otherwise. It follows that

$$p_r \circ h_n = \sum_\mu \sigma_r(\mu) s_\mu$$

summed over r-cores μ such that $l(\mu) \leqslant r$ and $|\mu| = nr$.

8. More generally, if ρ is any partition, let $h_n^{(\rho)} = p_\rho \circ h_n$, so that

$$h_n^{(\rho)} = \prod_{j \geqslant 1} (p_{\rho_j} \circ h_n) = \prod_{j \geqslant 1} h_n^{(\rho_j)}$$

is the coefficient of $t^{n\rho}=t_1^{n\rho_1}t_2^{n\rho_2}\dots$ in

(1) $$\prod_{i,j}(1-x_i^{\rho_j}t_j^{\rho_j})^{-1}=\prod_{i,j}\prod_{k=1}^{\rho_j}\left(1-x_i\omega_j^k t_j\right)^{-1},$$

where $\omega_j=\exp(2\pi i/\rho_j)$. By (4.3) this product is equal to

(2) $$\sum_\mu s_\mu(x)s_\mu(t_1y^{(1)},\dots,t_my^{(m)})$$

summed over partitions μ of length $\leqslant|\rho|$, where $y^{(j)}$ denotes the sequence $(\omega_j^k)_{1\leqslant k\leqslant\rho_j}$, and m is the length of ρ.

By (5.11) we have

(3) $$s_\mu(t_1y^{(1)},\dots,t_my^{(m)})=\sum\prod_{j=1}^m t_j^{|\nu^{(j)}-\nu^{(j-1)}|}s_{\nu^{(j)}/\nu^{(j-1)}}(y^{(j)})$$

summed over all sequences $(\nu^{(0)},\nu^{(1)},\dots,\nu^{(m)})$ of partitions such that $0=\nu^{(0)}\subset\nu^{(1)}\subset\dots\subset\nu^{(m)}=\mu$. Now from §5, Example 24(b) we have

(4) $$s_{\nu^{(j)}/\nu^{(j-1)}}(y^{(j)})=\begin{cases}\sigma_{\rho_j}(\nu^{(j)}/\nu^{(j-1)}) & \text{if } \nu^{(j)}\approx_{\rho_j}\nu^{(j-1)},\\ 0 & \text{otherwise.}\end{cases}$$

From (1)–(4) we can pick out the coefficient of each s_μ in $h_n^{(\rho)}$. The result may be stated as follows: define a *generalized tableau* of type ρ, shape μ, and weight $n\rho=(n\rho_1,n\rho_2,\dots,n\rho_m)$ to be a sequence $T=(\nu^{(0)},\dots,\nu^{(m)})$ of partitions satisfying the following conditions:

(i) $0=\nu^{(0)}\subset\nu^{(1)}\subset\dots\subset\nu^{(m)}=\mu$;
(ii) $|\nu^{(j)}-\nu^{(j-1)}|=n\rho_j$ for $1\leqslant j\leqslant m$;
(iii) $\nu^{(j)}\approx_{\rho_j}\nu^{(j-1)}$ $(1\leqslant j\leqslant m)$. (§5, Example 24.)
(When $\rho=(1^m)$ these are tableaux in the usual sense, of weight (n^m).)

For such a tableau T let

$$\sigma(T)=\prod_{j=1}^m\sigma_{\rho_j}(\nu^{(j)}/\nu^{(j-1)})=\pm1,$$

and define

$$K_{\mu,n\rho}^{(\rho)}=\sum_T\sigma(T)$$

summed over all generalized tableaux T of type ρ, shape μ, and weight $n\rho$. The integers $K_{\mu,n\rho}^{(\rho)}$ may be regarded as generalized Kostka numbers.

With these definitions, we have

$$p_\rho\circ h_n=h_n^{(\rho)}=\sum_\mu K_{\mu,n\rho}^{(\rho)}s_\mu,$$

summed over partitions μ such that $|\mu|=n|\rho|$ and $l(\mu)\leqslant|\rho|$.

9. Since

$$s_\lambda = \sum_\rho z_\rho^{-1} \chi_\rho^\lambda p_\rho$$

it follows from Example 8 that

$$s_\lambda \circ h_n = \sum_\rho z_\rho^{-1} \chi_\rho^\lambda p_\rho \circ h_n$$

$$= \sum_\mu \left(\sum_\rho z_\rho^{-1} \chi_\rho^\lambda K_{\mu, n\rho}^{(\rho)} \right) s_\mu$$

the outer sum being over partitions μ such that $|\mu| = n|\lambda|$ and $l(\mu) \leqslant |\lambda|$.

(a) When $|\lambda| = 2$ we have (Example 7)

$$h_n^{(2)} = \sum_\mu \sigma_2(\mu) s_\mu$$

summed over $|\mu| = 2n$, $l(\mu) \leqslant 2$, so that $\mu = (2n - j, j)$; $\sigma_2(\mu) = s_\mu(1, -1) = (-1)^j$, and therefore

(1) $$h_n^{(2)} = \sum_{j=0}^{n} (-1)^j s_{(2n-j, j)}.$$

On the other hand,

(2) $$h_n^{(1^2)} = h_n^2 = \sum_{j=0}^{n} s_{(2n-j, j)}.$$

From (1) and (2) it follows that

$$h_2 \circ h_n = \sum_{j \text{ even}} s_{(2n-j, j)},$$

$$e_2 \circ h_n = \sum_{j \text{ odd}} s_{(2n-j, j)}.$$

By duality (Example 1) we obtain

$$h_2 \circ e_n = \sum_{k \text{ even}} s_{(n+k, n-k)'},$$

$$e_2 \circ e_n = \sum_{k \text{ odd}} s_{(n+k, n-k)'}.$$

(b) When $|\lambda| = 3$ we have

$$K_{\mu(n^3)}^{(1^3)} = \text{number of tableaux of shape } \mu \text{ and weight } (n^3)$$

$$= 1 + m(\mu)$$

where $m(\mu) = \min(\mu_1 - \mu_2, \mu_2 - \mu_3)$, and $l(\mu) \leqslant 3$.

Next, since $h_n^{(3)} \equiv h_n^3 \pmod 3$, it follows that

$$K^{(3)}_{\mu(3n)} \equiv 1 + m(\mu) \pmod 3$$

and since $K^{(3)}_{\mu(3n)} = 0$ or ± 1, it is determined by this congruence.

Finally,

$$h_n^{(21)} = h_n^{(2)} h_n = \sum_{j=0}^{n} (-1)^j s_{(2n-j,j)} h_n,$$

from which we obtain

$$K^{(21)}_{\mu(2n,n)} = \begin{cases} 0 & \text{if } m(\mu) \text{ is odd,} \\ (-1)^{\mu_2} & \text{if } m(\mu) \text{ is even.} \end{cases}$$

From these values we obtain

$$h_3 \circ h_n = \sum_\mu ([\tfrac{1}{6}m(\mu)] + \varepsilon(\mu)) s_\mu,$$

$$s_{21} \circ h_n = \sum_\mu \{\tfrac{1}{3}m(\mu)\} s_\mu,$$

$$e_3 \circ h_n = \sum_\mu ([\tfrac{1}{6}m(\mu)] + \varepsilon(\mu) - (-1)^{\mu_2}) s_\mu,$$

summed in each case over partitions μ such that $|\mu| = 3n$ and $l(\mu) \leqslant 3$, where $\varepsilon(\mu) = 1$ if $m(\mu)$ and μ_2 are even or if $m(\mu) \equiv 3$ or $5 \pmod 6$, and $\varepsilon(\mu) = 0$ otherwise; and where $\{x\} = -[-x]$ is the least integer $\geqslant x$.

10. Foulkes conjectured that $h_m \circ h_n \leqslant h_n \circ h_m$ whenever $m \leqslant n$, with respect to the partial ordering on Λ defined in §7, Example 9(b). The results of Example 6 and Example 9(a) show that this is true for $m = 2$ and all $n \geqslant 2$.

11. From §5, Example 4 it follows that

$$\sum_\lambda s_\lambda = \sum_{r,s} h_r (h_s \circ e_2) = \sum_{r,s} e_r (h_s \circ h_2).$$

On passing to representations of S_n, these formulas give

$$\sum_{|\lambda|=n} \chi^\lambda = \sum \eta_r (\eta_s \circ \varepsilon_2) = \sum \varepsilon_r (\eta_s \circ \eta_2) \tag{1}$$

where $r + 2s = n$ in the second and third sums. Now $\eta_s \circ \varepsilon_2$ is the character of S_{2s} induced from the sign character of the wreath product $S_2 \sim S_s$ (the hyperoctahedral group of rank s), hence $\eta_r(\eta_s \circ \varepsilon_2)$ is the character of S_n induced from the character $\eta_r \times (\eta_s \circ \varepsilon_2)$ of the group $S_r \times (S_2 \sim S_s)$, which is the centralizer in S_n of an involution of cycle-type $(2^s 1^r)$, i.e. a product of s disjoint transpositions. It follows that

$$\sum_{|\lambda|=n} \chi^\lambda = \sum_c \xi_c$$

where c runs over the conjugacy classes of elements $w \in S_n$ such that $w^2 = 1$, and ξ_c is the induced representation from the centralizer of an element $w \in c$ just described. The other sum in (1) may be interpreted analogously.

Notes and references

Plethysm was introduced by D. E. Littlewood [L9]. His notation for our $s_\lambda \circ s_\mu$ is $\{\mu\} \otimes \{\lambda\}$. Many authors have computed (or have described algorithms to compute) $s_\lambda \circ s_\mu$ for particular choices of either λ or μ. For their work we refer to the bibliographies in Littlewood [L9] and Robinson [R7], and to [C2]. Examples 9(a) and 9(b) are due to R. M. Thrall [T5]. For the next case ($|\lambda| = 4$) see H. O. Foulkes [F6] and R. Howe [H10].

9. The Littlewood–Richardson rule

If μ and ν are partitions, the product $s_\mu s_\nu$ is an integral linear combination of Schur functions:

$$s_\mu s_\nu = \sum_\lambda c^\lambda_{\mu\nu} s_\lambda$$

or equivalently

$$s_{\lambda/\mu} = \sum_\nu c^\lambda_{\mu\nu} s_\nu \,. \tag{9.1}$$

The coefficients $c^\lambda_{\mu\nu}$ are non-negative integers, because by (7.3) and (7.5) $c^\lambda_{\mu\nu} = \langle \chi^\lambda, \chi^\mu . \chi^\nu \rangle$ is the multiplicity of χ^λ in the character $\chi^\mu . \chi^\nu$; also we have $c^\lambda_{\mu\nu} = 0$ unless $|\lambda| = |\mu| + |\nu|$ and $\mu, \nu \subset \lambda$.

This section is devoted to the statement and proof of a combinatorial rule for computing $c^\lambda_{\mu\nu}$, due to Littlewood and Richardson [L13].

Let T be a tableau. From T we derive a *word* or sequence $w(T)$ by reading the symbols in T from right to left (as in Arabic) in successive rows, starting with the top row. For example, if T is the tableau

	1	1	2	3
	2	3		
1	4			

$w(T)$ is the word 32113241.

If a word w arises in this way from a tableau of shape $\lambda - \mu$, we shall say that w is *compatible* with $\lambda - \mu$.

A word $w = a_1 a_2 \dots a_N$ in the symbols $1, 2, \dots, n$ is said to be a *lattice permutation* if for $1 \leqslant r \leqslant N$ and $1 \leqslant i \leqslant n-1$, the number of occurrences of the symbol i in $a_1 a_2 \dots a_r$ is not less than the number of occurrences of $i+1$.

We can now state the Littlewood–Richardson rule:

(9.2) *Let λ, μ, ν be partitions. Then $c^{\lambda}_{\mu\nu}$ is equal to the number of tableaux T of shape $\lambda - \mu$ and weight ν such that $w(T)$ is a lattice permutation.*

The proof we shall give of (9.2) depends on the following proposition. For any partitions λ, μ, π such that $\lambda \supset \mu$, let $\mathrm{Tab}(\lambda - \mu, \pi)$ denote the set of tableaux T of shape $\lambda - \mu$ and weight π, and let $\mathrm{Tab}^0(\lambda - \mu, \pi)$ denote the subset of those T such that $w(T)$ is a lattice permutation. From (5.14) we have

$$(9.3) \qquad |\mathrm{Tab}(\lambda - \mu, \pi)| = K_{\lambda - \mu, \pi} = \langle s_{\lambda/\mu}, h_\pi \rangle.$$

We shall prove that

(9.4) *There exists a bijection*

$$\mathrm{Tab}(\lambda - \mu, \pi) \xrightarrow{\sim} \coprod_{\nu} (\mathrm{Tab}^0(\lambda - \mu, \nu) \times \mathrm{Tab}(\nu, \pi)).$$

Before proving (9.4), let us deduce (9.2) from it. From (9.4) and (9.3), we have

$$\langle s_{\lambda/\mu}, h_\pi \rangle = \sum_{\nu} |\mathrm{Tab}^0(\lambda - \mu, \nu)| \langle s_\nu, h_\pi \rangle$$

for all partitions π, and therefore

$$s_{\lambda/\mu} = \sum_{\nu} |\mathrm{Tab}^0(\lambda - \mu, \nu)| s_\nu .$$

Comparison of this identity with (9.1) shows that $c^{\lambda}_{\mu\nu} = |\mathrm{Tab}^0(\lambda - \mu, \nu)|$.

To construct a bijection as required for (9.4), we shall follow the method of Littlewood and Robinson [R5], which consists in starting with a tableau T of shape $\lambda - \mu$ and successively modifying it until the word $w(T)$ becomes a lattice permutation, and simultaneously building up a tableau M, which serves to record the sequence of moves made.

If $w = a_1 a_2 \dots a_N$ is any word in the symbols $1, 2, \dots$, let $m_r(w)$ denote the number of occurrences of the symbol r in w. For $1 \leqslant p \leqslant N$ and $r \geqslant 2$, the difference $m_r(a_1 \dots a_p) - m_{r-1}(a_1 \dots a_p)$ is called the *r-index* of a_p in w. Observe that w is a lattice permutation if and only if all indices are $\leqslant 0$.

Let m be the maximum value of the r-indices in w, and suppose that $m > 0$. Take the first element of w at which this maximum is attained (clearly this element will be an r), and replace it by $r-1$. Denote the result of this operation by $S_{r-1,r}(w)$ (substitution for $r-1$ for r). Observe that $S_{r-1,r}(w)$ has maximum r-index $m-1$ (unless $m = 1$, in which case it can be -1).

(9.5) *The operation* $S_{r-1,r}$ *is one-to-one.*

Proof. Let $w' = S_{r-1,r}(w)$. To reconstruct w from w', let m' be the maximum r-index in w'. If $m' \geqslant 0$, take the last symbol in w' with r-index m', and convert the next symbol (which must be an $r-1$) into r. If $m' < 0$, the first symbol in w' must be an $r-1$, and this is converted into r. In either case the result is w, which is therefore uniquely determined by w' and r. |

(9.6) *Let* $w' = S_{r-1,r}(w)$. *Then* w' *is compatible with* $\lambda - \mu$ *if and only if* w *is compatible with* $\lambda - \mu$.

Proof. Let $w = w(T)$, $w' = w(T')$, where T and T' are arrays of shape $\lambda - \mu$. They differ in only one square, say x, which in T is occupied by r and in T' by $r-1$.

Suppose that T is a tableau. If T' is not a tableau there are two possibilities: either (a) the square y immediately to the left of x in T is occupied by r, or (b) the square immediately above x is occupied by $r-1$.

In case (a) the symbol r in square y would have a higher r-index in $w(T)$ than the r in square x, which is impossible. In case (b) the square x in T will be the left-hand end of a string of say s squares occupied by the symbol r, and immediately above this string there will be a string of s squares occupied by the symbol $r-1$. It follows that $w(T)$ contains a segment of the form

$$(r-1)^s \ldots r^s$$

where the unwritten symbols in between the two strings are all either $> r$ or $< r-1$, and the last r is the one to be replaced by $r-1$ to form w'. But the r-index of this r is equal to that of the element of w immediately preceding the first of the string of $(r-1)$'s, and this again is impossible. Hence if T is a tableau, so also is T'.

The reverse implication is proved similarly, using the recipe of (9.5) for passing back from w' to w. |

Suppose now that the word w has the lattice permutation property with respect to $(1, 2, \ldots, r-1)$ but not with respect to $(r-1, r)$, or in other words that all the s-indices are $\leqslant 0$ for $2 \leqslant s \leqslant r-1$ but not for $s = r$. This is the only situation in which we shall use the operator $S_{r-1,r}$. The effect of replacing r by $r-1$ in w as required by $S_{r-1,r}$ may destroy the lattice permutation property with respect to $(r-2, r-1)$, i.e. it may produce some $(r-1)$-indices equal to $+1$. In this case we operate with $S_{r-2,r-1}$ to produce

$$S_{r-2,r}(w) = S_{r-2,r-1} S_{r-1,r}(w).$$

At this stage the $(r-1)$-indices will all be $\leqslant 0$, but there may be some $(r-2)$-indices equal to $+1$, and so on. Eventually this process will stop, and we have then say

$$S_{a,r}(w) = S_{a,a+1} \ldots S_{r-1,r}(w)$$

for some a such that $1 \leqslant a \leqslant r-1$, and the word $S_{a,r}(w)$ again has the lattice property with respect to $(1, 2, \ldots, r-1)$, and maximal r-index strictly less than that of w.

At this point the following lemma is crucial:

(9.7) *If* w, $S_{a,r}(w) = w'$ *and* $S_{b,r}(w') = w''$ *all have the lattice property with respect to* $(1, 2, \ldots, r-1)$, *then* $b \leqslant a$.

Proof. Let $w = x_1 x_2 x_3 \ldots$. We have to study in detail the process of passing from w to w'. This starts by applying $S_{r-1,r}$, i.e. by replacing the first symbol r in w with r-index m, where m is the maximum of the r-indices, by $r-1$. Suppose that this happens at x_{p_0}. Then for each $s \geqslant 1$, the $(r-1)$-index of x_s is unaltered if $s < p_0$, and is increased by 1 if $s \geqslant p_0$. The element on which $S_{r-2,r-1}$ operates is therefore in the p_1th place, where p_1 is the first integer $\geqslant p_0$ for which x_{p_1} has $(r-1)$-index in w equal to 0. Likewise the element on which $S_{r-3,r-2}$ operates is in the p_2th place, where p_2 is the first integer $\geqslant p_1$ for which x_{p_2} has $(r-2)$-index zero, and so on.

In this way we obtain a sequence

$$p_0 \leqslant p_1 \leqslant \ldots \leqslant p_{r-a-1}$$

with the property that, for each $i \geqslant 1$, x_{p_i} is the first element not preceding $x_{p_{i-1}}$ for which the $(r-i)$-index is 0. Observe that in w' the element in the p_ith place still has $(r-i)$-index zero, for each $i \geqslant 1$ (though it will no longer be the first with this property).

Now consider the passage from $w' = y_1 y_2 y_3 \ldots$ to w''. In w' the maximum r-index is $m-1$ (which by assumption is still positive) and occurs first at say y_{q_0}, where $q_0 < p_0$. (This is because the r-index can by its definition only go up or down in single steps, and therefore the r-index $m-1$ occurs first in w at some element to the left of x_{p_0}; and the elements to the left of the p_0th are the same in w' as in w.) In w' the $(r-1)$-index of y_{p_1} is zero, and is therefore $+1$ in $S_{r-1,r}(w')$. Hence $S_{r-1,r}(w')$ admits the substitution $S_{r-2,r-1}$, which will operate on the element in the q_1th place, where q_1 is the first integer $\geqslant q_0$ for which the $(r-1)$-index of y_{q_1} in w' is 0, so that $q_0 \leqslant q_1 < p_1$. Continuing in this way we get a sequence

$$q_0 \leqslant q_1 \leqslant q_2 \leqslant \ldots \leqslant q_{r-a-1}$$

with $q_i < p_i$ for all $i \geqslant 0$, and w' admits the operator $S_{a,r}$.

If $S_{a,r}(w') = w''$, then $b = a$; if not, then $S_{a,r}(w')$ admits further substitutions $S_{a-1,a}, \ldots$, until $w'' = S_{b,r}(w')$ is attained, so that $b < a$ in this case. In either case we have $b \leqslant a$, as required. |

We shall now describe the algorithm of Littlewood and Robinson which constructs from a tableau T of shape $\lambda - \mu$ and weight π, where λ, μ, π are partitions, a pair (L, M) where $L \in \mathrm{Tab}^0(\lambda - \mu, \nu)$ for some partition ν, and $M \in \mathrm{Tab}(\nu, \pi)$.

If A is any array—not necessarily a tableau—and a, r are positive integers such that $a < r$, we denote by $R_{a,r}(A)$ the result of raising the right-hand element of the rth row of A up to the right-hand end of the ath row.

The algorithm begins with the word $w_1 = w(T)$ and the array M_1 consisting of π_1 1's in the first row, π_2 2's in the second row, and so on (i.e. M_1 is the unique tableau of shape π and weight π).

Operate on w_1 with S_{12} until there are no positive 2-indices, and simultaneously on M_1 with R_{12} the same number of times: say

$$w_2 = S_{12}^m(w_1), \qquad M_2 = R_{12}^m(M_1).$$

Next operate on w_2 with S_{23} or S_{13} as appropriate until there are no positive 2- or 3-indices, and simultaneously operate on M_2 with R_{23} or R_{13}: say

$$w_3 = \ldots S_{a_2,3} S_{a_1,3}(w_2), \qquad M_3 = \ldots R_{a_2,3} R_{a_1,3}(M_2)$$

where each a_1, a_2 is 1 or 2.

Continue in this way until we reach (w_l, M_l), where $l = l(\pi)$. Clearly from our construction w_l is a lattice permutation. From (9.6) it follows that w_l is compatible with $\lambda - \mu$, so that $w_l = w(L)$ where $L \in \mathrm{Tab}^0(\lambda - \mu, \nu)$ for some partition ν. Next, it is clear from the construction that at each stage the length $l_i(M_r)$ of the ith row of the array M_r is equal to the multiplicity $m_i(w_r)$ of the symbol i in the corresponding word w_r, so that the final array $M = M_l$ has shape ν and weight π.

We have to show moreover that M_l is a *tableau*. For this, we shall prove by induction on r that the first r rows of M_r form a tableau. This is clear if $r = 1$, so assume that $r > 1$ and the result is true for $r - 1$.

Consider the steps that lead from M_{r-1} to M_r: we have, say,

$$M_r = R_{a_m,r} \ldots R_{a_1,r}(M_{r-1});$$

let us put

$$M_{r-1,i} = R_{a_i,r} \ldots R_{a_1,r}(M_{r-1}) \qquad (1 \leqslant i \leqslant m)$$

and likewise

$$w_{r-1,i} = S_{a_i,r} \ldots S_{a_1,r}(w_{r-1}),$$

where each word $w_{r-1,i}$ has the lattice property with respect to $(1,2,\ldots, r-1)$. Each array $M_{r-1,i}$ is obtained from its predecessor $M_{r-1,i-1}$ (or M_{r-1} if $i = 1$) by moving up a single symbol r from the rth row to the a_ith row. By our construction the length $l_j(M_{r-1,i})$ of the jth row of $M_{r-1,i}$ is equal to the multiplicity $m_j(w_{r-1,i})$ of j in $w_{r-1,i}$, for each $j \geqslant 1$; and since each word $w_{r-1,i}$ has the lattice property with respect to $(1,2,\ldots,r-1)$, it follows that

$$l_1(M_{r-1,i}) \geqslant \ldots \geqslant l_{r-1}(M_{r-1,i}).$$

Also, by (9.7), the integers a_i satisfy $a_1 \geqslant \ldots \geqslant a_m$. It follows that no two symbols r can appear in the same column at any stage, and consequently the first r rows of M_r form a tableau.

The algorithm therefore provides a mapping

$$\mathrm{Tab}(\lambda - \mu, \pi) \to \coprod_{\nu} \mathrm{Tab}^0(\lambda - \mu, \nu) \times \mathrm{Tab}(\nu, \pi).$$

To complete the proof of (9.4) we have to show that this mapping is a bijection. For this purpose it is enough to show that, for each $r \geqslant 1$, we can unambiguously trace our steps back from (w_r, M_r) to (w_{r-1}, M_{r-1}). With the notation used above, we have

$$w_r = S_{a_m,r} \ldots S_{a_1,r}(w_{r-1}),$$

and the sequence $(a_1,\ldots,a_m)$ can be read off from the array M_r, since the a_i are the indices $< r$ of the rows in which the symbols r are located in M_r, arranged in descending order: $a_1 \geqslant a_2 \geqslant \ldots \geqslant a_m$ (by virtue of (9.7)). Since by (9.5) each $S_{a,r}$ is reversible, it follows that (w_{r-1}, M_{r-1}) is uniquely determined by (w_r, M_r). Finally, by (9.6), if w_r is compatible with $\lambda - \mu$, then so also is w_{r-1}, and the proof is complete. Q.E.D.

Remark. A lattice permutation $w = a_1 a_2 \ldots a_N$ of weight ν may be described by a standard tableau $T(w)$ of shape ν, in which the symbol r occurs in the a_rth row, for $1 \leqslant r \leqslant N$ (the fact that w is lattice ensures that $T(w)$ is a tableau). Hence the algorithm described above constructs from a word w a pair of tableaux $T(w_l)$ and M_l of the same shape ν, the first of which is standard and the second of weight π. It may be verified that this algorithm coincides with one described by Burge [B9] (see also Gansner [G1]).

Notes and references

The Littlewood–Richardson rule (9.2) was first stated, but not proved, in [L13] (p. 119). The proof subsequently published by Robinson [R5], and reproduced in Littlewood's book ([L9], pp. 94–6) is incomplete, and it is this proof that we have endeavoured to complete.

Complete proofs of the rule first appeared in the 1970s ([S7], [T4]).† Since then, many other formulations, proofs and generalizations have appeared, some of which are covered by the following references: Bergeron and Garsia [B4]; James [J7]; James and Peel [J10]; James and Kerber [J9]; Kerov [K8]; Littelmann [L7], [L8]; White [W3]; and Zelevinsky [Z2], [Z3].

† Gordon James [J8] reports that he was once told that 'the Littlewood–Richardson rule helped to get men on the moon, but it was not proved until after they had got there. The first part of this story might be an exaggeration.'

APPENDIX A: Polynomial functors and polynomial representations

1. Introduction

Let k be a field of characteristic 0 and let $\mathfrak{V}$ denote the category whose objects are finite-dimensional k-vector spaces and whose morphisms are k-linear maps. A (covariant) functor $F: \mathfrak{V} \to \mathfrak{V}$ will be said to be a *polynomial functor* if, for each pair of k-vector spaces X, Y, the mapping $F: \mathrm{Hom}(X,Y) \to \mathrm{Hom}(FX, FY)$ is a polynomial mapping. This condition may be expressed as follows:

(1.1) *Let* $f_i: X \to Y$ $(1 \leqslant i \leqslant r)$ *be morphisms in* $\mathfrak{V}$, *and let* $\lambda_1, \dots, \lambda_r \in k$. *Then* $F(\lambda_1 f_1 + \dots + \lambda_r f_r)$ *is a polynomial function of* $\lambda_1, \dots, \lambda_r$, *with coefficients in* $\mathrm{Hom}(FX, FY)$ *(depending on* $f_1, \dots, f_r$*).*

If $F(\lambda_1 f_1 + \dots + \lambda_r f_r)$ is homogeneous of degree n, for all choices of $f_1, \dots, f_r$, then F is said to be *homogeneous of degree n*. For example, the nth exterior power $\wedge^n$ and the nth symmetric power $\mathbf{S}^n$ are homogeneous polynomial functors of degree n.

Each polynomial functor F is a direct sum $\oplus_{n \geqslant 0} F_n$, where F_n is homogeneous of degree n (§2). We shall show that each F_n determines a representation of the symmetric group S_n on a finite-dimensional k-vector space E_n, such that

$$F_n(X) \cong (E_n \otimes X^{\otimes n})^{S_n}$$

functorially in X, and that $F_n \mapsto E_n$ defines an equivalence of the category of homogeneous polynomial functors of degree n with the category of finite-dimensional $k[S_n]$-modules. In particular, the irreducible polynomial functors correspond to the irreducible representations of symmetric groups, hence are indexed by partitions.

The connection with symmetric functions is the following. Let $u: k^m \to k^m$ be a semisimple endomorphism, with eigenvalues $\lambda_1, \dots, \lambda_m$ (in some extension of k). Then trace $F(u)$ is a symmetric polynomial function of $\lambda_1, \dots, \lambda_m$, say $\chi_m(F)(\lambda_1, \dots, \lambda_m)$, where $\chi_m(F) \in \Lambda_m$. As $m \to \infty$, the $\chi_m(F)$ determine an element $\chi(F) \in \Lambda$. If $F = F_\mu$ is irreducible (where μ is a partition), it will appear that $\chi(F_\mu)$ is the Schur function s_μ.

Notation. If $X \in \mathfrak{V}$ and $\lambda \in k$, we shall denote by λ_X (or just λ if the context permits) multiplication by λ in X.

2. Homogeneity

Let F be a polynomial functor on $\mathfrak{V}$, and let $\lambda \in k$. By (1.1), $F(\lambda_X)$ is a polynomial function of λ with coefficients in $\mathrm{End}(F(X))$, say

$$F(\lambda_X) = \sum_{n \geqslant 0} u_n(X)\lambda^n. \tag{2.1}$$

Since $F((\lambda\mu)_X) = F(\lambda_X \mu_X) = F(\lambda_X)F(\mu_X)$, we have

$$\sum_{n \geqslant 0} u_n(X)(\lambda\mu)^n = \Big(\sum_{n \geqslant 0} u_n(X)\lambda^n\Big)\Big(\sum_{n \geqslant 0} u_n(X)\mu^n\Big)$$

for all $\lambda, \mu \in k$, and therefore (because k is an infinite field) $u_n(X)^2 = u_n(X)$ for all $n \geqslant 0$, and $u_m(X)u_n(X) = 0$ if $m \neq n$. Also

$$\sum_{n \geqslant 0} u_n(X) = F(1_X) = 1_{F(X)}$$

by taking $\lambda = 1$ in (2.1). It follows that the $u_n(X)$ determine a direct sum decomposition

$$F(X) = \bigoplus_{n \geqslant 0} F_n(X) \tag{2.2}$$

where $F_n(X)$ is the image of $u_n(X)\colon F(X) \to F(X)$. Since $F(X)$ is finite-dimensional, all but a finite number of the summands $F_n(X)$ in (2.2) will be zero, for any given X.

Moreover, if $f\colon X \to Y$ is a k-linear map, we have $f\lambda_X = \lambda_Y f$ for all $\lambda \in k$, and hence $F(f)F(\lambda_X) = F(\lambda_Y)F(f)$. From (2.1) it follows that $F(f)u_n(X) = u_n(Y)F(f)$ for all $n \geqslant 0$, so that each u_n is an endomorphism of the functor F. Hence $F(f)$ defines by restriction k-linear maps $F_n(f)\colon F_n(X) \to F_n(Y)$, and therefore each F_n is a *functor*, which is clearly polynomial. Consequently we have a direct decomposition

$$F = \bigoplus_{n \geqslant 0} F_n \tag{2.3}$$

in which each F_n is a homogeneous polynomial functor of degree n.

Remarks. 1. The direct sum (2.3) may well have infinitely many non-zero components, although for any given $X \in \mathfrak{V}$ we must have $F_n(X) = 0$ for all sufficiently large n. An example is the exterior algebra functor $\wedge$.

If $F_n = 0$ for all sufficiently large n, we shall say that F has *bounded degree*.

2. F_0 is homogeneous of degree 0, so that $F_0(\lambda) = 1$ for all $\lambda \in k$, and in particular $F_0(0) = F_0(1)$. It follows that for all morphisms $f\colon X \to Y$ we have $F_0(f) = F_0(0)$, which is therefore independent of f and is an isomorphism of $F_0(X)$ onto $F_0(Y)$. Hence all the objects $F_0(X)$ are canonically isomorphic.

More generally, let r be a positive integer and $\mathfrak{B}^r = \mathfrak{B} \times \ldots \times \mathfrak{B}$ the category in which the objects are sequences $X = (X_1, \ldots, X_r)$ of length r of objects of $\mathfrak{B}$, and $\mathrm{Hom}(X,Y) = \prod_i \mathrm{Hom}(X_i, Y_i)$. As before, a functor $F: \mathfrak{B}^r \to \mathfrak{B}$ will be said to be *polynomial* if $F: \mathrm{Hom}(X,Y) \to \mathrm{Hom}(FX, FY)$ is a polynomial mapping for all $X, Y \in \mathfrak{B}^r$. If F is polynomial and $(\lambda) = (\lambda_1, \ldots, \lambda_r) \in k^r$, then $F((\lambda)_X) = F((\lambda_1)_{X_1}, \ldots, (\lambda_r)_{X_r})$ will be a polynomial function of $\lambda_1, \ldots, \lambda_r$ with coefficients in $\mathrm{End}(F(X))$, say

$$F((\lambda)_X) = \sum_{m_1, \ldots, m_r} u_{m_1 \ldots m_r}(X_1, \ldots, X_r) \lambda_1^{m_1} \ldots \lambda_r^{m_r}. \tag{2.4}$$

Exactly as before, we see that the $u_{m_1 \ldots m_r}$ are endomorphisms of F, and that if we denote the image of $u_{m_1 \ldots m_r}(X_1, \ldots, X_r)$ by $F_{m_1 \ldots m_r}(X_1, \ldots, X_r)$, then the $F_{m_1 \ldots m_r}$ are subfunctors of F which give rise to a direct decomposition

$$F = \bigoplus_{m_1, \ldots, m_r} F_{m_1 \ldots m_r}. \tag{2.5}$$

Each $F_{m_1 \ldots m_r}$ is homogeneous of multidegree $(m_1, \ldots, m_r)$, i.e. we have $F_{m_1 \ldots m_r}(\lambda_1, \ldots, \lambda_r) = \lambda_1^{m_1} \ldots \lambda_r^{m_r}$.

3. Linearization

Again let $F: \mathfrak{B} \to \mathfrak{B}$ be a polynomial functor. In view of the decomposition (2.3), we shall assume from now on that F is *homogeneous* of degree $n > 0$. The considerations at the end of §2 apply to the functor $F': \mathfrak{B}^n \to \mathfrak{B}$ defined by $F'(X_1, \ldots, X_n) = F(X_1 \oplus \ldots \oplus X_n)$, and show that there exists a direct sum decomposition, functorial in each variable,

$$F(X_1 \oplus \ldots \oplus X_n) = \oplus F'_{m_1 \ldots m_n}(X_1, \ldots, X_n)$$

where the direct sum on the right is over all $(m_1, \ldots, m_n) \in \mathbf{N}^n$ such that $m_1 + \ldots + m_n = n$.

Our main interest will be in the functor $F'_{1 \ldots 1}$, the image of the morphism $u_{1 \ldots 1}$. For brevity, we shall write L_F and v in place of $F'_{1 \ldots 1}$ and $u_{1 \ldots 1}$, respectively. We call L_F the *linearization* of F: it is homogeneous of degree 1 in each variable.

To recapitulate the definitions of L_F and v, let $Y = X_1 \oplus \ldots \oplus X_n$. Then there are monomorphisms $i_\alpha: X_\alpha \to Y$ and epimorphisms $p_\alpha: Y \to X_\alpha$ $(1 \leqslant \alpha \leqslant n)$, satisfying

$$p_\alpha i_\alpha = 1_{X_\alpha}, \qquad p_\alpha i_\beta = 0 \quad \text{if} \quad \alpha \neq \beta, \qquad \sum_\alpha i_\alpha p_\alpha = 1_Y. \tag{3.1}$$

For each $\lambda = (\lambda_1, \ldots, \lambda_n) \in k^n$, let $(\lambda)_Y$ or (λ) denote the morphism $\sum \lambda_\alpha i_\alpha p_\alpha: Y \to Y$, so that (λ) acts as scalar multiplication by λ_α on the component X_α. Then $v(X_1,\ldots,X_n)$ is by definition the coefficient of $\lambda_1\ldots\lambda_n$ in $F((\lambda)_Y)$, and $L_F(X_1,\ldots,X_n)$ is the image of $v(X_1,\ldots,X_n)$, and is a direct summand of $F(X_1\oplus\ldots\oplus X_n)$.

(3.2) *Example.* If F is the nth exterior power Λ^n, we have

$$F(X_1\oplus\ldots\oplus X_n)\cong\bigoplus_{m_1,\ldots,m_n}\Lambda^{m_1}(X_1)\oplus\ldots\oplus\Lambda^{m_n}(X_n)$$

summed over all $(m_1,\ldots,m_n)\in\mathbf{N}^n$ such that $m_1+\ldots+m_n=n$, and hence $L_F(X_1,\ldots,X_n)\cong X_1\otimes\ldots\otimes X_n$.

4. The action of the symmetric group

Let F as before be homogeneous of degree $n>0$, and let

$$L_F^{(n)}(X)=L_F(X,\ldots,X).$$

For each element s of the symmetric group S_n, let s_X or s denote the morphism $\sum i_{s(\alpha)}p_\alpha: X^n\to X^n$, where $X^n=X\oplus\ldots\oplus X$, so that s_X permutes the summands of X^n. For any $\lambda=(\lambda_1,\ldots,\lambda_n)\in k^n$ we have from (3.1)

$$s_X(\lambda)=\sum_\alpha\lambda_\alpha i_{s(\alpha)}p_\alpha=(s\lambda).s_X$$

where $s\lambda=(\lambda_{s^{-1}(1)},\ldots,\lambda_{s^{-1}(n)})$, and hence

$$F(s)F((\lambda))=F((s\lambda)).F(s). \tag{4.1}$$

By picking out the coefficient of $\lambda_1\ldots\lambda_n$ on either side, we see that

$$F(s)v=vF(s) \tag{4.2}$$

from which it follows that $F(s)$ defines by restriction an endomorphism $\tilde{F}(s)$ of $L_F^{(n)}$. Explicitly, if

$$j=j_X: L_F^{(n)}(X)\to F(X^n),\qquad q=q_X: F(X^n)\to L_F^{(n)}(X) \tag{4.3}$$

are the injection and projection associated with the direct summand $L_F^{(n)}(X)$ of $F(X^n)$, so that $qj=1$ and $jq=v$, then

$$\tilde{F}(s)=qF(s)j. \tag{4.4}$$

From (4.2) and (4.4) it follows that $\tilde{F}(st)=\tilde{F}(s)\tilde{F}(t)$ for $s,t\in S_n$, so that $s\mapsto\tilde{F}(s)$ is a representation of S_n on the vector space $L_F^{(n)}(X)$, functorial in X.

We shall now show that this representation of S_n determines the functor F up to isomorphism, and more precisely that there exists a functorial isomorphism of $F(X)$ onto the subspace of S_n-invariants of $L_F^{(n)}(X)$.

Example. In the example (3.2) we have $L_F^{(n)}(X) = \mathbf{T}^n(X)$, the nth tensor power of X over k, and the action of S_n on $L_F^{(n)}(X)$ in this case is given by

$$\tilde{F}(s)(x_1 \otimes \ldots \otimes x_n) = \varepsilon(s) x_{s^{-1}(1)} \otimes \ldots \otimes x_{s^{-1}(n)}$$

where $\varepsilon(s)$ is the sign of $s \in S_n$. Hence $L_F^{(n)}(X)^{S_n}$ is the space of skew-symmetric tensors in $\mathbf{T}^n(X)$, which is isomorphic to $\Lambda^n(X)$ since k has characteristic 0.

Let $i = \sum i_\alpha : X \to X^n$, $p = \sum p_\alpha : X^n \to X$. Then we have

(4.5) $$vF(ip)v = \sum_{s \in S_n} F(s)v.$$

Proof. Consider linear transformations $f: X^n \to X^n$ of the form $f = \sum_{\alpha,\beta} \xi_{\alpha\beta} i_\alpha p_\beta$, with $\xi_{\alpha\beta} \in k$; $F(f)$ will be a homogeneous polynomial of degree n in the n^2 variables $\xi_{\alpha\beta}$, with coefficients in $\mathrm{End}(F(X^n))$ depending only on X (and F). For each $s \in S_n$, let w_s denote the coefficient of $\xi_{s(1)1} \cdots \xi_{s(n)n}$ in $F(f)$.

We have $F(s)v = vF(s)v$ by (4.2) (since $v^2 = v$), hence $F(s)v$ is the coefficient of $\lambda_1 \ldots \lambda_n \mu_1 \ldots \mu_n$ in

$$F((\lambda))F(s)F((\mu)) = F((\lambda)s(\mu)) = F\Big(\sum_\alpha \lambda_{s(\alpha)} \mu_\alpha i_{s(\alpha)} p_\alpha\Big)$$

and therefore $F(s)v = w_s$.

On the other hand, $vF(ip)v$ is the coefficient of $\lambda_1 \ldots \lambda_n \mu_1 \ldots \mu_n$ in

$$F((\lambda))F(ip)F((\mu)) = F((\lambda)ip(\mu)) = F\Big(\sum_{\alpha,\beta} \lambda_\alpha \mu_\beta i_\alpha p_\beta\Big)$$

and this coefficient is clearly

$$\sum_{s \in S_n} w_s = \sum_s F(s)v. \quad |$$

We now define two morphisms of functors:

$$\xi = qF(i): F \to L_F^{(n)}, \qquad \eta = F(p)j: L_F^{(n)} \to F.$$

(4.6) *We have* $\eta\xi = n!$ (i.e. scalar multiplication by $n!$) *and* $\xi\eta = \sum_{s \in S_n} \tilde{F}(s)$.

Proof. $\eta\xi = F(p)jqF(i) = F(p)vF(i)$ is the coefficient of $\lambda_1 \dots \lambda_n$ in $F(p)F((\lambda))F(i) = F(p(\lambda)i)$. Now $p(\lambda)i: X \to Y$ is scalar multiplication by $\lambda_1 + \dots + \lambda_n$, so that $F(p(\lambda)i)$ is scalar multiplication by $(\lambda_1 + \dots + \lambda_n)^n$, and the coefficient of $\lambda_1 \dots \lambda_n$ is therefore $n!$.

Next, we have $\xi\eta = qF(i)F(p)j$, so that by (4.3) and (4.5)

$$j\xi\eta q = vF(ip)v = \sum_s F(s)v$$

and hence $\xi\eta = \Sigma_s qF(s)j = \Sigma_s \tilde{F}(s)$ by (4.4). |

Let $L_F^{(n)}(X)^{S_n}$ denote the subspace of S_n-invariants in $L_F^{(n)}(X)$. From (4.6) it follows that $\sigma = (n!)^{-1}\xi\eta$ is idempotent, with image $L_F^{(n)}(X)^{S_n}$. Let

$$\varepsilon: L_F^{(n)}(X)^{S_n} \to L_F^{(n)}(X), \qquad \pi: L_F^{(n)}(X) \to L_F^{(n)}(X)^{S_n}$$

be the associated injection and projection, so that $\pi\varepsilon = 1$ and $\varepsilon\pi = \sigma$. From (4.6) we have immediately

(4.7) *The morphisms*

$$\xi' = \pi\xi: F(X) \to L_F^{(n)}(X)^{S_n},$$

$$\eta' = \eta\varepsilon: L_F^{(n)}(X)^{S_n} \to F(X)$$

are functorial isomorphisms such that $\xi'\eta' = n!$ *and* $\eta'\xi' = n!$.

5. Classification of polynomial functors

It follows from (4.7) that every homogeneous polynomial functor of degree n is of the form $X \mapsto L(X, \dots, X)^{S_n}$, where $L: \mathfrak{V}^n \to \mathfrak{V}$ is homogeneous of degree 1 in each variable. The next step, therefore, is to find all such functors.

We shall begin with the case $n = 1$, so that $L: \mathfrak{V} \to \mathfrak{V}$ is homogeneous of degree 1. From (1.1), L is clearly additive: $L(f_1 + f_2) = L(f_1) + L(f_2)$.

(5.1) *There exists a functorial isomorphism*

$$L(X) \cong L(k) \otimes X.$$

Proof. For each $x \in X$ let $e(x): k \to X$ denote the mapping $\lambda \mapsto \lambda x$. Let Y be any k-vector space, and define

$$\psi_X: \mathrm{Hom}(L(X), Y) \to \mathrm{Hom}(X, \mathrm{Hom}(L(k), Y))$$

by $\psi_X(f)(x) = f \circ L(e(x))$. Clearly ψ_X is functorial in X, and to prove (5.1) it is enough to show that ψ_X is an isomorphism.

Since L is additive we have $L(X_1 \oplus X_2) \cong L(X_1) \oplus L(X_2)$. It follows that if ψ_{X_1} and ψ_{X_2} are isomorphisms, $\psi_{X_1 \oplus X_2}$ is also an isomorphism. Hence it is enough to verify that ψ_k is an isomorphism, and this is obvious. |

Now let $L: \mathfrak{V}^n \to \mathfrak{V}$ be homogeneous and linear in each variable.

(5.2) *There exists a functorial isomorphism*

$$L(X_1,\ldots,X_n) \cong L^{(n)}(k) \otimes X_1 \otimes \ldots \otimes X_n$$

where $L^{(n)}(k) = L(k,\ldots,k)$.

Proof. By repeated applications of (5.1),

$$\begin{aligned} L(X_1,\ldots,X_n) &\cong L(X_1,\ldots,X_{n-1},k) \otimes X_n \\ &\cong L(X_1,\ldots,X_{n-2},k,k) \otimes X_{n-1} \otimes X_n \\ &\quad \cdots\cdots \quad \cdots\cdots \\ &\cong L(k,\ldots,k) \otimes X_1 \otimes X_2 \otimes \ldots \otimes X_n. \end{aligned}$$ |

From (5.2) and (4.7) we have immediately

(5.3) *Let F be a homogeneous polynomial functor of degree n. Then there exists an isomorphism of functors*

$$F(X) \cong (L_F^{(n)}(k) \otimes \mathbf{T}^n(X))^{S_n},$$

where $\mathbf{T}^n(X) = X \otimes \ldots \otimes X$ is the nth tensor power of X. |

Let $\mathfrak{F}_n$ denote the category of homogeneous polynomial functors of degree n, and $\mathfrak{V}_{S_n}$ the category of finite-dimensional $k[S_n]$-modules.

(5.4) *The functors $\alpha: \mathfrak{F}_n \to \mathfrak{V}_{S_n}$, $\beta: \mathfrak{V}_{S_n} \to \mathfrak{F}_n$ defined by*

$$\alpha(F) = L_F^{(n)}(k), \qquad \beta(M)(X) = (M \otimes \mathbf{T}^n(X))^{S_n}$$

constitute an equivalence of categories.

Proof. We have $\beta\alpha \cong 1_{\mathfrak{F}_n}$ by (5.3), and we have to verify that $\alpha\beta \cong 1_{\mathfrak{V}_{S_n}}$. If $M \in \mathfrak{V}_{S_n}$ and $\beta(M) = F$, then

$$F(X_1 \oplus \ldots \oplus X_n) = (M \otimes \mathbf{T}^n(X_1 \oplus \ldots \oplus X_n))^{S_n}$$

and therefore

$$L_F(X_1,\dots,X_n) = \left(M \otimes \left(\bigoplus_{s\in S_n} X_{s(1)} \otimes \dots \otimes X_{s(n)}\right)\right)^{S_n}.$$

Hence

$$\alpha\beta(M) = L_F^{(n)}(k) \cong (M \otimes k[S_n])^{S_n} \cong M. \quad |$$

It follows that α and β establish a one–one correspondence between the isomorphism classes of homogeneous polynomial functors of degree n, and the isomorphism classes of finite-dimensional $k[S_n]$-modules.

In particular, the irreducible polynomial functors (which by (2.3) are necessarily homogeneous) correspond to the irreducible representations of the symmetric groups S_n, and are therefore naturally indexed by partitions.

For each partition λ of n, let M_λ be an irreducible $k[S_n]$-module with character χ^λ. From (5.4) the irreducible polynomial functor F_λ indexed by λ is given by

(5.5) $$F_\lambda(X) = (M_\lambda \otimes \mathbf{T}^n(X))^{S_n}.$$

Now if G is any finite group and U, V are finite-dimensional $k[G]$-modules, there is a canonical isomorphism $(U^* \otimes V)^G \cong \operatorname{Hom}_{k[G]}(U,V)$, functorial in both U and V, where $U^* = \operatorname{Hom}_k(U,k)$ is the contragredient of U. In the present context we have $M_\lambda^* \cong M_\lambda$, and therefore

(5.5′) $$F_\lambda(X) \cong \operatorname{Hom}_{k[S_n]}(M_\lambda, \mathbf{T}^n(X)).$$

Consider $\mathbf{T}^n(X)$ as a $k[S_n]$-module. Its decomposition into isotypic components is

$$\mathbf{T}^n(X) \cong \bigoplus_\lambda M_\lambda \otimes \operatorname{Hom}_{k[S_n]}(M_\lambda, \mathbf{T}^n(X))$$

so that

(5.6) $$\mathbf{T}^n(X) \cong \bigoplus_\lambda M_\lambda \otimes F_\lambda(X)$$

functorially in X.

6. Polynomial functors and $k[S_n]$-modules

We shall need the following facts. Let G, H be finite groups, M a finite-dimensional $k[G]$-module, and N a finite-dimensional $k[H]$-module. Then $M \otimes N$ is a $k[G \times H]$-module, and we have

(A) $$M^G \otimes N^H = (M \otimes N)^{G\times H}$$

as subspaces of $M \otimes N$.

Suppose now that H is a subgroup of G, and let $\mathrm{ind}_H^G N = k[G] \otimes_{k[H]} N$ be the $k[G]$-module induced by N, and $\mathrm{res}_G^H M$ the module M regarded as a $k[H]$-module by restriction of scalars. Then

$$\text{(B)} \qquad N^H \cong (\mathrm{ind}_H^G N)^G.$$

This is a particular case of Frobenius reciprocity:

$$\mathrm{Hom}_{k[H]}(\mathrm{res}_G^H M, N) \cong \mathrm{Hom}_{k[G]}(M, \mathrm{ind}_H^G N)$$

in which M is taken to be the trivial one-dimensional $k[G]$-module.

Suppose again that H is a subgroup of G. Then

$$\text{(C)} \qquad \mathrm{ind}_H^G(N \otimes \mathrm{res}_G^H M) \cong (\mathrm{ind}_H^G N) \otimes M.$$

For both sides are canonically isomorphic to $k[G] \otimes_{k[H]} N \otimes_k M$.

Finally, suppose that H is a normal subgroup of G, that M is a finite-dimensional $k[G]$-module and L a finite-dimensional $k[G/H]$-module, hence a $k[G]$-module on which H acts trivially. Then

$$\text{(D)} \qquad (L \otimes M^H)^{G/H} = (L \otimes M)^G$$

as subspaces of $L \otimes M$.

For $L \otimes M^H = (L \otimes M)^H$ since H acts trivially on L, hence

$$(L \otimes M^H)^{G/H} = ((L \otimes M)^H)^{G/H} = (L \otimes M)^G.$$

Now let E, F be homogeneous polynomial functors, of degrees m and n respectively. Then

$$E \otimes F : X \mapsto E(X) \otimes F(X)$$

is a homogeneous polynomial functor of degree $m + n$, hence corresponds as in §5 to a representation of S_{m+n}.

Suppose that $E = \beta(M)$, $F = \beta(N)$ in the notation of (5.4). Then

$$\begin{aligned}(E \otimes F)(X) &= (M \otimes \mathbf{T}^m(X))^{S_m} \otimes (N \otimes \mathbf{T}^n(X))^{S_n} \\ &\cong (M \otimes N \otimes \mathbf{T}^{m+n}(X))^{S_m \times S_n} \quad \text{by (A)} \\ &\cong \left(\mathrm{ind}_{S_m \times S_n}^{S_{m+n}}(M \otimes N) \otimes \mathbf{T}^{m+n}(X)\right)^{S_{m+n}} \quad \text{by (B) and (C)}\end{aligned}$$

so that $E \otimes F$ corresponds to the $k[S_{m+n}]$-module

$$\text{(6.1)} \qquad M \,.\, N = \mathrm{ind}_{S_m \times S_n}^{S_{m+n}}(M \otimes N)$$

which we call the *induction product* of M and N.

Since tensor products are commutative and associative up to isomorphism, so is the induction product.

Consider next the composition of polynomial functors. With E, F as before we have

$$\begin{aligned}(E \circ F)(X) &= E((N \otimes \mathbf{T}^n(X))^{S_n}) \\ &= \left(M \otimes \mathbf{T}^m((N \otimes \mathbf{T}^n(X))^{S_n})\right)^{S_m} \\ &\cong \left(M \otimes (\mathbf{T}^m(N) \otimes \mathbf{T}^{mn}(X))^{S_n^m}\right)^{S_m} \quad \text{by (A)}\end{aligned}$$

where $S_n^m = S_n \times \ldots \times S_n$ as a subgroup of S_{mn}. Now the normalizer of S_n^m in S_{mn} is the semidirect product $S_n^m \times S_m$, in which S_m acts by permuting the factors of S_n^m: this is the *wreath product* of S_n with S_m, denoted by $S_n \sim S_m$. Using (D) it follows that

$$\begin{aligned}(E \circ F)(X) &\cong (M \otimes \mathbf{T}^m(N) \otimes \mathbf{T}^{mn}(X))^{S_n \sim S_m} \\ &\cong \left(\operatorname{ind}_{S_n \sim S_m}^{S_{mn}}(M \otimes \mathbf{T}^m(N)) \otimes \mathbf{T}^{mn}(X)\right)^{S_{mn}}\end{aligned}$$

by (B) and (C). Hence $E \circ F$ corresponds to the $k[S_{mn}]$-module

$$M \circ N = \operatorname{ind}_{S_n \sim S_m}^{S_{mn}}(M \otimes \mathbf{T}^m(N)), \tag{6.2}$$

which we call the *composition product* or *plethysm* (Chapter I, §8) of M with N. Plethysm is linear in M:

$$(M_1 \oplus M_2) \circ N \cong (M_1 \circ N) \oplus (M_2 \circ N) \tag{6.3}$$

and distributive over the induction product:

$$(M_1 \circ N).(M_2 \circ N) \cong (M_1 . M_2) \circ N. \tag{6.4}$$

For the corresponding relation for the functors is

$$(E_1 \circ F) \otimes (E_2 \circ F) = (E_1 \otimes E_2) \circ F$$

which is obvious.

7. The characteristic map

If $\mathfrak{A}$ is any abelian category, let $K(\mathfrak{A})$ denote the Grothendieck group of $\mathfrak{A}$.

Let $\mathfrak{F}$ denote the category of polynomial functors of bounded degree (§2) on $\mathfrak{V}$. From (2.3) and (5.4) it follows that $\mathfrak{F}$ is abelian and semisimple, and that

$$K(\mathfrak{F}) \cong \bigoplus_{n \geqslant 0} K(\mathfrak{F}_n) \cong \bigoplus_{n \geqslant 0} K(\mathfrak{V}_{S_n}).$$

Moreover, $K(\mathfrak{V}_{S_n}) = R^n$ in the notation of Chapter I, §7.

The tensor product (§6) defines a structure of a commutative, associative, graded ring with identity element on $K(\mathfrak{F})$. In view of (6.1), this ring structure agrees with that defined on $R = \oplus R^n$ in Chapter I, §7, so that we may identify $K(\mathfrak{F})$ with R.

$K(\mathfrak{F})$ also carries a scalar product. If E, F are polynomial functors, then $\mathrm{Hom}(E, F)$ is a finite-dimensional k-vector space, and we define

$$\langle E, F \rangle = \dim_k \mathrm{Hom}(E, F).$$

Again by (5.4), it is clear that this scalar product is the same as that defined on R in Chapter I, §7.

We shall now give an intrinsic description of the characteristic map $\mathrm{ch}: R \to \Lambda$, defined in Chapter I, §7. Let F be a polynomial functor on $\mathfrak{V}$. For each $\lambda = (\lambda_1, \dots, \lambda_r) \in k^r$, let (λ) as before denote the diagonal endomorphism of k^r with eigenvalues $\lambda_1, \dots, \lambda_r$. Then trace $F((\lambda))$ is a polynomial function of $\lambda_1, \dots, \lambda_r$, which is *symmetric* because by (4.1)

$$\begin{aligned} \operatorname{trace} F((s\lambda)) = \operatorname{trace} F(s(\lambda)s^{-1}) &= \operatorname{trace} F(s)F((\lambda))F(s)^{-1} \\ &= \operatorname{trace} F((\lambda)) \end{aligned}$$

for all $s \in S_r$. Since the trace is additive, it determines a mapping

$$\chi_r: K(\mathfrak{F}) \to \Lambda_r,$$

namely $\chi_r(F)(\lambda_1, \dots, \lambda_r) = \operatorname{trace} F((\lambda))$; and since the trace is multiplicative with respect to tensor products, χ_r is a homomorphism of graded rings. Moreover, it is clear from the definitions that $\chi_r = \rho_{q,r} \circ \chi_q$ for $q \geqslant r$, in the notation of Chapter I, §2, and hence the χ_r determine a homomorphism of graded rings

(7.1) $$\chi: K(\mathfrak{F}) \to \Lambda.$$

To see that this homomorphism coincides with the characteristic map $\mathrm{ch}: R \to \Lambda$ defined in Chapter I, §7, we need only observe that $\chi(\wedge^n) = e_n$ and that $\wedge^n$ corresponds to the sign representation ε_n of S_n in the correspondence (5.4).

Hence from Chapter I, (7.5) and (7.6) it follows that

(7.2) *If* $F_\lambda: \mathfrak{V} \to \mathfrak{V}$ *is the irreducible polynomial functor corresponding to the partition* λ, *then* $\chi(F_\lambda)$ *is the Schur function* s_λ. |

If F is any polynomial functor, the decomposition (2.5) applied to the functor $F':(X_1,\dots,X_r) \mapsto F(X_1 \oplus \dots \oplus X_r)$ shows that the eigenvalues of $F((\lambda))$ are monomials $\lambda_1^{m_1}\dots\lambda_r^{m_r}$, with corresponding eigenspaces $F'_{m_1\dots m_r}(k,\dots,k)$, and therefore

$$\chi_r(F) = \sum_{m_1,\dots,m_r} \dim F'_{m_1\dots m_r}(k,\dots,k) x_1^{m_1}\dots x_r^{m_r}.$$

From this and the definition of plethysm in Chapter I, §8, it follows that

$$\chi(E \circ F) = \chi(E) \circ \chi(F) \tag{7.3}$$

for any two polynomial functors E, F.

In particular, if λ and μ are partitions, the functor $F_\lambda \circ F_\mu$ is a direct sum of irreducible functors F_π, so that in $K(\mathfrak{F})$ we have

$$F_\lambda \circ F_\mu = \sum_\pi a_{\lambda\mu}^\pi F_\pi$$

with non-negative integral coefficients $a_{\lambda\mu}^\pi$. By (7.2) and (7.3) it follows that

$$s_\lambda \circ s_\mu = \sum_\pi a_{\lambda\mu}^\pi s_\pi \tag{7.4}$$

with coefficients $a_{\lambda\mu}^\pi \geqslant 0$.

Example

It follows from §7 that a polynomial functor F is determined up to isomorphism by its trace $\chi(F)$. Hence identities in the ring Λ may be interpreted as statements about polynomial functors. Consider for example the identities of Chapter I, §5, Examples 5, 7, and 9. From Example 5(a) we obtain

$$\mathbf{S}^k(\mathbf{S}^2 V) \cong \bigoplus_\mu F_\mu(V) \tag{1}$$

summed over *even* partitions μ of $2k$, and from Example 5(b)

$$\mathbf{S}^k(\Lambda^2 V) \cong \bigoplus_\nu F_\nu(V) \tag{2}$$

summed over partitions ν of $2k$ such that ν' is even (i.e. all columns of ν have even length). Likewise Example 7 gives

$$\mathbf{S}^k(V \oplus \Lambda^2 V) \cong \bigoplus_\lambda F_\lambda(V) \tag{3}$$

summed over partitions λ such that $|\lambda|+c(\lambda)=2k$, where $c(\lambda)$ (*loc. cit.*) is the number of columns of odd length in λ, or equivalently the number of odd parts of λ'. Finally from Example 9(a) and 9(b) we obtain

(4) $$\bigwedge^p(\bigwedge^2 V)\cong \bigoplus_{\pi} F_\pi(V)$$

summed over partitions $\pi=(\alpha_1-1,\dots,\alpha_p-1\,|\,\alpha_1,\dots,\alpha_p)$ in Frobenius notation, and

(5) $$\bigwedge^p(\mathbf{S}^2 V)\cong \bigoplus_{\rho} F_\rho(V)$$

summed over partitions $\rho=(\alpha_1+1,\dots,\alpha_p+1\,|\,\alpha_1,\dots,\alpha_p)$.

8. The polynomial representations of $GL_m(k)$

Let G be any group (finite or not) and let R be a matrix representation of G of degree d over an algebraically closed field k of characteristic 0, the representing matrices being $R(g)=(R_{ij}(g))$, $g\in G$. Thus R determines d^2 functions $R_{ij}: G\to k$, called the *matrix coefficients* of R. If we replace R by an equivalent representation $g\mapsto AR(g)A^{-1}$, where A is a fixed matrix, the space of functions on G spanned by the matrix coefficients is unaltered. It follows that if R is reducible, the R_{ij} are linearly dependent over k, because in an equivalent representation some of them will be zero. Again, if R and S are equivalent irreducible representations of G, the matrix coefficients of S are linearly dependent on those of R.

(8.1) *Let $R^{(1)}, R^{(2)},\dots$ be a sequence of matrix representations over k of a group G. Then the following are equivalent:*

(i) *All the matrix coefficients $R_{ij}^{(1)}, R_{ij}^{(2)},\dots$ are linearly independent;*

(ii) *The representations $R^{(1)}, R^{(2)},\dots$ are irreducible and pairwise inequivalent.*

We have just seen that (i) implies (ii). The reverse implication is a theorem of Frobenius and Schur (see [C5], p. 183).

Now let $V=k^m$ and let $G=GL(V)=GL_m(k)$. Let x_{ij} $(1\leqslant i,j\leqslant m)$ be the coordinate functions on G, so that $x_{ij}(g)$ is the (i,j) element of the matrix $g\in G$. Let

$$P=\bigoplus_{n\geqslant 0} P^n = k[x_{ij}: 1\leqslant i,j\leqslant m]$$

be the algebra of polynomial functions of G, where P^n consists of the polynomials in the x_{ij} that are homogeneous of degree n. A matrix representation R of G is said to be *polynomial* if its matrix coefficients are polynomials in the x_{ij}. Clearly, each polynomial functor F_λ such that

$l(\lambda) \leqslant m$ (so that $F_\lambda(V) \neq 0$) gives rise to an equivalence class of polynomial representations R^λ of G, in which $g \in G$ acts as $F_\lambda(g)$ on $F_\lambda(V)$.

We shall show that

(8.2) *The representations R^λ such that $l(\lambda) \leqslant m$ are inequivalent irreducible polynomial representations of G, and conversely every irreducible polynomial representation of G is equivalent to some R^λ.*

Proof. By (7.2), the dimension of $F_\lambda(V)$ is equal to $s_\lambda(x_1, \dots, x_m)$ evaluated at $x_1 = \dots = x_m = 1$, and hence by Chapter I, (4.3)

(1) $$d_n = \sum_{|\lambda| = n} (\dim F_\lambda(V))^2$$

is equal to the coefficient of t^n in $(1-t)^{-m^2}$, so that

(2) $$d_n = \dim_k P^n.$$

On the other hand, the decomposition (5.6)

$$\mathbf{T}^n(V) \cong \bigoplus_{|\lambda| = n} M_\lambda \otimes F_\lambda(V)$$

shows that the representation of G on $\mathbf{T}^n(V)$ is equivalent to the diagonal sum of $\dim M_\lambda$ copies of R^λ, for each partition λ of n of length $\leqslant m$. Now the matrix coefficients of $\mathbf{T}^n$ are the monomials of degree n in the x_{ij}, hence span P_n. Consequently the R_{ij}^λ also span P^n; but from (1) and (2) above it follows that the total number of these matrix coefficients is $d_n = \dim_k P^n$. Hence the R_{ij}^λ such that $|\lambda| = n$ and $l(\lambda) \leqslant m$ form a *k-basis* of P^n, and it follows from (8.1) that the R^λ are irreducible and pairwise inequivalent.

Finally, if R is any polynomial representation of G, the argument of §2 shows that R is a direct sum of homogeneous polynomial representations. So if R is irreducible, its matrix coefficients R_{ij} are homogeneous of degree say n, i.e. $R_{ij} \in P^n$. Hence by (8.1) R is equivalent to some R^λ. |

In the course of the above proof we have shown that

(8.3) *The matrix coefficients R_{ij}^λ, where λ ranges over all partitions of length $\leqslant m$, form a k-basis of P.* |

If $f \in \Lambda$ is any symmetric function, we may regard f as a function on G by the rule

$$f(x) = f(\xi_1, \dots, \xi_m)$$

where $\xi_1, \dots, \xi_m$ are the eigenvalues of $x \in G$. For example, $e_r(x) = \operatorname{trace}(\wedge^r x)$ is the sum of the principal $r \times r$ minors of the matrix x, and hence is a polynomial function of x. Since each $f \in \Lambda$ is a polynomial in the e_r, it follows that f is a polynomial function on G.

(8.4)(i) *The character of the representation R^λ of G is the Schur function s_λ.*
(ii) *A polynomial representation of G is determined up to equivalence by its character.*

Proof. (i) From (7.2), $s_\lambda(x)$ is the trace of $F_\lambda(x)$ if $x \in G$ is a diagonal matrix, and hence also if x is diagonalizable. Since the diagonalizable matrices are Zariski-dense in G (because x is diagonalizable whenever $D(x) \neq 0$, where $D(x)$ is the discriminant of the characteristic polynomial of x) and both $s_\lambda(x)$ and trace $F_\lambda(x)$ are polynomial functions of x, it follows that $s_\lambda(x) = \text{trace } F_\lambda(x)$ for all $x \in G$.

(ii) This follows from (i), since the Schur functions s_λ such that $l(\lambda) \leqslant m$ are linearly independent. |

The group G acts on P (and on each P^n) by the rule

$$(gp)(x) = p(xg)$$

for $p \in P$ and $g, x \in G$. Hence

$$gR_{ij}^\lambda(x) = R_{ij}^\lambda(xg) = \sum_r R_{ir}^\lambda(x) R_{rj}^\lambda(g),$$

so that

$$gR_{ij}^\lambda = \sum_r R_{rj}^\lambda(g) R_{ir}^\lambda.$$

This equation shows that for each $i = 1, 2, \dots, d_\lambda$, where $d_\lambda = \dim F_\lambda(V)$ is the degree of R^λ, the subspace of P spanned by the R_{ij}^λ $(1 \leqslant j \leqslant d_\lambda)$ is an irreducible G-module affording the representation R^λ, and hence is isomorphic to $F_\lambda(V)$. Consequently, if P_λ is the subspace of P spanned by all the R_{ij}^λ, we have

(8.5)
$$P = \bigoplus_\lambda P_\lambda$$

where λ ranges over all partitions of length $\leqslant m$, and $P_\lambda \cong F_\lambda(V)^{d_\lambda}$.

Examples

1. (a) Let λ be a partition of n and let M_λ (§7, Example 15) denote the Specht module corresponding to λ. It has a basis (f_t) indexed by the standard tableaux t of shape λ, where (*loc. cit.*) f_t is a product of Vandermonde determinants on the columns of t, so that

(1)
$$wf_t = \varepsilon(w) f_t$$

if $w \in S_n$ lies in the column stabilizer C_t of t.

If $v \in S_n$, then vf_t is a $\mathbf{Z}$-linear combination of the f_s, say

(2) $$vf_t = \sum_s \rho_{st}^\lambda(v) f_s$$

summed over standard tableaux s of shape λ, and the mapping $v \mapsto (\rho_{st}^\lambda(v))$ is an irreducible representation of S_n by integer matrices.

(b) Let X be a k-vector space with basis $x_1, \dots, x_q$. Then $\mathbf{T}^n(X)$ has a basis consisting of the tensors

$$x_\tau = x_{\tau(1)} \otimes \dots \otimes x_{\tau(n)}$$

where τ runs through all mappings $[1,n] \to [1,q]$. If $w \in S_n$ we have $w(x_\tau) = x_{\tau w^{-1}}$.

Hence $M_\lambda \otimes \mathbf{T}^n(X)$ has as basis the products $f_t \otimes x_\tau$, where t is a standard tableau of shape λ, and $\tau: [1,n] \to [1,q]$. Now consider $F_\lambda(X) = (M_\lambda \otimes \mathbf{T}^n(X))^{S_n}$: clearly it will be spanned by the elements

(3) $$x_{t,\tau} = \sum_{w \in S_n} w(f_t \otimes x_\tau) = \sum_{w \in S_n} wf_t \otimes x_{\tau w^{-1}}.$$

Given τ as above, we may choose $v \in S_n$ so that $\sigma = \tau v^{-1}$ is an *increasing* mapping of $[1,n]$ into $[1,q]$. We then have

$$x_{t,\tau} = \sum_{w \in S_n} wf_t \otimes x_{\sigma v w^{-1}}$$

$$= \sum_{w \in S_n} wvf_t \otimes x_{\sigma w^{-1}}$$

and hence by (2) and (3) we have

(4) $$x_{t,\tau} = \sum_s \rho_{st}^\lambda(v) x_{s,\sigma}$$

and therefore $F_\lambda(X)$ is spanned by the elements $x_{t,\tau}$ such that t is a standard tableau of shape λ and $\tau: [1,n] \to [1,q]$ is an increasing mapping.

(c) We regard a standard tableau t of shape λ as a bijective mapping of (the shape of) λ onto $[1,n]$, so that $t(i,j)$ is the integer occupying the square $(i,j) \in \lambda$. Then $T = \tau \circ t$ is a mapping $\lambda \to [1,q]$, i.e. it is a filling of the squares of λ with the numbers $1, 2, \dots, q$. If τ is increasing, then T is increasing (in the weak sense) along rows and down columns, and we have $x_{t,\tau} = 0$ unless T is a (column-strict) tableau. For if there are two squares a, b in the same column of λ such that $T(a) = T(b)$, i.e. $\tau(t(a)) = \tau(t(b))$, let $w \in S_n$ be the transposition that interchanges $t(a)$ and $t(b)$. Since $w \in C_t$ it follows from (1) above that $wf_t = -f_t$, and since also $wx_\tau = x_\tau$ we have $w(f_t \otimes x_\tau) = -f_t \otimes x_\tau$, and consequently $x_{t,\tau} = 0$.

(d) On the other hand, if $T = \tau \circ t$ is a tableau, then τ is uniquely determined by T, because the sequence $(\tau(1), \dots, \tau(n))$ is the weight of T (Chapter I, §1). In general the standard tableau t is not uniquely determined by T: for example, if

$T = {}^{1\,2}_{2}$ then t can be either ${}^{1\,2}_{3}$ or ${}^{1\,3}_{2}$. However, if $T = \tau \circ t = \tau \circ t_1$ with t, t_1 standard tableaux, then $t_1 = vt$ for some $v \in S_n$ such that $\tau = \tau \circ v$, and hence

$$x_{t_1,\tau} = \sum_{w \in S_n} w(f_{t_1} \otimes x_\tau)$$

$$= \sum_{w \in S_n} wv(f_t \otimes x_\tau) = x_{t,\tau}$$

and we may therefore unambiguously define $x_T = x_{t,\tau}$ when T is a column-strict tableau.

(e) It follows from (b) and (c) above that $F_\lambda(X)$ is spanned by the elements x_T, where $T: \lambda \to [1, q]$ is a column-strict tableau. But now the dimension of $F_\lambda(X)$ is by (7.2) equal to the Schur function s_λ evaluated at $(1^q) = (1, \dots, 1)$, which by Chapter I, (5.12) is equal to the number of tableaux T as above. Hence the x_T from a *k-basis* of $F_\lambda(X)$.

2. In continuation of Example 1, let Y be a k-vector space with basis $y_1, \dots, y_p$, and let $\alpha: X \to Y$ be a k-linear mapping, say

$$\alpha x_j = \sum_{i=1}^{p} a_{ij} y_i,$$

so that α is represented by the $p \times q$ matrix $A = (a_{ij})$ over k, relative to the given bases of X and Y. Then α induces $\mathbf{T}^n(\alpha): \mathbf{T}^n(X) \to \mathbf{T}^n(Y)$, given by

$$\mathbf{T}^n(\alpha) x_\tau = \sum_\sigma a_{\sigma\tau} y_\sigma \tag{5}$$

where $a_{\sigma\tau} = a_{\sigma(1)\tau(1)} \cdots a_{\sigma(n)\tau(n)}$ and in the sum σ runs through all mappings $[1, n] \to [1, p]$.

If now λ is a partition of n, we have a k-linear mapping $F_\lambda(\alpha): F_\lambda(X) \to F_\lambda(Y)$, say

$$F_\lambda(\alpha) x_T = \sum_S a_{ST} y_S \tag{6}$$

summed over column-strict tableaux $S: \lambda \to [1, p]$. We shall now compute the matrix coefficients a_{ST}.

If $T = \tau \circ t$ as in Example 1, then

$$F_\lambda(\alpha) x_T = \sum_{w \in S_n} w(f_t) \otimes \mathbf{T}^n(\alpha) x_{\tau w^{-1}}$$

$$= \sum_{w \in S_n} w(f_t) \otimes \sum_\sigma a_{\sigma, \tau w^{-1}} y_\sigma$$

by (3) and (5) above. By replacing σ by σw^{-1} in the inner sum above and observing that $a_{\sigma w^{-1}, \tau w^{-1}} = a_{\sigma,\tau}$, we obtain

$$F_\lambda(\alpha) x_T = \sum_\sigma a_{\sigma\tau} y_{t,\sigma} \tag{7}$$

summed over $\sigma:[1,n]\to[1,p]$. For each such σ, choose $v\in S_n$ such that $\sigma_0=\sigma v^{-1}$ is an increasing mapping; then by (4) of Example 1 we have

$$(8)\qquad \begin{aligned} y_{t,\sigma} &= \sum_s \rho_{st}^{\lambda}(v) y_{s,\sigma_0} \\ &= \sum_s \rho_{st}^{\lambda}(v) y_S \end{aligned}$$

summed over standard tableaux s such that $S=\sigma_0\circ s$ is a tableau. From (7) and (8) we obtain

$$F_\lambda(\alpha)x_T=\sum \rho_{st}^{\lambda}(v)a_{\sigma_0 v,\tau}y_S$$

summed over all triples (σ_0,s,v) such that $\sigma_0:[1,n]\to[1,p]$ is increasing, s is a standard tableau of shape λ, $S=\sigma_0\circ s$ is a column-strict tableau, and $v\in S_n$ runs through a set of representatives of the cosets Gv of the stabilizer G of σ_0 in S_n. Hence finally

$$a_{ST}=\sum_{s,v}\rho_{st}^{\lambda}(v)a_{\sigma_0 v,\tau}$$

summed over s, v as above. In particular, this formula shows that a_{ST} is a polynomial in the a_{ij} with integer coefficients.

3. As in Chapter I, §7, Example 15, let λ be a partition of n and let T be any numbering of the diagram of λ with the numbers $1,2,\dots,n$. Let R (resp. C) be the subgroup of S_n that stabilizes each row (resp. column) of T, and let

$$e=e_T=\sum_{u,v}\varepsilon(u)uv\in k[S_n],$$

summed over $(u,v)\in C\times R$. Then (*loc. cit.*) $k[S_n]e_T$ is a minimal left ideal of the group algebra $k[S_n]$, isomorphic to M_λ. Deduce from (5.5′) that

$$F_\lambda(V)\cong e_T\mathbf{T}^n(V)$$

as G-modules.

4. Let $e_1,\dots,e_m$ be the standard basis of $V=k^m$, so that

$$ge_j=\sum_{i=1}^m g_{ij}e_i$$

if $g=(g_{ij})\in G$. Then $\mathbf{T}=\mathbf{T}^n(V)$ has a basis consisting of the tensors $e_\tau=e_{\tau(1)}\otimes\dots\otimes e_{\tau(n)}$, where τ runs through all mappings of $[1,n]$ into $[1,m]$. Let $\langle u,v\rangle$ be the scalar product on $\mathbf{T}$ defined by $\langle e_\sigma,e_\tau\rangle=\delta_{\sigma\tau}$, and for each $u\in\mathbf{T}$ let $\varphi=\varphi_u:\mathbf{T}\to P^n$ be the linear mapping defined by

$$\varphi(v)(g)=\langle u,gv\rangle$$

for $v\in\mathbf{T}$ and $g\in G$. Verify that φ is a G-module homomorphism and deduce that (in the notation of Example 3) $\varphi(e_T\mathbf{T})$ is either zero or is a G-submodule of P^n isomorphic to $F_\lambda(V)$.

5. If $g = (g_{ij}) \in G$ let

$$g^{(r)} = (g_{ij})_{1 \leqslant i, j \leqslant r}$$

for $1 \leqslant r \leqslant m$ (so that $g^{(m)} = g$), and for any partition λ of length $\leqslant m$ let

$$\Delta_\lambda(g) = \prod_{i \geqslant 1} \det(g^{(\lambda'_i)}).$$

Then Δ_λ is a polynomial function on G. If $b = (b_{ij}) \in G$ is upper triangular, we have

$$\Delta_\lambda(b) = b_{11}^{\lambda_1} \dots b_{mm}^{\lambda_m},$$

and

$$\Delta_\lambda(gb) = \Delta_\lambda(b'g) = \Delta_\lambda(g)\,\Delta_\lambda(b)$$

for all $g \in G$.

Show that the G-submodule of P generated by Δ_λ is irreducible and isomorphic to $F_\lambda(V)$. (This is a particular case of the construction of Example 4. Take T to be the standard tableau of shape λ in which the numbers $1, 2, \dots, n = |\lambda|$ occur in order in successive rows, and let $u = e_\tau$ where $(\tau(1), \dots, \tau(n))$ is the weakly increasing sequence in which i occurs λ_i times, for each $i = 1, 2, \dots, m$. The subgroup of S_n that fixes u is the row-stabilizer $R = S_{\lambda_1} \times \dots \times S_{\lambda_m}$ of T, so that $\sum_{w \in R} w(u) = ru$, where $r = |R| = \lambda_1! \dots \lambda_m!$. Hence $\varphi(e_T u)$ is the function on G whose value at $g \in G$ is

$$r . \sum_{w \in C} \varepsilon(w) \langle e_\tau, gwe_\tau \rangle = r\Delta_\lambda(g)$$

where C is the column stabilizer of T. Hence $\Delta_\lambda \in \varphi(e_T \mathbf{T})$; since $\Delta_\lambda \neq 0$, it follows that $\varphi(e_T \mathbf{T})$ is irreducible and isomorphic to $F_\lambda(V)$.)

6. The *Schur algebra* of degree n of G (or V) is

$$\mathfrak{S}^n = \mathrm{End}_{k[S_n]} \mathbf{T}^n(V).$$

For each $g \in G$ we have $\mathbf{T}^n(g) \in \mathfrak{S}^n$, and G acts on $\mathfrak{S}^n$ by the rule $g\alpha = \mathbf{T}^n(g) \circ \alpha$ for $g \in G$ and $\alpha \in \mathfrak{S}^n$.

If $\alpha \in \mathfrak{S}^n$, let p_α be the function on G defined by

$$p_\alpha(g) = \mathrm{trace}(g\alpha). \tag{1}$$

(a) Show that $p_\alpha \in P^n$ and that $\alpha \mapsto p_\alpha : \mathfrak{S}^n \to P^n$ is an isomorphism of G-modules. (Relative to the basis (e_τ) of $\mathbf{T}^n(V)$ (Example 4), each $\alpha \in \mathfrak{S}^n$ is represented by a matrix $(a_{\sigma\tau})$ such that $a_{\sigma w, \tau w} = a_{\sigma\tau}$ for all $w \in S_n$. We have $\mathrm{trace}(g\sigma) = \sum_{\sigma,\tau} a_{\tau\sigma} g_{\sigma\tau}$ and hence

$$p_\alpha = \sum_{\sigma,\tau} a_{\tau\sigma} x_{\sigma\tau}$$

where $x_{\sigma\tau} = \prod_{i=1}^n x_{\sigma(i), \tau(i)} = x_{\sigma w, \tau w}$ for $w \in S_n$. It follows that $\alpha \mapsto p_\alpha$ is a linear isomorphism of $\mathfrak{S}^n$ onto P^n, and it is clear from (1) that $hp_\alpha = p_{h\alpha}$ for $h \in G$.)

(b) The $\mathsf{T}^n(g)$, $g \in G$, span $\mathfrak{S}^n$ as a k-vector space. (Suppose not, then there exists $\alpha \neq 0$ in $\mathfrak{S}^n$ such that $\operatorname{trace}(\mathsf{T}^n(g)\alpha) = 0$ for all $g \in G$, since the bilinear form $(\alpha, \beta) \mapsto \operatorname{trace}(\alpha\beta)$ on $\mathfrak{S}^n$ is nondegenerate. But then $p_\alpha(g) = 0$ for all $g \in G$, so that $p_\alpha = 0$ and therefore $\alpha = 0$.)

Hence the G-submodules of $\mathfrak{S}^n$ are precisely the left ideals.

(c) We have

$$\mathfrak{S}^n \cong \prod_\lambda \operatorname{End}_k(F_\lambda(V))$$

as k-algebras, where the product is taken over all partitions λ of n of length $\leqslant m$. (Use (5.6).)

Notes and references

Practically everything in this Appendix is contained, explicitly or implicitly, in Schur's thesis [S4] and his subsequent 1927 paper [S6]. We have assumed throughout that the ground field k has characteristic zero; for a study of the polynomial representations of $GL_n(k)$ when k has positive characteristic, see [G14], and for polynomial functors over k (or indeed over any commutative ring) see [A2].

APPENDIX B: Characters of wreath products

1. Notation

If G is any finite group, let G_* denote the set of conjugacy classes in G, and let G^* denote the set of irreducible complex characters of G. If $c \in G_*$ and $\gamma \in G^*$ we shall denote by $\gamma(c)$ the value of γ at an element $x \in c$, and by ζ_c the order of the centralizer of x in G, so that the number of elements in c is $|c| = |G|/\zeta_c$.

Let $R(G)$ denote the complex vector space spanned by G^*, or equivalently the space of complex-valued class functions on G. We have $\dim R(G) = |G^*| = |G_*|$. On $R(G)$ we have a hermitian scalar product

$$\langle u, v\rangle_G = \frac{1}{|G|} \sum_{x \in G} u(x)\overline{v(x)} \tag{1.1}$$

relative to which G^* is an orthonormal basis of $R(G)$, i.e. $\langle \beta, \gamma\rangle_G = \delta_{\beta\gamma}$ for $\beta, \gamma \in G^*$.

Later we shall encounter families of partitions indexed by G_* and by G^*. In general, if X is a finite set and $\rho = (\rho(x))_{x\in X}$ a family of partitions indexed by X, or equivalently a mapping $\rho: X \to \mathscr{P}$ (where $\mathscr{P}$ is the set of all partitions), we denote by $\|\rho\|$ the sum

$$\|\rho\| = \sum_{x \in X} |\rho(x)|$$

and by $\mathscr{P}(X)$ (resp. $\mathscr{P}_n(X)$) the set of all $\rho: X \to \mathscr{P}$ (resp. the set of all $\rho: X \to \mathscr{P}$ such that $\|\rho\| = n$). Finally, if $\rho, \sigma \in \mathscr{P}(X)$, then $\rho \cup \sigma$ is the function $x \mapsto \rho(x) \cup \sigma(x)$ for $x \in X$.

2. The wreath product $G \sim S_n$

Let $G^n = G \times \ldots \times G$ be the direct product of n copies of G. The symmetric group S_n acts on G^n by permuting the factors: $s(g_1, \ldots, g_n) = (g_{s^{-1}(1)}, \ldots, g_{s^{-1}(n)})$. The *wreath product* $G_n = G \sim S_n$ is the semidirect product of G^n with S_n defined by this action, that is to say it is the group whose underlying set is $G^n \times S_n$, with multiplication defined by $(g, s)(h, t) = (g \,.\, s(h), st)$, where $g, h \in G^n$ and $s, t \in S_n$. More concretely, the elements of G_n may be thought of as permutation matrices with entries in G, the matrix corresponding to (g, s) having (i, j) element $g_i\, \delta_{i, s(j)}$, where $g = (g_1, \ldots, g_n)$.

When $n=1$, G_1 is just G. When $n=0$, G_0 is the group of one element. The order of G_n is $|G|^n . n!$ for all $n \geqslant 0$.

An embedding of $S_m \times S_n$ in S_{m+n} gives rise to an embedding of $G_m \times G_n$ in G_{m+n}, and any two such embeddings are conjugate in G_{m+n}.

3. Conjugacy classes and types

Let $x=(g,s) \in G_n$, where $g=(g_1,\dots,g_n) \in G^n$ and $s \in S_n$. The permutation s can be written as a product of disjoint cycles: if $z=(i_1 i_2 \dots i_r)$ is one of these cycles, the element $g_{i_r} g_{i_{r-1}} \dots g_{i_1} \in G$ is determined up to conjugacy in G by g and z, and is called the *cycle-product* of x corresponding to the cycle z. For each conjugacy class $c \in G_*$ and each integer $r \geqslant 1$, let $m_r(c)$ denote the number of r-cycles in s whose cycle-product lies in c. In this way each element $x \in G_n$ determines an array $(m_r(c))_{r \geqslant 1, c \in G_*}$ of non-negative integers such that $\Sigma_{r,c}\, r m_r(c) = n$. Equivalently, if $\rho(c)$ denotes the partition having $m_r(c)$ parts equal to r, for each $r \geqslant 1$, then $\rho = (\rho(c))_{c \in G_*}$ is a partition-valued function on G_* such that $\|\rho\| = n$, i.e. $\rho \in \mathscr{P}_n(G_*)$ in the notation introduced in §1. This function ρ is called the *type* of $x=(g,s) \in G_n$. Note that the cycle-type of s in S_n is $\sigma = \bigcup_{c \in G_*} \rho(c)$.

We shall show that two elements of G_n are conjugate if and only if they have the same type.

(a) If $w \in S_n$, let $\overline{w} = (1,w) \in G_n$. If $x=(g,s) \in G_n$ we have $\overline{w} x \overline{w}^{-1} = (wg, wsw^{-1})$. If $z=(i_1 \dots i_r)$ is a cycle in s, as above, then $wzw^{-1} = (w(i_1) \dots w(i_r))$ is a cycle of wsw^{-1}; moreover $(wg)_{w(i_\alpha)} = g_{i_\alpha}$, so that the cycle-product is unaltered, and therefore x and $\overline{w} x \overline{w}^{-1}$ have the same type.

(b) Let $h=(h_1,\dots,h_n) \in G^n$. Then $hxh^{-1} = (hgs(h^{-1}), s)$, and $hgs(h^{-1})$ has ith component $h_i g_i h_{s^{-1}(i)}^{-1}$. Hence the cycle-product of hxh^{-1} corresponding to the cycle z in s is

$$\left(h_{i_r} g_{i_r} h_{i_{r-1}}^{-1}\right)\left(h_{i_{r-1}} g_{i_{r-1}} h_{i_{r-2}}^{-1}\right) \dots \left(h_{i_1} g_{i_1} h_{i_r}^{-1}\right) = h_{i_r}(g_{i_r} \dots g_{i_1}) h_{i_r}^{-1}$$

which is conjugate in G to $g_{i_r} \dots g_{i_1}$. It follows that x and hxh^{-1} have the same type.

(c) From (a) and (b) it follows that conjugate elements in G_n have the same type. Conversely, suppose that $x=(g,s)$ and $y=(h,t)$ in G_n have the same type. Then $s,t \in S_n$ have the same cycle-type $\sigma = (\sigma_1, \sigma_2, \dots)$ and are therefore conjugate in S_n. Hence by conjugating y by a suitable permutation $w \in S_n$ we may assume that $t=s$; then both x and y lie in the same subgroup $G_\sigma = G_{\sigma_1} \times G_{\sigma_2} \times \dots$ of G_n, and it is enough to show that they are conjugate in G_σ. This effectively reduces the question to the case where $s \in S_n$ is an n-cycle, say $s = (12\ldots n)$, and the products $\bar{g} = g_n g_{n-1} \cdots g_1$, $\bar{h} = h_n h_{n-1} \cdots h_1$ are conjugate in G. Now choose $u_n \in G$ such that $\bar{g} = u_n \bar{h} u_n^{-1}$, and define $u_1, \ldots, u_{n-1} \in G$ successively by the equations

$$g_1 = u_1 h_1 u_n^{-1},\ g_2 = u_2 h_2 u_1^{-1}, \ldots, g_{n-1} = u_{n-1} h_{n-1} u_{n-2}^{-1}.$$

A simple calculation now shows that $uyu^{-1} = x$, where $u = (u_1, \ldots, u_n)$; hence x and y are conjugate in G_n, and the proof is complete.

(d) We shall now compute the order of the centralizer in G_n of an element $x = (g, s)$ of type $\rho \in \mathscr{P}_n(G_*)$. First, the number of possibilities for $s \in S_n$ is

$$n! \Big/ \prod_{r \geq 1} r^{m_r} . m_r! \tag{1}$$

where $m_r = \Sigma_c m_r(c)$. Next, for each such s and each $r \geq 1$ there are

$$m_r! \Big/ \prod_c m_r(c)! \tag{2}$$

ways of distributing the m_r r-cycles among the conjugacy classes of G. Finally, for each cycle $z = (i_1 \ldots i_r)$ in s such that $g_{i_r} \cdots g_{i_1} \in c$, there are

$$|G|^{r-1} . |c| = |G|^r / \zeta_c \tag{3}$$

choices for $(g_{i_1}, \ldots, g_{i_r})$. From (1), (2), and (3) it follows that the number of elements in G_n of type ρ is

$$\frac{n!}{\prod_c z_{\rho(c)}} \cdot \frac{|G|^n}{\prod_c \zeta_c^{l(\rho(c))}}$$

and hence that the order of the centralizer in G_n of an element of type ρ is

$$Z_\rho = \prod_{c \in G_*} z_{\rho(c)} \zeta_c^{l(\rho(c))}. \tag{3.1}$$

4. The algebra $R(G)$

Let

$$R(G) = \bigoplus_{n \geq 0} R(G_n).$$

We define a multiplication on $R(G)$ as follows. Let $u \in R(G_m)$, $v \in R(G_n)$, and embed $G_m \times G_n$ in G_{m+n}. Then $u \times v$ is an element of $R(G_m \times G_n)$, and we define

$$uv = \operatorname{ind}_{G_m \times G_n}^{G_{m+n}} (u \times v) \tag{4.1}$$

which is an element of $R(G_{m+n})$. Thus we have defined a bilinear multiplication $R(G_m)\times R(G_n)\to R(G_{m+n})$, and just as in Chapter I, §7 (which deals with the case $G=\{1\}$) it is not difficult to verify that with this multiplication $R(G)$ is a commutative, associative, graded C-algebra with identity element.

In addition, $R(G)$ carries a hermitian scalar product; if $u,v\in R(G)$, say $u=\Sigma u_n$ and $v=\Sigma v_n$ with $u_n,v_n\in R(G_n)$, we define

(4.2) $$\langle u,v\rangle=\sum_{n\geqslant 0}\langle u_n,v_n\rangle_{G_n}$$

where the scalar product on the right is that defined by (1.1), with G_n replacing G.

5. The algebra $\Lambda(G)$

Let $p_r(c)$ $(r\geqslant 1, c\in G_*)$ be independent indeterminates over C, and let

$$\Lambda(G)=\mathbf{C}[p_r(c):r\geqslant 1,c\in G_*].$$

For each $c\in G_*$ we may think of $p_r(c)$ as the rth power sum in a sequence of variables $x_c=(x_{ic})_{i\geqslant 1}$. We assign degree r to $p_r(c)$, and then $\Lambda(G)$ is a graded C-algebra.

If $\sigma=(\sigma_1,\sigma_2,\dots)$ is any partition, let $p_\sigma(c)=p_{\sigma_1}(c)p_{\sigma_2}(c)\dots$. If now ρ is any partition-valued function on G_*, we define

(5.1) $$P_\rho=\prod_{c\in G_*}p_{\rho(c)}(c).$$

Clearly the P_ρ form a C-basis of $\Lambda(G)$. If $f\in\Lambda(G)$, say $f=\Sigma_\rho f_\rho P_\rho$ (where all but a finite number of the coefficients $f_\rho\in\mathbf{C}$ are zero), let

(5.2) $$\bar f=\sum_\rho \bar f_\rho P_\rho$$

so that in particular $\overline{P}_\rho=P_\rho$.

Next, we define a hermitian scalar product on $\Lambda(G)$ as follows: if $f=\Sigma_\rho f_\rho P_\rho$, $g=\Sigma_\rho g_\rho P_\rho$ then

(5.3) $$\langle f,g\rangle=\sum_\rho f_\rho\bar g_\rho Z_\rho$$

with Z_ρ given by (3.1). Equivalently,

(5.3′) $$\langle P_\rho,P_\sigma\rangle=\delta_{\rho\sigma}Z_\rho.$$

Finally, let $\Psi:G_m\to\Lambda(G)$ be the mapping defined by $\Psi(x)=P_\rho$ if

$x \in G_m$ has type ρ. If also $y \in G_n$ has type σ, then $x \times y \in G_m \times G_n$ is well-defined up to conjugacy in G_{m+n}, and has type $\rho \cup \sigma$, so that

$$\Psi(x \times y) = \Psi(x)\Psi(y). \tag{5.4}$$

6. The characteristic map

This is a **C**-linear mapping

$$\mathrm{ch}: R(G) \to \Lambda(G)$$

defined as follows: if $f \in R(G_n)$ then

$$\begin{aligned} \mathrm{ch}(f) &= \langle f, \Psi \rangle_{G_n} \\ &= \frac{1}{|G_n|} \sum_{x \in G_n} f(x)\Psi(x). \end{aligned} \tag{6.1}$$

If f_ρ is the value of f at elements of type ρ, then

$$\mathrm{ch}(f) = \sum_\rho Z_\rho^{-1} f_\rho P_\rho. \tag{6.2}$$

In particular, if φ_ρ is the characteristic function of the set of elements $x \in G_n$ of type ρ, we have $\mathrm{ch}(\varphi_\rho) = Z_\rho^{-1} P_\rho$, from which it follows that ch is a linear isomorphism.

Let $f, g \in G_n$. Then from (5.3) and (6.2)

$$\begin{aligned} \langle \mathrm{ch}(f), \mathrm{ch}(g) \rangle &= \sum_\rho Z_\rho^{-1} f_\rho \bar{g}_\rho \\ &= \langle f, g \rangle_{G_n} \end{aligned}$$

from which it follows that ch is an *isometry* for the scalar products on $R(G)$ and $\Lambda(G)$ defined by (4.2) and (5.3).

If $u \in R(G_m)$, $v \in R(G_n)$, then by (4.2) and (6.1)

$$\begin{aligned} \mathrm{ch}(uv) &= \langle \mathrm{ind}_{G_m \times G_n}^{G_{m+n}}(u \times v), \Psi \rangle_{G_{m+n}} \\ &= \langle u \times v, \Psi|_{G_m \times G_n} \rangle_{G_m \times G_n} \end{aligned}$$

by Frobenius reciprocity,

$$\begin{aligned} &= \langle u, \Psi \rangle_{G_m} \langle v, \Psi \rangle_{G_n} \qquad \text{by (5.4)} \\ &= \mathrm{ch}(u)\mathrm{ch}(v). \end{aligned}$$

Hence we have proved that

(6.3) ch: $R(G) \to \Lambda(G)$ *is an isometric isomorphism of graded* **C**-*algebras.*

7. Change of variables

For each irreducible character $\gamma \in G^*$ and each $r \geqslant 1$ let

$$p_r(\gamma) = \sum_{c \in G_*} \zeta_c^{-1} \gamma(c) p_r(c) \tag{7.1}$$

so that (by the orthogonality of the characters of G) we have

$$p_r(c) = \sum_{\gamma \in G^*} \overline{\gamma(c)} p_r(\gamma) = \sum_{\gamma \in G^*} \gamma(c) \overline{p_r(\gamma)}. \tag{7.1'}$$

The $p_r(\gamma)$ are algebraically independent and generate $\Lambda(G)$ as **C**-algebra. From (7.1') we have

$$\begin{aligned} \sum_c \zeta_c^{-1} p_r(c) \otimes p_r(c) &= \sum_c \sum_{\beta, \gamma} \zeta_c^{-1} \bar{\gamma}(c) \beta(c) p_r(\gamma) \otimes \overline{p_r(\beta)} \\ &= \sum_{\beta, \gamma} \langle \beta, \gamma \rangle_G p_r(\gamma) \otimes \overline{p_r(\beta)} \\ &= \sum_\gamma p_r(\gamma) \otimes \overline{p_r(\gamma)} \end{aligned} \tag{7.2}$$

in $\Lambda(G) \otimes \Lambda(G)$.

We may regard $p_r(\gamma)$ as the rth power sum of a new sequence of variables $y_\gamma = (y_{i\gamma})_{i \geqslant 1}$, and we may then define, for example, Schur functions $s_\mu(\gamma) = s_\mu(y_\gamma)$ for any partition μ, and more generally

$$S_\lambda = \prod_{\gamma \in G^*} s_{\lambda(\gamma)}(\gamma) \tag{7.3}$$

for any $\lambda \in \mathscr{P}(G^*)$.

(7.4) $(S_\lambda)_{\lambda \in \mathscr{P}(G^*)}$ *is an orthonormal basis of* $\Lambda(G)$.

Proof. In view of the definition (5.3') of the scalar product on $\Lambda(G)$, it is enough to show that

$$\sum_\lambda S_\lambda \otimes \bar{S}_\lambda = \sum_\rho Z_\rho^{-1} P_\rho \otimes P_\rho, \tag{1}$$

where the sum on the left is over all $\lambda \in \mathscr{P}(G^*)$, and that on the right is over all $\rho \in \mathscr{P}(G_*)$. Now the left-hand side of (1) is equal to

$$\prod_{\gamma \in G^*} \left(\sum_{\mu} s_\mu(\gamma) \otimes \overline{s_\mu(\gamma)} \right) = \prod_{\gamma \in G^*} \left(\exp \sum_{r \geq 1} \frac{1}{r} p_r(\gamma) \otimes \overline{p_r(\gamma)} \right)$$

$$(2) \qquad = \exp\left(\sum_{r \geq 1} \frac{1}{r} \sum_{\gamma \in G^*} p_r(\gamma) \otimes \overline{p_r(\gamma)} \right).$$

On the other hand, the right-hand side of (1) is by virtue of (3.1) equal to

$$\prod_{c \in G_*} \left(\sum_{\sigma} \zeta_c^{-l(\sigma)} z_\sigma^{-1} p_\sigma(c) \otimes p_\sigma(c) \right) = \prod_{c \in G_*} \left(\exp \sum_{r \geq 1} \frac{1}{r} \zeta_c^{-1} p_r(c) \otimes p_r(c) \right)$$

$$(3) \qquad = \exp\left(\sum_{r \geq 1} \frac{1}{r} \sum_{c \in G_*} \zeta_c^{-1} p_r(c) \otimes p_r(c) \right)$$

Now (7.2) shows that (2) and (3) are equal. |

8. The characters $\eta_{n,\gamma}$

Let E_γ be a G-module with character $\gamma \in G^*$. The group G_n acts on the nth tensor power $\mathbf{T}^n(E_\gamma) = E_\gamma \otimes \ldots \otimes E_\gamma$ as follows: if $u_1, \ldots, u_n \in E_\gamma$ and $(g, s) \in G_n$, where $g = (g_1, \ldots, g_n) \in G^n$ and $s \in S_n$, then

$$(g, s)(u_1 \otimes \ldots \otimes u_n) = g_1 u_{s^{-1}(1)} \otimes \ldots \otimes g_n u_{s^{-1}(n)}.$$

We wish to compute the character $\eta_n = \eta_n(\gamma)$ of this representation of G_n. First, if $x \in G_r$ and $y \in G_{n-r}$, so that x acts on the first r factors of $\mathbf{T}^n(E_\gamma)$ and y on the last $n - r$ factors, it is clear that

$$(8.1) \qquad \eta_n(x \times y) = \eta_r(x)\eta_{n-r}(y).$$

Hence it is enough to compute $\eta_n(g, s)$ when $g \in G^n$ and s is an n-cycle, say $s = (12 \ldots n)$. For this purpose let $e_1, \ldots, e_d$ be a basis of E_γ and let

$$g e_j = \sum_i a_{ij}(g) e_i$$

so that $g \mapsto a(g) = (a_{ij}(g))$ is the matrix representation of G defined by this basis. Then we have

$$(g, s)(e_{j_1} \otimes \ldots \otimes e_{j_n}) = g_1 e_{j_n} \otimes g_2 e_{j_1} \otimes \ldots \otimes g_n e_{j_{n-1}}$$

in which the coefficient of $e_{j_1} \otimes \ldots \otimes e_{j_n}$ is

$$a_{j_1 j_n}(g_1) a_{j_2 j_1}(g_2) \ldots a_{j_n j_{n-1}}(g_n). \tag{1}$$

To obtain the trace we must sum (1) over all $j_1, \ldots, j_n$, which gives

$$\begin{aligned} \eta_n(g, s) &= \text{trace}\, a(g_n) a(g_{n-1}) \ldots a(g_1) \\ &= \text{trace}\, a(g_n g_{n-1} \cdots g_1) = \gamma(c) \end{aligned}$$

if the cycle-product $g_n g_{n-1} \cdots g_1$ lies in the conjugacy class c. From (8.1) it now follows that if $x \in G_n$ has type ρ, then

$$\eta_n(\gamma)(x) = \prod_c \gamma(c)^{l(\rho(c))}. \tag{8.2}$$

We now calculate:

$$\begin{aligned} \sum_{n \geqslant 0} \text{ch}(\eta_n(\gamma)) &= \sum_\rho Z_\rho^{-1} \prod_{c \in G_*} \gamma(c)^{l(\rho(c))} p_{\rho(c)}(c) \\ &= \prod_{c \in G_*} \left(\sum_\sigma (\zeta_c^{-1} \gamma(c))^{l(\sigma)} z_\sigma^{-1} p_\sigma(c) \right) \\ &= \prod_{c \in G_*} \exp\left(\sum_{r \geqslant 1} \frac{1}{r} \zeta_c^{-1} \gamma(c) p_r(c) \right) \\ &= \exp\left(\sum_{r \geqslant 1} \frac{1}{r} \sum_{c \in G_*} \zeta_c^{-1} \gamma(c) p_r(c) \right) \\ &= \exp\left(\sum_{r \geqslant 1} \frac{1}{r} p_r(\gamma) \right) \qquad \text{by (7.1)} \\ &= \sum_{n \geqslant 0} h_n(\gamma). \end{aligned}$$

Hence

$$\text{ch}(\eta_n(\gamma)) = h_n(\gamma) \tag{8.3}$$

for all $\gamma \in G^*$ and $n \geqslant 0$.

9. The irreducible characters of G_n

For each partition μ of m and each $\gamma \in G^*$, let

$$\chi^\mu(\gamma) = \det\left(\eta_{\mu_i - i + j}(\gamma) \right). \tag{9.1}$$

This is an alternating sum of induction products of characters, hence is a (perhaps virtual) *character* of G_m. From (8.3) and (6.3) we have

$$\text{ch}(\chi^\mu(\gamma)) = \det\left(h_{\mu_i-i+j}(\gamma)\right) = s_\mu(\gamma). \tag{9.2}$$

Next, for each $\lambda \in \mathscr{P}_n(G^*)$, define

$$X^\lambda = \prod_{\gamma \in G^*} \chi^{\lambda(\gamma)}(\gamma) \tag{9.3}$$

which for the same reason is a character of G_n. From (9.2) it follows that

$$\text{ch}(X^\lambda) = \prod_{\gamma \in G^*} s_{\lambda(\gamma)}(\gamma) = S_\lambda; \tag{9.4}$$

and since ch is an isometry (6.3) it follows from (7.4) that

$$\langle X^\lambda, X^\mu \rangle_{G_n} = \langle S_\lambda, S_\mu \rangle = \delta_{\lambda\mu}$$

for all $\lambda, \mu \in \mathscr{P}_n(G^*)$. Hence the X^λ are, up to sign, distinct irreducible characters of G_n; and since $|\mathscr{P}_n(G^*)| = |\mathscr{P}_n(G_*)| = |(G_n)_*|$, there are as many of them as there are conjugacy classes in G_n, so that they are *all* the irreducible characters of G_n. It remains to settle the question of sign, which we shall do by computing the degree of each character X^λ.

Let X_ρ^λ denote the value of X^λ at elements of G_n of type $\rho \in \mathscr{P}_n(G_*)$. From (9.4) and (6.2) we have

$$S_\lambda = \sum_\rho Z_\rho^{-1} X_\rho^\lambda P_\rho \tag{9.5}$$

or equivalently

$$X_\rho^\lambda = \langle S_\lambda, P_\rho \rangle, \tag{9.5$'$}$$

$$P_\rho = \sum_\lambda X_\rho^\lambda \overline{S}_\lambda = \sum_\lambda \overline{X}_\rho^\lambda S_\lambda. \tag{9.5$''$}$$

Let $c_0 \in G_*$ be the class consisting of the identity element. The type of the identity element of G_n is ρ where $\rho(c_0) = (1^n)$, $\rho(c) = 0$ if $c \neq c_0$. Hence $P_\rho = p_1(c_0)^n$, and since $p_1(c_0) = \Sigma_\gamma d_\gamma p_1(\gamma)$, where $d_\gamma = \gamma(c_0)$ is the degree of γ, it follows from (9.5′) that the degree of X^λ is equal to the coefficient of S_λ in $(\Sigma_\gamma d_\gamma p_1(\gamma))^n$, i.e. to the coefficient of $\Pi_\gamma s_{\lambda(\gamma)}(\gamma)$ in

$$\frac{n!}{\prod_\gamma |\lambda(\gamma)|!} \prod_\gamma \left(d_\gamma p_1(\gamma)\right)^{|\lambda(\gamma)|}$$

which by Chapter I, (7.6) is equal to

$$n! \prod_{\gamma} \left(d_{\gamma}^{|\lambda(\gamma)|} / h(\lambda(\gamma)) \right) \tag{9.6}$$

where $h(\lambda(\gamma))$ is the product of the hook-lengths of the partition $\lambda(\gamma)$. Since this number is positive, it follows that X^{λ} (and not $-X^{\lambda}$) is an irreducible character. So, finally,

(9.7) *The irreducible complex characters of* $G_n = G \sim S_n$ *are the* X^{λ} ($\lambda \in \mathscr{P}_n(G^*)$) *defined by* (9.3), *and the value of* X^{λ} *at elements of type* $\rho \in \mathscr{P}_n(G_*)$ *is*

$$X_{\rho}^{\lambda} = \langle S_{\lambda}, P_{\rho} \rangle.$$

Moreover the degree of X^{λ} *is given by* (9.6).

Example

The simplest nontrivial case of this theory is that in which the group G has two elements ± 1. Then G_n is the group of signed permutation matrices, of order $2^n n!$, or equivalently the hyperoctahedral group of rank n. The conjugacy classes and the irreducible characters of G_n are each indexed by pairs of partitions (λ, μ) such that $|\lambda| + |\mu| = n$.

If the power sums corresponding to the identity (resp. non-identity) conjugacy class are denoted by $p_r(a)$ (resp. $p_r(b)$), and those corresponding to the trivial (resp. nontrivial) character by $p_r(x)$ (resp. $p_r(y)$), the change of variables formula (7.1′) reads

$$p_r(a) = p_r(x) + p_r(y), \qquad p_r(b) = p_r(x) - p_r(y)$$

and the formula (9.5″) reads

$$p_{\rho}(a) p_{\sigma}(b) = \sum_{\lambda, \mu} X_{\rho, \sigma}^{\lambda, \mu} s_{\lambda}(x) s_{\mu}(y)$$

where $X_{\rho,\sigma}^{\lambda,\mu}$ is the value of the character indexed by (λ, μ) at the class indexed by (ρ, σ) (where $|\lambda| + |\mu| = |\rho| + |\sigma| = n$).

Notes and references

The characters of the wreath products $G \sim S_n$, where G is any finite group, were first worked out by W. Specht in his dissertation [S17], and our account does not differ materially from his. See also [M3].

II

HALL POLYNOMIALS

1. Finite $\mathfrak{o}$-modules

Let $\mathfrak{o}$ be a (commutative) discrete valuation ring, $\mathfrak{p}$ its maximal ideal, $k = \mathfrak{o}/\mathfrak{p}$ the residue field. Later we shall require k to be a finite field, but for the present this restriction is unnecessary. We shall be concerned with *finite* $\mathfrak{o}$-modules M, that is to say, modules M which possess a finite composition series, or equivalently finitely-generated $\mathfrak{o}$-modules M such that $\mathfrak{p}^r M = 0$ for some $r \geqslant 0$. If k is finite, the finite $\mathfrak{o}$-modules are precisely those which have a finite number of elements.

Two examples to bear in mind are

(1.1) *Example*. Let p be a prime number, M a finite abelian p-group. Then $p^r M = 0$ for large r, so that M may be regarded as a module over the ring $\mathbf{Z}/p^r\mathbf{Z}$ for all large r, and hence as a module over the ring $\mathfrak{o} = \mathbf{Z}_p$ of p-adic integers. The residue field is $k = \mathbf{F}_p$.

(1.2) *Example*. Let k be a field, M a finite-dimensional vector space over k, and let T be a nilpotent endomorphism of M. Then M may be regarded as a $k[t]$-module, where t is an indeterminate, by defining $tx = Tx$ for all $x \in M$. Since T is nilpotent we have $t^r M = 0$ for all large r, and hence M may be regarded as a module over the power series ring $\mathfrak{o} = k[[t]]$, which is a discrete valuation ring with residue field k.

Remarks. 1. In both these examples the ring $\mathfrak{o}$ is a *complete* discrete valuation ring. In general, if M is a finite $\mathfrak{o}$-module as at the beginning, we have $\mathfrak{p}^r M = 0$ for all sufficiently large r, so that M is an $\mathfrak{o}/\mathfrak{p}^r$-module and hence a module over the $\mathfrak{p}$-adic completion $\hat{\mathfrak{o}}$ of $\mathfrak{o}$, which has the same residue field k as $\mathfrak{o}$. Hence there would be no loss of generality in assuming at the outset that $\mathfrak{o}$ is complete.

2. Suppose now that k is finite. The complete discrete valuation rings with finite residue field are precisely the rings of integers of $\mathfrak{p}$-adic fields (Bourbaki, *Alg. Comm.*, Chapter VI, §9), and a $\mathfrak{p}$-adic field K is either a finite extension of the field $\mathbf{Q}_p$ of p-adic numbers (if char. $K = 0$) or is a field of formal power series $k((t))$ over a finite field (if char. $K > 0$). The two examples (1.1) and (1.2) (with k finite) are therefore typical.

3. The results of this chapter will all be valid under the wider hypothesis that $\mathfrak{o}$ is the ring of integers (i.e. the unique maximal order) of a division algebra of finite rank over a $\mathfrak{p}$-adic field (Deuring, *Algebren*, Ch. VI, §11).

Since $\mathfrak{o}$ is a principal ideal domain, every finitely generated $\mathfrak{o}$-module is a direct sum of cyclic $\mathfrak{o}$-modules. For a finite $\mathfrak{o}$-module M, this means that M has a direct sum decomposition of the form

$$(1.3) \qquad M \cong \bigoplus_{i=1}^{r} \mathfrak{o}/\mathfrak{p}^{\lambda_i}$$

where the λ_i are positive integers, which we may assume are arranged in descending order: $\lambda_1 \geqslant \lambda_2 \geqslant \ldots \geqslant \lambda_r > 0$. In other words, $\lambda = (\lambda_1, \ldots, \lambda_r)$ is a *partition*.

(1.4) *Let* $\mu_i = \dim_k(\mathfrak{p}^{i-1}M/\mathfrak{p}^iM)$. *Then* $\mu = (\mu_1, \mu_2, \ldots)$ *is the conjugate of the partition* λ.

Proof. Let x_j be a generator of the summand $\mathfrak{o}/\mathfrak{p}^{\lambda_i}$ in (1.3), and let π be a generator of $\mathfrak{p}$. Then $\mathfrak{p}^{i-1}M$ is generated by those of the $\pi^{i-1}x_j$ which do not vanish, i.e. those for which $\lambda_j \geqslant i$. Hence μ_j is equal to the number of indices j such that $\lambda_j \geqslant i$, and therefore $\mu_i = \lambda'_i$. |

From (1.4) it follows that the partition λ is determined uniquely by the module M, and we call λ the *type* of M. Clearly two finite $\mathfrak{o}$-modules are isomorphic if and only if they have the same type, and every partition λ occurs as a type. If λ is the type of M, then $|\lambda| = \sum \lambda_i$ is the *length* $l(M)$ of M, i.e. the length of a composition series of M. The length is an additive function of M: this means that if

$$0 \to M' \to M \to M'' \to 0$$

is a short exact sequence of finite $\mathfrak{o}$-modules, then

$$l(M') - l(M) + l(M'') = 0.$$

If N is a submodule of M, the *cotype* of N in M is defined to be the type of M/N.

Cyclic and elementary modules

A finite $\mathfrak{o}$-module M is *cyclic* (i.e. generated by one element) if and only if its type is a partition (r) consisting of a single part $r = l(M)$, and M is *elementary* (i.e. $\mathfrak{p}M = 0$) if and only if the type of M is (1^r). If M is elementary of type (1^r), then M is a vector space over k, and $l(M) = \dim_k M = r$.

Duality

Let π be a generator of the maximal ideal $\mathfrak{p}$. If $m \leqslant n$, multiplication by

π^{n-m} is an injective $\mathfrak{o}$-homomorphism of $\mathfrak{o}/\mathfrak{p}^m$ into $\mathfrak{o}/\mathfrak{p}^n$. Let E denote the direct limit:

$$E = \varinjlim \mathfrak{o}/\mathfrak{p}^n.$$

Then E is an injective $\mathfrak{o}$-module containing $\mathfrak{o}/\mathfrak{p} = k$, and is 'the' injective envelope of k, i.e. the smallest injective $\mathfrak{o}$-module which contains k as a submodule.

If now M is any finite $\mathfrak{o}$-module, the *dual* of M is defined to be

$$\hat{M} = \mathrm{Hom}_{\mathfrak{o}}(M, E).$$

$\hat{M}$ is a finite $\mathfrak{o}$-module isomorphic to M, hence of the same type as M. (To see this, observe that $M \mapsto \hat{M}$ commutes with direct sums; hence it is enough to check that $\hat{M} \cong M$ when M is cyclic (and finite), which is easy.) Since E is injective, an exact sequence

$$0 \to N \to M \to M/N \to 0$$

(where N is a submodule of M) gives rise to an exact sequence

$$0 \leftarrow \hat{N} \leftarrow \hat{M} \leftarrow (M/N)\hat{} \leftarrow 0$$

and $(M/N)\hat{}$ is the annihilator N^0 of N in $\hat{M}$, i.e. the set of all $\xi \in \hat{M}$ such that $\xi(N) = 0$. The natural mapping $M \to \hat{\hat{M}}$ is an isomorphism for all finite $\mathfrak{o}$-modules M, and identifies N with N^{00}. Hence

(1.5) *$N \leftrightarrow N^0$ is a one-one correspondence between the submodules of M, $\hat{M}$ respectively, which maps the set of all $N \subset M$ of type ν and cotype μ onto the set of all $N^0 \subset \hat{M}$ of type μ and cotype ν.* |

Automorphisms

Suppose that the residue field k is finite, with q elements. If M is a finite $\mathfrak{o}$-module and x is a non-zero element of M, we shall say that x has *height* r if $\mathfrak{p}^r x = 0$ and $\mathfrak{p}^{r-1} x \neq 0$. The zero element of M is assigned height 0. We denote by M_r the submodule of M consisting of elements of height $\leqslant r$, so that M_r is the annihilator of $\mathfrak{p}^r$ in M.

(1.6) *The number of automorphisms of a finite $\mathfrak{o}$-module M of type λ is*

$$a_\lambda(q) = q^{|\lambda| + 2n(\lambda)} \prod_{i \geqslant 1} \varphi_{m_i(\lambda)}(q^{-1}),$$

where as usual $\varphi_m(t) = (1-t)(1-t^2)\dots(1-t^m)$.

The number of automorphisms of M is equal to the number of sequences $(x_1, \dots, x_r)$ of elements of M such that x_i has height $\lambda_i (1 \leqslant i \leqslant r)$

and M is the direct sum of the cyclic submodules $\mathfrak{o}x_i$. To enumerate such sequences we shall use the following lemma:

(1.7) *Let N be a submodule of M, generated by elements of height $\geqslant r$, and let $x \in M$. Then the following conditions on x are equivalent:*
(i) *x has height r and $\mathfrak{o}x \cap N = 0$;*
(ii) *$x \in M_r - (M_{r-1} + N_r)$.*
Moreover the number of $x \in M$ satisfying these equivalent conditions is

$$q^{\lambda'_1 + \dots + \lambda'_r}(1 - q^{\nu'_r - \lambda'_r}) \tag{1.8}$$

if ν is the type of N.

Proof. If x satisfies (i), clearly $x \in M_r$. If $x \in M_{r-1} + N_r$, then $0 \neq \mathfrak{p}^{r-1}x \subset N_r \subset N$, so that $\mathfrak{o}x \cap N \neq 0$, contrary to assumption. Conversely, if x satisfies (ii), it is clear that height $(x) = r$. If $\mathfrak{o}x \cap N \neq 0$, then for some $m < r$ we shall have $\mathfrak{p}^m x \subset N$, and therefore $\mathfrak{p}^{r-1}x$ is contained in the socle N_1 of N. Since N is generated by elements of height $\geqslant r$, it follows that $\mathfrak{p}^{r-1}x = \mathfrak{p}^{r-1}y$ for some $y \in N$; hence $x - y \in M_{r-1}$ and therefore $x \in (M_{r-1} + N) \cap M_r = M_{r-1} + N_r$.

We have $M/M_r \cong \mathfrak{p}^r M$, so that

$$l(M_r) = l(M) - l(\mathfrak{p}^r M) = \sum_{i=1}^{r} l(\mathfrak{p}^{i-1}M/\mathfrak{p}^i M) = \lambda'_1 + \dots + \lambda'_r$$

by (1.4). Also

$$\begin{aligned} l(M_{r-1} + N_r) &= l(M_{r-1}) + l((M_{r-1} + N_r)/M_{r-1}) \\ &= l(M_{r-1}) + l(N_r/N_{r-1}) \\ &= \lambda'_1 + \dots + \lambda'_{r-1} + \nu'_r \end{aligned}$$

which proves (1.8). |

The number of automorphisms of M is therefore the product of the numbers (1.8) for $r = \lambda_1, \lambda_2, \dots$, where $\nu = (\lambda_1, \dots, \lambda_{k-1})$ if $r = \lambda_k$. It is not hard to see that this product is equal to $a_\lambda(q)$ as defined in (1.6).

Example

Let M, N be finite $\mathfrak{o}$-modules of types λ, μ respectively. Then $M \oplus N$ has type $\lambda \cup \mu$, and $M \otimes N$, $\mathrm{Hom}(M, N)$, $\mathrm{Tor}_1(M, N)$, and $\mathrm{Ext}^1(M, N)$ all have type $\lambda \times \mu$ (Chapter I, §1).

2. The Hall algebra

In this section the residue field k of $\mathfrak{o}$ is assumed to be *finite*.

Let $\lambda, \mu^{(1)}, \dots, \mu^{(r)}$ be partitions, and let M be a finite $\mathfrak{o}$-module of type λ. We define

$$G^{\lambda}_{\mu^{(1)}\dots\mu^{(r)}}(\mathfrak{o})$$

to be the number of chains of submodules of M:

$$M = M_0 \supset M_1 \supset \dots \supset M_r = 0$$

such that M_{i-1}/M_i has type $\mu^{(i)}$, for $1 \leqslant i \leqslant r$. In particular, $G^{\lambda}_{\mu\nu}(\mathfrak{o})$ is the number of submodules N of M which have type ν and cotype μ. Since $l(M) = l(M/N) + l(N)$, it is clear that

(2.1) $G^{\lambda}_{\mu\nu}(\mathfrak{o}) = 0$ *unless* $|\lambda| = |\mu| + |\nu|$.

Philip Hall had the idea of using the numbers $G^{\lambda}_{\mu\nu}(\mathfrak{o})$ as the multiplication constants of a ring, as follows. Let $H = H(\mathfrak{o})$ be a free $\mathbf{Z}$-module on a basis (u_λ) indexed by all partitions λ. Define a product in H by the rule

$$u_\mu u_\nu = \sum_\lambda G^{\lambda}_{\mu\nu}(\mathfrak{o}) u_\lambda.$$

By (2.1) the sum on the right has only finitely many non-zero terms.

(2.2) $H(\mathfrak{o})$ *is a commutative and associative ring with identity element.*

Proof. The identity element is u_0, where 0 is the empty partition. Associativity follows from the fact that the coefficient of u_λ in either $u_\mu(u_\nu u_\rho)$ or $(u_\mu u_\nu)u_\rho$ is just $G^{\lambda}_{\mu\nu\rho}$. Commutativity follows from (1.5), which shows that $G^{\lambda}_{\mu\nu} = G^{\lambda}_{\nu\mu}$. ∎

The ring $H(\mathfrak{o})$ is the *Hall algebra* of $\mathfrak{o}$.

(2.3) *The ring* $H(\mathfrak{o})$ *is generated (as* $\mathbf{Z}$*-algebra) by the elements* $u_{(1^r)}$ $(r \geqslant 1)$, *and they are algebraically independent over* $\mathbf{Z}$.

Proof. For convenience let us write v_r in place of $u_{(1^r)}$, and for any partition λ consider the product

$$v_{\lambda'} = v_{\lambda'_1} v_{\lambda'_2} \dots v_{\lambda'_s}$$

where $\lambda' = (\lambda'_1, \dots, \lambda'_s)$ is as usual the conjugate of λ. This product $v_{\lambda'}$ will be a linear combination of the u_μ, say

(1) $$v_{\lambda'} = \sum_\mu a_{\lambda\mu} u_\mu$$

in which the coefficient $a_{\lambda\mu}$ is by definition equal to the number of chains

$$M = M_0 \supset M_1 \supset \ldots \supset M_s = 0 \tag{2}$$

in a fixed finite $\mathfrak{o}$-module M of type μ, such that M_{i-1}/M_i is of type $(1^{\lambda'_i})$, i.e. elementary of length λ'_i, for $1 < i < s$. If such a chain (2) exists (that is, if $a_{\lambda\mu} \neq 0$) we must have $\mathfrak{p}M_{i-1} \subset M_i$ $(1 \leqslant i \leqslant s)$ and therefore $\mathfrak{p}^i M \subset M_i$ for $1 < i < s$. Hence

$$l(M/\mathfrak{p}^i M) \geqslant l(M/M_i)$$

which by virtue of (1.4) gives the inequality

$$\mu'_1 + \ldots + \mu'_i \geqslant \lambda'_1 + \ldots + \lambda'_i$$

for $1 \leqslant i \leqslant s$. Hence $\mu' \geqslant \lambda'$ and therefore by Chapter I, (1.11), $\mu \leqslant \lambda$. Moreover, the same reasoning shows that if $\mu = \lambda$ there is only one possible chain (2), namely $M_i = \mathfrak{p}^i M$.

Consequently we have $a_{\lambda\mu} = 0$ unless $\mu \leqslant \lambda$, and $a_{\lambda\lambda} = 1$. In other words, the matrix $(a_{\lambda\mu})$ is strictly upper unitriangular (Chapter I, §6), and so the equations (1) can be solved to give the u_μ as integral linear combinations of the $v_{\lambda'}$. Hence the $v_{\lambda'}$ form a $\mathbf{Z}$-basis of $H(\mathfrak{o})$, which proves (2.3). |

From (2.3) it follows that the Hall algebra $H(\mathfrak{o})$ is isomorphic to the ring Λ of symmetric functions (Chapter I). The obvious choice of isomorphism would be that which takes each $u_{(1^r)}$ to the rth elementary symmetric function e_r; however, as we shall see in the next chapter, a more intelligent choice is to map $u_{(1^r)}$ to $q^{-r(r-1)/2}e_r$, where q is the number of elements in the residue field of $\mathfrak{o}$. Thus each generator u_λ of $H(\mathfrak{o})$ is mapped to a symmetric function. We shall identify and study these symmetric functions in the next chapter; the remainder of the present chapter will be devoted to computing the structure constants $G^\lambda_{\mu\nu}(\mathfrak{o})$.

3. The *LR*-sequence of a submodule

Let T be a tableau (Chapter I, §1) of shape $\lambda - \mu$ and weight $\nu = (\nu_1, \ldots, \nu_r)$. Then T determines (and is determined by) a sequence of partitions

$$S = (\lambda^{(0)}, \lambda^{(1)}, \ldots, \lambda^{(r)})$$

such that $\lambda^{(0)} = \mu$, $\lambda^{(r)} = \lambda$, and $\lambda^{(i)} \supset \lambda^{(i-1)}$ for $1 \leqslant i \leqslant r$, by the condition that $\lambda^{(i)} - \lambda^{(i-1)}$ is the skew diagram consisting of the squares occupied by the symbol i in T (and hence is a horizontal strip, because T is a tableau).

A sequence of partitions S as above will be called a *LR-sequence of type* $(\mu, \nu; \lambda)$ if

(LR1) $\lambda^{(0)} = \mu$, $\lambda^{(r)} = \lambda$, and $\lambda^{(i)} \supset \lambda^{(i-1)}$ for $1 \leqslant i \leqslant r$;

(LR2) $\lambda^{(i)} - \lambda^{(i-1)}$ is a horizontal strip of length ν_i, for $1 \leqslant i \leqslant r$. (These two conditions ensure that S determines a tableau T.)

(LR3) The word $w(T)$ obtained by reading T from right to left in successive rows, starting at the top, is a lattice permutation (Chapter I, §9).

For (LR3) to be satisfied it is necessary and sufficient that, for $i \geqslant 1$ and $k \geqslant 0$, the number of symbols i in the first k rows of T should be not less than the number of symbols $i+1$ in the first $k+1$ rows of T. In other words, a condition equivalent to (LR3) is

$$\text{(LR3}'\text{)} \qquad \sum_{j=1}^{k} \left(\lambda_j^{(i)} - \lambda_j^{(i-1)}\right) \geqslant \sum_{j=1}^{k+1} \left(\lambda_j^{(i+1)} - \lambda_j^{(i)}\right)$$

for all $i \geqslant 1$ and $k \geqslant 0$.

We shall show in this section that every submodule N of a finite $\mathfrak{o}$-module M gives rise to a LR-sequence of type $(\mu', \nu'; \lambda')$, where λ, μ, ν are the types of M, M/N, and N respectively. Before we come to the proof, a few lemmas are required. We do not need to assume in this section that the residue field of $\mathfrak{o}$ is finite.

(3.1) *Let M be a finite $\mathfrak{o}$-module of type λ, and let N be a submodule of type ν and cotype μ in M. Then $\mu \subset \lambda$ and $\nu \subset \lambda$.*

Proof. Since

$$\frac{\mathfrak{p}^{i-1}(M/N)}{\mathfrak{p}^i(M/N)} \cong \frac{\mathfrak{p}^{i-1}M + N}{\mathfrak{p}^i M + N} \cong \frac{\mathfrak{p}^{i-1}M}{\mathfrak{p}^{i-1}M \cap (\mathfrak{p}^i M + N)}$$

and since also $\mathfrak{p}^{i-1}M \cap (\mathfrak{p}^i M + N) \supset \mathfrak{p}^i M$, it follows that

$$l(\mathfrak{p}^{i-1}(M/N)/\mathfrak{p}^i(M/N)) \leqslant l(\mathfrak{p}^{i-1}M/\mathfrak{p}^i M)$$

and hence that $\mu_i' \leqslant \lambda_i'$ by (1.4). Consequently $\mu \subset \lambda$. By duality (1.5), it follows that $\nu \subset \lambda$ also. ∎

Let M be an $\mathfrak{o}$-module of type λ, N an elementary submodule of M. Then $\mathfrak{p}N = 0$, so that $N \subset S$ where

$$S = \{x \in M : \mathfrak{p}x = 0\}$$

is the *socle* of M, i.e. the unique largest elementary submodule of M.

(3.2) *The type of M/S is* $\tilde{\lambda} = (\lambda_1 - 1, \lambda_2 - 1, \ldots)$.

Proof. If $M = \oplus\, \mathfrak{o}/\mathfrak{p}^{\lambda_i}$, then clearly $S = \oplus\, \mathfrak{p}^{\lambda_i - 1}/\mathfrak{p}^{\lambda_i}$, whence $M/S \cong \oplus\, \mathfrak{o}/\mathfrak{p}^{\lambda_i - 1}$. ∎

(3.3) *Let M be a finite $\mathfrak{o}$-module of type λ, and N an elementary submodule of M, of cotype μ. Then $\lambda - \mu$ is a vertical strip (i.e. $\lambda_i - \mu_i = 0$ or 1 for each i).*

Proof. We have $N \subset S$, hence $M/S \cong (M/N)/(S/N)$, and therefore $\tilde{\lambda} \subset \mu \subset \lambda$ by (3.1) and (3.2). Hence

$$0 \leqslant \lambda_i - \mu_i \leqslant \lambda_i - \tilde{\lambda}_i = 1$$

and therefore $\lambda - \mu$ is a vertical strip. ∎

Notice that if $\mu \subset \lambda$ then $\lambda - \mu$ is a vertical strip if and only if $\tilde{\lambda} \subset \mu$.

(3.4) *Let M be a finite $\mathfrak{o}$-module of type λ, and let N be a submodule of M, of type ν and cotype μ. For each $i \geqslant 0$, let $\lambda^{(i)}$ be the cotype of $\mathfrak{p}^i N$. Then the sequence*

$$S(N) = (\lambda^{(0)\prime}, \lambda^{(1)\prime}, \ldots, \lambda^{(r)\prime})$$

(where $\mathfrak{p}^r N = 0$) is an LR-sequence of type $(\mu', \nu'; \lambda')$.

Proof. Clearly $\lambda^{(0)} = \mu$ and $\lambda^{(r)} = \lambda$, and $\lambda^{(i)} \supset \lambda^{(i-1)}$ by (3.1) applied to the module $M/\mathfrak{p}^i N$ and the submodule $\mathfrak{p}^{i-1}N/\mathfrak{p}^i N$. Hence (LR1) is satisfied. Since $\mathfrak{p}^{i-1}N/\mathfrak{p}^i N$ is an elementary $\mathfrak{o}$-module, it follows from (3.3) that $\lambda^{(i)} - \lambda^{(i-1)}$ is a vertical strip, and hence that $\lambda^{(i)\prime} - \lambda^{(i-1)\prime}$ is a horizontal strip, of length equal to $l(\mathfrak{p}^{i-1}N/\mathfrak{p}^i N) = \nu_i'$ (by (1.4)). Hence (LR2) is satisfied.

As to (LR3′), we have

$$\lambda_j^{(i)\prime} = l(\mathfrak{p}^{j-1}(M/\mathfrak{p}^i N)/\mathfrak{p}^j(M/\mathfrak{p}^i N))$$

(again by (1.4)), so that

$$\sum_{j=1}^{k} \lambda_j^{(i)\prime} = l((M/\mathfrak{p}^i N)/\mathfrak{p}^k(M/\mathfrak{p}^i N)) = l(M/(\mathfrak{p}^k M + \mathfrak{p}^i N))$$

and therefore

$$\sum_{j=1}^{k} \left(\lambda_j^{(i)\prime} - \lambda_j^{(i-1)\prime}\right) = l(V_{ki})$$

where $V_{ki} = (\mathfrak{p}^k M + \mathfrak{p}^{i-1}N)/(\mathfrak{p}^k M + \mathfrak{p}^i N)$. Likewise

$$\sum_{j=1}^{k+1} \left(\lambda_j^{(i+1)\prime} - \lambda_j^{(i)\prime}\right) = l(V_{k+1,i+1}).$$

Since multiplication by a generator of $\mathfrak{p}$ induces a homomorphism of V_{ki} onto $V_{k+1,i+1}$, it follows that $l(V_{ki}) \geqslant l(V_{k+1,i+1})$, and hence (LR3′) is satisfied. |

Example

Let V be an n-dimensional vector space over an algebraically closed field k. A *flag* in V is a sequence $F = (V_0, V_1, \dots, V_n)$ of subspaces of V, such that $0 = V_0 \subset V_1 \subset \dots \subset V_n = V$ and $\dim V_i = i$ for $0 \leqslant i \leqslant n$. Let X denote the set of all flags in V. The group $G = GL(V)$ acts transitively on X, so that X may be identified with G/B, where B is the subgroup which fixes a given flag, and therefore X is a (non-singular projective) algebraic variety, the *flag manifold* of V.

Now let $u \in G$ be a unipotent endomorphism of V. Then as in (1.2) V becomes a $k[t]$-module of finite length, with t acting on V as the nilpotent endomorphism $u-1$. Let λ be the type of V, so that λ is a partition of n which describes the Jordan canonical form of u, and let $X_\lambda \subset X$ be the set of all flags $F \in X$ fixed by u. These flags F are the composition series of the $k[t]$-module V. The set X_λ is a closed subvariety of X.

For each $F = (V_0, V_1, \dots, V_n) \in X_\lambda$, let $\lambda^{(i)}$ be the cotype of the submodule V_{n-i} of V. Then by (3.1) we have

$$0 = \lambda^{(0)} \subset \lambda^{(1)} \subset \dots \subset \lambda^{(n)} = \lambda$$

and $|\lambda^{(i)} - \lambda^{(i-1)}| = 1$ for $1 \leqslant i \leqslant n$, so that F determines in this way a *standard tableau* of T of shape λ. Hence we have a partition of X_λ into subsets X_T indexed by the standard tableaux T of shape λ.

These subsets X_T have the following properties (Spaltenstein [S16]):

(a) X_T is a smooth irreducible locally closed subvariety of X_λ.

(b) $\dim X_T = n(\lambda)$.

(c) X_T is a disjoint union $\bigcup_{j=1}^m X_{T,j}$ such that each $X_{T,j}$ is isomorphic to an affine space and $\bigcup_{k \leqslant j} X_{T,k}$ is closed in X_T, for $j = 1, 2, \dots, m$.

From these results it follows that the closures $\bar{X}_T$ of the X_T are the irreducible components of X_λ, and all have the same dimension $n(\lambda)$. The number of irreducible components is therefore equal to the degree of the irreducible character χ^λ of S_n (Chapter I, §7).

If k contains the finite field $\mathbf{F}_q$ of q elements, the number $X_\lambda(q)$ of $\mathbf{F}_q$-rational points of X_λ is equal to $Q^\lambda_{(1^n)}(q)$ (Chapter III, §7).

More general results concerning the partial flags may be found in [H8], [S14].

4. The Hall polynomial

In this section we shall compute the structure constants $G^\lambda_{\mu\nu}(\mathfrak{o})$ of the Hall algebra. (The residue field of $\mathfrak{o}$ is assumed to be finite, as in §2.) Let S be an LR-sequence of type $(\mu', \nu'; \lambda')$, and let M be a finite $\mathfrak{o}$-module of type λ. Denote by $G_S(\mathfrak{o})$ the number of submodules N of M whose associated LR-sequence $S(N)$ is S. By (3.4), each such N has type ν and cotype μ.

Let q denote the number of elements in the residue field of $\mathfrak{o}$, and recall that $n(\lambda) = \Sigma(i-1)\lambda_i$, for any partition λ. Then:

(4.1) *For each LR-sequence S of type $(\mu', \nu'; \lambda')$, there exists a monic polynomial $g_S(t) \in \mathbf{Z}[t]$ of degree $n(\lambda) - n(\mu) - n(\nu)$, independent of $\mathfrak{o}$, such that*

$$g_S(q) = G_S(\mathfrak{o})$$

(In other words, $G_S(\mathfrak{o})$ is a 'polynomial in q'.)

Now define, for any three partitions λ, μ, ν

$$g^{\lambda}_{\mu\nu}(t) = \sum_S g_S(t) \tag{4.2}$$

summed over all LR-sequences S of type $(\mu', \nu'; \lambda')$. This polynomial is the *Hall polynomial* corresponding to λ, μ, ν. Recall from Chapter I (§§5 and 9) that $c^{\lambda}_{\mu\nu}$ denotes the coefficient of the Schur function s_λ in the product $s_\mu s_\nu$; that $c^{\lambda}_{\mu\nu} = c^{\lambda'}_{\mu'\nu'}$; and that $c^{\lambda'}_{\mu'\nu'}$ is the number of LR-sequences of type $(\mu', \nu'; \lambda')$. Then from (4.1) it follows that

(4.3) (i) *If $c^{\lambda}_{\mu\nu} = 0$, the Hall polynomial $g^{\lambda}_{\mu\nu}(t)$ is identically zero. (In particular, $g^{\lambda}_{\mu\nu}(t) = 0$ unless $|\lambda| = |\mu| + |\nu|$ and $\mu, \nu \subset \lambda$.)*
(ii) *If $c^{\lambda}_{\mu\nu} \neq 0$, then $g^{\lambda}_{\mu\nu}(t)$ has degree $n(\lambda) - n(\mu) - n(\nu)$ and leading coefficient $c^{\lambda}_{\mu\nu}$.*
(iii) *In either case, $G^{\lambda}_{\mu\nu}(\mathfrak{o}) = g^{\lambda}_{\mu\nu}(q)$.*
(iv) $g^{\lambda}_{\mu\nu}(t) = g^{\lambda}_{\nu\mu}(t)$.

The only point that requires comment is (iv). From (2.2) we have $G^{\lambda}_{\mu\nu}(\mathfrak{o}) = G^{\lambda}_{\nu\mu}(\mathfrak{o})$ for all $\mathfrak{o}$, hence $g^{\lambda}_{\mu\nu}(q) = g^{\lambda}_{\nu\mu}(q)$ for all prime-powers q, and so $g^{\lambda}_{\mu\nu}(t) = g^{\lambda}_{\nu\mu}(t)$. ∎

The starting point of the proof of (4.1) is the following proposition:

(4.4) *Let M be a finite $\mathfrak{o}$-module of type λ and let N be an elementary submodule of cotype α in M (so that $\lambda - \alpha$ is a vertical strip, by (3.3)). Let β be a partition such that $\alpha \subset \beta \subset \lambda$ and let $H_{\alpha\beta\lambda}(\mathfrak{o})$ denote the number of submodules $P \subset N$ of cotype β in M. Then*

$$H_{\alpha\beta\lambda}(\mathfrak{o}) = h_{\alpha\beta\lambda}(q)$$

where $h_{\alpha\beta\lambda}(t) \in \mathbf{Z}[t]$ is the polynomial

$$h_{\alpha\beta\lambda}(t) = t^{d(\alpha,\beta,\lambda)} \prod_{i \geqslant 1} [\lambda'_i - \beta'_i, \beta'_i - \alpha'_i](t^{-1}), \tag{4.4.1}$$

in which

$$d(\alpha, \beta, \lambda) = \sum_{r<s} (\beta_r - \alpha_r)(\lambda_s - \beta_s) \tag{4.4.2}$$

and $[r, s]$ is the Gaussian polynomial $\begin{bmatrix} r+s \\ r \end{bmatrix}$ (Chapter I, §2, Example 1) *if $r,s \geqslant 0$, and is zero otherwise.*

Proof. Let $\theta = \lambda - \beta$, $\varphi = \beta - \alpha$. Also let $N_i = N \cap \mathfrak{p}^i M$. Since

$$\mathfrak{p}^i M / N_i \cong (\mathfrak{p}^i M + N)/N \cong \mathfrak{p}^i (M/N)$$

we have

$$\begin{aligned} l(N_i) &= l(\mathfrak{p}^i M) - l(\mathfrak{p}^i(M/N)) \\ &= \sum_{j>i} (\lambda'_j - \alpha'_j) \end{aligned}$$

by (1.4), or equivalently

$$n_i = l(N_i) = \sum_{j>i} (\theta'_j + \varphi'_j). \tag{1}$$

Now let P be a submodule of N, of cotype β in M, and let $P_i = P \cap \mathfrak{p}^i M = P \cap N_i$. Then

$$P/P_i \cong (P + \mathfrak{p}^i M)/\mathfrak{p}^i M$$

and therefore

$$\begin{aligned} l(P_{i-1}) - l(P_i) &= l(P/P_i) - l(P/P_{i-1}) \\ &= l((P + \mathfrak{p}^i M)/\mathfrak{p}^i M) - l((P + \mathfrak{p}^{i-1} M)/\mathfrak{p}^{i-1} M) \\ &= l(\mathfrak{p}^{i-1} M/\mathfrak{p}^i M) - l((P + \mathfrak{p}^{i-1} M)/(P + \mathfrak{p}^i M)) \\ &= \lambda'_i - \beta'_i = \theta'_i \end{aligned}$$

by (1.4). Conversely, if P is a submodule of N such that

$$l(P_{i-1}/P_i) = \theta'_i \qquad (i \geqslant 1) \tag{2}$$

then the preceding calculation shows that P has cotype β in M.

Suppose that $i \geqslant 1$ and P_i are given. We ask for the number of submodules P_{i-1} of N_{i-1} which satisfy (2) and

$$P_{i-1} \cap N_i = P_i. \tag{3}$$

The number of sequences $x = (x_1, \ldots, x_{\theta_i'})$ in N_{i-1} which are linearly independent modulo N_i is

$$(q^{n_{i-1}} - q^{n_i})(q^{n_{i-1}} - q^{n_i+1}) \ldots (q^{n_{i-1}} - q^{n_i+\theta_i'-1}). \tag{4}$$

For each such sequence x, the submodule P_{i-1} generated by P_i and x satisfies (2) and (3). Conversely, any such P_{i-1} can be obtained in this way from any sequence x of θ_i' elements of P_{i-1} which are linearly independent modulo P_i. The number of such sequences is

$$(q^{p_{i-1}} - q^{p_i})(q^{p_{i-1}} - q^{p_i+1}) \ldots (q^{p_{i-1}} - q^{p_i+\theta_i'-1}) \tag{5}$$

where

$$p_i = l(P_i) = \sum_{j>i} \theta_j' \tag{6}$$

by virtue of (2). So the number of submodules P_{i-1} of N_{i-1} which satisfy (2) and (3) is the quotient of (4) by (5), namely

$$q^{\theta_i'(n_{i-1}-p_{i-1})}[\theta_i', \varphi_i'](q^{-1}).$$

Taking the product of these for all $i \geqslant 1$, and observing that $n_{i-1} - p_{i-1} = \sum_{j \geqslant i} \varphi_j'$ (from (1) and (6)) we obtain

$$H_{\alpha\beta\lambda}(\mathfrak{o}) = q^d \prod_{i \geqslant 1} [\lambda_i' - \beta_i', \beta_i' - \alpha_i'](q^{-1})$$

where

$$d = \sum_{i \leqslant j} \theta_i' \varphi_j'.$$

Now θ_i' is the number of squares of the skew diagram $\theta = \lambda - \beta$ in the ith column, and $\sum_{j \geqslant i} \varphi_j'$ is the number of squares of $\varphi = \beta - \alpha$ in the same or later colums, hence in *higher* rows. It follows that

$$d = \sum_{r \leqslant s} \varphi_r \theta_s = d(\alpha, \beta, \lambda)$$

which completes the proof of (4.4). |

Suppose in particular that N in (4.4) is the socle S of M, so that $\alpha = \tilde{\lambda}$ by (3.2). Then $H_{\tilde{\lambda}\beta\lambda}(\mathfrak{o})$ is the number of elementary submodules P of cotype β in M, so that

$$H_{\tilde{\lambda}\beta\lambda}(\mathfrak{o}) = G^{\lambda}_{\beta(1^m)}(\mathfrak{o})$$

where $m = |\lambda - \beta| = |\theta|$. In this case we have $\varphi = \beta - \tilde{\lambda}$, so that $\varphi_i = \beta_i - \lambda_i + 1 = 1 - \theta_i$, and the exponent $d = d(\tilde{\lambda}, \beta, \lambda)$ is therefore

$$\begin{aligned} d &= \sum_{r \leqslant s} (1 - \theta_r)\theta_s \\ &= \sum_{r < s} (1 - \theta_r)\theta_s = \sum_{r < s} \theta_s - \sum_{r < s} \theta_r \theta_s . \end{aligned}$$

because $\theta_r = \theta_r^2$ for all r. Since $\Sigma\theta_r = m$ we have

$$\sum_{r < s} \theta_r \theta_s = \tfrac{1}{2}m^2 - \tfrac{1}{2}\sum \theta_r^2 = \tfrac{1}{2}m(m-1) \tag{4.5}$$

and hence

$$d = \sum (s-1)\theta_s - \tfrac{1}{2}m(m-1) = n(\lambda) - n(\beta) - n(1^m).$$

So from (4.4) we have the formula

$$G^{\lambda}_{\beta(1^m)}(\mathfrak{o}) = q^{n(\lambda) - n(\beta) - n(1^m)} \prod_{i \geqslant 1} [\lambda'_i - \beta'_i, \beta'_i - \lambda'_{i+1}](q^{-1})$$

(because $\tilde{\lambda}'_i = \lambda'_{i+1}$ for all $i \geqslant 1$). This is valid for any partitions λ, β, where $m = |\lambda| - |\beta|$. (If $\lambda - \beta$ is not a vertical strip, both sides are zero.)

Equivalently, if we define

$$g^{\lambda}_{\beta(1^m)}(t) = t^{n(\lambda) - n(\beta) - n(1^m)} \prod_{i \geqslant 1} [\lambda'_i - \beta'_i, \beta'_i - \lambda'_{i+1}](t^{-1}) \tag{4.6}$$

then we have $g^{\lambda}_{\beta(1^m)}(q) = G^{\lambda}_{\beta(1^m)}(\mathfrak{o})$.

The next stage in the proof of (4.1) is

(4.7) *Let $R = (\alpha', \beta', \lambda')$ be a three-term LR-sequence. Then there exists a monic polynomial $F_R(t) \in \mathbf{Z}[t]$, depending only on R, of degree $n(\beta) - n(\alpha) - \binom{n}{2}$ where $n = |\beta - \alpha|$, with the following property: if M is a finite $\mathfrak{o}$-module of type λ, and P is an elementary submodule of cotype β in M, then the number of submodules N of cotype α in M such that $\mathfrak{p}N = P$ is equal to $F_R(q)$.*

(Observe that when $\beta=\lambda$ (so that $\lambda-\alpha$ is a vertical strip) we have $P=0$, so that N is elementary and therefore $F_R(t)=g^{\lambda}_{\alpha(1^n)}(t)$ in this case.)

Proof. Let Q be a submodule of P, and let γ be the cotype of Q in M, so that we have $\alpha\subset\beta\subset\gamma\subset\lambda$.

Let $f(P,Q)$ (resp. $g(P,Q)$) denote the number of submodules N of cotype α in M such that $N\supset P\supset Q\supset\mathfrak{p}N$ (resp. $N\supset P\supset Q=\mathfrak{p}N$). The number we want to calculate is $g(P,P)$. Now $f(P,Q)$ is easily obtained, as we shall see in a moment, and then $g(P,Q)$ can be obtained by Möbius inversion [R8].

First of all let us calculate $f(P,Q)$. We have

$$\begin{aligned} N\supset P\supset Q\supset\mathfrak{p}N &\Leftrightarrow N\supset P \text{ and } N/Q \text{ is elementary}\\ &\Leftrightarrow S\supset N\supset P, \quad \text{where } S/Q \text{ is the socle of } M/Q\\ &\Leftrightarrow N/P\subset S/P. \end{aligned}$$

By (3.2) S has cotype $\tilde{\gamma}$ in M, and so by (4.4) (applied to the module M/P and its elementary submodule S/P) we have

(i) $$f(P,Q)=H_{\tilde{\gamma}\alpha\beta}(\mathfrak{o})$$

if $\tilde{\gamma}\subset\alpha$, i.e. if $\gamma-\alpha$ is a vertical strip, and $f(P,Q)=0$ otherwise.

Now it is clear that

(ii) $$f(P,Q)=\sum_{R\subset Q} g(P,R)$$

summed over all submodules (i.e. vector subspaces) R of Q. The equations (ii) for fixed P and varying $Q\subset P$ can be solved by Möbius inversion: the solution is

$$g(P,Q)=\sum_{R\subset Q} f(P,R)\mu(R,Q)$$

where μ is the Möbius function on the lattice of subspaces of the vector space P, which (*loc. cit.*) is given by

$$\mu(R,Q)=(-1)^d q^{d(d-1)/2}$$

where $d=\dim_k(Q/R)$. Hence in particular we have

(iii) $$g(P,P)=\sum_{R\subset P}(-1)^m q^{m(m-1)/2} f(P,R)$$

where $m=\dim_k(P/R)$ and the summation is over all subspaces R of P.

Now for each partition δ such that $\beta \subset \delta \subset \lambda$, the number of $R \subset P$ of cotype δ in M is $H_{\beta\delta\lambda}(\mathfrak{o})$. Hence from (i) and (iii) it follows that

(iv) $$g(P,P) = \sum_{\delta} (-1)^m q^{m(m-1)/2} H_{\beta\delta\lambda}(\mathfrak{o}) H_{\tilde{\delta}\alpha\beta}(\mathfrak{o})$$

where the sum is over all δ such that $\beta \subset \delta \subset \lambda$ and such that $\delta - \alpha$ is a vertical strip, and $m = |\delta - \beta|$. So if we define the polynomial $F_R(t)$ by

(4.8) $$F_R(t) = \sum_{\delta} (-1)^m t^{m(m-1)/2} h_{\beta\delta\lambda}(t) h_{\tilde{\delta}\alpha\beta}(t)$$

summed over partitions δ as above, then it follows from (iv) and (4.4) that $g(P,P) = F_R(q)$.

To complete the proof it remains to be shown that this polynomial $F_R(t)$ is monic of degree $n(\beta) - n(\alpha) - \binom{n}{2}$, where $n = |\beta - \alpha|$.

The degree of the summand corresponding to δ in (4.8) is by (4.4.2) equal to

$$d = \tfrac{1}{2}m(m-1) + \sum_{r<s} (\varphi_r \psi_s + (1 - \theta_r - \varphi_r)\theta_s),$$

where $\theta = \beta - \alpha$, $\varphi = \delta - \beta$ and $\psi = \lambda - \delta$, and $m = |\varphi|$. Each of θ, φ, ψ is a vertical strip, so that each θ_r, φ_r, and ψ_r is 0 or 1: in particular we have $\frac{1}{2}m(m-1) = \sum_{r<s} \varphi_r \varphi_s - m$ by (4.5), so that

(v) $$d = \sum_{r<s} \varphi_r(\varphi_s + \psi_s - \theta_s) + \sum_{r<s} (1 - \theta_r)\theta_s - |\varphi|.$$

Now $(\alpha', \beta', \lambda')$ is an LR-sequence. This implies that, for each $r \geqslant 1$, the number of squares of $\beta' - \alpha' = \theta'$ in the columns with indices $\geqslant r$ is not less than the number of squares of $\lambda' - \beta'$ in the same columns: that is to say, we have

(vi) $$\sum_{s \geqslant r} \theta_s \geqslant \sum_{s \geqslant r} (\varphi_s + \psi_s)$$

for each $r \geqslant 1$. From (v) and (vi) it follows that

$$d \leqslant \sum_{r<s} (1 - \theta_r)\theta_s$$

with equality if and only if $\varphi = 0$, i.e. if and only if $\delta = \beta$. Hence the dominant term of the sum (4.8) is $h_{\tilde{\beta}\alpha\beta}(t) = g^{\beta}_{\alpha(1^n)}(t)$, which by (4.6) is monic of degree $n(\beta) - n(\alpha) - \binom{n}{2}$. This completes the proof. |

The proof of (4.1) can now be rapidly completed. Let M be an $\mathfrak{o}$-module of type λ, and let $S = (\lambda^{(0)\prime}, \ldots, \lambda^{(r)\prime})$ be an LR-sequence of type $(\mu', \nu'; \lambda')$. Let N be a submodule of M such that $S(N) = S$, and let

$N_1 = \mathfrak{p}N$. Then clearly $S(N_1) = (\lambda^{(1)\prime}, \dots, \lambda^{(r)\prime}) = S_1$ say. Conversely, if we are given a submodule N_1 of M such that $S(N_1) = S_1$, the number of submodules N such that $S(N) = S$ and $\mathfrak{p}N = N_1$ is equal to $F_{R_1}(q)$, where $R_1 = (\lambda^{(0)\prime}, \lambda^{(1)\prime}, \lambda^{(2)\prime})$, as we see by (4.7) applied to the module $M/\mathfrak{p}N_1$ and its elementary submodule $N_1/\mathfrak{p}N_1$. Consequently

$$G_S(\mathfrak{o}) = G_{S_1}(\mathfrak{o}) \cdot F_{R_1}(q)$$

and therefore

$$G_S(\mathfrak{o}) = \prod_{i=1}^{r} F_{R_i}(q)$$

where $R_i = (\lambda^{(i-1)\prime}, \lambda^{(i)\prime}, \lambda^{(i+1)\prime})$. (When $i = r$ we take $\lambda^{(r+1)\prime} = \lambda^{(r)\prime}$, so that as remarked earlier $F_{R_r}(q) = g^{\lambda}_{\alpha(1^m)}(q)$, where $\alpha = \lambda^{(r-1)\prime}$ and $m = |\lambda - \alpha|$.)

Hence if we define

$$g_S(t) = \prod_{i=1}^{r} F_{R_i}(t) \tag{4.9}$$

we have $g_S(q) = G_S(\mathfrak{o})$, and by (4.7) the polynomial $g_S(t)$ is monic of degree

$$\sum_{i=1}^{r} \left(n(\lambda^{(i)}) - n(\lambda^{(i-1)}) - \binom{\nu_i'}{2} \right) = n(\lambda) - n(\mu) - n(\nu). \quad \text{Q.E.D.}$$

The proof just given provides an explicit (if complicated) expression for $g^{\lambda}_{\mu\nu}(t)$, via the formulas (4.2), (4.4), (4.8), and (4.9). If a, b, c, N are non-negative integers such that $b \leqslant c$, let us define

$$\Phi(a,b,c;N;t) = \sum_{r \geqslant 0} (-1)^r t^{Nr + r(r+1)/2} \begin{bmatrix} a \\ r \end{bmatrix} \begin{bmatrix} c-r \\ b-r \end{bmatrix} \tag{4.10}$$

where $\begin{bmatrix} a \\ r \end{bmatrix}$ and $\begin{bmatrix} c-r \\ b-r \end{bmatrix}$ are Gaussian polynomials in t. (The sum on the right of (4.10) is finite, since the term written is zero as soon as $r > \min(a,b)$.)

With this notation we have

(4.11) *Let $S = (\alpha^{(0)}, \alpha^{(1)}, \dots, \alpha^{(r)})$ be an LR-sequence of type $(\mu', \nu'; \lambda')$. Then*

$$g_S(t) = t^{n(\lambda) - n(\mu) - n(\nu)} \prod_{i > j \geqslant 1} \Phi\left(a_{ij}, b_{ij}, c_{ij}; N_{ij}; t^{-1}\right)$$

where

$$a_{ij} = \alpha_{i+1}^{(j+1)} - \alpha_{i+1}^{(j)}, \qquad b_{ij} = \alpha_i^{(j-1)} - \alpha_{i+1}^{(j)},$$

$$c_{ij} = \alpha_i^{(j)} - \alpha_{i+1}^{(j)}, \qquad N_{ij} = \sum_{h \leqslant i} (a_{h-1,j-1} - a_{h,j}).$$

(Thus a_{ij} is the number of symbols $j+1$ in the $(i+1)$th row of the tableau defined by S; c_{ij} is the number of columns of $\alpha^{(j)}$ of length i, and $c_{ij} - b_{ij} = a_{i-1,j-1}$ is the number of j's in the ith row of S. Finally, N_{ij} is the excess of the number of j's in the first i rows of S over the number of $(j+1)$'s in the first $i+1$ rows, and hence is $\geqslant 0$ by virtue of the lattice permutation property.)

Proof. With the notation of (4.7) let $R = (\alpha', \beta', \lambda')$ be a three-term LR-sequence, and let

$$a_i = \lambda'_{i+1} - \beta'_{i+1}, \quad b_i = \alpha'_i - \beta'_{i+1}, \quad c_i = \beta'_i - \beta'_{i+1}.$$

As in (4.8) let δ be a partition such that $\tilde{\delta} \subset \alpha \subset \beta \subset \delta \subset \lambda$, where $\tilde{\delta}'_i = \delta'_{i+1}$ for all $i \geqslant 1$, and let $r_i = \delta'_{i+1} - \beta'_{i+1}$, so that $m = |\delta - \beta| = \Sigma r_i$. (Since R is an LR-sequence we have $\beta'_1 = \delta'_1 = \lambda'_1$, so that $a_0 = r_0 = 0$; also $0 \leqslant r_i \leqslant a_i \leqslant c_i$ and $0 \leqslant r_i \leqslant b_i \leqslant c_i$.)

We shall use (4.8) to calculate $F_R(t)$. From (4.4) we have

$$(1) \qquad t^{m(m-1)/2} h_{\beta\delta\lambda}(t) h_{\tilde{\delta}\alpha\beta}(t) = t^A \prod_{i \geqslant 1} \begin{bmatrix} a_i \\ r_i \end{bmatrix} (t^{-1}) \begin{bmatrix} c_i - r_i \\ b_i - r_i \end{bmatrix} (t^{-1}),$$

in which the exponent of t is

$$(2) \qquad A = \tfrac{1}{2}\left(\textstyle\sum r_j\right)^2 - \tfrac{1}{2}\textstyle\sum r_j + \displaystyle\sum_{i<j} \left((a_i - r_i) r_j + (c_i - b_i)(b_j - r_j)\right)$$

$$= -\tfrac{1}{2}\textstyle\sum r_j(r_j+1) - \sum N_j r_j + \displaystyle\sum_{i<j} (c_i - b_i) b_j$$

where $N_j = \Sigma_{i<j}(c_i - a_i - b_i)$. Moreover,

$$\sum_{i<j} (c_i - b_i) b_j = \sum_{i<j} (\beta'_i - \alpha'_i)(\alpha'_j - \beta'_{j+1})$$

which reduces to

$$(3) \qquad n(\beta) - n(\alpha) - \tfrac{1}{2} s(s-1)$$

where $s = |\beta - \alpha|$.

From (1), (2), (3), and (4.8) we obtain

$$F_R(t) = t^{n(\beta)-n(\alpha)-s(s-1)/2} \prod_{i \geq 1} \Phi(a_i, b_i, c_i; N_i; t^{-1}). \tag{4}$$

Since by (4.9)

$$g_S(t) = \prod_{j=1}^{r} F_{R_j}(t),$$

where $R_j = (\alpha^{(j-1)}, \alpha^{(j)}, \alpha^{(j+1)})$, the formula (4) leads directly to (4.11). (Since $(j+1)$'s appear for the first time in the $(j+1)$th row of S, it follows that $a_{ij} = 0$ if $i < j$, whence the restriction $i \geq j$ in the product on the right-hand side of (4.11). |

As an example (which we shall make use of later) we shall compute $g^{\lambda}_{\mu\nu}(t)$ when $\nu = (r)$ is a partition with only one part $r = |\lambda - \mu|$. (The case when $\nu = (1^r)$ is given by (4.6).) First we have

(4.12) $g^{\lambda}_{\mu(r)}(t) = 0$ *unless* $\lambda - \mu$ *is a horizontal strip of length* r.

Proof. Let M be a finite $\mathfrak{o}$-module of type λ, and let N be a cyclic submodule of cotype μ in M. Let $N_i = N \cap \mathfrak{p}^i M$. Then since $\mathfrak{p}^i M / N_i \cong (\mathfrak{p}^i M + N)/N \cong \mathfrak{p}^i (M/N)$ we have

$$l(N_{i-1}/N_i) = \lambda'_i - \mu'_i$$

and since $\mathfrak{p} N_{i-1} = \mathfrak{p}(N \cap \mathfrak{p}^{i-1} M) \subset N \cap \mathfrak{p}^i M = N_i$, it follows that

$$0 \leq \lambda'_i - \mu'_i \leq l(N_{i-1}/\mathfrak{p} N_{i-1}).$$

Since N is cyclic, so is N_i, and hence $l(N_{i-1}/\mathfrak{p} N_{i-1}) \leq 1$. Hence $\lambda'_i - \mu'_i = 0$ or 1 for each $i \geq 1$, and therefore $\lambda - \mu$ is a horizontal strip. So unless $\lambda - \mu$ is a horizontal strip we have $g^{\lambda}_{\mu(r)}(q) = 0$ for all prime-powers q, and therefore $g^{\lambda}_{\mu(r)}(t) = 0$. |

Remark. Alternatively (and more rapidly) (4.12) follows from (4.3): we have $g^{\lambda}_{\mu(r)}(t) = 0$ unless $c^{\lambda}_{\mu(r)} \neq 0$, i.e. unless s_λ occurs in the product $s_\mu h_r$, which by Chapter I, (5.16) requires $\lambda - \mu$ to be a horizontal strip.

Assume then that $\lambda - \mu$ is a horizontal r-strip. Then $\lambda' - \mu'$ is a vertical r-strip, and there is only one LR-sequence $S = (\alpha^{(0)}, \alpha^{(1)}, \ldots, \alpha^{(r)})$ of type $(\mu', (1^r); \lambda')$, namely that obtained by filling in the squares of the vertical strip $\lambda' - \mu'$ consecutively, starting at the top. For this S we have $a_{ij} = 0$ or 1 for each pair (i, j), and also $c_{ij} - b_{ij} = a_{i-1, j-1} = 0$ or 1, since there is at most one square of $\lambda' - \mu'$ in any row. Moreover, if $a_{ij} = 1$ (so that there is a square labelled $j+1$ in the $(i+1)$th row) we have $N_{ij}=0$. Hence, from the definition (4.10) of Φ we find that $\Phi(a_{ij}, b_{ij}, c_{ij}; N_{ij}; t)$ is equal to

$$\begin{array}{ll} 1 & \text{if } a_{i-1,j-1}=a_{ij}=0 \text{ or } 1; \\ 1-t & \text{if } a_{i-1,j-1}=0,\ a_{ij}=1; \\ (1-t^{c_{ij}})/(1-t) & \text{if } a_{i-1,j-1}=1,\ a_{ij}=0. \end{array}$$

Moreover, in the latter case (i.e. when there is a square labelled j in the ith row, but none labelled j in the row below), c_{ij} is the number of columns of λ' of length i, that is $c_{ij}=m_i(\lambda)$. Putting these facts together, we obtain

(4.13) *Let* $\sigma=\lambda-\mu$ *be a horizontal strip of total length* r, *and let* I *be the set of integers* $i\geqslant 1$ *such that* $\sigma_i'=1$ *and* $\sigma_{i+1}'=0$. *Then*

$$g^{\lambda}_{\mu(r)}(t)=\frac{t^{n(\lambda)-n(\mu)}}{1-t^{-1}}\prod_{i\in I}(1-t^{-m_i(\lambda)})$$

where $m_i(\lambda)$ *is the multiplicity of* i *in* λ. |

Examples

1. Let $(x;t)_r=(1-x)(1-xt)\dots(1-xt^{r-1})$ for all $r\geqslant 0$. Then $\Phi(a,b,c;N;t)$ is equal to the coefficient of x^b in $(xt^{N+1};t)_a/(x;t)_{c-b+1}$. (Use Chapter I, §2, Example 3.) Deduce that if $N=c-a-b$ we have

$$\Phi(a,b,c;N;t)=\begin{bmatrix} c-a \\ b \end{bmatrix}$$

This applies in particular to the terms corresponding to $i=j$ in the product (4.11), since $N_{ii}=c_{ii}-a_{ii}-b_{ii}$.

2. Let

$${}_2\varphi_1(\alpha,\beta;\gamma;t,x)=\sum_{r\geqslant 0}\frac{(\alpha;t)_r(\beta;t)_r}{(\gamma;t)_r(t;t)_r}x^r$$

in the standard notation for basic hypergeometric series (see e.g. [G5]). In this notation we have

$$\Phi(a,b,c;N;t)=\begin{bmatrix} c \\ b \end{bmatrix}{}_2\varphi_1(t^a,t^b;t^c;t^{-1},t^N)$$

so that $g_S(t)$ is a product of Gaussian polynomials and terminating ${}_2\varphi_1$'s.

Notes and references

The contents of §§1 and 2, together with the theorem (4.3), are due to Philip Hall, who did not publish anything more than a summary of his theory [H3]. The contents of §3, in particular (3.4), are due to J. A. Green [G13]. Theorem (4.1) was first proved by T. Klein [K10], a student of Green. Our proof is different from hers.

It should be pointed out that Hall was in fact anticipated by more than half a century by E. Steinitz, who in 1900 defined what we have called the Hall polynomials and the Hall algebra, recognized their connection with Schur functions, and conjectured Hall's theorem (4.3). Steinitz's note [S26] is a summary of a lecture given at the annual meeting of the Deutsche Mathematiker-Vereinigung in Aachen in 1900; it gives neither proofs nor indications of method, and remained forgotten until brought to light by K. Johnsen in 1982 [J13].

For generalizations of the notion of the Hall algebra, see the papers by C. M. Ringel [R2], [R3], [R4].

APPENDIX*: Another proof of Hall's theorem

This Appendix is devoted to a simple proof of the following slightly weakened form of (4.3) (we shall freely use the terminology and notation of Chapter II):

(AZ.1) *For any three partitions λ, μ, ν there exists a polynomial $g^{\lambda}_{\mu\nu}(t) \in \mathbf{Z}[t]$ such that $G^{\lambda}_{\mu\nu}(\mathfrak{o}) = g^{\lambda}_{\mu\nu}(q)$. Moreover, $g^{\lambda}_{\mu\nu}(t)$ has degree $\leqslant n(\lambda) - n(\mu) - n(\nu)$, and the coefficient of $t^{n(\lambda)-n(\mu)-n(\nu)}$ is equal to $c^{\lambda}_{\mu\nu}$.*

Our proof does not use the Littlewood–Richardson rule (Chapter I, (9.2)). Therefore, combining it with the arguments of §§3 and 4 of Chapter II we shall obtain a new proof of the Littlewood–Richardson rule which makes the appearance of the lattice permutations more natural.

Our proof is based on a combinatorial interpretation of the coefficients $a_{\lambda\mu}$ from the formula (1) of Chapter II, §2. We need some definitions.

A *composition* is a sequence $\alpha = (\alpha_1, \alpha_2, \dots)$ of non-negative integers with only a finite number of non-zero terms. So a partition is a composition such that $\alpha_1 \geqslant \alpha_2 \geqslant \dots$. The group S_∞ of finite permutations of $\mathbf{N}^+ = \{1, 2, 3, \dots\}$ acts on compositions by $w\alpha = (\alpha_{w^{-1}(1)}, \alpha_{w^{-1}(2)}, \dots)$. We shall write $\alpha \sim \beta$ if α and β are conjugate under this action: clearly , each S_∞-orbit contains exactly one partition.

As well as partitions the compositions have diagrams: the diagram of α is formally defined as the set $\{(i,j) \in \mathbf{Z}^2 : 1 \leqslant j \leqslant \alpha_i\}$ and is graphically displayed as the set of squares containing α_i squares in the ith row.

If α and β are two compositions, an *array* of shape α and weight β is a numbering of the squares of the diagram of α by positive integers such that for any $i \geqslant 1$ there are β_i squares numbered by i (more formally, an array is a function $A: \alpha \to \mathbf{N}^+$ such that Card $A^{-1}(i) = \beta_i$ for all $i \geqslant 1$; it will be convenient for us to assume that A is defined on all of $\mathbf{N}^+ \times \mathbf{N}^+$ so that $A(i,j) = +\infty$ when $j > \alpha_i$). For any $x = (i,j) \in \mathbf{N}^+ \times \mathbf{N}^+$ let $x^{\rightarrow} = (i, j+1)$. An array α will be called *row-ordered* (resp. *row-strict*) if $A(x^{\rightarrow}) \geqslant A(x)$ (resp. $A(x^{\rightarrow}) > A(x)$) for any $x \in \alpha$. Likewise we define column-ordered and column-strict arrays.

* This Appendix was written by A. Zelevinsky for the Russian version of the first edition of this book, and is reproduced here (in English, for the reader's convenience) with his permission.

We define a total ordering on $\mathbf{N}^+ \times \mathbf{N}^+$ by

$$(i,j) <_L (i',j') \Leftrightarrow \text{either } j<j', \text{ or } j=j' \text{ and } i>i'.$$

Finally, for each row-strict array A of shape α we let

$$d(A) = \text{Card}\{(x,y) \in \alpha \times \alpha : y <_L x, A(x) < A(y) < A(x^{\rightarrow})\}.$$

(AZ.2) *Let λ, μ be partitions and α, β compositions such that $\alpha \sim \mu$ and $\beta \sim \lambda'$. Then the coefficient $a_{\lambda\mu}$* (Chapter II, §2) *is equal to*

$$a_{\lambda\mu} = \sum_A q^{d(A)}$$

summed over all row-strict arrays A of shape α and weight β.

Before proving (AZ.2) we shall derive from it (AZ.1). From (AZ.2) the polynomial

$$\sum_A t^{d(A)} \in \mathbf{Z}[t],$$

where the sum is the same as in (AZ.2), depends only on λ and μ; we denote it by $a_{\lambda\mu}(t)$.

(AZ.3) (a) *$a_{\lambda\mu}(t)$ has coefficients $\geqslant 0$.*
(b) *$a_{\lambda\mu}(1)$ is equal to the number of* $(0,1)$*-matrices with row sums $\mu_1, \mu_2, \dots$ and column sums $\lambda'_1, \lambda'_2, \dots$.*
(c) *$a_{\lambda\mu}(t) = 0$ unless $\mu \leqslant \lambda$. Moreover, $a_{\lambda\lambda}(t) = 1$.*

Proof. (a) is evident. To prove (b) it is enough to establish a bijection between row-strict arrays of shape μ and weight λ', and $(0,1)$-matrices with row sums $\mu_1, \mu_2, \dots$ and column sums $\lambda'_1, \lambda'_2, \dots$. To do this we assign to an array A the matrix (c_{ij}), where $c_{ij} = 1$ if the ith row of A contains j, and $c_{ij} = 0$ otherwise; this is the required bijection. Finally, (c) follows at once from (a), (b), and the Gale–Ryser theorem (Chapter I, §7, Example 9). ∎

From (AZ.3)(c), $(a_{\lambda\mu}(t))$ is a strictly upper unitriangular matrix over $\mathbf{Z}[t]$. Hence it is invertible, and its inverse has the same form. Therefore, the entries in the transition matrices between the bases $(v_{\lambda'})$ and (u_λ) in $H(\mathfrak{o})$ are integer polynomials in q. Since the multiplication law in $H(\mathfrak{o})$ with respect to the basis $(v_{\lambda'})$ does not depend on q, it follows that the structure constants in the basis (u_λ) are integer polynomials in q. This proves the existence of the Hall polynomials $g^\lambda_{\mu\nu}(t) \in \mathbf{Z}[t]$. It remains to find their degrees and leading coefficients.

(AZ.4) *$a_{\lambda\mu}(t)$ has degree $\leqslant n(\mu) - n(\lambda)$, and the coefficient of $t^{n(\mu)-n(\lambda)}$ is equal to $K_{\mu'\lambda'}$* (Chapter I, §5).

This follows at once from the next combinatorial lemma:

(AZ.5) *For any row-strict array A of shape μ and weight λ' let $\tilde{d}(A)$ denote the number of pairs $(x,y)\in\mu\times\mu$ such that y lies above x (in the same column) and $A(x)<A(y)<A(x^{\rightarrow})$. Then*
(a) $d(A)+\tilde{d}(A)=n(\mu)-n(\lambda)$;
(b) $\tilde{d}(A)=0$ *if and only if A is column-ordered.*

Proof. (a) Let

$$D(A)=\{(x,y)\in\mu\times\mu: y<_L x, A(x)\leqslant A(y)<A(x^{\rightarrow})\},$$

$$N(\mu)=\{(x,y)\in\mu\times\mu: y \text{ lies above } x\},$$

$$\tilde{D}(A)=\{(x,y)\in N(\mu): A(x)<A(y)<A(x^{\rightarrow})\};$$

then we have

$$\operatorname{Card} D(A)=d(A)+\sum_{i\geqslant 1}\binom{\lambda'_i}{2}=d(A)+n(\lambda),$$

$$\operatorname{Card} N(\mu)=\sum_{i\geqslant 1}\binom{\mu'_i}{2}=n(\mu),$$

and

$$\operatorname{Card}\tilde{D}(A)=\tilde{d}(A).$$

We shall construct a mapping $\varphi: D(A)\to N(\mu)$. Let $(x,y)\in D(A)$ and suppose that $x=(i_1,j_1)$, $y=(i_2,j_2)$. Clearly $i_1\neq i_2$; let $i=\max(i_1,i_2)$, $i'=\min(i_1,i_2)$, $j=\min(j_1,j_2)$ and finally $\varphi(x,y)=((i,j),(i',j))$. The definitions readily imply that φ is a bijection of $D(A)$ onto $N(\mu)-\tilde{D}(A)$, whence our assertion.

(b) The 'if' part is evident. Now suppose that A is not column-ordered: that is to say, there are $x=(i,j)$, $y=(i',j)\in\mu$ such that $i>i'$ and $A(x)<A(y)$. Choose such a pair with least possible j; clearly it belongs to $\tilde{D}(A)$, hence $\tilde{d}(A)\neq 0$. |

Now let $\tilde{a}_{\lambda\mu}(t)=t^{n(\mu)-n(\lambda)}a_{\lambda\mu}(t^{-1})$. From (AZ.4) and (AZ.5) we have

(AZ.6) $$\tilde{a}_{\lambda\mu}(t)=\sum_A t^{\tilde{d}(A)}$$

summed over all row-strict arrays of shape μ and weight λ'; in particular, $\tilde{a}_{\lambda\mu}(t)\in\mathbf{Z}[t]$. Moreover, $\tilde{a}_{\lambda\mu}(0)=K_{\mu'\lambda'}$. |

Now consider the ring $\Lambda[t]=\Lambda_{\mathbf{Z}[t]}$ of polynomials in t with coefficients from Λ; we shall write down the elements of $\Lambda[t]$ as $P(x;t)$. Clearly, $\Lambda[t]$

is a free $\mathbf{Z}[t]$-module with basis (e_λ). From (AZ.3)(c), the matrix $(\tilde{a}_{\lambda\mu}(t))$ is strictly upper unitriangular. Therefore, the equations

$$e_{\lambda'} = \sum_\mu \tilde{a}_{\lambda\mu}(t) P_\mu(x;t) \tag{1}$$

uniquely determine the elements $P_\mu(x;t) \in \Lambda[t]$, and they form a $\mathbf{Z}[t]$-basis of $\Lambda[t]$. Let $f^\lambda_{\mu\nu}(t)$ be the structure constants of $\Lambda[t]$ with respect to this basis, i.e.

$$P_\mu(x;t)P_\nu(x;t) = \sum_\lambda f^\lambda_{\mu\nu}(t) P_\lambda(x;t);$$

it is clear that $f^\lambda_{\mu\nu}(t) \in \mathbf{Z}[t]$ for all λ, μ, ν.

(AZ.7)(a) $g^\lambda_{\mu\nu}(t) = t^{n(\lambda)-n(\mu)-n(\nu)} f^\lambda_{\mu\nu}(t^{-1})$.
(b) $P_\lambda(x;0) = s_\lambda(x)$. *In particular,* $f^\lambda_{\mu\nu}(0) = c^\lambda_{\mu\nu}$ *are the structure constants of* Λ *in the basis* (s_λ).

Proof. (a) follows at once from the definitions. To prove (b) it is enough to observe that

$$M(e,s)_{\lambda'\mu} = K_{\mu'\lambda'} = \tilde{a}_{\lambda\mu}(0)$$

by (AZ.6) and the results of Chapter I, §6. |

From (AZ.7), $g^\lambda_{\mu\nu}(t)$ has degree $\leqslant n(\lambda) - n(\mu) - n(\nu)$, and the coefficient of this power of t is equal to $f^\lambda_{\mu\nu}(0) = c^\lambda_{\mu\nu}$. This completes the proof of (AZ.1).

It remains to prove (AZ.2). For this we reformulate the definitions of an array and of $d(A)$ in terms of sequences of compositions. If α and β are two compositions, we shall write $\beta \dashv \alpha$ if $\alpha_i - 1 \leqslant \beta_i \leqslant \alpha_i$ for any $i \geqslant 1$. If $\beta \dashv \alpha$ then we define $d(\alpha, \beta)$ to be the number of pairs (i,j) such that $\beta_i = \alpha_i$, $\beta_j = \alpha_j - 1$ and $(j, \alpha_j) <_L (i, \alpha_i)$.

(AZ.8) *Let* α *and* β *be two compositions and suppose that* $\beta_i = 0$ *for* $i > r$. *There is a natural one-to-one correspondence between row-strict arrays A of shape* α *and weight* β, *and sequences of compositions* $(\alpha^{(0)}, \alpha^{(1)}, \dots, \alpha^{(r)})$ *such that* $0 = \alpha^{(0)} \dashv \alpha^{(1)} \dashv \dots \dashv \alpha^{(r)} = \alpha$ *and* $|\alpha^{(i)}| - |\alpha^{(i-1)}| = \beta_i$ *for* $i \geqslant 1$. *Moreover, this correspondence transforms* $d(A)$ *into* $\sum_{i \geqslant 1} d(\alpha^{(i)}, \alpha^{(i-1)})$.

Proof. We attach to an array A the sequence $(\alpha^{(i)})$, where $\alpha^{(i)} = A^{-1}(\{1,2,\dots,i\})$. All our assertions are verified directly. |

Remembering the definition of $a_{\lambda\mu}$ (Chapter II, §2) we see that an evident induction reduces the proof of (AZ.2) to the next statement:

(AZ.9) *Let* λ, μ *be partitions with* $|\lambda| = |\mu| + r$, *and* α *a composition such that* $\alpha \sim \lambda$. *Then*

$$G^{\lambda}_{\mu(1^r)}(\mathfrak{o}) = \sum_{\beta} q^{d(\alpha,\beta)}$$

summed over all compositions β *such that* $\beta \dashv \alpha$ *and* $\beta \sim \mu$.

Proof. Let M be a finite $\mathfrak{o}$-module of type λ. Recall that $G^{\lambda}_{\mu(1^r)}(\mathfrak{o})$ is the number of submodules $N \subset M$ of type (1^r) and cotype μ. The condition that N has type (1^r) means that N is an r-dimensional vector k-subspace of the socle S of M. Let $G_r(S)$ denote the set of these subspaces. We shall use the decomposition of $G_r(S)$ into Schubert cells. Recall that if a basis $(v_i)_{i \in I}$ of S is given, where I is a totally ordered set, then the corresponding Schubert cells C_J in $G_r(S)$ are parametrized by r-subsets $J \subset I$. The elements of C_J have coordinates $(c_{ij} \in k: j \in J, i \in I - J, j < i)$; the subspace corresponding to (c_{ij}) has the basis $(v_j + \Sigma_i c_{ij} v_i)_{j \in J}$. It is known (and easy to prove) that $G_r(S)$ is the disjoint union of the C_J. Moreover, we have Card $(C_J) = q^{d(J)}$, where $d(J)$ is the number of pairs (i, j) such that $j \in J$, $i \in I - J$ and $j < i$.

Now we express M in the form $M = \bigoplus_{i \geqslant 1} \mathfrak{o} x_i$, where $\mathrm{Ann}(x_i) = \mathfrak{p}^{\alpha_i}$. Let I denote the set of indices $i \geqslant 1$ such that $\alpha_i > 0$, and for each $i \in I$ let $v_i = \pi^{\alpha_i - 1} x_i$, where π is a generator of $\mathfrak{p}$. It is clear that the $v_i (i \in I)$ form a k-basis of S. We order I by requiring that j precedes i if and only if $(j, \alpha_j) <_L (i, \alpha_i)$. Consider the corresponding decomposition of $G_r(S)$ into Schubert cells. The subsets $J \subset I$ are in natural one-to-one correspondence with compositions $\beta \dashv \alpha$: to a subset J there corresponds the composition β such that $\beta_i = \alpha_i - 1$ for $i \in J$, and $\beta_i = \alpha_i$ for $i \in I - J$. It is clear that this correspondence transforms $d(J)$ into $d(\alpha, \beta)$. Finally, it is easy to prove that all submodules $N \in C_J$ have the same cotype μ, where μ is the partition such that $\mu \sim \beta$. This completes the proof of (AZ.9). |

Remarks. 1. The polynomials $P_\lambda(x; t)$ defined by (1) are called the Hall–Littlewood functions. They will be studied in detail in Chapter III.

2. It is easy to see that (AZ.9) is equivalent to Chapter II, (4.6): each of these statements follows at once from the other by means of the following well-known expression for Gaussian polynomials:

(2)
$$\begin{bmatrix} n \\ r \end{bmatrix} = \sum_J t^{d(J)}$$

where the sum is over all r-subsets of a totally ordered n-set I, and $d(J)$ is defined in the proof of (AZ.9) above. The formula (2) follows e.g. from Chapter I, §2, Example 3. On the other hand, one of the standard ways to prove (2) is to count the number of points on a Grassmannian over a finite field in two different ways, which was essentially done above.

III

HALL–LITTLEWOOD SYMMETRIC FUNCTIONS

1. The symmetric polynomials R_λ

Let $x, \ldots, x_n$ and t be independent indeterminates over $\mathbf{Z}$, and let λ be a partition of length $\leqslant n$. Define

$$R_\lambda(x_1, \ldots, x_n; t) = \sum_{w \in S_n} w\left(x_1^{\lambda_1} \ldots x_n^{\lambda_n} \prod_{i<j} \frac{x_i - tx_j}{x_i - x_j}\right).$$

The denominator in each term of the sum on the right-hand side is, up to sign, the Vandermonde polynomial

$$a_\delta = \prod_{i<j}(x_i - x_j)$$

(Chapter I, §3), and therefore we have

$$(1.1) \quad R_\lambda(x_1, \ldots, x_n; t) = a_\delta^{-1} \sum_{w \in S_n} \varepsilon(w) . w\left(x_1^{\lambda_1} \ldots x_n^{\lambda_n} \prod_{i<j}(x_i - tx_j)\right)$$

where as usual $\varepsilon(w)$ is the sign of the permutation w. The sum on the right-hand side of (1.1) is skew-symmetric in $x_1, \ldots, x_n$, hence is divisible by a_δ in the ring $\mathbf{Z}[x_1, \ldots, x_n, t]$, and consequently R_λ is a homogeneous *symmetric polynomial* in $x_1, \ldots, x_n$, of degree $|\lambda|$, with coefficients in $\mathbf{Z}[t]$. Hence R_λ can be expressed as a linear combination of the Schur functions $s_\mu(x_1, \ldots, x_n)$, with coefficients in $\mathbf{Z}[t]$. In fact we have

$$(1.2) \qquad R_\lambda(x_1, \ldots, x_n; t) = \sum_\mu u_{\lambda\mu}(t) s_\mu(x_1, \ldots, x_n)$$

where $u_{\lambda\mu}(t) \in \mathbf{Z}[t]$, *and* $u_{\lambda\mu}(t) = 0$ *unless* $|\lambda| = |\mu|$ *and* $\lambda \geqslant \mu$.

Moreover, the polynomial $u_{\lambda\lambda}(t)$ can be explicitly computed. For each integer $m \geqslant 0$ let

$$v_m(t) = \prod_{i=1}^{m} \frac{1 - t^i}{1 - t} = \varphi_m(t)/(1-t)^m$$

and for a partition $\lambda = (\lambda_1, \dots, \lambda_n)$ of length $\leqslant n$ (in which some of the λ_i may be zero) we define

$$v_\lambda(t) = \prod_{i \geqslant 0} v_{m_i}(t)$$

where m_i is the number of λ_j equal to i, for each $i \geqslant 0$. Then we have

(1.3) $$u_{\lambda\lambda}(t) = v_\lambda(t).$$

Proof of (1.2) and (1.3). The product $\prod_{i<j}(x_i - tx_j)$, when multiplied out, is a sum of terms of the form

$$\prod_{i<j} x_i^{r_{ij}}(-tx_j)^{r_{ji}}$$

where (r_{ij}) is any $n \times n$ matrix of 0's and 1's such that

(i) $$r_{ii} = 0, \qquad r_{ij} + r_{ji} = 1 \text{ if } i \neq j.$$

For such a matrix (r_{ij}), let

(ii) $$\alpha_i = \lambda_i + \sum_j r_{ij},$$

(iii) $$d = \sum_{i<j} r_{ji}.$$

Then from (1.1) it is clear that R_λ is a sum of terms

$$(-t)^d a_\alpha a_\delta^{-1}$$

where as in Chapter I, §3, a_α is the skew-symmetric polynomial generated by $x^\alpha = x_1^{\alpha_1} \dots x_n^{\alpha_n}$. Since $a_\alpha = 0$ if any two of the α_i are equal, we may assume that $\alpha_1, \dots, \alpha_n$ are all distinct. We rearrange them in descending order, say

(iv) $$a_{w(i)} = \mu_i + n - i \qquad (1 \leqslant i \leqslant n)$$

for some permutation $w \in S_n$ and some partition $\mu = (\mu_1, \dots, \mu_n)$. Then $a_\alpha a_\delta^{-1}$ is equal to $\varepsilon(w) s_\mu$, and to prove (1.2) it is enough to show that $\mu \leqslant \lambda$.

Let $s_{ij} = r_{w(i), w(j)}$. The matrix (s_{ij}) satisfies the same conditions (i) as the matrix (r_{ij}), and from (ii) and (iv) we have

$$\mu_i + n - i = \lambda_{w(i)} + \sum_j s_{ij}.$$

Hence for $1 \leqslant k \leqslant n$

(v) $$\mu_1 + \ldots + \mu_k = \lambda_{w(1)} + \ldots + \lambda_{w(k)} + \sum_{i=1}^{k} \sum_{j=1}^{n} s_{ij} - \sum_{i=1}^{k} (n-i).$$

Now

$$\begin{aligned} \sum_{i=1}^{k} \sum_{j=1}^{n} s_{ij} &= \sum_{i,j=1}^{k} s_{ij} + \sum_{i=1}^{k} \sum_{j=k+1}^{n} s_{ij} \\ &= \tfrac{1}{2}k(k-1) + \sum_{i=1}^{k} \sum_{j>k} s_{ij} \\ &\leqslant \tfrac{1}{2}k(k-1) + k(n-k) \\ &= \sum_{i=1}^{k} (n-i). \end{aligned}$$

Hence it follows from (v) that

$$\begin{aligned} \mu_1 + \ldots + \mu_k &\leqslant \lambda_{w(1)} + \ldots + \lambda_{w(k)} \\ &\leqslant \lambda_1 + \ldots + \lambda_k \end{aligned}$$

and therefore $\mu \leqslant \lambda$. This proves (1.2).

Furthermore, these calculations show that $\lambda = \mu$ if and only if $\lambda_{w(i)} = \lambda_i$ for $1 \leqslant i \leqslant n$, and $s_{ij} = 1$ for all pairs $i < j$. It follows that

(vi) $$u_{\lambda\lambda}(t) = \sum_{w} \varepsilon(w)(-t)^d$$

summed over all $w \in S_n$ which fix λ, where

$$d = \sum_{i<j} r_{ji} = \sum_{i<j} s_{w^{-1}(j), w^{-1}(i)}$$

is equal to the number $n(w)$ of pairs $i < j$ in $\{1, 2, \ldots, n\}$ such that $w^{-1}(j) < w^{-1}(i)$; this number $n(w)$ is also the number of pairs $l < k$ such that $w(k) < w(l)$, and the signature $\varepsilon(w)$ is equal to $(-1)^{n(w)}$. Hence from (vi) we have

(vii) $$u_{\lambda\lambda}(t) = \sum_{w \in S_n^{\lambda}} t^{n(w)}$$

where S_n^{λ} is the subgroup of permutations $w \in S_n$ such that $\lambda_{w(i)} = \lambda_i$ for $1 \leqslant i \leqslant n$. Clearly $S_n^{\lambda} = \prod_{i \geqslant 0} S_{m_i}$ where as before m_i is the number of λ_j

equal to i, and hence to prove (1.3) it is enough to show that

(viii) $$\sum_{w \in S_m} t^{n(w)} = v_m(t).$$

We prove (viii) by induction on m. Let w_i denote the transposition (i, m), for $1 \leqslant i \leqslant m$ (so that w_m is the identity). The w_i are coset representatives of S_{m-1} in S_m, and we have

$$n(w'w_i) = n(w') + m - i$$

for $w' \in S_{m-1}$, because in the sequence $(w'w_i(1), \ldots, w'w_i(m))$ the number m occurs in the ith place and is therefore followed by $m - i$ numbers less than m. Consequently

$$\sum_{w \in S_m} t^{n(w)} = \left(\sum_{w' \in S_{m-1}} t^{n(w')} \right) (1 + t + \ldots + t^{m-1})$$

from which (viii) follows immediately. This completes the proof of (1.3). |

By taking $\lambda = 0$ in (1.2) and (1.3) we have

(1.4) $$\sum_{w \in S_n} w \left(\prod_{i<j} \frac{x_i - tx_j}{x_i - x_j} \right) = v_n(t),$$

which is independent of $x_1, \ldots, x_n$.

Next we shall show that

(1.5) *$R_\lambda(x_1, \ldots, x_n; t)$ is divisible by $v_\lambda(t)$ (i.e. all the coefficients of R_λ are divisible by $v_\lambda(t)$ in $\mathbf{Z}[t]$).*

Proof. Suppose for example that $\lambda_1 = \ldots = \lambda_m > \lambda_{m+1}$. Then any $w \in S_n$ which permutes only the digits $1, 2, \ldots, m$ will fix the monomial $x_1^{\lambda_1} \ldots x_n^{\lambda_n}$, and by (1.4) we can extract a factor $v_m(t)$ from R_λ. It follows that

$$R_\lambda(x_1, \ldots, x_n; t) = v_\lambda(t) \sum_{w \in S_n/S_n^\lambda} w \left(x_1^{\lambda_1} \ldots x_n^{\lambda_n} \prod_{\lambda_i > \lambda_j} \frac{x_i - tx_j}{x_i - x_j} \right)$$

$$= v_\lambda(t) P_\lambda(x_1, \ldots, x_n; t), \qquad \text{say.}$$

Since R_λ is a polynomial in $x_1, \ldots, x_n$, so also is P_λ; and since t occurs only in the numerators of the terms in the sum P_λ, it follows that P_λ is a symmetric polynomial in $x_1, \ldots, x_n$ with coefficients in $\mathbf{Z}[t]$. |

It is these polynomials P_λ, rather than the R_λ, which are the subject of this chapter.

2. Hall–Littlewood functions

The polynomials $P_\lambda(x_1,\ldots,x_n;t)$ just defined are called the *Hall–Littlewood polynomials*. They were first defined indirectly by Philip Hall, in terms of the Hall algebra (Chapter II) and then directly by D. E. Littlewood [L12], essentially as we have defined them. From the proof of (1.5) we have two equivalent definitions:

$$(2.1)\qquad P_\lambda(x_1,\ldots,x_n;t)=\frac{1}{v_\lambda(t)}\sum_{w\in S_n} w\left(x_1^{\lambda_1}\ldots x_n^{\lambda_n}\prod_{i<j}\frac{x_i-tx_j}{x_i-x_j}\right),$$

$$(2.2)\qquad P_\lambda(x_1,\ldots,x_n;t)=\sum_{w\in S_n/S_n^\lambda} w\left(x_1^{\lambda_1}\ldots x_n^{\lambda_n}\prod_{\lambda_i>\lambda_j}\frac{x_i-tx_j}{x_i-x_j}\right).$$

The P_λ serve to interpolate between the Schur functions s_λ and the monomial symmetric functions m_λ, because

$$(2.3)\qquad P_\lambda(x_1,\ldots,x_n;0)=s_\lambda(x_1,\ldots,x_n)$$

as is clear from (2.1), and

$$(2.4)\qquad P_\lambda(x_1,\ldots,x_n;1)=m_\lambda(x_1,\ldots,x_n)$$

as is clear from (2.2).

As with the other types of symmetric functions studied in Chapter I, the number of variables $x_1,\ldots,x_n$ is immaterial, provided only that it is not less than the length of the partition λ. For we have

(2.5) *Let λ be a partition of length $\leqslant n$. Then*

$$P_\lambda(x_1,\ldots,x_n,0;t)=P_\lambda(x_1,\ldots,x_n;t).$$

Proof. From (2.2) we have

$$P_\lambda(x_1,\ldots,x_{n+1};t)=\sum_{w\in S_{n+1}/S_{n+1}^\lambda} w\left(x_1^{\lambda_1}\ldots x_{n+1}^{\lambda_{n+1}}\prod_{\lambda_i>\lambda_j}\frac{x_i-tx_j}{x_i-x_j}\right).$$

When we set x_{n+1} equal to 0, the only terms on the right-hand side which survive will be those which correspond to permutations $w\in S_{n+1}$ which send $n+1$ to some r such that $\lambda_r=0$; modulo S_{n+1}^λ, such a permutation fixes $n+1$, so that the summation is effectively over S_n/S_n^λ. ∎

Remark. The polynomials $R_\lambda(x_1,\ldots,x_n;t)$ defined in §1 do not have this stability property, and are therefore of little interest.

By virtue of (2.5) we may pass to the limit and define $P_\lambda(x;t)$ to be the element of $\Lambda[t]$ whose image in $\Lambda_n[t]$ for each $n \geqslant l(\lambda)$ is $P_\lambda(x_1,\dots,x_n;t)$. The symmetric function $P_\lambda(x;t)$ is the *Hall–Littlewood function* corresponding to the partition λ. It is homogeneous of degree $|\lambda|$.

From (1.2), (1.3), and (1.5) it follows that

$$P_\lambda(x_1,\dots,x_n;t) = \sum_\mu w_{\lambda\mu}(t) s_\mu(x_1,\dots,x_n)$$

with coefficients $w_{\lambda\mu}(t) \in \mathbf{Z}[t]$ such that $w_{\lambda\mu}(t) = 0$ unless $|\lambda| = |\mu|$ and $\lambda \geqslant \mu$, and $w_{\lambda\lambda}(t) = 1$. Hence

(2.6) *The transition matrix $M(P,s)$ which expresses the P_λ in terms of the s_μ is strictly upper unitriangular* (Chapter I, §6). |

Since the s_μ form a **Z**-basis of the ring Λ of symmetric functions, and therefore also a $\mathbf{Z}[t]$-basis of $\Lambda[t]$, it follows from (2.6) that the same is true of the P_λ:

(2.7) *The symmetric functions $P_\lambda(x;t)$ form a $\mathbf{Z}[t]$-basis of $\Lambda[t]$.* |

Next we consider P_λ when $\lambda = (1^r)$ and when $\lambda = (r)$. In the first case we have

$$P_{(1^r)}(x;t) = e_r(x), \tag{2.8}$$

the rth elementary symmetric function of the x_i.

Proof. By stability (2.5) $P_{(1^r)}$ is uniquely determined by its image in $\Lambda_r[t]$: in other words, we may assume that the number of variables is r. But then it is clear from (2.2) that $P_{(1^r)}(x_1,\dots,x_r;t) = x_1 \dots x_r = e_r$. |

We now define

$$q_r = q_r(x;t) = (1-t)P_{(r)}(x;t) \qquad (r \geqslant 1),$$

$$q_0 = q_0(x;t) = 1.$$

From (2.2) we have, for $r \geqslant 1$,

$$q_r(x_1,\dots,x_n;t) = (1-t)\sum_{i=1}^n x_i^r \prod_{j \neq i} \frac{x_i - tx_j}{x_i - x_j}. \tag{2.9}$$

The generating function for the q_r is

$$Q(u) = \sum_{r=0}^\infty q_r(x;t)u^r = \prod_i \frac{1 - x_i tu}{1 - x_i u} \tag{2.10}$$

$$= H(u)/H(tu)$$

in the notation of Chapter I.

Proof. Suppose first that the number of variables x_i is finite, and put $z = u^{-1}$. By the usual rule for partial fractions we have

$$\prod_{i=1}^{n} \frac{z - tx_i}{z - x_i} = 1 + \sum_{i=1}^{n} \frac{(1-t)x_i}{z - x_i} \prod_{j \neq i} \frac{x_i - tx_j}{x_i - x_j},$$

so that

$$\prod_{i=1}^{n} \frac{1 - tux_i}{1 - ux_i} = 1 + (1-t) \sum_{i=1}^{n} \frac{x_i u}{1 - x_i u} \prod_{j \neq i} \frac{x_i - tx_j}{x_i - x_j}$$

in which the coefficient of u^r, for $r \geqslant 0$, is equal to $q_r(x_1, \ldots, x_n; t)$ by (2.9). Now let $n \to \infty$ as usual. ∎

It will be convenient to introduce another family of symmetric functions $Q_\lambda(x;t)$, which are scalar multiples of the $P_\lambda(x;t)$. They are defined as follows:

$$Q_\lambda(x;t) = b_\lambda(t) P_\lambda(x;t) \tag{2.11}$$

where

$$b_\lambda(t) = \prod_{i \geqslant 1} \varphi_{m_i(\lambda)}(t) \tag{2.12}$$

Here, as usual, $m_i(\lambda)$ denotes the number of times i occurs as a part of λ, and $\varphi_r(t) = (1-t)(1-t^2)\ldots(1-t^r)$. In particular,

$$Q_{(r)}(x;t) = q_r(x;t). \tag{2.13}$$

We shall refer to the Q_λ as well as the P_λ as Hall–Littlewood functions. They may also be defined inductively, as follows. If f is any polynomial or formal power series in $x_1, \ldots, x_n$ (and possibly other variables), and $1 \leqslant i \leqslant n$, let $f^{(i)}$ denote the result of setting $x_i = 0$ in f. Then if we write Q_λ for $Q_\lambda(x_1, \ldots, x_n; t)$, we have

$$Q_\lambda = \sum_{i=1}^{n} x_i^{\lambda_1} g_i Q_\mu^{(i)}, \tag{2.14}$$

where $\mu = (\lambda_2, \lambda_3, \ldots)$ is the partition obtained by deleting the largest part of λ, and

$$g_i = (1-t) \prod_{j \neq i} \frac{x_i - tx_j}{x_i - x_j}.$$

Proof. Let l be the length of λ. From the definitions of $b_\lambda(t)$ and $v_\lambda(t)$ we have

$$v_\lambda(t) = v_{n-l}(t) b_\lambda(t)/(1-t)^l,$$

and therefore

$$\begin{aligned} Q_\lambda(x_1,\dots,x_n;t) &= \frac{(1-t)^l}{v_{n-l}(t)} R_\lambda(x_1,\dots,x_n;t) \\ &= (1-t)^l \sum_{w \in S_n/S_{n-l}} w\left(x_1^{\lambda_1} \dots x_l^{\lambda_l} \prod_{i=1}^{l} \prod_{j>i} \frac{x_i - tx_j}{x_i - x_j} \right) \end{aligned}$$

by (1.4), where S_{n-l} acts on $x_{l+1},\dots,x_n$. It follows that

$$Q_\lambda(x_1,\dots,x_n;t) = \sum_{w \in S_n/S_{n-1}} w\left(x_1^{\lambda_1} g_1 Q_\mu(x_2,\dots,x_n;t)\right)$$

which is equivalent to (2.14). ∎

Let

$$\begin{aligned} F(u) &= \frac{1-u}{1-tu} \\ &= 1 + \sum_{r \geq 1} (t^r - t^{r-1}) u^r \end{aligned}$$

as a power series in u. Then we have

(2.15) *Let $u_1, u_2, \dots$ be independent indeterminates. Then $Q_\lambda(x;t)$ is the coefficient of $u^\lambda = u_1^{\lambda_1} u_2^{\lambda_2} \dots$ in*

$$Q(u_1,u_2,\dots) = \prod_{i \geq 1} Q(u_i) \prod_{i<j} F\left(u_i^{-1} u_j\right).$$

Proof. We proceed by induction on the length l of λ. When $l=1$, (2.15) follows from (2.13). So let $l > 1$ and assume the result true for $\mu = (\lambda_2, \lambda_3, \dots)$.

As before, let $Q^{(i)}(u)$ (resp. $Q^{(i)}(u_1,u_2,\dots)$) denote the result of setting $x_i = 0$ in $Q(u)$ (resp. $Q(u_1,u_2,\dots)$). From (2.10) we have

$$Q^{(i)}(u) = F(x_i u) Q(u)$$

and hence

(1)
$$Q^{(i)}(u_1,u_2,\dots) = Q(u_1,u_2,\dots) \prod_{j \geq 1} F(x_i u_j).$$

From (2.14), $Q_\lambda(x_1,\dots,x_n;t)$ is the coefficient of u^λ in

$$\sum_{\lambda_1>0} u_1^{\lambda_1} \sum_{i=1}^{n} x_i^{\lambda_1} g_i Q^{(i)}(u_2,u_3,\dots)$$

$$= Q(u_2,u_3,\dots) \sum_{\lambda_1>0} u_1^{\lambda_1} \sum_{i=1}^{n} x_i^{\lambda_1} g_i \prod_{j>2} F(x_i u_j) \tag{2}$$

by (1) above. Let us expand the product $\prod_{j>2} F(x_i u_j)$ as a power series in x_i, say

$$\prod_{j>2} F(x_i u_j) = \sum_{m>0} f_m x_i^m$$

where the coefficients f_m are polynomials in $t, u_2, u_3, \dots$. Then the expression (2) is equal to $Q(u_2,u_3,\dots)$ multiplied by

$$\begin{aligned}\sum_{\lambda_1>0} u_1^{\lambda_1} \sum_{m>0} f_m \sum_{i=1}^{n} x_i^{\lambda_1+m} g_i &= \sum_{\lambda_1,m} u_1^{\lambda_1} q_{\lambda_1+m} f_m \\ &= Q(u_1) - \sum_{m\geqslant 0} f_m u_1^{-m} - X, \text{ say} \\ &= Q(u_1) - \prod_{j\geqslant 2} F(u_1^{-1}u_j) - X\end{aligned}$$

where X involves only negative powers of u_1, and therefore does not contribute to the coefficient of u^λ. Hence finally Q_λ is the coefficient of u^λ in

$$Q(u_1)\prod_{j>2} F(u_1^{-1}u_j).Q(u_2,u_3,\dots) = Q(u_1,u_2,u_3,\dots) \quad \|$$

For any finite sequence $\alpha = (\alpha_1, \alpha_2, \dots)$ of integers let

$$q_\alpha = q_\alpha(x;t) = \prod_{i>1} q_{\alpha_i}(x;t)$$

with the convention that $q_r = 0$ if $r<0$ (so that $q_\alpha = 0$ if any α_i is negative). The raising operators $R_{ij}, i<j$ (Chapter I, §1) act as follows: $R_{ij}q_\alpha = q_{R_{ij}\alpha}$. Then the coefficient of u^α in

$$F(u_i^{-1}u_j)\prod_{k>1} Q(u_k) = \left(1 + \sum_{r>1} (t^r - t^{r-1}) u_i^{-r} u_j^r\right) \sum_\beta q_\beta u^\beta$$

is

$$1 + \sum_{r>1} (t^r - t^{r-1}) R_{ij}^r q_\alpha = \frac{1-R_{ij}}{1-tR_{ij}} q_\alpha .$$

It follows that (2.15) may be restated in terms of raising operators:

(2.15′)
$$Q_\lambda = \prod_{i<j} \frac{1-R_{ij}}{1-tR_{ij}} q_\lambda$$

for all partitions λ.

Hence

$$Q_\lambda = \prod_{i<j} \left(1+(t-1)R_{ij}+(t^2-t)R_{ij}^2+\dots\right)q_\lambda$$

and it follows from Chapter I, (1.14) that

$$Q_\lambda = \sum_\mu a_{\lambda\mu}(t) q_\mu$$

where the polynomials $a_{\lambda\mu}(t) \in \mathbf{Z}[t]$ are such that $a_{\lambda\mu}(t)=0$ unless $\lambda \leqslant \mu$, and $a_{\lambda\lambda}(t)=1$. By (2.7), the Q_λ form a $\mathbf{Q}(t)$-basis of $\Lambda \otimes_{\mathbf{Z}} \mathbf{Q}(t)$; and therefore

(2.16) *The symmetric functions* q_λ *form a* $\mathbf{Q}(t)$*-basis of* $\Lambda \otimes_{\mathbf{Z}} \mathbf{Q}(t)$*, and the transition matrix* (Chapter I, § 6) $M(Q,q)$ *is strictly lower unitriangular.*

Examples

1. If we set $x_i = t^{i-1} (1 \leqslant i \leqslant n)$ we have

$$R_\lambda(1,t,\dots,t^{n-1};t) = t^{n(\lambda)} v_n(t)$$

directly from the definition (1.1), because the only term in the sum which does not vanish is that corresponding to $w=1$. Hence

$$Q_\lambda(1,t,\dots,t^{n-1};t) = t^{n(\lambda)} \varphi_n(t)/\varphi_{m_0}(t)$$

where $m_0 = n - l(\lambda)$. Let $n \to \infty$, then also $m_0 \to \infty$ and so when $x_i = t^{i-1}$ for all $i \geqslant 1$ we have

$$Q_\lambda(1,t,t^2,\dots;t) = t^{n(\lambda)}.$$

2. We can use the inductive definition (2.14) to define Q_λ for any finite sequence $(\lambda_1,\lambda_2,\dots,\lambda_l)$ of non-negative integers, not necessarily in descending order. The formula (2.15′) will still be valid and can be used to extend the definition of Q_λ to *any* sequence $(\lambda_1,\lambda_2,\dots\lambda_l)$ of integers, positive or negative. If $\lambda_l<0$, clearly $Q_\lambda=0$.

To reduce Q_λ when the λ_i are not in descending order to a linear combination of the Q_μ where μ is a partition, we may proceed as follows: since interchange of x_1 and x_2 transforms

$$\frac{x_1^r x_2^s (x_2 - tx_1)(x_1 - tx_2)}{x_1 - x_2}$$

into

$$-\frac{x_1^s x_2^r (x_2 - tx_1)(x_1 - tx_2)}{x_1 - x_2},$$

it follows that

$$Q_{(r,s+1)} - tQ_{(r+1,s)} = -(Q_{(s,r+1)} - tQ_{(s+1,r)})$$

or, replacing r by $r-1$, that

$$Q_{(s,r)} = tQ_{(r,s)} - Q_{(r-1,s+1)} + tQ_{(s+1,r-1)}.$$

Assuming $s < r$, this relation enables us to express $Q_{(s,r)}$ in terms of the $Q_{(r-i,s+i)}$ where $0 \leqslant i \leqslant [\frac{1}{2}(r-s)] = m$ say. The result is

$$Q_{(s,r)} = tQ_{(r,s)} + \sum_{i=1}^{m} (t^{i+1} - t^{i-1}) Q_{(r-i,s+i)}$$

if $r - s = 2m + 1$, whereas if $r - s = 2m$ the last term in the sum must be replaced by $(t^m - t^{m-1})\, Q_{(r-m,\, s+m)}$.

For simplicity we have stated these formulas for a two-term sequence (s, r): but the same holds for any two consecutive terms of a sequence λ.

3. The definitions (2.1) and (2.2) of P_λ can be written in terms of lowering operations $R_{ji} (i < j)$:

$$P_\lambda = v_\lambda(t)^{-1} \prod_{i<j} (1 - tR_{ji}) s_\lambda$$

$$= \prod_{\lambda_i > \lambda_j} (1 - tR_{ji}) s_\lambda$$

Hence, for example,

$$P_{(n)} = \prod_{j=2}^{n} (1 - tR_{j1}) s_{(n)}$$

$$= \sum_J (-t)^{|J|} \prod_{j \in J} R_{j1} s_{(n)}$$

summed over all subsets J of $\{2, 3, \dots, n\}$. The only J for which $\prod_{j \in J} R_{j1} s_{(n)} \neq 0$ are $J = \{2, 3, \dots, i\}$ $(2 \leqslant i \leqslant n)$, and therefore

$$P_{(n)} = \sum_{r=0}^{n-1} (-t)^r s_{(n-r,1^r)}.$$

3. The Hall algebra again

By (2.7), the product $P_\mu P_\nu$ of two Hall–Littlewood functions will be a linear combination of the P_λ, where $|\lambda| = |\mu| + |\nu|$ (λ, μ, ν being partitions) with coefficients in $\mathbf{Z}[t]$: that is to say, there exist polynomials $f^\lambda_{\mu\nu}(t) \in \mathbf{Z}[t]$ such that

$$P_\mu(x;t)P_\nu(x;t) = \sum_\lambda f^\lambda_{\mu\nu}(t)P_\lambda(x;t).$$

When $t = 0$ we have by (2.3)

$$f^\lambda_{\mu\nu}(0) = c^\lambda_{\mu\nu}, \tag{3.1}$$

the coefficient of s_λ in $s_\mu s_\nu$. Likewise, from (2.4), $f^\lambda_{\mu\nu}(1)$ is the coefficient of m_λ in the product $m_\mu m_\nu$ when expressed as a sum of monomial symmetric functions.

In order to connect the symmetric functions P_λ with the Hall algebra, we need to compute the polynomial $f^\lambda_{\mu\nu}(t)$ in the case $\nu = (1^m)$.

Recall that

$$\begin{bmatrix} n \\ r \end{bmatrix} = \varphi_n(t)/\varphi_r(t)\varphi_{n-r}(t)$$

if $0 \leqslant r \leqslant n$ and that $\begin{bmatrix} n \\ r \end{bmatrix} = 0$ otherwise, where

$$\varphi_r(t) = (1-t)(1-t^2)\dots(1-t^r).$$

(3.2) *We have*

$$f^\lambda_{\mu(1^m)}(t) = \prod_{i \geqslant 1} \begin{bmatrix} \lambda'_i - \lambda'_{i+1} \\ \lambda'_i - \mu'_i \end{bmatrix}$$

(*and therefore* $f^\lambda_{\mu(1^m)}(t) = 0$ *unless* $\lambda - \mu$ *is a vertical m-strip*).

Proof. We shall work with a finite set of variables $x_1, \dots, x_n$, where $n \geqslant l(\mu) + m$. We have to multiply P_μ, given by the formula (2.2), by $P_{(1^m)}$, which by (2.7) is equal to the mth elementary symmetric function e_m. Let $k = \mu_1$ be the largest part of μ, and split up the set $\{x_1, \dots, x_n\}$ into subsets $X_0, \dots, X_k$ (some possibly empty) such that $x_j \in X_i$ if and only if $\mu_j = i$, so that $|X_i| = m_i$, the multiplicity of i as a part of μ. Let $e_r(X_i)$ denote the elementary symmetric functions of the set of variables X_i. Since

$$\prod_{i=1}^n (1 + x_i t) = \prod_{j=0}^k \prod_{x_i \in X_j} (1 + x_i t)$$

it follows that

(1) $$e_m(x_1,\dots,x_n)=\sum_r e_{r_0}(X_0)\dots e_{r_k}(X_k)$$

summed over all $r=(r_0,\dots,r_k)\in\mathbf{N}^{k+1}$ such that $r_i\leqslant m_i$ for each i and $\sum r_i=m$.

To each such r there corresponds a partition λ defined by

$$\lambda'_i=\mu'_i+r_{i-1}\qquad(1\leqslant i\leqslant k+1).$$

Clearly $\lambda-\mu$ is a vertical m-strip, and conversely every partition λ such that $\lambda-\mu$ is a vertical m-strip arises in this way.

Each $e_{r_i}(X_i)$ on the right-hand side of (1) is equal to $P_{(1^{r_i})}(X_i;t)$ by (2.7), and hence by (2.1) can be written as a sum over the symmetric group S_{m_i} acting on the set X_i. In this way (1) takes the form

(2) $$e_m(x_1,\dots,x_n)=\sum_\lambda c_{\lambda,\mu}(t)^{-1}\Phi_{\lambda,\mu}(x_1,\dots,x_n;t)$$

summed over all partitions $\lambda\supset\mu$ such that $\lambda-\mu$ is a vertical m-strip, where

$$\Phi_{\lambda,\mu}=\sum_{w\in S_n^\mu}w\left(x_1^{\lambda_1-\mu_1}\dots x_n^{\lambda_n-\mu_n}\prod_{\substack{i<j\\ \mu_i=\mu_j}}\frac{x_i-tx_j}{x_i-x_j}\right)$$

and

$$c_{\lambda,\mu}(t)=\prod_{i=0}^{k}v_{r_i}(t)v_{m_i-r_i}(t).$$

If we now multiply both sides of (2) by $P_\mu(x_1,\dots,x_n;t)$ as given by (2.2) we shall obtain

$$P_\mu e_m=\sum_\lambda c_{\lambda,\mu}(t)^{-1}R_\lambda$$

from which it follows that the coefficient of P_λ in $P_\mu e_m$ is

$$f^\lambda_{\mu(1^m)}(t)=v_\lambda(t)c_{\lambda,\mu}(t)^{-1}$$

which easily reduces to the expression given. |

If we now compare (3.2) with Chapter II, (4.6) we see that

(3.3) $$G^\lambda_{\mu(1^m)}(\mathfrak{o})=q^{n(\lambda)-n(\mu)-n(1^m)}f^\lambda_{\mu(1^m)}(q^{-1}).$$

From this we deduce the following structure theorem for the Hall algebra $H(\mathfrak{o})$:

(3.4) *Let* $\psi: H(\mathfrak{o}) \otimes_{\mathbf{Z}} \mathbf{Q} \to \Lambda_{\mathbf{Q}}$ *be the* $\mathbf{Q}$*-linear mapping defined by*

$$\psi(u_\lambda) = q^{-n(\lambda)} P_\lambda(x; q^{-1}).$$

Then ψ *is an isomorphism of rings.*

Proof. By (2.7), ψ is a linear isomorphism. Since $H(\mathfrak{o})$ is freely generated (as $\mathbf{Z}$-algebra) by the $u_{(1^r)}$ (Chapter II, (2.3)), we may define a ring homomorphism $\psi': H(\mathfrak{o}) \otimes_{\mathbf{Z}} \mathbf{Q} \to \Lambda_{\mathbf{Q}}$ by

$$\psi'(u_{(1^r)}) = q^{-r(r-1)/2} e_r.$$

We shall show that $\psi' = \psi$, which will prove (3.4). To do this we shall prove by induction on the partition λ that $\psi'(u_\lambda) = \psi(u_\lambda)$. When $\lambda = 0$, this is clear from the definitions. Now assume that $\lambda \neq 0$, and let μ be the partition obtained from λ by deleting the last column. Suppose that this last column has m elements. Then from Chapter II, (4.6) we have $G^\lambda_{\mu(1^m)}(\mathfrak{o}) = 1$, and $G^\nu_{\mu(1^m)}(\mathfrak{o}) = 0$ unless $\nu \supset \mu$ and $\nu - \mu$ is a vertical m-strip. Moreover, if $\nu - \mu$ is a vertical m-strip, it is easy to see that $\nu \leqslant \lambda$. Hence

$$u_\mu u_{(1^m)} = u_\lambda + \sum_{\nu < \lambda} G^\nu_{\mu(1^m)}(\mathfrak{o}) u_\nu \tag{1}$$

and likewise

$$P_\mu P_{(1^m)} = P_\lambda + \sum_{\nu < \lambda} f^\nu_{\mu(1^m)}(q^{-1}) P_\nu \tag{2}$$

where P_μ stands for $P_\mu(x; q^{-1})$, and so on. If we now apply ψ' to both sides of (1) and recall that $P_{(1^m)} = e_m$ by (2.7), and compare the result with (2), taking as inductive hypothesis that $\psi'(u_\nu) = q^{-n(\nu)} P_\nu$ for all $\nu < \lambda$, then it follows from (3.3) that $\psi'(u_\lambda) = q^{-n(\lambda)} P_\lambda = \psi(u_\lambda)$. |

Remark. This proof depends on the identity (3.3), which appears as merely a happy accident. It is possible to give a proof of (3.4) which does not depend on an apparent miracle: see the notes at the end of Chapter V.

From (3.4) it follows that

$$G^\lambda_{\mu\nu}(\mathfrak{o}) = q^{n(\lambda) - n(\mu) - n(\nu)} f^\lambda_{\mu\nu}(q^{-1}) \tag{3.5}$$

for any three partitions λ, μ, ν, and therefore by Chapter II, (4.3) the polynomials $f^\lambda_{\mu\nu}$ are related to the Hall polynomials $g^\lambda_{\mu\nu}$ of Chapter II as follows:

$$g^\lambda_{\mu\nu}(t) = t^{n(\lambda) - n(\mu) - n(\nu)} f^\lambda_{\mu\nu}(t^{-1}). \tag{3.6}$$

In particular:

(3.7) (i) *If* $f^{\lambda}_{\mu\nu}(0)=0$, *then* $f^{\lambda}_{\mu\nu}(t)$ *is identically zero.*
(ii) $f^{\lambda}_{\mu\nu}(t)=0$ *unless* $|\lambda|=|\mu|+|\nu|$ *and* $\mu,\ \nu\subset\lambda$.

The results of Chapter II, §4 (especially (4.2) and (4.11)) provide an explicit formula for $f^{\lambda}_{\mu\nu}(t)$, namely

(3.8) $$f_{\lambda\mu}(t)=\sum_S f_S(t)$$

summed over LR-sequences of type $(\mu',\nu';\lambda')$, where in the notation of Chapter II, (4.10) and (4.11)

(3.9) $$f_S(t)=\prod_{i>j}\Phi(a_{ij},b_{ij},c_{ij};N_{ij};t).$$

In particular:

(3.10) *If* $\lambda\supset\mu$ *and* $\theta=\lambda-\mu$ *is a horizontal r-strip, then*

$$f^{\lambda}_{\mu(r)}(t)=(1-t)^{-1}\prod_{i\in I}(1-t^{m_i(\lambda)})$$

where I is the set of positive integers i such that $\theta'_i>\theta'_{i+1}$ *(so that* $\theta'_i=1$ *and* $\theta'_{i+1}=0$*) and* $m_i(\lambda)$ *is the multiplicity of i as a part of* λ. *In all other cases* $f^{\lambda}_{\mu(r)}(t)=0$.

In view of (3.6), this is simply a transcription of Chapter II, (4.13).

Remark. We have arrived at this result (which we shall later make use of) by an indirect route, via the Hall algebra. It is also possible to derive (3.10) directly from the definition of the $P_\lambda(x;t)$: see Morris [M13].

Examples

1. Let λ be a partition. Then for each $m\geqslant 0$,

$$\sum_\mu G^{\lambda}_{\mu(1^m)}(\mathfrak{o})$$

is the number of elementary submodules E of type (1^m) in a fixed $\mathfrak{o}$-module M of type λ. All these submodules E lie in the socle S of M, which is a k-vector space of dimension $l=l(\lambda)$, and the sum above is therefore equal to the number of m-dimensional subspaces of an l-dimensional vector space over a finite field with q elements, which is $\begin{bmatrix} l \\ m \end{bmatrix}(q)$. Hence by (3.3) we obtain

$$\sum_\mu t^{n(\mu)}f^{\lambda}_{\mu(1^m)}(t)=t^{n(\lambda)-m(m-1)/2}\begin{bmatrix} l \\ m \end{bmatrix}(t^{-1}).$$

Now let y be an indeterminate. Then

$$\left(\sum_{\mu} t^{n(\mu)}P_{\mu}\right)\left(\sum_{m} e_m y^m\right) = \sum_{\lambda,\mu,m} y^m t^{n(\mu)} f^{\lambda}_{\mu(1^m)}(t) P_{\lambda}$$

$$= \sum_{\lambda} t^{n(\lambda)}P_{\lambda} \sum_{m} y^m t^{-m(m-1)/2} \begin{bmatrix} l(\lambda) \\ m \end{bmatrix}(t^{-1})$$

$$(1) \qquad = \sum_{\lambda} t^{n(\lambda)}P_{\lambda} \prod_{j=1}^{l(\lambda)} (1+t^{1-j}y)$$

by the identity of Chapter I, §2, Example 3. In particular, when $y = -1$ we obtain

$$\left(\sum_{\mu} t^{n(\mu)}P_{\mu}\right)\left(\sum_{m} (-1)^m e_m\right) = 1$$

and therefore

$$(2) \qquad \sum_{|\mu|=n} t^{n(\mu)}P_{\mu} = h_n.$$

Hence the identity (1) takes the form

$$(3) \qquad \prod_{i\geqslant 1}(1+x_i y)/(1-x_i) = \sum_{\lambda} t^{n(\lambda)} \prod_{j=1}^{l(\lambda)}(1+t^{1-j}y) . P_{\lambda}(x;t).$$

2. (a) Let

$$\tilde{P}_{\lambda} = q^{-n(\lambda)}P_{\lambda}(x;q^{-1}), \qquad \tilde{Q}_{\lambda} = q^{|\lambda|+n(\lambda)}Q_{\lambda}(x;q^{-1})$$

so that $\tilde{Q}_{\lambda} = a_{\lambda}(q)\tilde{P}_{\lambda}$, where by Chapter II, (1.6) $a_{\lambda}(q)$ is the order of the automorphism group Aut M_{λ} of an $\mathfrak{o}$-module of type λ. From §2, Example 1 it follows that the specialization $x_i \mapsto q^{-i} (i \geqslant 1)$ maps each $\tilde{Q}_{\lambda}$ to 1 and hence $\tilde{P}_{\lambda}$ to $a_{\lambda}(q)^{-1}$. Hence from (3.4) the mapping $u_{\lambda} \mapsto |\text{Aut } M_{\lambda}|^{-1}$ of the Hall algebra $H(\mathfrak{o})$ into $\mathbf{Q}$ is multiplicative, i.e. is a *character* of $H(\mathfrak{o})$.

(b) From Example 1 above we have

$$\sum_{|\lambda|=n} \tilde{P}_{\lambda} = h_n$$

and therefore, specializing as above,

$$\sum_{|\lambda|=n} |\text{Aut } M_{\lambda}|^{-1} = h_n(q^{-1}, q^{-2}, \ldots).$$

Hence

$$\sum_{\lambda} |\text{Aut } M_{\lambda}|^{-1} = \prod_{i\geqslant 1}(1-q^{-i})^{-1} = \sum_{\lambda} q^{-|\lambda|}$$

or equivalently

$$\sum_{\lambda} |\mathrm{Aut}\, M_\lambda|^{-1} = \sum_{\lambda} |M_\lambda|^{-1}$$

as convergent infinite series.

(c) More generally, the identity (3) of Example 1 above leads to

$$\sum_{\lambda} \frac{(x;q)_{l(\lambda)}}{|\mathrm{Aut}\, M_\lambda|} = \prod_{i \geqslant 1} \frac{1-xq^{-i}}{1-q^{-i}}$$

where $(x;q)_n = (1-x)(1-qx)\dots(1-q^{n-1}x)$.

3. (a) Let $\mu = (\mu^{(1)},\dots,\mu^{(r)})$ be any sequence of partitions, and let

$$v_\mu(M_\lambda) = g^{\lambda}_{\mu^{(1)}\dots\mu^{(r)}}(q)$$

be the number of chains of submodules

$$M_\lambda = M_0 \supset M_1 \supset \dots \supset M_r = 0$$

such that M_{i-1}/M_i has type $\mu^{(i)}$, for $1 \leqslant i \leqslant r$. Show that

$$\sum_{\lambda} \frac{v_\mu(M_\lambda)}{|\mathrm{Aut}\, M_\lambda|} = \prod_{i=1}^{r} |\mathrm{Aut}\, M_\mu^{(i)}|^{-1}. \tag{1}$$

In particular, if $v(M_\lambda)$ is the number of composition series of M_λ, then

$$\sum_{|\lambda|=r} \frac{v(M_\lambda)}{|\mathrm{Aut}\, M_\lambda|} = \frac{1}{(q-1)^r}.$$

(Specialize the identity

$$\sum_{\lambda} g^{\lambda}_{\mu^{(1)}\dots\mu^{(r)}}(q) u_\lambda = u_{\mu^{(1)}}\dots u_{\mu^{(r)}}$$

as in Example 2 above.)

(b) Let $H = (H_1,\dots,H_r)$ be any sequence of finite abelian groups, and for each finite abelian group G let $v_H(G)$ denote the number of chains of subgroups

$$G = G_0 \geqslant G_1 \geqslant \dots \geqslant G_r = 0$$

such that $G_{i-1}/G_i \cong H_i$ for $1 \leqslant i \leqslant r$. Show that

$$\sum_{(G)} \frac{v_H(G)}{|\mathrm{Aut}\, G|} = \prod_{i=1}^{r} |\mathrm{Aut}\, H_i|^{-1}$$

where the sum on the left is over all isomorphism classes of finite abelian groups. (Split G and the H_i into their p-primary components, and use (a) above.)

4. The fact (Example 2 (a)) that $u_\lambda \mapsto a_\lambda(q)^{-1}$ is a character of the Hall algebra $H(\mathfrak{o})$ is equivalent to the following relation:

$$\sum_\lambda a_\lambda(q)^{-1} a_\mu(q) a_\nu(q) g^\lambda_{\mu\nu}(q) = 1 \tag{1}$$

for all partitions μ, ν. This may be proved directly as follows.

Fix $\mathfrak{o}$-modules L, M, N of respective types λ, μ, ν, and let E^L_{MN} denote the set of all exact sequences

$$E(\alpha, \beta)\colon 0 \to N \overset{\alpha}{\to} L \overset{\beta}{\to} M \to 0.$$

In such an exact sequence, $N' = \alpha(N)$ is a submodule of L of type ν, with $L/N' \cong M$ of type μ; hence the number of choices for N' is $g^\lambda_{\mu\nu}(q)$. Given N', there are $a_\nu(q)$ choices for α, and $a_\mu(q)$ choices for β, and therefore

$$e^L_{MN} = |E^L_{MN}| = a_\mu(q) a_\nu(q) g^\lambda_{\mu\nu}(q).$$

Hence (1) is equivalent to

$$\sum_{(L)} a_L^{-1} e^L_{MN} = 1 \tag{2}$$

where $a_L = |\mathrm{Aut}\, L|$ and the sum is over isomorphism classes of finite $\mathfrak{o}$-modules L.

The group $\mathrm{Aut}\, L$ acts on E^L_{MN} by $\varphi E(\alpha, \beta) = E(\varphi\alpha, \beta\varphi^{-1})$, and it is not difficult to verify that $\varphi \in \mathrm{Aut}\, L$ fixes $E(\alpha, \beta)$ if and only if $\varphi = 1 + \alpha\theta\beta$, where $\theta \in \mathrm{Hom}(M, N)$. Hence the orbits are all of size a_L/h, where $h = |\mathrm{Hom}(M, N)|$, and the number of them is therefore $a_L^{-1} e^L_{MN} h$.

On the other hand, the orbits of $\mathrm{Aut}\, L$ in E^L_{MN}, for all L, are the equivalence classes of extensions of M by N (see e.g. [C1], Chapter XIV) and are in one–one correspondence with the elements of $\mathrm{Ext}^1(M, N)$. Now $\mathrm{Ext}^1(M, N)$ and $\mathrm{Hom}(M, N)$ are finite $\mathfrak{o}$-modules of the same type $\mu \times \nu$ (Chapter II, §1, Example) and in particular have the same order $h (= q^{|\mu \times \nu|})$. Hence we have

$$\sum_{(L)} a_L^{-1} e^L_{MN} h = h$$

which proves (2) and therefore also (1).

5. Let $\mu = (\mu_1, \dots, \mu_r)$ be a partition of length r, and for a finite $\mathfrak{o}$-module M let $w_\mu(M)$ denote the number of chains of submodules

$$M = M_0 \supset M_1 \supset \dots \supset M_r = 0$$

such that $|M_{i-1}/M_i| = q^{\mu_i}$ for $1 \leqslant i \leqslant r$. Thus

$$w_\mu(M_\lambda) = \sum g^\lambda_{\nu^{(1)} \dots \nu^{(r)}}(q)$$

summed over all sequences $(\nu^{(1)}, \dots, \nu^{(r)})$ of partitions such that $|\nu^{(i)}| = \mu_i$ for $1 \leqslant i \leqslant r$. From Example 2 it follows that

$$\sum_\lambda w_\mu(M_\lambda) \tilde{P}_\lambda = h_\mu$$

and hence that

(1) $$\sum_\lambda \frac{w_\mu(M_\lambda)}{|\mathrm{Aut}\, M_\lambda|} = h_\mu(q^{-1}, q^{-2}, \dots).$$

Since (Chapter I, §2, Example 4)

$$\begin{aligned} h_r(q^{-1}, q^{-2}, \dots) &= q^{-r}/\varphi_r(q^{-1}) \\ &= q^{r(r-1)/2}/(q-1)\dots(q^r-1), \end{aligned}$$

we may rewrite (1) in the form

(2) $$\sum_\lambda \frac{w_\mu(M_\lambda)}{|\mathrm{Aut}\, M_\lambda|} = q^{n(\mu')} \prod_{i \geq 1} (q^i - 1)^{-\mu'_i}.$$

In particular, when $r = 2$ and $\mu = (n-k, k)$, $w_\mu(M_\lambda)$ is the number of submodules of M_λ of order q^k. More particularly still, when $\mu = (n-1, 1)$, $w_\mu(M_\lambda)$ is the number of submodules of M_λ of order q. These all lie in the socle, which is a k-vector space of dimension $l(\lambda)$, so that $w_\mu(M_\lambda) = (q^{l(\lambda)} - 1)/(q-1)$ when $\mu = (n-1, 1)$. Deduce that

$$\sum_\lambda \frac{q^{l(\lambda)} t^{|\lambda|}}{|\mathrm{Aut}\, M_\lambda|} = \frac{1+t}{\prod_{i \geq 1} (1 - q^{-i} t)}$$

where the sum on the left is over all partitions λ.

Notes and References

The identities of Examples 2(b) and 3(a) are due to P. Hall [H1], [H2]. See also [M4].

4. Orthogonality

We shall now generalize the developments of Chapter I, §4 by giving three series expansions for the product

$$\prod_{i,j} (1 - t x_i y_j)/(1 - x_i y_j).$$

The first of these is

(4.1) $$\prod_{i,j} (1 - t x_i y_j)/(1 - x_i y_j) = \sum_\lambda z_\lambda(t)^{-1} p_\lambda(x) p_\lambda(y)$$

summed over all partitions λ, where

$$z_\lambda(t) = z_\lambda \cdot \prod_{i \geq 1} (1 - t^{\lambda_i})^{-1}.$$

Proof. We have

$$\log \prod_{i,j} (1 - tx_i y_j)/(1 - x_i y_j) = \sum_{i,j} \left(\log(1 - tx_i y_j) - \log(1 - x_i y_j)\right)$$
$$= \sum_{i,j} \sum_{m=1}^{\infty} \frac{1-t^m}{m} (x_i y_j)^m$$
$$= \sum_{m=1}^{\infty} \frac{1-t^m}{m} p_m(x) p_m(y),$$

and therefore

$$\prod_{i,j} (1 - tx_i y_j)/(1 - x_i y_j) = \prod_{m=1}^{\infty} \exp\left(\frac{1-t^m}{m} p_m(x) p_m(y)\right)$$
$$= \prod_{m=1}^{\infty} \sum_{r_m=0}^{\infty} \frac{(1-t^m)^{r_m}}{m^{r_m} . r_m!} p_m(x)^{r_m} p_m(y)^{r_m}$$
$$= \sum_{\lambda} z_\lambda(t)^{-1} p_\lambda(x) p_\lambda(y). \quad |$$

Next we have

(4.2)
$$\prod_{i,j} (1 - tx_i y_j)/(1 - x_i y_j) = \sum_{\lambda} q_\lambda(x;t) m_\lambda(y)$$
$$= \sum_{\lambda} m_\lambda(x) q_\lambda(y;t)$$

summed over all partitions λ.

Proof. From (2.10) it follows that

$$\prod_{i,j} (1 - tx_i y_j)/(1 - x_i y_j) = \prod_j \sum_{r_j=0}^{\infty} q_{r_j}(x;t) y_j^{r_j}$$

and the product on the right, when multiplied out, is equal to

$$\sum_{\lambda} q_\lambda(x;t) m_\lambda(y).$$

Likewise with the x's and y's interchanged. |

From (4.2) it follows that

(4.3) *The transition matrix* $M(q, m)$ *is symmetric.* |

Next, in generalization of Chapter I, (4.3) we have

(4.4)
$$\prod_{i,j}(1-tx_iy_j)/(1-x_iy_j) = \sum_\lambda P_\lambda(x;t)Q_\lambda(y;t)$$
$$= \sum_\lambda b_\lambda(t)P_\lambda(x;t)P_\lambda(y;t).$$

Proof. Consider the transition matrices

$$A = M(q,Q), \qquad B = M(m,Q), \qquad C = M(q,m)$$

and let $D = B'A$. By (2.16), A^{-1} is lower unitriangular, hence so is A. Also B is upper triangular, because

$$B^{-1} = M(Q,P)M(P,s)M(s,m)$$

and $M(Q,P)$ is diagonal by (2.11), $M(P,s)$ is upper triangular by (2.6), and $M(s,m)$ is upper triangular by Chapter I, (6.5). It follows that D is lower triangular. On the other hand, $D = B'CB$, and C is symmetric by (4.3); hence D is symmetric as well as triangular, and is therefore a diagonal matrix, with diagonal elements equal to those of B, so that $D = M(P,Q)$ and therefore $D_{\lambda\lambda} = b_\lambda(t)^{-1}$.

Hence

$$\sum_\lambda q_\lambda(x;t)m_\lambda(y) = \sum_{\lambda,\mu,\nu} A_{\lambda\mu}B_{\lambda\nu}Q_\mu(x;t)Q_\nu(y;t)$$
$$= \sum_\mu b_\mu(t)^{-1}Q_\mu(x;t)Q_\mu(y;t)$$
$$= \sum_\mu P_\mu(x;t)Q_\mu(y;t)$$

and so (4.4) follows from (4.2). |

Remark. There is yet another expansion for $\prod(1-tx_iy_j)/(1-x_iy_j)$. Let us define

(4.5)
$$S_\lambda(x;t) = \det\left(q_{\lambda_i-i+j}(x;t)\right)$$

for any partition λ, so that

(4.6)
$$S_\lambda(x;t) = \prod_{i<j}(1-R_{ij})q_\lambda = \prod_{i<j}(1-tR_{ij})Q_\lambda$$

by Chapter I, (3.4″) and (2.15′). If we introduce a set of (fictitious) variables ξ_i by means of

$$\prod_i(1-tx_iy)/(1-x_iy) = \prod_i(1-\xi_iy)^{-1}$$

then $q_r(x;t)=h_r(\xi)$ and therefore $S_\lambda(x;t)=s_\lambda(\xi)$ by Chapter I, (3.4). Hence from Chapter I, (4.3) we have

(4.7) $$\prod_{i,j}(1-tx_iy_j)/(1-x_iy_j)=\sum_\lambda S_\lambda(x;t)s_\lambda(y)$$
$$=\sum_\lambda s_\lambda(x)S_\lambda(y;t).$$

We now define a scalar product on $\Lambda[t]$ (with values in $\mathbf{Q}(t)$) by requiring that the bases $(q_\lambda(x;t))$ and $(m_\lambda(x))$ be dual to each other:

(4.8) $$\langle q_\lambda(x;t),m_\mu(x)\rangle=\delta_{\lambda\mu}.$$

The same considerations as in Chapter I, §4, applied to the identities (4.1), (4.4), and (4.7) show that

(4.9) $$\langle P_\lambda(x;t),Q_\mu(x;t)\rangle=\delta_{\lambda\mu},$$

(4.10) $$\langle S_\lambda(x;t),s_\mu(x)\rangle=\delta_{\lambda\mu},$$

(4.11) $$\langle p_\lambda(x),p_\mu(x)\rangle=z_\lambda(t)\delta_{\lambda\mu}.$$

(4.12) *The bilinear form $\langle u,v\rangle$ on $\Lambda[t]$ is symmetric.*

When $t=0$, this scalar product specializes to that of Chapter I, because $P_\lambda(x;0)=Q_\lambda(x;0)=s_\lambda(x)$. When $t=1$ it collapses, because $b_\lambda(1)=0$ and therefore $Q_\lambda(x;1)=0$ for all partitions $\lambda\neq 0$.

Remark. Since $z_\lambda(t)=z_\lambda p_\lambda(1,t,t^2,\dots)$, it follows from (4.11) and Chapter I, (7.12) that the scalar product (4.8) is given by

(4.13) $$\langle f,g\rangle=(f*g)(1,t,t^2,\dots)$$

for all $f,g\in\Lambda[t]$, where $f*g$ is the internal product of f and g defined in Chapter I, §7.

Examples

1. By taking $y_i=t^{i-1}$ for all $i\geqslant 1$ in (4.4) and making use of §2, Example 1, we obtain

$$\sum_\lambda t^{n(\lambda)}P_\lambda(x;t)=\prod(1-x_i)^{-1},$$

so that

$$\sum_{|\lambda|=n}t^{n(\lambda)}P_\lambda(x;t)=h_n(x).$$

(See also §3, Example 1.)

2. By applying the involution ω (Chapter I, §2) to the y-variables in (4.7) we obtain

$$\sum_{\lambda} S_{\lambda}(x;t)s_{\lambda'}(y) = \prod_{i,j}(1+x_i y_j)/(1+tx_i y_j).$$

Now take $y_i = t^{i-1}$ for all $i \geqslant 1$; then (Chapter I, §3, Example 2)

$$s_{\lambda'}(y) = t^{n(\lambda')}H_{\lambda}(t)^{-1}$$

and therefore

$$\sum_{\lambda} S_{\lambda}(x;t)t^{n(\lambda')}H_{\lambda}(t)^{-1} = \prod(1+x_i).$$

3. For each integer $n \geqslant 0$, under the specialization

$$p_r \mapsto (1-t^{nr})/(1-t^r) \qquad (r \geqslant 1)$$

$Q_{\lambda}(x;t)$ specializes to

$$t^{n(\lambda)}\prod_{i=1}^{l(\lambda)}(1-t^{n-i+1})$$

(§2, Example 1). Since this is true for all $n \geqslant 0$, we may replace t^n by an indeterminate u: under the specialization

$$p_r \mapsto (1-u^r)/(1-t^r) \qquad (r \geqslant 1)$$

$Q_{\lambda}(x;t)$ specializes to

$$t^{n(\lambda)}\prod_{i=1}^{l(\lambda)}(1-t^{1-i}u).$$

By applying this specialization to the y-variables in (4.4), we obtain another proof of the identity (3) of §3, Example 1.

Notes and references

The identity (4.4) is equivalent to the orthogonality relations for Green's polynomials (§7). The proof given here is due to Littlewood [L12].

5. Skew Hall–Littlewood functions

Since the symmetric functions P_{λ} form a basis of $\Lambda[t]$, any symmetric function u is uniquely determined by its scalar products with the P_{λ}: for by (4.9) we have

$$u = \sum_{\lambda}\langle u, P_{\lambda}\rangle Q_{\lambda}.$$

In particular, for each pair of partitions λ, μ we may define a symmetric function $Q_{\lambda/\mu}$ by

(5.1) $$\langle Q_{\lambda/\mu}, P_\nu \rangle = \langle Q_\lambda, P_\mu P_\nu \rangle = f^\lambda_{\nu\mu}(t)$$

or equivalently

(5.2) $$Q_{\lambda/\mu} = \sum_\nu f^\lambda_{\mu\nu}(t) Q_\nu .$$

Since $f^\lambda_{\mu\nu}(t) = 0$ unless $\mu \subset \lambda$ (3.7), it follows that

(5.3) $Q_{\lambda/\mu} = 0$ *unless* $\mu \subset \lambda$.

From (5.1) we have $\langle Q_{\lambda/\mu}, u \rangle = \langle Q_\lambda, P_\mu u \rangle$ for any symmetric function u. In particular, when $\mu = 0$ it follows that $Q_{\lambda/0} = Q_\lambda$. When $t = 0$, $Q_{\lambda/\mu}$ reduces to the skew Schur function $s_{\lambda/\mu}$.

Likewise we define $P_{\lambda/\mu}$ by interchanging the P's and Q's in (5.1):

(5.1') $$\langle P_{\lambda/\mu}, Q_\nu \rangle = \langle P_\lambda, Q_\mu Q_\nu \rangle$$

for all partitions ν. Since $Q_\lambda = b_\lambda(t) P_\lambda$ it follows that

(5.4) $$Q_{\lambda/\mu} = b_{\lambda/\mu}(t) P_{\lambda/\mu}$$

where $b_{\lambda/\mu}(t) = b_\lambda(t)/b_\mu(t)$.

From (5.2) it follows that

$$\sum_\lambda Q_{\lambda/\mu}(x;t) P_\lambda(y;t) = \sum_{\lambda,\nu} f^\lambda_{\mu\nu}(t) Q_\nu(x;t) P_\lambda(y;t)$$
$$= \sum_\nu Q_\nu(x;t) P_\mu(y;t) P_\nu(y;t)$$

and therefore by (4.4)

$$\sum_\lambda Q_{\lambda/\mu}(x;t) P_\lambda(y;t) = P_\mu(y,t) \prod_{i,j} \frac{1 - tx_i y_j}{1 - x_i y_j} .$$

From this we have

$$\sum_{\lambda,\mu} Q_{\lambda/\mu}(x;t) P_\lambda(y;t) Q_\mu(z;t)$$
$$= \sum_\mu P_\mu(y;t) Q_\mu(z;t) \prod_{i,j} \frac{1 - tx_i y_j}{1 - x_i y_j}$$
$$= \prod_{i,j} \frac{1 - tx_i y_j}{1 - x_i y_j} \prod_{j,k} \frac{1 - ty_j z_k}{1 - y_j z_k}$$

which by (4.4) again is equal to

$$\sum_{\lambda} Q_\lambda(x,z;t)P_\lambda(y;t).$$

Consequently,

(5.5)
$$Q_\lambda(x,z;t) = \sum_{\mu} Q_{\lambda/\mu}(x;t)Q_\mu(z;t)$$

where by (5.3) the summation is over partitions $\mu \subset \lambda$. Likewise,

(5.5′)
$$P_\lambda(x,z;t) = \sum_{\mu} P_{\lambda/\mu}(x;t)P_\mu(z;t).$$

Just as in Chapter I, §5, these formulas enable us to express the symmetric functions $P_{\lambda/\mu}$ and $Q_{\lambda/\mu}$ as sums of monomials. Since (m_ν) and (q_ν) are dual bases for the scalar product (4.8), it follows that

(5.6)
$$\begin{aligned} Q_{\lambda/\mu} &= \sum_{\nu} \langle Q_{\lambda/\mu}, q_\nu \rangle m_\nu \\ &= \sum_{\nu} \langle Q_\lambda, P_\mu q_\nu \rangle m_\nu . \end{aligned}$$

Now from (3.10) we have

(5.7)
$$P_\mu q_r = \sum_{\lambda} \varphi_{\lambda/\mu}(t) P_\lambda$$

summed over all partitions λ such that $\theta = \lambda - \mu$ is a horizontal r-strip, where

(5.8)
$$\varphi_{\lambda/\mu}(t) = \prod_{i \in I} (1 - t^{m_i(\lambda)})$$

in which I is the set of integers $i \geqslant 1$ such that $\theta'_i > \theta'_{i+1}$ (i.e. $\theta'_i = 1$ and $\theta'_{i+1} = 0$).

We shall use (5.7) to express $P_\mu q_\nu$, where μ and ν are any partitions, as a linear combination of the P_λ. Let T be a tableau (Chapter I, §1) of shape $\lambda - \mu$ and weight ν. Then T determines (and is determined by) a sequence of partitions $(\lambda^{(0)}, \dots, \lambda^{(r)})$ such that $\mu = \lambda^{(0)} \subset \lambda^{(1)} \subset \dots \subset \lambda^{(r)} = \lambda$ and such that each $\lambda^{(i)} - \lambda^{(i-1)}$ is a horizontal strip. Let

(5.9)
$$\varphi_T(t) = \prod_{i=1}^{r} \varphi_{\lambda^{(i)}/\lambda^{(i-1)}}(t),$$

then we have

(5.10)
$$P_\mu q_\nu = \sum_{\lambda} \Big(\sum_{T} \varphi_T(t) \Big) P_\lambda$$

summed over all partitions $\lambda \supset \mu$ such that $|\lambda - \mu| = |\nu|$, the inner sum being over all tableaux T of shape $\lambda - \mu$ and weight ν. This is a direct consequence of (5.7) and induction on the length of ν.

From (5.6) and (5.10) it follows that

(5.11) $$Q_{\lambda/\mu} = \sum_T \varphi_T(t) x^T$$

summed over all tableaux T of shape $\lambda - \mu$, where as in Chapter I, (5.13) x^T *is the monomial defined by the tableau T.*

There is an analogous result for $P_{\lambda/\mu}$. First, when $\lambda - \mu = \theta$ is a horizontal strip, we define

(5.8′) $$\psi_{\lambda/\mu}(t) = \prod_{j \in J} (1 - t^{m_j(\mu)})$$

where J is the set of integers $j \geqslant 1$ such that $\theta'_j < \theta'_{j+1}$ (i.e. $\theta'_j = 0$ and $\theta'_{j+1} = 1$), and then we define

(5.9′) $$\psi_T(t) = \prod_{i=1}^{r} \psi_{\lambda^{(i)}/\lambda^{(i-1)}}(t)$$

for a tableau T as above. It is easily verified that

(5.12) $$\varphi_{\lambda/\mu}(t) / \psi_{\lambda/\mu}(t) = b_\lambda(t) / b_\mu(t)$$

and hence that

(5.13) $$\varphi_T(t) / \psi_T(t) = b_\lambda(t) / b_\mu(t)$$

if T has shape $\lambda - \mu$. From (5.4), (5.11), and (5.13) it follows that

(5.11′) $$P_{\lambda/\mu} = \sum_T \psi_T(t) x^T$$

summed over all tableaux T of shape $\lambda - \mu$.

Also, in place of (5.7) we have

(5.7′) $$Q_\mu q_r = \sum_\lambda \psi_{\lambda/\mu}(t) Q_\lambda$$

summed over all $\lambda \supset \mu$ such that $\lambda - \mu$ is a horizontal r-strip, by virtue of (5.7) and (5.12).

Remarks. 1. Whereas the skew Schur function $s_{\lambda/\mu}$ depends only on the difference $\lambda - \mu$, the symmetric functions $P_{\lambda/\mu}$ and $Q_{\lambda/\mu}$ depend on both λ and μ; this is clear from (5.11) and (5.11′).

2. In the case where there is only one variable x, we have

$$Q_{\lambda/\mu}(x;t) = \varphi_{\lambda/\mu}(t)x^{|\lambda-\mu|} \tag{5.14}$$

if $\lambda - \mu$ is a horizontal strip, and $Q_{\lambda/\mu}(x;t) = 0$ otherwise. Similarly

$$P_{\lambda/\mu}(x;t) = \psi_{\lambda/\mu}(t)x^{|\lambda-\mu|} \tag{5.14'}$$

if $\lambda - \mu$ is a horizontal strip, and $P_{\lambda/\mu}(x;t) = 0$ otherwise.

These are special cases of (5.11) and (5.11').

Examples

Several of the formal identities given in the Examples in Chapter I, §5 can be generalized.

1. $$\sum_\lambda P_\lambda(x;t) = \prod_i (1-x_i)^{-1} \prod_{i<j} (1-tx_ix_j)/(1-x_ix_j),$$

summed over all partitions λ.

It is enough to prove this identity when $x = \{x_1, \dots, x_n\}$, the sum on the left being over all partitions of length $\leqslant n$. By induction on n, it is therefore enough to prove that

$$\sum_\lambda P_\lambda(x,y;t) = \frac{1}{1-y} \prod_{i=1}^n \frac{1-tx_iy}{1-x_iy} \sum_\nu P_\nu(x;t)$$

where we have written y in place of x_{n+1}. Now by (5.5') and (5.14') we have

$$\begin{aligned} P_\lambda(x,y;t) &= \sum_{\mu\subset\lambda} P_{\lambda/\mu}(y;t)P_\mu(x;t) \\ &= \sum_{\mu\subset\lambda} \psi_{\lambda/\mu}(t)y^{|\lambda-\mu|}P_\mu(x;t), \end{aligned}$$

summed over partitions $\mu \subset \lambda$ such that $\lambda - \mu$ is a horizontal strip. On the other hand, by (2.9),

$$\prod_{i=1}^n \frac{1-tx_iy}{1-x_iy} \sum_\nu P_\nu(x;t) = \sum_{\nu,r} P_\nu(x;t)q_r(x;t)y^r,$$

and

$$P_\nu(x;t)q_r(x;t)y^r = \sum_{\mu\supset\nu} \varphi_{\mu/\nu}(t)P_\mu(x;t)y^{|\mu-\nu|},$$

by (5.7), the summation being over all $\mu \supset \nu$ such that $\mu - \nu$ is a horizontal r-strip. Hence we are reduced to proving that

$$\sum_{\lambda\supset\mu} \psi_{\lambda/\mu}(t)y^{|\lambda-\mu|} = (1-y)^{-1} \sum_{\nu\subset\mu} \varphi_{\mu/\nu}(t)y^{|\mu-\nu|} \tag{1}$$

for any partition μ, where $\lambda - \mu$ and $\mu - \nu$ are horizontal strips.

For each subset I of $[1, \mu_1]$ consider the partitions $\lambda \supset \mu$ and $\nu \subset \mu$ such that

(2)
$$\psi_{\lambda/\mu}(t) = \varphi_{\mu/\nu}(t) = \prod_{i \in I} (1 - t^{m_i(\mu)}).$$

Let $i_1, i_1 + i_2, \ldots, i_1 + \ldots + i_r$ be the elements of I in ascending order. Then it is easily seen that the contribution to the left-hand side of (1) from the partitions λ satisfying (2) is

(3) $$(1 + y + \ldots + y^{i_1 - 1})(y + y^2 + \ldots + y^{i_2 - 1}) \ldots (y + \ldots + y^{i_r - 1}) . y(1 - y)^{-1}$$

and likewise that the contribution to the right-hand side of (1) from the partitions ν satisfying (2) is

$$(1 - y)^{-1}(y + y^2 + \ldots + y^{i_1})(y + y^2 + \ldots + y^{i_2 - 1}) \ldots (y + \ldots + y^{i_r - 1})$$

which is visibly equal to (3). Hence the two sides of (1) are indeed equal, and the proof is complete.

When $t = 0$, this identity reduces to that of Chapter I, §5, Example 4.

2.
$$\sum_\mu P_\mu(x;t) = \prod_i (1 - x_i^2)^{-1} \prod_{i<j} (1 - tx_i x_j)/(1 - x_i x_j)$$

summed over all partitions μ with all parts *even*.

In view of Example 1 it is enough to show that

$$\sum_\mu P_\mu(x;t) \sum_{m \geq 0} e_m(x) = \sum_\lambda P_\lambda(x;t)$$

where the sum on the right is over all partitions λ. Now each partition λ determines uniquely an even partition μ by decreasing each odd part of λ by 1, so that $\lambda_i' - \mu_i' = 0$ if i is even and $= m_i(\lambda)$ if i is odd. It follows from (3.2) that $f^\lambda_{\mu(1^m)}(t) = 1$, where $m = |\lambda - \mu|$, and hence that the coefficient of P_λ in $P_\mu e_m$ is 1.

3.
$$\sum_\nu c_\nu(t) P_\nu(x;t) = \prod_{i<j} (1 - tx_i x_j)/(1 - x_i x_j)$$

where the sum is over all partitions ν with all columns of even length, and $c_\nu(t) = \prod_{i \geq 1} (1 - t)(1 - t^3) \ldots (1 - t^{m_i - 1})$, where $m_i = m_i(\nu)$.

From Example 1 it follows that

$$\prod_{i<j} (1 - tx_i x_j)/(1 - x_i x_j) = \left(\sum_\lambda P_\lambda(x;t) \right) \left(\sum_m (-1)^m e_m(x) \right)$$

$$= \sum_{\lambda, m} (-1)^m \sum_\mu f^\mu_{\lambda(1^m)}(t) P_\mu(x;t),$$

from which and (3.2) the coefficient of $P_\mu(x;t)$ (μ any partition) in $\prod_{i<j} (1 - tx_i x_j)/(1 - x_i x_j)$ is equal to

$$\prod_{i \geq 1} \sum_{r_i = 0}^{m_i(\mu)} (-1)^{r_i} \begin{bmatrix} m_i(\mu) \\ r_i \end{bmatrix}.$$

Now (see [A2], Theorem 3.4) the inner sum is 0 if $m_i(\mu)$ is odd, and is equal to $(1-t)(1-t^3)\dots(1-t^{m_i(\mu)-1})$ if $m_i(\mu)$ is even. Hence the result.

Since $c_\nu(t) = b_\nu(t)/b_{\nu/2}(t^2)$, where $\nu/2$ is the partition defined by $m_i(\nu/2) = \frac{1}{2}m_i(\nu)$ for all i, the identity just proved can be restated in the form

$$\sum_\nu b_{\nu/2}(t^2)^{-1}Q_\nu(x;t) = \prod_{i<j}(1-tx_ix_j)/(1-x_ix_j)$$

again summed over partitions ν with all columns of even length, i.e. partitions in which each part occurs an even number of times.

4.
$$\prod_i \frac{1-tx_i}{1-x_i}\prod_{i<j}\frac{1-tx_ix_j}{1-x_ix_j} = \sum_\lambda d_\lambda(t)^{-1}Q_\lambda(x;t)$$

summed over all partitions λ, where

$$d_\lambda(t) = \prod_{i\geqslant 1}\prod_{j=1}^{[m_i/2]}(1-t^{2j})$$

in which $m_i = m_i(\lambda)$.

From Example 3 and (2.10) we have

$$\prod_i \frac{1-tx_i}{1-x_i}\prod_{i<j}\frac{1-tx_ix_j}{1-x_ix_j} = \left(\sum_{r=0}^{\infty} q_r(x;t)\right)\left(\sum_\nu b_{\nu/2}(t^2)^{-1}Q_\nu(x;t)\right)$$
$$= \sum_{\lambda,\nu}\frac{\psi_{\lambda/\nu}(t)}{b_{\nu/2}(t^2)}Q_\lambda(x;t)$$

by (5.7′), where the summation is over all ν with all columns even and all $\lambda\supset\nu$ such that $\lambda-\nu$ is a horizontal strip. Each partition λ determines a unique such ν, by removing the bottom square from each column of odd length in λ. If $\lambda-\nu=\theta$, this means that $\theta_i'=1$ or 0 according as λ_i' is odd or even. We have $m_i(\nu) = m_i(\lambda)-\theta_i'+\theta_{i+1}'$, and $\psi_{\lambda/\nu}(t)$ is the product $\prod(1-t^{m_i(\lambda)-\theta_i'+\theta_{i+1}'})$ taken over those $i\geqslant 1$ for which $\theta_i'=0$ and $\theta_{i+1}'=1$. It follows that $b_{\nu/2}(t^2)/\psi_{\lambda/\nu}(t) = d_\lambda(t)$ as defined above.

5. Let

$$\Phi(x_1,\dots,x_n;t) = \prod_{i=1}^n(1-x_i)^{-1}\prod_{i<j}(1-tx_ix_j)/(1-x_ix_j)$$
$$= \sum_\lambda P_\lambda(x_1,\dots,x_n;t)$$

(Example 1). Let u be another indeterminate, then the formal power series

$$S(u) = \sum P_\lambda(x_1,\dots,x_n;t)u^{\lambda_0}$$

where the sum is over all $\lambda_0 \geqslant \lambda_1 \geqslant \ldots \geqslant \lambda_n \geqslant 0$, and $\lambda = (\lambda_1, \ldots, \lambda_n)$, is equal to

$$\sum_{\varepsilon} \frac{\Phi(x_1^{\varepsilon_1}, \ldots, x_n^{\varepsilon_n}; t)}{1 - u\prod x_i^{(1-\varepsilon_i)/2}}$$

summed over all $\varepsilon = (\varepsilon_1, \ldots, \varepsilon_n)$ where each ε_i is ± 1 (hence a sum of 2^n terms).

Let X denote the set of variables $x_1, \ldots, x_n$, and for any subset E of X, let $p(E)$ denote the product of the elements of E. (In particular, $p(\varnothing) = 1$.)

Suppose $\lambda = (\lambda_1, \ldots, \lambda_n)$ is of the form $(\mu_1^{r_1}, \mu_2^{r_2}, \ldots, \mu_k^{r_k})$, where $\mu_1 > \mu_2 > \ldots > \mu_k \geqslant 0$, and the r_i are positive inegers whose sum is n. Then from (2.2) we have

$$P_\lambda(x_1, \ldots, x_n; t) = \sum_f p(f^{-1}(1))^{\mu_1} \ldots p(f^{-1}(k))^{\mu_k} \prod_{f(x_i)<f(x_j)} \frac{x_i - tx_j}{x_i - x_j} \tag{1}$$

summed over all surjective mappings $f: X \to \{1, 2, \ldots, k\}$ such that $|f^{-1}(i)| = r_i$ for $1 \leqslant i \leqslant k$.

Each such f determines a *filtration* of X:

$$\mathscr{F}: \varnothing = F_0 \subset F_1 \subset \ldots \subset F_k = X$$

(all inclusions strict) according to the rule

$$x \in F_r \Leftrightarrow f(x) \leqslant r.$$

Conversely, such a filtration $\mathscr{F}$ of length k determines a surjection $f: X \to \{1, 2, \ldots, k\}$ uniquely. From (1) it follows that

$$S(u) = \sum_{\mathscr{F}} \pi_{\mathscr{F}} \sum u^{\mu_0} \prod_{i=1}^{k} p(F_i - F_{i-1})^{\mu_i}, \tag{2}$$

where the outer sum is over all filtrations $\mathscr{F} = (F_0, \ldots, F_k)$ of X (with k arbitrary) and the inner sum is over all integers $\mu_0, \mu_1, \ldots, \mu_k$ such that $\mu_0 \geqslant \mu_1 > \mu_2 > \ldots > \mu_k \geqslant 0$, and finally

$$\pi_{\mathscr{F}} = \prod_{f(x_i)<f(x_j)} \frac{x_i - tx_j}{x_i - x_j}$$

where f is the function $X \to \{1, 2, \ldots, k\}$ defined by $\mathscr{F}$.

Now let $\nu_i = \mu_i - \mu_{i+1}$ $(0 \leqslant i \leqslant k-1)$, $\nu_k = \mu_k$, so that $\nu_0 \geqslant 0$, $\nu_k \geqslant 0$ and $\nu_i > 0$ for $1 \leqslant i \leqslant k-1$. Then the inner sum in (2) is

$$\sum_{\nu_0, \ldots, \nu_k} u^{\nu_0 + \ldots + \nu_k} \prod_{i=1}^{k} p(F_i)^{\nu_i} = \frac{1}{1-u} \prod_{i=1}^{k-1} \frac{p(F_i)u}{1 - p(F_i)u} \frac{1}{1 - p(X)u}$$

$$= (1-u)^{-1} A_{\mathscr{F}}(u), \text{ say,}$$

and hence from (2) we have

$$S(u) = (1-u)^{-1} \sum_{\mathscr{F}} \pi_{\mathscr{F}} A_{\mathscr{F}}(u), \tag{3}$$

summed over all filtrations $\mathscr{F}$ of X as before.

Observe that

$$S(u)(1-u)=\sum_{\lambda} P_\lambda(x_1,\dots,x_n;t)u^{\lambda_1}$$

and hence that $S(u)(1-u)|_{u=1}$ is equal to $\Phi(x_1,\dots,x_n;t)$, so that

$$\Phi(x_1,\dots,x_n;t)=\sum_{\mathscr{F}} \pi_{\mathscr{F}} A_{\mathscr{F}}(1). \tag{4}$$

The formula (3) shows that $S(u)$ is a rational function of u (with coefficients in $\mathbf{Q}(x_1,\dots,x_n,t)$). The denominators on the right-hand side of (3) are products of factors of the form $(1-p(Y)u)$, where $Y\subset X$. We may therefore express $S(u)$ as a sum of partial fractions: say

$$S(u)=\sum_{Y\subset X}\frac{a(Y)}{1-p(Y)u} \tag{5}$$

To compute $a(Y)$, we have to decompose $(1-u)^{-1}A_{\mathscr{F}}(u)$ into a sum of partial fractions in the usual way, for each filtration $\mathscr{F}$ which contains Y, and then add together the results. The reader will not find it difficult to verify, using (4), that for each $Y\subset X$ we have

$$a(Y)=\Phi(x_1^{\varepsilon_1},\dots,x_n^{\varepsilon_n};t) \tag{6}$$

where $\varepsilon_i=-1$ or $+1$ according as $x_i\in Y$ or $x_i\notin Y$. Since

$$p(Y)=\prod_{x_i\in Y}x_i=\prod_{i=1}^{n}x_i^{(1-\varepsilon_i)/2},$$

this gives the stated result.

6. Show that

$$\sum_{\lambda}\frac{q^{n(\lambda)}}{|\mathrm{Aut}\,M_\lambda|}=(1-q^{-1})\prod_{i\geq 1}\frac{1+q^{-i}}{1-q^{-i}},$$

$$\sum_{\mu\text{ even}}\frac{q^{n(\mu)}}{|\mathrm{Aut}\,M_\mu|}=\prod_{i\geq 2}\frac{1}{1-q^{-i}},$$

where (as in Chapter II) M_λ is a finite $\mathfrak{o}$-module of type λ, and q is the cardinality of the residue field of $\mathfrak{o}$. (Apply the specialization $x_i\mapsto q^{-i}(i\geq 1)$ to the identities of Examples 1 and 2 above, bearing in mind §3, Example 2(a).)

7. (a) Let $Q'_\lambda(x;t)=Q_\lambda(x';t)$ where the x' variables are the products $x_it^{j-1}(i,j\geq 1)$. (In λ-ring notation, $Q'_\lambda(x;t)=Q_\lambda(x/(1-t);t)$. From (4.4) it follows that

$$\sum_{\lambda}P_\lambda(x;t)Q'_\lambda(y;t)=\prod_{i,j}(1-x_iy_j)^{-1} \tag{1}$$

or equivalently that (Q'_λ) is the basis of $\Lambda[t]$ dual to (P_λ), relative to the scalar product of Chapter I, §4.

Now let n be a positive integer and $\zeta = \exp 2\pi i/n$, and set $x_j = \zeta^{j-1}$ $(1 \leqslant j \leqslant n)$, $x_j = 0$ for $j > n$. From §2, Example 1, if λ is a partition of length $\leqslant n$ we have

(2) $$P_\lambda(1, \zeta, \ldots, \zeta^{n-1}; \zeta) = \zeta^{n(\lambda)}\varphi_n(\zeta)/\prod_{i \geqslant 0} \varphi_{m_i}(\zeta),$$

where $m_j = m_j(\lambda)$ for $j \geqslant 1$, and $m_0 = n - l(\lambda)$. The right-hand side of (2) will contain $1 - \zeta^n$ in the numerator, and hence will vanish, unless $m_s = n$ for some s, that is to say unless $\lambda = (s^n)$; and in that case, since $n(\lambda) = \frac{1}{2}n(n-1)s$, we have

(3) $$P_\lambda(1, \zeta, \ldots, \zeta^{n-1}; \zeta) = \zeta^{n(\lambda)} = (-1)^{s(n-1)}.$$

From (1), (2), and (3) it follows that

$$\sum_{s \geqslant 0} (-1)^{s(n-1)} Q'_{(s^n)}(y; \zeta) = \prod_{j \geqslant 1} (1 - y_j^n)^{-1}$$

so that

(4) $$Q'_{(s^n)}(y; \zeta) = (-1)^{s(n-1)} h_s(y^n)$$

where $y^n = (y_1^n, y_2^n, \ldots)$.

(b) From (5.7′) we have

(5) $$Q'_\mu(x; \zeta) q'_r(x; \zeta) = \sum_\lambda \psi_{\lambda/\mu}(\zeta) Q'_\lambda(x; \zeta)$$

(where $q'_r = Q'_{(r)}$) summed over partitions $\lambda \supset \mu$ such that $\lambda - \mu$ is a horizontal strip, where $\psi_{\lambda/\mu}$ is given by (5.8′). It follows that, for each $s \geqslant 1$,

(6) $$Q'_{\mu \cup (s^n)}(x; \zeta) q'_r(x; \zeta) = \sum_\lambda \psi_{\lambda/\mu}(\zeta) Q'_{\lambda \cup (s^n)}(x; \zeta)$$

with the same coefficients $\psi_{\lambda/\mu}$ as before.

Let $\varphi_s: \Lambda_{\mathbf{C}} \to \Lambda_{\mathbf{C}}$ denote the linear mapping defined by

$$\varphi_s(Q'_\mu(x; \zeta)) = Q'_{\mu \cup (s^n)}(x; \zeta).$$

From (4) we have $\varphi_s(1) = (-1)^{s(n-1)} h_s(x^n)$. The equations (5) and (6) show that φ_s commutes with multiplication by q'_r, for each $r \geqslant 1$, and hence that φ_s is $\Lambda_{\mathbf{C}}$-*linear*. It follows that $\varphi_s(f) = f\varphi_s(1)$ for all $f \in \Lambda_{\mathbf{C}}$, i.e. that

(7) $$Q'_{\mu \cup (s^n)}(x; \zeta) = (-1)^{s(n-1)} Q'_\mu(x; \zeta) h_s(x^n).$$

(c) Deduce that the symmetric functions $Q'_\lambda(x; \zeta)$ enjoy the following factorization property: let $m_i(\lambda) = nq_i + r_i$ where $0 \leqslant r_i \leqslant n-1$, and let $\mu = (1^{q_1}2^{q_2}\dots)$, $\rho = (1^{r_1}2^{r_2}\dots)$. Then

$$Q'_\lambda(x;\zeta) = (-1)^{(n-1)|\mu|}Q'_\rho(x;\zeta)h_\mu(x^n).$$

In particular, when $n = 1$ (so that $\zeta = 1$) we have $P_\lambda(x;\zeta) = m_\lambda$, and $\rho = 0$, $\mu = \lambda$: so we recover the fact that (h_λ) is the basis of Λ dual to (m_λ).

8. As in Chapter I, §5, Example 25, let Δ denote the diagonal map $(\Delta f = f(x,y))$. Also, for each $f \in \Lambda[t]$, let $f^\perp$ denote the adjoint of multiplication of f, relative to the scalar product (4.8), so that $\langle f^\perp g, h\rangle = \langle g, fh\rangle$ for all $g, h \in \Lambda[t]$.

(a) Show that Δ is the adjoint of the product map, i.e. that

$$(1) \qquad \langle \Delta f, g \otimes h\rangle = \langle f, gh\rangle$$

for all $f, g, h \in \Lambda[t]$. (Use (5.5) and (5.1).)

(b) We have

$$(2) \qquad \Delta q_n = \sum_{r+s=n} q_r \otimes q_s$$

and hence by (a) above

$$\langle q_n, gh\rangle = \sum_{r+s=n} \langle q_r, g\rangle\langle q_s, h\rangle$$

for all $g, h \in \Lambda[t]$. By taking $g = q_m$ deduce that

$$q_m^\perp q_n = \begin{cases} (1-t)q_{n-m} & \text{if } m \geqslant 1, \\ q_n & \text{if } m = 0. \end{cases}$$

If we define

$$Q^\perp(u) = \sum_{m \geqslant 0} q_m^\perp u^m, \qquad Q(v) = \sum_{n \geqslant 0} q_n v^n$$

where u, v are indeterminates, this last relation may be written in the form

$$(3) \qquad Q^\perp(u)(Q(v)) = Q(v)F(uv)^{-1}$$

where $F(uv) = (1-uv)/(1-tuv)$ as in §2.

(c) Show that $Q^\perp(u)\colon \Lambda[t] \to \Lambda[t,u]$ is a ring homomorphism. (We have

$$\begin{aligned} \langle q_n^\perp(fg), h\rangle &= \langle fg, q_n h\rangle \\ &= \langle f \otimes g, \Delta(q_n)\Delta(h)\rangle \end{aligned}$$

by (1) and the fact that Δ is a ring homomorphism,

$$= \sum_{r+s=n} \langle (q_r^\perp f) \otimes (q_s^\perp g), \Delta h \rangle$$

$$= \sum_{r+s=n} \langle (q_r^\perp f)(q_s^\perp g), h \rangle$$

by (2) and (1) again.

Hence

$$q_n^\perp(fg) = \sum_{r+s=n} (q_r^\perp f)(q_s^\perp g).)$$

It now follows from (3) that

$$Q^\perp(u) \circ Q(v) = F(uv)^{-1} Q(v) \circ Q^\perp(u) \tag{4}$$

as operators on $\Lambda[t]$, where $Q(v)$ is regarded as a multiplication operator, i.e. $Q(v)f = \Sigma_{n \geqslant 0} q_n f v^n$.

(d) Let $Q(u)^{-1} = \Sigma r_n u^n = R(u)$ and let

$$R^\perp(u) = \sum r_n^\perp u^n.$$

From (4) we have

$$R(v) \circ Q^\perp(u) = F(uv)^{-1} Q^\perp(u) \circ R(v)$$

and hence, on taking adjoints,

$$R^\perp(v) \circ Q(u) = F(uv) Q(u) \circ R^\perp(v). \tag{5}$$

(e) Now define operators B_n $(n \in \mathbf{Z})$ by

$$B_n = \sum_{i \geqslant 0} q_{n+i} r_i^\perp$$

or collectively

$$B(u) = \sum_n B_n u^n = Q(u) \circ R^\perp(u^{-1}). \tag{6}$$

Since

$$Q(u) = \exp\left(\sum_{n \geqslant 1} \frac{1-t^n}{n} p_n u^n \right)$$

it follows that

$$B(u) = \exp\left(\sum_{n \geqslant 1} \frac{1-t^n}{n} p_n u^n \right) \circ \exp\left(- \sum_{n \geqslant 1} \frac{1-t^n}{n} p_n^\perp u^{-n} \right).$$

Also define

(7) $$B(u_1,\dots,u_n) = Q(u_1)\dots Q(u_n)R^{\perp}(u_1^{-1})\dots R^{\perp}(u_n^{-1})$$

where $u_1,\dots,u_n$ are independent indeterminates. Deduce from (5) that

(8) $$B(u)B(v) = B(u,v)F(u^{-1}v)$$

and hence by iteration that

(9) $$B(u_1)\dots B(u_n) = B(u_1,\dots,u_n)\prod_{i<j} F(u_i^{-1}u_j).$$

From (9) it follows that

$$B(u_1)\dots B(u_n)(1) = Q(u_1)\dots Q(u_n)\prod_{i<j} F(u_i^{-1}u_j)$$

$$= Q(u_1,\dots,u_n)$$

in the notation of (2.15). Hence for all partitions $\lambda = (\lambda_1,\dots,\lambda_n)$ of length $\leqslant n$ we have

$$Q_\lambda(x;t) = B_{\lambda_1}\dots B_{\lambda_n}(1).$$

(f) Deduce from (8) that

$$B(u)B(v)(u-tv) = B(v)B(u)(tu-v)$$

and hence that

$$B_{m-1}B_n - tB_mB_{n-1} = tB_nB_{m-1} - B_{n-1}B_m$$

for all $m,n \in \mathbf{Z}$.

Notes and references

Example 7 is due to Lascoux, Leclerc, and Thibon [L6].
Example 8. The expression $B(u)$ is a 'vertex operator': see Jing [J12].

6. Transition matrices

We have met various (integral or rational) bases for the ring $\Lambda[t]$ of symmetric functions with coefficients in $\mathbf{Z}[t]$: in particular (§4)

(6.1) $(P_\lambda), (Q_\lambda)$ *are dual bases,*

$(q_\lambda), (m_\lambda)$ *are dual bases,*

$(S_\lambda), (s_\lambda)$ *are dual bases,*

with respect to the scalar product of §4.

By (2.6) the transition matrix

$$K(t) = M(s, P)$$

from the Schur functions $s_\lambda(x)$ to the Hall–Littlewood functions $P_\mu(x;t)$ is strictly upper unitriangular. For any t, u we have

$$M(P(x;t), P(x;u)) = K(t)^{-1}K(u); \tag{6.2}$$

also $K(0)$ is the identity matrix, and $K(1)$ is the Kostka matrix K of Chapter I, §6. Also from (6.1), (6.2), and Chapter I, (6.1) we have

$$M(Q,q) = M(P,m)^* = (K(t)^{-1}K)^* = K(t)'K^*$$

where * denotes transposed inverse. Since (Chapter I, §6) K^* is the matrix of the operator $\prod_{i<j}(1-R_{ij})$, it follows from (2.15) that

(6.3) *$K(t)'$ is the matrix of the operator* $\prod_{i<j}(1-tR_{ij})^{-1}$.

Moreover, the rule (5.11′) for expressing $P_\lambda(x;t)$ as a sum of monomials shows that

$$(K(t)^{-1}K)_{\lambda\mu} = \sum_T \psi_T(t) \tag{6.4}$$

summed over all tableaux T of shape λ and weight μ.

The table of transition matrices between the six bases listed in (6.1) is now easily calculated, and is as shown on p. 241 ($b(t)$ denotes the diagonal matrix whose entries are the $b_\lambda(t)$).

For $n \leqslant 6$ the matrices $K(t)$ are given below.

	2	1^2
2	1	t
1^2		1

	3	21	1^3
3	1	t	t^3
21		1	$t+t^2$
1^3			1

	4	31	2^2	21^2	1^4
4	1	t	t^2	t^3	t^6
31		1	t	$t+t^2$	$t^3+t^4+t^5$
2^2			1	t	t^2+t^4
21^2				1	$t+t^2+t^3$
1^4					1

	5	41	32	31^2	2^21	21^3	1^5
5	1	t	t^2	t^3	t^4	t^6	t^{10}
41		1	t	$t+t^2$	t^2+t^3	$t^3+t^4+t^5$	$t^6+t^7+t^8+t^9$
32			1	t	$t+t^2$	$t^2+t^3+t^4$	$t^4+t^5+t^6+t^7+t^8$
31^2				1	t	$t+t^2+t^3$	$t^3+t^4+2t^5+t^6+t^7$
2^21					1	$t+t^2$	$t^2+t^3+t^4+t^5+t^6$
21^3						1	$t+t^2+t^3+t^4$
1^5							1

	6	51	42	41^2	3^2	321	31^3	2^3	2^21^2	21^4	1^6
6	1	t	t^2	t^3	t^3	t^4	t^6	t^6	t^7	t^{10}	t^{15}
51		1	t	$t+t^2$	t^2	t^2+t^3	$t^3+t^4+t^5$	t^4+t^5	$t^4+t^5+t^6$	$t^6+t^7+t^8+t^9$	$t^{10}+t^{11}+t^{12}+t^{13}+t^{14}$
42			1	t	t	$t+t^2$	$t^2+t^3+t^4$	$t^2+t^3+t^4$	$2t^3+t^4+t^5$	$t^4+t^5+2t^6+t^7+t^8$	$t^7+t^8+2t^9+t^{10}+2t^{11}+t^{12}+t^{13}$
41^2				1	0	t	$t+t^2+t^3$	t^3	$t^2+t^3+t^4$	$t^3+t^4+2t^5+t^6+t^7$	$t^6+t^7+2t^8+2t^9+2t^{10}+t^{11}+t^{12}$
3^2					1	t	t^3	t^3	t^2+t^4	$t^4+t^5+t^7$	$t^6+t^8+t^9+t^{10}+t^{12}$
321						1	$t+t^2$	$t+t^2$	$t+2t^2+t^3$	$t^2+2t^3+2t^4+2t^5+t^6$	$t^4+2t^5+2t^6+3t^7+3t^8+2t^9+2t^{10}+t^{11}$
31^3							1	0	t	$t+t^2+t^3+t^4$	$t^3+t^4+2t^5+2t^6+2t^7+t^8+t^9$
2^3								1	t	t^2+t^4	$t^3+t^5+t^6+t^7+t^9$
2^21^2									1	$t+t^2+t^3$	$t^2+t^3+2t^4+t^5+2t^6+t^7+t^8$
21^4										1	$t+t^2+t^3+t^4+t^5$
1^6											1

	m	q	P	Q	s	S
m	1	$K^{-1}K(t)b(t)^{-1}K(t)'K^*$	$K^{-1}K(t)$	$K^{-1}K(t)b(t)^{-1}$	K^{-1}	$K^{-1}K(t)b(t)^{-1}K(t)'$
q	$K'K(t)^*b(t)K(t)^{-1}K$	1	$K'K(t)^*b(t)$	$K'K(t)^*$	$K'K(t)^*b(t)K(t)^{-1}$	K'
P	$K(t)^{-1}K$	$b(t)^{-1}K(t)'K^*$	1	$b(t)^{-1}$	$K(t)^{-1}$	$b(t)^{-1}K(t)'$
Q	$b(t)K(t)^{-1}K$	$K(t)'K^*$	$b(t)$	1	$b(t)K(t)^{-1}$	$K(t)'$
s	K	$K(t)b(t)^{-1}K(t)'K^*$	$K(t)$	$K(t)b(t)^{-1}$	1	$K(t)b(t)^{-1}K(t)'$
S	$K(t)^*b(t)K(t)^{-1}K$	K^*	$K(t)^*b(t)$	$K(t)^*$	$K(t)^*b(t)K(t)^{-1}$	1

These tables suggest that the polynomials $K_{\lambda\mu}(t)$ have all their coefficients $\geqslant 0$. Since $K_{\lambda\mu}(1) = K_{\lambda\mu}$ is (Chapter I, (6.4)) the number of tableaux of shape λ and weight μ, Foulkes [F7] conjectured that it should be possible to attach to each such tableau a positive integer $c(T)$ such that

$$K_{\lambda\mu}(t) = \sum_T t^{c(T)},$$

the sum being over all tableaux T of shape λ and weight μ. Such a rule has been found by Lascoux and Schützenberger ([L4]; see also [B10]), and goes as follows.

We consider *words* (or sequences) $w = a_1 \dots a_n$ in which each a_i is a positive integer. The *weight* of w is the sequence $\mu = (\mu_1, \mu_2, \dots)$, where μ_i is the number of a_j equal to i. Assume that $\mu_1 \geqslant \mu_2 \geqslant \dots$, i.e. that μ is a *partition*. If $\mu = (1^n)$ (so that w is a derangement of $12\dots n$) we call w a *standard* word.

(i) Suppose first that w is a standard word. Attach an *index* to each element of w as follows: the number 1 has index 0, and if r has index i, then $r+1$ has index i or $i+1$ according as it lies to the right or left of r. The *charge* $c(w)$ of w is defined to be the sum of the indices.

(ii) Now let w be any word, subject only to the condition that its weight should be a partition μ. We extract a standard word from w as follows. Reading from the left, choose the first 1 that occurs in w, then the first 2 to the right of the 1 chosen, and so on. If at any stage there is no $s+1$ to the right of the s chosen, go back to the beginning. This procedure selects a standard subword w_1 of w, of length μ'_1. Now erase w_1 from w and repeat the procedure to obtain a standard subword w_2 (of length μ'_2), and so on. (For example, if $w = 32214113$, the subword w_1 is 2143, consisting of the underlined symbols in w: $3\underline{2}2\underline{1}\underline{4}11\underline{3}$. When w_1 is erased, we are left with 3211, so that $w_2 = 321$ and $w_3 = 1$.)

The *charge* of w is defined to be the sum of the charges of the standard subwords: $c(w) = \sum c(w_i)$. (In the example above, the indices (attached as subscripts) are $2_1 1_0 4_2 3_1$ for w_1, so that $c(w_1) = 4$; $3_2 2_1 1_0$ for w_2, so that $c(w_2) = 3$; and 1_0 for w_3, so that $c(w_3) = 0$: hence $c(w) = 4 + 3 + 0 = 7$.)

(iii) Finally, if T is a tableau, by reading T from right to left in consecutive rows, starting from the top (as in Chapter I, §9), we obtain a word $w(T)$, and the *charge* $c(T)$ of T is defined to be $c(w(T))$.

The theorem of Lascoux and Schützenberger now states that

(6.5) (i) *We have*

$$K_{\lambda\mu}(t) = \sum_T t^{c(T)}$$

summed over all tableaux T of shape λ and weight μ;

(ii) *If* $\lambda \geqslant \mu$, $K_{\lambda\mu}(t)$ *is monic of degree* $n(\mu) - n(\lambda)$. (*If* $\lambda \not\geqslant \mu$, *we know already* (2.6) *that* $K_{\lambda\mu}(t) = 0$.)

For the proof of (6.5) we refer to [L4], [B10]. (6.5)(ii) is an easy consequence of (6.3) (see Example 4).

Examples

1. $K_{(n)\mu}(t) = t^{n(\mu)}$ for any partition μ of n. This follows from the identity

$$h_n = \sum_{|\mu|=n} t^{n(\mu)} P_\mu$$

(§3, Example 1; §4, Example 1). Alternatively, use (6.5).

2. $K_{\lambda(1^n)}(t) = t^{n(\lambda')}\varphi_n(t)H_\lambda(t)^{-1}$ for any partition λ of n, where

$$H_\lambda(t) = \prod_{x \in \lambda} (1 - t^{h(x)})$$

is the hook-length polynomial. This follows from §4, Example 2, since $K_{\lambda(1^n)}(t)$ is the coefficient of $S_\lambda(x;t)$ in $Q_{(1^n)}(x;t) = \varphi_n(t)e_n(x)$.

From Chapter I, §5, Example 14, it follows that

$$K_{\lambda(1^n)}(t) = \sum_T t^{c'(T)}$$

summed over all standard tableaux T of shape λ, where $c'(T)$ is the sum of the positive integers $i < n$ such that $i+1$ lies in a column to the right of i in T.

3. Tables of the polynomials $K_{\lambda\mu}(t)$ suggest the following conjecture: if the *lowest* power of t which is present in $K_{\lambda\mu}(t)$ is $t^{a(\lambda,\mu)}$, then $a(\lambda,\mu) = a(\mu',\lambda')$.

4. Let $\varepsilon_1, \dots, \varepsilon_n$ be the standard basis of $\mathbf{Z}^n$ and let R^+ denote the set of vectors $\varepsilon_i - \varepsilon_j$ such that $1 \leqslant i < j \leqslant n$. For any $\xi = (\xi_1, \dots, \xi_n) \in \mathbf{Z}^n$ such that $\sum \xi_i = 0$, let

$$P(\xi;t) = \sum_{(m_\alpha)} t^{\sum m_\alpha},$$

summed over all families $(m_\alpha)_{\alpha \in R^+}$ of non-negative integers such that $\xi = \sum m_\alpha \alpha$. The polynomial $P(\xi;t)$ is non-zero if and only if $\xi = \sum_1^{n-1} \eta_i(\varepsilon_i - \varepsilon_{i+1})$, where the η_i are $\geqslant 0$, and then it is monic of degree $\sum \eta_i$. Since $\eta_i = \xi_1 + \dots + \xi_i$ $(1 \leqslant i \leqslant n-1)$, the degree of $P(\xi;t)$ is

$$\sum (n-i)\xi_i = \langle \xi, \delta \rangle$$

where as usual $\delta = (n-1, n-2, \dots, 1, 0)$ and the scalar product is the standard one.

Now by (6.3), $K(t)_{\lambda\mu}$ is the coefficient of s_λ in

$$\prod_{1 \leqslant i < j \leqslant n} (1 + tR_{ij} + t^2R_{ij}^2 + \dots)s_\mu,$$

hence is the coefficient of $a_{\lambda+\delta}$ in

$$\sum_{(m_\alpha)} t^{\Sigma m_\alpha} a_{\mu+\delta+\Sigma m_\alpha a}$$

or equivalently the coefficient of $x^{\lambda+\delta}$ in

$$\sum_{w\in S_n} \varepsilon(w) \sum_{(m_\alpha)} t^{\Sigma m_\alpha} x^{w(\mu+\delta+\Sigma m_\alpha \alpha)}.$$

It follows from this that

$$K(t)_{\lambda\mu} = \sum_{w\in S_n} \varepsilon(w) P(w^{-1}(\lambda+\delta)-(\mu+\delta);t). \tag{1}$$

Now

$$\begin{aligned}\langle w^{-1}(\lambda+\delta)-(\mu+\delta),\delta\rangle &= \langle \lambda+\delta, w\delta\rangle - \langle \mu+\delta,\delta\rangle \\ &\leqslant \langle \lambda+\delta,\delta\rangle - \langle \mu+\delta,\delta\rangle \\ &= n(\mu)-n(\lambda)\end{aligned}$$

and equality holds if and only if $w=1$. Hence the dominant term in the sum (1) is that corresponding to $w=1$, and therefore $K(t)_{\lambda\mu}$ is monic of degree $n(\mu)-n(\lambda)$.

From (1) it follows that $K_{\lambda\mu}(t)$ is equal to the coefficient of x^μ in $\det(x_j^{\lambda_i-i+j})/\prod_{i<j}(1-tx_i^{-1}x_j)$ (P. Hoffman).

5. From the tables of this section and Chapter I, §6, we have

$$M(e,P)=M(e,s)M(s,P)=K'JK(t)$$

or equivalently

$$M(e,P)_{\lambda\mu} = \sum_\nu K(t)_{\nu\mu} K_{\nu'\lambda}. \tag{1}$$

On the other hand, the formula (1) in the Appendix to Chapter II shows that

$$M(e,P)_{\lambda\mu} = \sum_A t^{\tilde{d}(A)}, \tag{2}$$

where the sum is over all row-strict arrays of shape μ and weight λ, and $\tilde{d}(A)$ is defined in Chapter II, Appendix (AZ.5).

From (1) and (2) we conclude that Foulkes's conjecture is equivalent to the following combinatorial statement: for any two partitions λ,μ there exists a bijection φ between pairs of tableaux (T,T') of conjugate shapes and respective weights μ and λ, and row-strict arrays of shape μ and weight λ, such that $\tilde{d}(\varphi(T,T'))$ does not depend on T'.

Indeed, letting $c(T)=\tilde{d}(\varphi(T,T'))$, we have

$$K(t)_{\lambda\mu} = \sum_T t^{c(T)}$$

summed over all tableaux T of shape λ and weight μ. It would be interesting to give an explicit construction of such a bijection φ.

6. The polynomials $K_{\lambda\mu}(t)$ have the following geometrical interpretation, due to G. Lusztig [L15]. Let V be an n-dimensional vector space over an algebraically closed field k. The conjugacy classes of unipotent operators on V are parametrized by the partitions of n: the class c_λ consists of operators with Jordan blocks of sizes $\lambda_1, \lambda_2, \dots$. The closure $\bar{c}_\lambda$ of c_λ is an algebraic variety (singular, in general) of dimension $2n(\lambda')$, and we have

$$\bar{c}_\lambda = \bigcup_{\mu \leqslant \lambda} c_\mu .$$

Let $x \in c_\mu \subset \bar{c}_\lambda$, let $H^i(\bar{c}_\lambda)$ denote the Deligne–Goreski–MacPherson cohomology sheaves on $\bar{c}_\lambda$ (see, for example, [L15]), and let $H^i(\bar{c}_\lambda)_x$ be the stalk of $H^i(\bar{c}_\lambda)$ at x. Then $H^i(\bar{c}_\lambda) = 0$ for i odd, and

$$\sum_{i \geqslant 0} t^i \dim H^{2i}(\bar{c}_\lambda)_x = t^{n(\mu)-n(\lambda)} K_{\lambda\mu}(t^{-1}).$$

This gives another proof of (6.5) (ii) and of the fact that all coefficients of $K_{\lambda\mu}(t)$ are $\geqslant 0$. It would be interesting to give a geometrical interpretation of the Lascoux–Schützenberger theorem (6.5) (i).

7. If $\alpha = (\alpha^{(0)}, \alpha^{(1)}, \alpha^{(2)}, \dots)$ is any finite sequence of partitions, let

$$P_n^k(\alpha) = \sum_{i=1}^{n} (\alpha_i^{(k-1)} - 2\alpha_i^{(k)} + \alpha_i^{(k+1)})$$

for $k, n \geqslant 1$, and

$$C(\alpha) = \sum_{k,n \geqslant 1} \binom{\alpha_n^{(k-1)} - \alpha_n^{(k)}}{2}.$$

Also let $[a, b]$ denote the Gaussian polynomial $\begin{bmatrix} a+b \\ b \end{bmatrix}$. Then

$$K_{\lambda\mu}(t) = \sum_{\alpha} t^{C(\alpha)} \prod_{k,n \geqslant 1} [P_n^k(\alpha), \alpha_n^{(k)} - \alpha_n^{(k+1)}]$$

where α runs through all sequences of partitions such that $\alpha^{(0)} = \mu'$ and $|\alpha^{(k)}| = \lambda_{k+1} + \lambda_{k+2} + \dots$ for $k \geqslant 1$. (This is a consequence of (6.5)(i); for the proof, see [K9].)

8. Let $N_\lambda(\mathfrak{o})$ denote the number of submodules of a finite $\mathfrak{o}$-module of type λ, where (as in Chapter II) $\mathfrak{o}$ is a discrete valuation ring with finite residue field k. Then

$$N_\lambda(\mathfrak{o}) = q^{n(\lambda)} \sum_{\mu} (\mu_1 - \mu_2 + 1) K_{\mu\lambda}(q^{-1})$$

summed over partitions $\mu \geqslant \lambda$ of length $\leqslant 2$, where q is the cardinality of k.

(We have

$$
\begin{aligned}
N_\lambda(\mathrm{o}) &= \sum_{\mu,\nu} g^\lambda_{\mu\nu}(\mathrm{o}) \\
&= q^{n(\lambda)} \sum_{\mu,\nu} q^{-n(\mu)-n(\nu)} f^\lambda_{\mu\nu}(q^{-1}) \\
&= q^{n(\lambda)} \sum_{\mu,\nu} q^{-n(\mu)-n(\nu)} \langle Q_\lambda(q^{-1}), P_\mu(q^{-1}) P_\nu(q^{-1})\rangle \\
&= q^{n(\lambda)} \langle Q_\lambda(q^{-1}), \left(\sum h_n\right)^2\rangle.
\end{aligned}
$$

Now

$$
\begin{aligned}
\left(\sum h_n\right)^2 = \prod (1-x_i)^{-2} &= \sum_\mu s_\mu(x) s_\mu(1,1) \\
&= \sum_\mu (\mu_1 - \mu_2 + 1) s_\mu(x),
\end{aligned}
$$

summed over partitions μ of length $\leqslant 2$. Hence

$$
\begin{aligned}
N_\lambda(\mathrm{o}) &= q^{n(\lambda)} \sum_\mu (\mu_1 - \mu_2 + 1) \langle Q_\lambda(q^{-1}), s_\mu\rangle \\
&= q^{n(\lambda)} \sum_\mu (\mu_1 - \mu_2 + 1) K_{\mu\lambda}(q^{-1}).)
\end{aligned}
$$

Notes and references

Examples 5 and 6 were contributed by A. Zelevinsky. Example 7 is due to Kirillov and Reshetikhin [K9].

7. Green's polynomials

Let $X(t)$ denote the transition matrix $M(p, P)$ between the power-sum products p_ρ and the Hall–Littlewood functions P_λ:

(7.1) $$p_\rho(x) = \sum_\lambda X^\lambda_\rho(t) P_\lambda(x;t)$$

(so that ρ is the row-index and λ the column-index). By (2.7), $X^\lambda_\rho(t) \in \mathbf{Z}[t]$, and is zero unless $|\lambda| = |\rho|$. When $t = 0$ we have $P_\lambda(x;0) = s_\lambda(x)$, so that by Chapter I, (7.8)

(7.2) $$X^\lambda_\rho(0) = \chi^\lambda_\rho,$$

the value of the irreducible character χ^λ of S_n at elements of cycle-type ρ.

From (4.1) and (4.4) we have

$$\sum_{|\rho|=n} z_\rho(t)^{-1} p_\rho(x) p_\rho(y) = \sum_{|\lambda|=n} b_\lambda(t) P_\lambda(x;t) P_\lambda(y;t).$$

Substituting (7.1) in this relation, we obtain

$$X'(t) z(t)^{-1} X(t) = b(t) \tag{7.3}$$

where $X'(t)$ is the transpose of $X(t)$, and $b(t)$ (resp. $z(t)$) is the diagonal matrix with entries $b_\lambda(t)$ (resp. $z_\lambda(t)$). An equivalent form of (7.3) is

$$X(t) b(t)^{-1} X'(t) = z(t). \tag{7.4}$$

In other words we have the *orthogonality relations*

$$\sum_{|\rho|=n} z_\rho(t)^{-1} X_\rho^\lambda(t) X_\rho^\mu(t) = \delta_{\lambda\mu} b_\lambda(t), \tag{7.3'}$$

$$\sum_{|\lambda|=n} b_\lambda(t)^{-1} X_\rho^\lambda(t) X_\sigma^\lambda(t) = \delta_{\rho\sigma} z_\rho(t) \tag{7.4'}$$

which when $t=0$ reduce to the orthogonality relations for the characters of the symmetric group S_n.

Since $X(t)^{-1} = b(t)^{-1} X'(t) z(t)^{-1}$ by (7.3), we can solve the equations (7.1) to give

$$Q_\lambda(x;t) = \sum_\rho z_\rho(t)^{-1} X_\rho^\lambda(t) p_\rho(x). \tag{7.5}$$

Next, since $M(s,P) = M(p,s)^{-1} M(p,P)$, we have $K(t) = X(0)^{-1} X(t)$, so that

$$X(t) = X(0) K(t) \tag{7.6}$$

or more explicitly, using (7.2),

$$X_\rho^\mu(t) = \sum_\lambda \chi_\rho^\lambda K_{\lambda\mu}(t). \tag{7.6'}$$

Since $K_{\lambda\mu}(t)$ is monic of degree $n(\mu) - n(\lambda) \leqslant n(\mu)$ (6.5), it follows that the dominant term on the right-hand side of (7.6') is $\chi_\rho^{(n)} K_{(n)\mu}(t) = t^{n(\mu)}$ (§6, Example 1), so that

(7.7) *$X_\rho^\mu(t)$ is monic of degree $n(\mu)$.*

Green's polynomials $Q_\rho^\lambda(q)$ are defined by

$$Q_\rho^\lambda(q) = q^{n(\lambda)} X_\rho^\lambda(q^{-1}). \tag{7.8}$$

In terms of these polynomials, the orthogonality relations (7.3′) and (7.4′) take the forms

(7.9) $$\sum_{|\rho|=n} y_\rho(q)^{-1} Q_\rho^\lambda(q) Q_\rho^\mu(q) = \delta_{\lambda\mu} a_\lambda(q),$$

(7.10) $$\sum_{|\lambda|=n} a_\lambda(q)^{-1} Q_\rho^\lambda(q) Q_\sigma^\lambda(q) = \delta_{\rho\sigma} y_\rho(q),$$

where $a_\lambda(q) = q^{|\lambda|+2n(\lambda)} b_\lambda(q^{-1})$, $y_\rho(q) = q^{-|\rho|} z_\rho(q^{-1})$ (Chapter II, (1.6)).

Also (7.6′) now becomes

(7.11) $$Q_\rho^\mu(q) = \sum_\lambda \chi_\rho^\lambda \tilde{K}_{\lambda\mu}(q)$$

where $\tilde{K}_{\lambda\mu}(q) = q^{n(\mu)} K_{\lambda\mu}(q^{-1}) = q^{n(\lambda)} +$ terms of higher degree by (6.5).

Examples

1. $X_\rho^{(n)}(t) = 1$ for all partitions ρ of n. This is a particular case of (7.7). Equivalently, $Q_\rho^{(n)}(q) = 1$.

2. $X_{(n)}^\lambda(t) = t^{n(\lambda)} \varphi_{l(\lambda)-1}(t^{-1})$, or equivalently $Q_{(n)}^\lambda(q) = \varphi_{l(\lambda)-1}(q)$. This follows from the identity (3) of §3, Example 1:

$$\prod_{i \geq 1} (1 - x_i y)/(1 - x_i) = \sum_\lambda t^{n(\lambda)} \prod_{j=1}^{l(\lambda)} (1 - t^{1-j} y) . P_\lambda(x;t).$$

Divide both sides by $1 - y$ and let $y \to 1$: since

$$\lim_{y \to 1} \frac{1}{1-y} \left(\prod_i \frac{1 - x_i y}{1 - x_i} - 1 \right) = \sum_i \frac{x_i}{1 - x_i} = \sum_{n \geq 1} p_n(x)$$

we obtain

$$p_n(x) = \sum_{|\lambda|=n} t^{n(\lambda)} \prod_1^{l(\lambda)-1} (1 - t^{-i}) P_\lambda(x;t)$$

which gives the result.

3. $X_\rho^{(1^n)}(t) = \prod_{i=1}^n (t^i - 1)/\prod_{j \geq 1}(t^{\rho_j} - 1)$, or equivalently $Q_\rho^{(1^n)}(q) = \varphi_n(q)/\prod_{j \geq 1}(1 - q^{\rho_j})$. This follows from (7.5) with $\lambda = (1^n)$: since $Q_{(1^n)}(x;t) = \varphi_n(t) e_n(x)$ we have

$$\sum_\rho z_\rho(t)^{-1} X_\rho^{(1^n)}(t) p_\rho(x) = \varphi_n(t) e_n(x)$$

$$= \varphi_n(t) \sum_\rho \varepsilon_\rho z_\rho^{-1} p_\rho(x)$$

so that $X_\rho^{(1^n)}(t) = \varepsilon_\rho z_\rho^{-1} z_\rho(t) \varphi_n(t)$, which is equivalent to the result stated.

4. $X^\lambda_{(1^n)}(t) = (1-t)^{-n}\Sigma_T\, \varphi_T(t)$, summed over all standard tableaux T of shape λ. ($\varphi_T(t)$ was defined in (5.9).)

For by (7.1) we have

$$X^\lambda_{(1^n)}(t) = \langle Q_\lambda, p_1^n\rangle = (1-t)^{-n}\langle Q_\lambda, q_1^n\rangle$$
$$= (1-t)^{-n}\sum_T \varphi_T(t)$$

by (4.8) and (5.11).

5. By taking $x_i = t^{i-1}$ in (7.5) we obtain by §2, Example 1

$$t^{n(\lambda)} = \sum_\rho z_\rho^{-1} X^\lambda_\rho(t)$$

or, in terms of Green's polynomials,

$$\sum_{|\rho|=n} z_\rho^{-1} Q^\lambda_\rho(q) = 1$$

for all partitions λ of n.

6. $X^\lambda_\rho(1) = \langle p_\rho, h_\lambda\rangle$ (for the scalar product of Chapter I), hence is the value of the induced character $\mathrm{ind}_{S_\lambda}^{S_n}(1)$ at elements of type ρ.

7. Let ω be a primitive rth root of unity ($r \geqslant 2$), and consider the effect of replacing the parameter t by ω. Let $K_r = \mathbf{Q}(\omega)$ be the rth cyclotomic field, $\mathfrak{O}_r = \mathbf{Z}[\omega]$ its ring of integers. By (2.7) the $P_\lambda(x;\omega)$ form an $\mathfrak{O}_r$-basis of $\Lambda \otimes_{\mathbf{Z}} \mathfrak{O}_r$. On the other hand, the definition (2.12) of $b_\lambda(t)$ shows that $b_\lambda(\omega)$ and therefore also $Q_\lambda(x;\omega)$ is zero if $m_i(\lambda) \geqslant r$ for some i, that is to say if the partition λ has r or more equal parts. Let $\mathscr{P}_{(r)}$ denote the set of partitions λ such that $m_i(\lambda) < r$ for all i, and let $\Lambda_{(r)}$ denote the $\mathfrak{O}_r$-submodule of $\Lambda \otimes_{\mathbf{Z}} \mathfrak{O}_r$ generated by the $P_\lambda(x;\omega)$ with $\lambda \in \mathscr{P}_{(r)}$. Then $\Lambda_{(r)}$ is an $\mathfrak{O}_r$-subalgebra of $\Lambda \otimes_{\mathbf{Z}} \mathfrak{O}_r$.

Let $A_{(r)} = \Lambda_{(r)} \otimes_{\mathfrak{O}_r} K_r$, which is the K_r-vector space generated freely by the $Q_\lambda(x;\omega)$ such that $\lambda \in \mathscr{P}_{(r)}$. Since $z_\rho(\omega)^{-1} = z_\rho^{-1}\prod_{j\geqslant 1}(1-\omega^{\rho_j})$ is zero if any ρ_i is divisible by r, it follows from (7.5) that $A_{(r)}$ is contained in the K_r-algebra $B_{(r)}$ generated by the power sums p_n such that n is not divisible by r. Now if $A^n_{(r)}, B^n_{(r)}$ are the components of degree n of $A_{(r)}$, $B_{(r)}$ respectively, it follows that $\Sigma_{n\geqslant 0} \dim A^n_{(r)} \cdot u^n$ is the generating function for partitions λ with no part repeated more than $r-1$ times, so that

$$\sum_{n\geqslant 0} \dim A^n_{(r)} \cdot u^n = \prod_{i\geqslant 1}(1+u^i+\ldots+u^{(r-1)i})$$

$$= \prod_{i\geqslant 1}(1-u^{ri})/(1-u^i), \tag{1}$$

and $\Sigma_{n\geqslant 0} \det B^n_{(r)} \cdot u^n$ is the generating function for partitions ρ none of whose parts is divisible by r, so that

$$\sum_{n\geqslant 0} \dim B^n_{(r)} \cdot u^n = \prod_{\substack{i\geqslant 1\\ i\not\equiv 0(r)}} (1-u^i)^{-1}. \tag{2}$$

From (1) and (2) it follows that $\dim A^n_{(r)} = \dim B^n_{(r)}$ and hence that $A_{(r)} = B_{(r)}$; hence $A_{(r)}$ is a K_r*-algebra* and therefore $\Lambda_{(r)}$ is an $\mathfrak{O}_r$*-algebra*, as asserted.

The scalar product defined in §4 is non-degenerate on $\Lambda_{(r)}$, and the orthogonality relations (7.3′), (7.4′) are now orthogonality relations for the matrix $X^\lambda_\rho(\omega)$, where λ runs through partitions of n which have no part repeated r times or more, and ρ through partitions of n which have no part divisible by r.

8. Let us write

$$Q^\mu_\rho(q) = \sum_i \psi^{\mu,i}(\rho)q^i \qquad (|\mu| = |\rho| = n) \tag{1}$$

where the coefficients $\psi^{\mu,i}$ may be regarded as class-functions on S_n. Hotta and Springer [H9] have shown that $\psi^{\mu,i}$ is a *character* of S_n, for each partition μ of n: they define an action of S_n on the rational cohomology $H^*(X_\mu)$ of the variety X_μ (Chapter II, §3, Example) and show that

$$\psi^{\mu,i}(w) = \varepsilon(w)\text{trace}\left(w^{-1}, H^{2i}(X_\mu)\right)$$

where as usual $\varepsilon(w)$ is the sign of $w \in S_n$.

If

$$\psi^{\mu,i} = \sum_\lambda k^{(i)}_{\lambda\mu} \chi^\lambda \tag{2}$$

is the decomposition of $\psi^{\mu,i}$ as a sum of irreducible characters, it follows from (1) and (2) and (7.11) that

$$\tilde{K}_{\lambda\mu}(q) = \sum_i k^{(i)}_{\lambda\mu} q^i$$

with coefficients $k^{(i)}_{\lambda\mu}$ integers $\geqslant 0$.

When $\rho = (1^n)$, $Q^\mu_{(1^n)}(q)$ is the Poincaré polynomial of the variety X_μ. When q is a prime-power, we have $Q^\mu_{(1^n)}(q) = |X_\mu(q)|$, the number of $\mathbf{F}_q$-rational points of X_μ.

The representations $\psi^{\mu,i}$ and their generalizations are studied in a number of recent papers. We mention the paper of W. Borho and R. MacPherson [B6], where is explained their relationship with the Deligne–Goreski–MacPherson sheaves on closures of unipotent classes (§6, Example 6).

Notes and references

The polynomials $Q^\lambda_\rho(q)$ were introduced by Green [G11], who proved the orthogonality relations (7.9), (7.10). For tables of these polynomials see Green (*loc. cit.*) for $n \leqslant 5$, and Morris [M12] for $n = 6, 7$.

8. Schur's Q-functions

We have seen in §2 that when $t = 0$ the Hall–Littlewood function $P_\lambda(x;t)$ reduces to the Schur function s_λ, and when $t = 1$ to the monomial symmetric function m_λ; and in §3 that when $t = q^{-1}$, q a prime power, the $P_\lambda(x;t)$ have a combinatorial significance via the Hall algebra. There is at least one other value of t that is of particular interest, namely $t=-1$: the symmetric functions $Q_\lambda(x;-1)$ were introduced in Schur's 1911 paper [S5] on the projective representations of the symmetric and alternating groups, where they play a role analogous to that of the Schur functions in connection with the linear representations of the symmetric groups (Chapter I, §7).

To simplify notation, we shall write $q_r, P_\lambda, Q_\lambda$ for $q_r(x;-1)$, $P_\lambda(x;-1), Q_\lambda(x;-1)$, respectively. As in §2, let

$$Q(t)=\sum_{r\geqslant 0} q_r t^r$$

(the symbol t now being available again), then from (2.10) we have

$$Q(t)=\prod_i \frac{1+tx_i}{1-tx_i}=E(t)H(t) \tag{8.1}$$

from which it follows that

$$Q(t)Q(-t)=1$$

that is to say

$$\sum_{r+s=n}(-1)^r q_r q_s=0 \tag{8.2}$$

for all $n\geqslant 1$. When n is odd, (8.2) tells us nothing. When n is even, say $n=2m$, it becomes

$$q_{2m}=\sum_{r=1}^{m-1}(-1)^{r-1}q_r q_{2m-r}+\tfrac{1}{2}(-1)^{m-1}q_m^2, \tag{8.2$'$}$$

which shows that $q_{2m}\in\mathbf{Q}[q_1,q_2,\dots,q_{2m-1}]$ and hence (by induction on m) that

$$q_{2m}\in\mathbf{Q}[q_1,q_3,q_5,\dots,q_{2m-1}]. \tag{8.3}$$

Another consequence of (8.2′) is the following. Recall (Chapter I, §1, Example 9) that a partition is *strict* if all its parts are distinct.

(8.4) *Let λ be a partition of n. Either λ is strict, or $q_\lambda=q_{\lambda_1}q_{\lambda_2}\dots$ is a $\mathbf{Z}$-linear combination of the q_μ such that μ is strict and $\mu>\lambda$.*

Proof. We proceed by induction, assuming the result true for all $\mu>\lambda$. (The induction starts at $\lambda=(n)$, which is strict.) If λ is not strict then for some pair $i<j$ we have $\lambda_i=\lambda_j=m$, say. The relation (8.2′) then shows that q_λ is a $\mathbf{Z}$-linear combination of the q_μ where $\mu=R_{ij}^r\lambda$, $1\leqslant r\leqslant m$. By Chapter I, (1.14) each such μ is $>\lambda$, and so (8.3) is true for μ. This completes the induction step. |

Let Γ denote the subring of Λ generated by the q_r:

$$\Gamma = \mathbf{Z}[q_1, q_2, q_3, \ldots].$$

Γ is a graded ring: $\Gamma = \bigoplus_{n \geqslant 0} \Gamma^n$, where $\Gamma^n = \Gamma \cap \Lambda^n$ is spanned by the q_λ such that $|\lambda| = n$.

Also let $\Gamma_{\mathbf{Q}} = \Gamma \otimes \mathbf{Q} = \mathbf{Q}[q_1, q_2, q_3, \ldots]$. From (8.3) it follows that $\Gamma_{\mathbf{Q}}$ is generated over $\mathbf{Q}$ by the q_r with r odd. In fact these elements are algebraically independent. For from (8.1) we have

$$\frac{Q'(t)}{Q(t)} = \frac{E'(t)}{E(t)} + \frac{H'(t)}{H(t)} = P(t) + P(-t)$$

$$= 2 \sum_{r \geqslant 0} p_{2r+1} t^{2r},$$

and therefore

$$rq_r = 2(p_1 q_{r-1} + p_3 q_{r-3} + \ldots).$$

The series on the right terminates with $p_{r-1} q_1$ if r is even, and with p_r if r is odd. Hence these equations enable us to express the odd power sums in terms of the q's, and vice versa. Since (Chapter I, §2) $p_1, p_3, \ldots$ are algebraically independent over $\mathbf{Q}$, it follows that

(8.5) *We have*

$$\Gamma_{\mathbf{Q}} = \mathbf{Q}[p_r : r \text{ odd}] = \mathbf{Q}[q_r : r \text{ odd}]$$

and the q_r *(r odd) are algebraically independent over* $\mathbf{Q}$.

A partition λ is *odd* if all its parts are odd. For each $n \geqslant 1$, the number N_n of odd partitions of n is equal to the number of strict partitions of n, because the generating function for odd partitions is

$$\sum_{\lambda \text{ odd}} t^{|\lambda|} = \prod_{r \geqslant 1} \frac{1}{1 - t^{2r-1}} = \prod_{r \geqslant 1} \frac{1 - t^{2r}}{1 - t^r} = \prod_{r \geqslant 1} (1 + t^r),$$

which is the generating function for strict partitions.

(8.6) (i) *The* q_λ, λ *odd, form a* $\mathbf{Q}$*-basis of* $\Gamma_{\mathbf{Q}}$.
(ii) *The* q_λ, λ *strict, form a* $\mathbf{Z}$*-basis of* Γ.

Proof. (i) follows from (8.5), and shows that the dimension of $\Gamma_{\mathbf{Q}}^n$ is N_n. From (8.4) it follows that the q_λ, where λ is a strict partition of n, span Γ^n and hence also $\Gamma_{\mathbf{Q}}^n$. Since there are N_n of them, they form a $\mathbf{Q}$-*basis* of $\Gamma_{\mathbf{Q}}^n$. Hence they are linearly independent over $\mathbf{Q}$, and therefore also over $\mathbf{Z}$, which proves (ii). ∎

Consider now $P_\lambda = P_\lambda(x; -1)$ and $Q_\lambda = Q_\lambda(x; -1)$. The symmetric functions P_λ are well-defined and non-zero for all partitions λ, and since the transition matrix $M(P, m)$ is unitriangular, it follows that the P_λ form a $\mathbf{Z}$-basis of Λ. On the other hand, $b_\lambda(-1) = \prod_{i \geqslant 1} \varphi_{m_i(\lambda)}(-1)$ will vanish if any $m_i(\lambda) > 1$, that is to say if the partition λ is not strict. If λ is strict, we have $b_\lambda(-1) = 2^{l(\lambda)}$, and hence

(8.7)
$$Q_\lambda = \begin{cases} 2^{l(\lambda)} P_\lambda & \text{if } \lambda \text{ is strict,} \\ 0 & \text{otherwise.} \end{cases}$$

From (2.15), with $t = -1$, we have

(8.8) *Let* λ *be a strict partition of length* $\leqslant n$. *Then* Q_λ *is equal to the coefficient of* $t^\lambda = t_1^{\lambda_1} t_2^{\lambda_2} \dots$ *in*

$$Q(t_1, \dots, t_n) = \prod_{i=1}^{n} Q(t_i) \prod_{i<j} F(t_i^{-1} t_j)$$

where

$$F(y) = \frac{1-y}{1+y} = 1 + 2 \sum_{r \geqslant 1} (-1)^r y^r.$$

As in §2, (8.8) is equivalent to the raising operator formula

(8.8′)
$$Q_\lambda = \prod_{i<j} \frac{1 - R_{ij}}{1 + R_{ij}} q_\lambda.$$

From (8.8′) it follows that Q_λ is of the form

$$Q_\lambda = q_\lambda + \sum_{\mu > \lambda} a_{\lambda\mu} q_\mu$$

with coefficients $a_{\lambda\mu} \in \mathbf{Z}$. If λ is strict, it follows from (8.4) that we have

$$Q_\lambda = q_\lambda + \sum_{\mu > \lambda} a'_{\lambda\mu} q_\mu$$

where now the sum on the right is restricted to *strict* partitions $\mu > \lambda$. From (8.6) (ii) it now follows that

(8.9) *The* Q_λ, λ *strict, form a* $\mathbf{Z}$*-basis of* Γ. |

We come next to Schur's definition of Q_λ. From (8.8) it follows that

$Q_{(r,s)}$ (where $r > s \geqslant 0$) is the coefficient of $t^r u^s$ in $Q(t,u) = Q(t)Q(u)F(t^{-1}u)$, so that

(8.10)
$$Q_{(r,s)} = q_r q_s + 2\sum_{i=1}^{s}(-1)^i q_{r+i}q_{s-i}.$$

If $r \leqslant s$ we *define* $Q_{(r,s)}$ by this formula. Then it follows from (8.2) that $Q_{(s,r)} = -Q_{(r,s)}$.

Now let λ be a strict partition, which we may write in the form $\lambda = (\lambda_1, \lambda_2, \dots, \lambda_{2n})$ where $\lambda_1 > \dots > \lambda_{2n} \geqslant 0$. Then the $2n \times 2n$ matrix

$$M_\lambda = \left(Q_{(\lambda_i,\lambda_j)}\right)$$

is skew-symmetric, and Schur's definition of Q_λ (which for us is a theorem) reads:

(8.11)
$$Q_\lambda = \mathrm{Pf}(M_\lambda)$$

where $\mathrm{Pf}(M_\lambda)$ is the *Pfaffian* of the skew-symmetric matrix M_λ. (We recall that if $A = (a_{ij})$ is a skew symmetric matrix of even size $2n \times 2n$, its determinant is a perfect square: $\det(A) = \mathrm{Pf}(A)^2$, where

$$\mathrm{Pf}(A) = \sum_w \varepsilon(w) . a_{w(1)w(2)} \cdots a_{w(2n-1)w(2n)}$$

summed over $w \in S_{2n}$ such that $w(2r-1) < w(2r)$ for $1 \leqslant r \leqslant n$, and $w(2r-1) < w(2r+1)$ for $1 \leqslant r \leqslant n-1$.)

Proof of (8.11). From above, $\mathrm{Pf}(M_\lambda)$ is a sum of terms of the form

$$\pm Q_{(\mu_1,\mu_2)} \cdots Q_{(\mu_{2n-1},\mu_{2n})}$$

where $(\mu_1, \dots, \mu_{2n}) = (\lambda_{i_1}, \dots, \lambda_{i_{2n}})$ is a derangement of $(\lambda_1, \dots, \lambda_{2n})$. By (8.8) the term just written is equal to the coefficient of $t^\lambda = t_1^{\lambda_1} t_2^{\lambda_2} \dots$ in

$$\pm Q(t_1) \dots Q(t_{2n}) F\left(t_{i_1}^{-1} t_{i_2}\right) \dots F\left(t_{i_{2n-1}}^{-1} t_{i_{2n}}\right),$$

and consequently $\mathrm{Pf}(M_\lambda)$ is the coefficient of t^λ in

(1)
$$Q(t_1) \dots Q(t_{2n}) \mathrm{Pf}\left(\left(F\left(t_i^{-1} t_j\right)\right)\right).$$

Now $F(t_i^{-1} t_j) = (t_i - t_j)/(t_i + t_j)$, and it is well-known that

(2)
$$\mathrm{Pf}\left(\frac{t_i - t_j}{t_i + t_j}\right) = \prod_{i<j}\left(\frac{t_i - t_j}{t_i + t_j}\right).$$

From (1) and (2) it follows that $\mathrm{Pf}(M_\lambda)$ is the coefficient of t^λ in

$$Q(t_1)\dots Q(t_{2n})\prod_{i<j} F\left(t_i^{-1}t_j\right) = Q(t_1,\dots,t_{2n})$$

hence by (8.8) is equal to Q_λ. ∎

(For a proof of (2) (which is a classical result) see Example 5 below; and for an extension of (8.11) to skew Q-functions $Q_{\lambda/\mu}$, see Example 9.)

Next, the scalar product of §4 is not well-defined on all of Λ when $t=-1$, because $z_\lambda(t)$ has a pole at $t=-1$ if any part of λ is even. But if λ is odd it follows from the definition of $z_\lambda(t)$ that $z_\lambda(-1)=2^{-l(\lambda)}z_\lambda$, and hence the restriction to Γ of the scalar product is well defined and positive definite. We have

(8.12) (i) $\langle Q_\lambda, Q_\mu\rangle = 2^{l(\lambda)}\delta_{\lambda\mu}$ *if* λ, μ *are strict.*
(ii) $\langle p_\lambda, p_\mu\rangle = 2^{-l(\lambda)}z_\lambda\delta_{\lambda\mu}$ *if* λ, μ *are odd.*

These correspond to the series expansions

(8.13)
$$\prod_{i,j}\frac{1+x_iy_j}{1-x_iy_j} = \sum_{\lambda \text{ strict}} 2^{-l(\lambda)}Q_\lambda(x)Q_\lambda(y)$$
$$= \sum_{\lambda \text{ odd}} 2^{l(\lambda)}z_\lambda^{-1}p_\lambda(x)p_\lambda(y),$$

obtained by setting $t=-1$ in (4.4) and (4.1).

Let λ,μ be partitions such that μ is strict, $\lambda\supset\mu$ and $\lambda-\mu$ is a horizontal strip. Then $m_i(\lambda)\leqslant 2$ for each $i\geqslant 1$, and hence the formula (5.8) shows that

(8.14)
$$\varphi_{\lambda/\mu}(-1) = \begin{cases} 2^{a(\lambda-\mu)} & \text{if } \lambda \text{ is strict,}\\ 0 & \text{otherwise,}\end{cases}$$

where $a(\lambda-\mu)$ is the number of integers $i\geqslant 1$ such that $\lambda-\mu$ has a square in the ith column but not in the $(i+1)$st column. Hence from (5.7) we have

(8.15)
$$P_\mu q_r = \sum_\lambda 2^{a(\lambda-\mu)}P_\lambda$$

for a strict partition μ, where the sum is over strict $\lambda\supset\mu$ such that $\lambda-\mu$ is a horizontal strip.

This result may be more conveniently rephrased in terms of shifted diagrams. As in Chapter I, §1, Example 9, let $S(\lambda)$ be the shifted diagram of a strict partition λ, obtained from the usual diagram by shifting the ith row $(i-1)$ squares to the right, for each $i>1$. The effect of replacing

$\lambda - \mu$ by $S(\lambda - \mu) = S(\lambda) - S(\mu)$ is to convert the horizontal strip into a disjoint union of border strips, and $a(\lambda - \mu)$ is just the number of connected components of $S(\lambda - \mu)$, i.e. the number of border strips of which it is composed.

We are therefore led to define a *shifted tableau* T of shape $S(\lambda - \mu) = S(\lambda) - S(\mu)$, where λ and μ are strict partitions, to be a sequence of strict partitions $(\lambda^{(0)}, \lambda^{(1)}, \dots, \lambda^{(r)})$ such that $\mu = \lambda^{(0)} \subset \lambda^{(1)} \subset \dots \subset \lambda^{(r)} = \lambda$, and such that each $S(\lambda^{(i)} - \lambda^{(i-1)})$ is a disjoint union of border strips; or, equivalently, as a labelling of the squares of $S(\lambda - \mu)$ with the integers $1, 2, \dots, r$ which is weakly increasing along rows and down columns and is such that no 2×2 block of squares bears the same label, so that for each i the set of squares labelled i is a disjoint union of border strips. If we denote by $b(T)$ the number of border strips of which T is composed, then it follows from (8.14) and (5.11) that, for λ and μ strict partitions, we have

(8.16)
$$Q_{\lambda/\mu} = \sum_T 2^{b(T)} x^T$$

summed over shifted tableaux of shape $S(\lambda - \mu)$, where as usual x^T is the monomial defined by the tableau T.

Remarks. 1. The diagonal squares (i, i) in $S(\lambda) - S(\mu)$ must all lie in distinct border strips, so that $b(T) \geqslant l(\lambda) - l(\mu)$ for each tableau T of shape $S(\lambda) - S(\mu)$.
2. The formula (8.16) shows that $Q_{\lambda/\mu}$, like $s_{\lambda/\mu}$ but unlike $Q_{\lambda/\mu}(x; t)$ for arbitrary t, depends only on the skew diagram $\lambda - \mu$.

For later purposes it will be convenient to rephrase (8.16). Let $\mathbf{P}'$ denote the ordered alphabet $\{1' < 1 < 2' < 2 < \dots\}$. The symbols $1', 2', \dots$ are said to be *marked*, and we shall denote by $|a|$ the unmarked version of any $a \in \mathbf{P}'$. Let λ and μ again be strict partitions such that $\lambda \supset \mu$. A *marked shifted tableau* T of shape $S(\lambda - \mu)$ is a labelling of the squares of $S(\lambda - \mu)$ with symbols from $\mathbf{P}'$ such that (compare Chapter I, §5, Example 23):

(M1) The labels increase (in the weak sense) along each row and down each column.
(M2) Each column contains at most one k, for each $k \geqslant 1$.
(M3) Each row contains at most one k', for each $k \geqslant 1$.

The conditions (M2) and (M3) say that for each $k \geqslant 1$, the set of squares labelled k (resp. k') is a horizontal (resp. vertical) strip.

The *weight* (or content) of T is the sequence $(\alpha_1, \alpha_2, \dots)$ where α_k, for each $k \geqslant 1$, is the number of squares in T labelled either k or k'.

Let $|T|$ be the (unmarked) shifted tableau obtained from T by deleting all the marks, i.e., by replacing each $a \in \mathbf{P}'$ by its unmarked version $|a|$. The conditions (M1)–(M3) ensure that no 2×2 block of squares can bear the same label in $|T|$, so that $|T|$ is a shifted tableau as defined above. Moreover, in each border strip β in $|T|$, each square of which is labelled k, it is easy to see that the conditions (M1)–(M3) uniquely determine the assignment of marked and unmarked symbols (k' or k) to the squares of β, with the exception of the square $(i, j) \in \beta$ nearest the diagonal (i.e. such that $(i+1, j) \notin \beta$ and $(i, j-1) \notin \beta$), which can be marked or unmarked. Call this square the *free* square of β.

From this discussion it follows that the number of free squares in $|T|$ is $b(|T|)$, and hence that the number of marked shifted tableaux T arising from a given (unmarked) shifted tableau $|T|$ is $2^{b(|T|)}$. Consequently (8.16) may be restated as

(8.16′)
$$Q_{\lambda/\mu} = \sum_T x^{|T|}$$

summed over marked *shifted tableaux of shape* $S(\lambda - \mu)$.

Equivalently:

(8.16″)
$$Q_{\lambda/\mu} = \sum_\nu K'_{\lambda-\mu,\nu} m_\nu$$

where $K'_{\lambda-\mu,\nu}$ *is the number of marked shifted tableaux T of shape* $\lambda - \mu$ *and weight* ν.

Remarks. 1. From (5.2) we have

$$Q_{\lambda/\mu} = \sum_\nu f^\lambda_{\mu\nu}(-1) Q_\nu .$$

Since $f^\lambda_{\mu\nu}(t) \in \mathbf{Z}[t]$, this shows that $Q_{\lambda/\mu}$ is a $\mathbf{Z}$-linear combination of the Q_ν (where we may assume ν strict, otherwise $Q_\nu = 0$). Hence it follows from (8.8) that $Q_{\lambda/\mu} \in \Gamma$ for all λ, μ.

2. The formula (8.16) or (8.16′) may be taken as a combinatorial definition of the Q-functions, and the same remarks apply as in the case of the Schur functions (Chapter I, §5). In particular, it is not at all obvious from this definition that $Q_{\lambda/\mu}$ is in fact a *symmetric* function of the x_i.

To conclude, we shall mention without proof two positivity results, and their combinatorial interpretations.

(8.17) (i) *If* μ, ν *are strict partitions and*

$$P_\mu P_\nu = \sum_\lambda f^\lambda_{\mu\nu} P_\lambda$$

then $f^\lambda_{\mu\nu} \geqslant 0$ *for all* λ.

(ii) *If λ is a strict partition and*

$$P_\lambda = \sum_\mu g_{\lambda\mu} s_\mu$$

then $g_{\lambda\mu} \geqslant 0$ for all μ.

Since the P_λ (for all partitions λ) form a **Z**-basis of Λ, the coefficients $f^\lambda_{\mu\nu} (= f^\lambda_{\mu\nu}(-1)$, in the notation of §3) are integers; and since $P_\mu, P_\nu \in \Gamma_\mathbf{Q}$ by virtue of (8.7) and (8.9), we have $f^\lambda_{\mu\nu} = 0$ if λ is not strict.

In (8.17)(ii), the matrix $(g_{\lambda\mu})$ is strictly upper unitriangular, and is the inverse of the matrix $K(-1)$ (§6). Hence (8.17)(ii) says that the matrix $K(-1)^{-1}$ has non-negative (integer) entries; or again, in the notation of Chapter I, §7, Example 9, that $P_\lambda \in \Lambda_+$ for all strict partitions λ.

There are at least three ways of proving both parts of (8.17). One is to interpret the coefficients $f^\lambda_{\mu\nu}$ and $g_{\lambda\mu}$ in terms of the projective representations of the symmetric groups, in the same sort of way that the Littlewood–Richardson coefficients $c^\lambda_{\mu\nu}$ can be shown to be non-negative by reference to the linear representations of the symmetric groups. Another way is to interpret the Q_λ as cohomology classes dual to Schubert cycles in the Grassmannian of maximal isotropic subspaces in $\mathbf{C}^{2n}$ relative to a skew-symmetric bilinear form of rank n. The third method, which we shall describe below, provides a shifted analogue of the Littlewood–Richardson rule for $f^\lambda_{\mu\nu}$, and likewise for $g_{\lambda\mu}$. A fourth possibility, which as far as I know has not been investigated, would be to interpret the P_λ (λ strict) as traces of polynomial functors on an appropriate category, in analogy with Schur's theory (Chapter I, Appendix A).

Let T be a marked (shifted or unshifted) tableau, that is to say a labelling of the shape of T with symbols from the alphabet $\mathbf{P}'$, satisfying the conditions (M1)–(M3) above. As in Chapter I, §9, T gives rise to a *word* $w = w(T)$ by reading the symbols in T from right to left in successive rows, starting with the top row. Let $\hat{w}$ be the word obtained from w by reading w from right to left, and then replacing each k by $(k+1)'$ and each k' by k, for all $k \geqslant 1$. For example, if $w = 11'2'12'2$, then $\hat{w} = 3'22'212'$.

Suppose T contains altogether n symbols, so that w is of length n; then $w\hat{w}$ is of length $2n$, say $w\hat{w} = a_1 a_2 \dots a_{2n}$. For each $k \geqslant 1$ and $0 \leqslant p \leqslant 2n$, let $m_k(a_1 \dots a_p)$ denote the number of k's among $a_1, \dots, a_p$. The word $w = w(T)$ is said to have the *lattice property* if, for each k and p as above, whenever $m_{k+1}(a_1 \dots a_p) = m_k(a_1 \dots a_p)$, we have $|a_{p+1}| \neq k+1$ (i.e. a_{p+1} is neither $k+1$ nor $(k+1)'$).

With this explained we can now state

(8.18) (i) *Let λ, μ, ν be strict partitions. Then the coefficient $f^{\lambda}_{\mu\nu}$ of P_λ in $P_\mu P_\nu$ is equal to the number of marked shifted tableaux T of shape $S(\lambda-\mu)$ ad weight ν such that*
(a) *$w(T)$ has the lattice property;*
(b) *for each $k \geqslant 1$, the last* (i.e. *rightmost*) *occurrence of k' in $w(T)$ precedes the last occurrence of k.*

(ii) *Let λ, μ be partitions, λ strict. Then the coefficient $g_{\lambda\mu}$ of s_μ in P_λ is equal to the number of* (*unshifted*) *marked tableaux T of shape μ and weight λ satisfying* (a) *and* (b) *above.*

These theorems are due to J. Stembridge [S28].

Examples

1. The formula (2.1) for $P_\lambda(x_1,\dots,x_n;t)$ no longer makes sense at $t=-1$, even if λ is strict (unless $l(\lambda)=n$ or $n-1$), because $v_\lambda(t)$ vanishes at $t=-1$ if $l(\lambda)<n-1$. But the alternative formula (2.2) does make sense at $t=-1$, and for a strict partition λ of length $l \leqslant n$ it gives

$$P_\lambda(x_1,\dots,x_n)=\sum_{w\in S_n/S_{n-l}} w\left(x^\lambda \prod \frac{x_i+x_j}{x_i-x_j}\right),$$

where S_{n-l} acts on $x_{l+1},\dots,x_n$, and the product on the right is over pairs (i,j) such that $1\leqslant i\leqslant l$ and $1\leqslant i<j\leqslant n$. This leads to the following expression for $Q_\lambda(x_1,\dots,x_n)$: let π_n denote the linear operator on the ring $L_n=\mathbf{Z}[x_1^{\pm 1},\dots,x_n^{\pm 1}]$ of Laurent polynomials defined by

$$\pi_n f=\sum_{w\in S_n} w(fx^\delta/a_\delta)$$

where $\delta=(n-1,n-2,\dots,1,0)$ and $a_\delta=\prod_{i<j}(x_i-x_j)$. Then we have, for a strict partition λ of length $l\leqslant n$,

$$Q_\lambda(x_1,\dots,x_n)=2^l\pi_n(x^\lambda f_l),$$

where $f_l=\prod(1+x_i^{-1}x_j)$, the product being taken over pairs (i,j) such that $1\leqslant i\leqslant l$ and $j>i$.

2. Suppose λ is a strict partition of length n or $n-1$. Then in n variables $x_1,\dots,x_n$ we have $P_\lambda=s_{\lambda-\delta}s_\delta$, where $\delta=(n-1,n-2,\dots,1,0)$. (Use (2.1) with $t=-1$, and observe that $\prod_{i<j}(x_i+x_j)=\prod_{i<j}(x_i^2-x_j^2)/(x_i-x_j)=s_\delta$.)

3. (a) Let ω be the involution on Λ defined in Chapter I, §2. From (8.1) we have $\omega q_n=q_n$ for all $n\geqslant 1$, hence ω fixes each element of the ring Γ, and in particular $\omega Q_\lambda=Q_\lambda$ and $\omega P_\lambda=P_\lambda$ for each strict partition λ.

(b) We have $P_\delta=s_\delta$ for any 'staircase' partition $\delta=(n-1,n-2,\dots,1,0)$. (For P_δ is of the form

$$P_\delta=\sum_{\mu\leqslant\delta} g_{\delta\mu}s_\mu$$

with $g_{\delta\delta} = 1$; from (a) above it follows that $g_{\delta\mu} = g_{\delta\mu'}$, and since $\mu \leqslant \delta$ implies $\mu' \geqslant \delta$ (Chapter I, (1.11)) it follows that $g_{\delta\mu} = 0$ unless $\mu = \delta$.

4. Let $A = (a_{ij})$ be a 'marked matrix', i.e. with entries $a_{ij} \in \mathbf{P}' \cup \{0\}$. The matrix $|A| = (|a_{ij}|)$ obtained by deleting the marks is a matrix of non-negative integers. Deduce from (8.13) and (8.16) that for any two partitions μ and ν the number of marked matrices A such that $|A|$ has row sums μ_i and column sums ν_i is equal to the number of pairs (S, T) of marked shifted tableaux of the same (shifted) shape, such that S has weight μ, T has weight ν, and the main diagonal in T is unmarked.

5. Let $P(t_1, \dots, t_m)$, where m is even, denote the Pfaffian of the $m \times m$ matrix M whose (i, j) element is $(t_i - t_j)/(t_i + t_j)$. From the definition of the Pfaffian it follows that

$$(1) \qquad P(t_1, \dots, t_m) = \sum_{i=2}^{m} (-1)^i P(t_1, t_i) P(t_2, \dots, \hat{t}_i, \dots, t_m).$$

Also it is clear that

$$U(t_1, \dots, t_m) = P(t_1, \dots, t_m) \prod_{i<j} (t_i + t_j)$$

is a homogeneous polynomial in the t_i, of degree $\frac{1}{2}m(m-1)$. Now $P^2 = \det M$ vanishes whenever two of the t_i are equal, hence the same is true of the polynomial U, and therefore U is divisible by $V = \prod_{i<j}(t_i - t_j)$. Since U and V have the same degree, we conclude that $U = c_m V$ for some constant c_m depending only on m, and hence that

$$\mathrm{Pf}\left(\frac{t_i - t_j}{t_i + t_j}\right) = c_m \prod_{i<j} \frac{t_i - t_j}{t_i + t_j}.$$

It remains to show that $c_m = 1$. From the relation (1), on multiplying both sides by $\prod_{i<j}(t_i + t_j)$ and then comparing the coefficients of $t_1^{m-1} t_2^{m-2} \dots t_{m-1}$ on either side, we obtain

$$c_m = \sum_{i=2}^{m} (-1)^i c_{m-2} = c_{m-2}$$

and hence $c_m = c_{m-2} = \dots = c_2 = 1$.

6. (a) Show that

$$Q(t) = \exp\left(\sum_{r \text{ odd}} 2 p_r t^r / r \right)$$

and hence that

$$q_n = \sum_{\rho \text{ odd}} z_\rho^{-1} 2^{l(\rho)} p_\rho.$$

(b) Let V be a vector space (over a field of characteristic zero) with basis $v_1, \dots, v_n$, and let S_n act on V by permuting the v_i. Then S_n acts on the exterior algebra $\bigwedge V$: if the character of this representation is φ_n, then

$$\operatorname{ch}(\varphi_n) = q_n.$$

(We have $\varphi_n(w) = \prod_{i=1}^n (1 + \xi_i)$, where the ξ_i are the eigenvalues of $w \in S_n$ acting on V (i.e. of the permutation matrix corresponding to w). To each r-cycle in w there correspond r eigenvalues ξ_i, namely the rth roots of unity. The corresponding product $\prod(1 + \xi_i)$ is zero if r is even, and is equal to 2 if r is odd. It follows that $\varphi_n(w) = 0$ if w has any cycles of even length, and that $\varphi_n(w) = 2^l$ if w consists of l cycles of odd length. The result now follows from (a) above.)

(c) From (8.1) we have

$$q_n = \sum_{r+s=n} e_r h_s = \sum_{a+b=n-1} 2 s_{(a|b)}$$

by Pieri's formula (Chapter I, (5.16)).

7. (a) Let

$$S_\lambda = S_\lambda(x; -1) = \det(q_{\lambda_i - i + j})$$

as in (4.5). In the notation of Chapter I, §5, Example 23, $S_\lambda = s_\lambda(x/-x)$. Hence (*loc. cit.*) we have

$$S_\lambda = \sum_T x^T$$

summed over all (unshifted) marked tableaux of shape λ.

(b) From the definition, S_λ is of the form

(1) $$S_\lambda = q_\lambda + \sum_{\mu > \lambda} c_{\lambda\mu} q_\mu$$

with integer coefficients $c_{\lambda\mu}$. (The transition matrix $M(S, q)$ is the same as $M(s, h)$, hence is the transposed inverse of the Kostka matrix K (Chapter I, §6).) Hence $S_\lambda \in \Gamma$ for all partitions λ, and if λ is strict it follows from (8.4) that

(2) $$S_\lambda = q_\lambda + \sum_{\mu > \lambda} c'_{\lambda\mu} q_\mu$$

where the sum is now over *strict* partitions $\mu > \lambda$. Hence from (8.6) (ii) it follows that the S_λ, λ strict, form another $\mathbf{Z}$-basis of the ring Γ.

(c) Show that

$$S_\lambda = \sum_{\rho \text{ odd}} z_\rho^{-1} 2^{l(\rho)} \chi_\rho^\lambda p_\rho$$

$$= s_\lambda * q_n = \operatorname{ch}(\chi^\lambda \varphi_n)$$

where (Example 4 (b)) φ_n is the character of the exterior algebra representation of S_n.

8. For each $f \in \Gamma$, let $f^{\perp}: \Gamma \to \Gamma$ denote the adjoint of multiplication by f, so that

$$\langle f^{\perp} g, h\rangle = \langle g, fh\rangle$$

for all $g, h \in \Gamma$. Let

$$Q^{\perp}(t) = \sum_{n \geqslant 0} q_n^{\perp} t^n,$$

then (§5, Example 8) we have

$$Q^{\perp}(t)Q(u) = F(-tu)Q(u)Q^{\perp}(t) \tag{1}$$

as operators on Γ, where t and u are commuting indeterminates, $Q(u)$ is the operator of multiplication by $Q(u)$, and $F(y) = (1-y)/(1+y)$ as in (8.8).

(a) Define operators $B_m: \Gamma \to \Gamma$ for each $m \in \mathbf{Z}$ by

$$B_m = q_m - q_{m+1}q_1^{\perp} + q_{m+2}q_2^{\perp} - \dots$$

(where $q_m = 0$ if $m < 0$, and $q_0 = 0$). Collectively we have

$$B(t) = \sum_{m \in \mathbf{Z}} B_m t^m = Q(t)Q^{\perp}(-t^{-1}).$$

More generally, if $t_1, \dots, t_n$ are commuting indeterminates, let

$$B(t_1, \dots, t_n) = \prod_{i=1}^{n} Q(t_i) \prod_{i=1}^{n} Q^{\perp}(-t_i^{-1}). \tag{2}$$

From (1) it follows that

$$B(t)B(u) = F(t^{-1}u)B(t,u) \tag{3}$$

and hence by iteration that

$$B(t_1)\dots B(t_n) = B(t_1, \dots, t_n) \prod_{i<j} F\left(t_i^{-1}t_j\right). \tag{4}$$

Since $Q^{\perp}(t)(1) = 1$, it follows from (2) and (4) that

$$B(t_1)\dots B(t_n)(1) = Q(t_1, \dots, t_n)$$

in the notation of (8.8), and hence that

$$Q_\lambda = B_{\lambda_1} \dots B_{\lambda_l}(1) \tag{5}$$

if λ is a strict partition of length l.

(b) Let

$$\begin{aligned} \varphi(t,u) = \varphi(u,t) &= F(t^{-1}u) + F(tu^{-1}) \\ &= 2 \sum_{n \in \mathbf{Z}} (-1)^n t^{-n} u^n, \end{aligned}$$

a formal Laurent series in t and u.

Show that

(6) $$Q(t)Q(u)\varphi(t,u)=\varphi(t,u).$$

(The coefficient of $t^r u^s$ on the left-hand side is

$$a_{rs}=\sum(-1)^c q_a q_b$$

summed over $a,b\in\mathbf{N}$ and $c\in\mathbf{Z}$ such that $a+c=r$ and $b-c=s$, so that $a+b=r+s$; hence

$$(-1)^r a_{rs}=\sum_{a+b=r+s}(-1)^a q_a q_b=\delta_{0,r+s}$$

since $Q(t)Q(-t)=1$.)

From (6) it follows by taking adjoints that

$$Q^{\perp}(t)Q^{\perp}(u)\varphi(t,u)=\varphi(t,u)$$

and hence that

(7) $$B(t,u)\varphi(t,u)=\varphi(t,u).$$

Finally, from (3) and (7) we have

(8) $$B(t,u)+B(u,t)=B(t,u)\varphi(t,u)=\varphi(t,u).$$

Explicitly, (8) says that for all $r,s\in\mathbf{Z}$:

(C1) $B_rB_s=-B_sB_r$ if $r+s\neq 0$, and in particular $B_r^2=0$ if $r\neq 0$;

(C2) $B_rB_{-r}+B_{-r}B_r=2(-1)^r$ if $r\neq 0$, and $B_0^2=1$.

(c) Let $\alpha=(\alpha_1,\dots,\alpha_n)\in\mathbf{Z}^n$ be any sequence of n integers. We define Q_α to be the coefficient of $t^\alpha=t_1^{\alpha_1}\dots t_n^{\alpha_n}$ in the Laurent series $Q(t_1,\dots,t_n)$, or equivalently

$$Q_\alpha=B_{\alpha_1}\dots B_{\alpha_n}(1).$$

Use the commutation rules (C1), (C2) to show that

(i) if $|\alpha_1|,\dots,|\alpha_n|$ are all distinct, then Q_α is skew-symmetric in α (i.e. $Q_{w\alpha}=\varepsilon(w)Q_\alpha$ for all $w\in S_n$), and $Q_\alpha=0$ if any α_i is negative.

(ii) If $\alpha_i=\alpha_{i+1}$ for some i, then $Q_\alpha=0$.

(iii) If $\alpha_i=-\alpha_{i+1}=r\neq 0$, then $Q_\alpha=0$ if $r>0$, and $Q_\alpha=2(-1)^rQ_\beta$ if $r<0$, where β is the sequence obtained from α by deleting α_i and α_{i+1}.

Deduce that, for each $\alpha\in\mathbf{Z}^n$, Q_α is either zero or is equal to $\pm 2^rQ_\lambda$ for some integer $r\geqslant 0$ and strict partition λ [H7].

9. Let $\lambda=(\lambda_1,\dots,\lambda_m)$ and $\mu=(\mu_1,\dots,\mu_n)$ be strict partitions such that $\lambda_1>\dots>\lambda_m>0$ and $\mu_1>\dots>\mu_n\geqslant 0$. Since we allow $\mu_n=0$ we may assume that $m+n$ is *even*. Then we have

$$Q_{\lambda/\mu}=\mathrm{Pf}(M_{\lambda,\mu})$$

where $M_{\lambda,\mu}$ is the skew-symmetric matrix of $m+n$ rows and columns

$$M_{\lambda,\mu}=\begin{pmatrix} M_\lambda & N_{\lambda,\mu}\\ -N_{\lambda,\mu}' & 0\end{pmatrix}$$

in which $M_\lambda = (Q_{(\lambda_i, \lambda_j)})$ as in (8.11), and $N_{\lambda,\mu}$ is the $m \times n$ matrix $(q_{\lambda_i - \mu_j})$.
(With the notation of Example 8, we have

$$Q_{\lambda/\mu} = 2^{-l(\mu)} Q_\mu^\perp(Q_\lambda)$$

and by (8.8) $Q_\mu^\perp(Q_\lambda)$ is the coefficient of $t^\lambda u^\mu$ in

$$\prod_k Q^\perp(u_k)\Big(\prod_i Q(t_i)\Big)\prod_{i<j} F\big(t_i^{-1}t_j\big)\prod_{k<l} F(u_k^{-1}u_l)$$

which by use of Example 8, formula (1) is equal to

$$(1)\qquad \prod_i Q(t_i)\prod_{i<j} F\big(t_i^{-1}t_j\big)\prod_{k<l} F(u_k^{-1}u_l)\prod_{i,k} F(-t_i u_k).$$

The product of all the F's in this expression can be written in the form

$$\prod_{r<s} F(v_r^{-1}v_s)$$

where $(v_1, v_2, \dots, v_{m+n}) = (-u_n^{-1}, \dots, -u_2^{-1}, -u_1^{-1}, t_1, \dots, t_m)$. Hence the same argument as in the proof of (8.11) shows that the expression (1) is the Pfaffian of the matrix

$$(a_r a_s F(v_r^{-1}v_s))$$

of $m+n$ rows and columns, where $a_r = 1$ if $r \leqslant n$, and $a_r = Q(v_r)$ if $r > n$. On picking out the coefficient of $t^\lambda u^\mu$, we obtain the result.)

10. Define a ring homomorphism $\varphi: \Lambda \to \Gamma$ by

$$\varphi(e_n) = q_n \qquad (n \geqslant 1).$$

(a) Show that
 (i) $\varphi(h_n) = q_n$,
 (ii) $\varphi(p_n) = 2p_n$ if n is odd, and $\varphi(p_n) = 0$ if n is even,
 (iii) φ commutes with the involution ω,
 (iv) $\varphi(s_{(a|b-1)}) = \frac{1}{2}(q_a q_b + Q_{(a,b)})$.

(b) Let λ, μ be strict partitions of lengths m, n respectively, such that $\lambda \supset \mu$. Let $D(\lambda) = (\lambda_1, \dots, \lambda_m | \lambda_1 - 1, \dots, \lambda_m - 1)$ be the 'double' of λ (Chapter I, §1, Example 9), and define $D(\mu)$ similarly as the double of μ. Then we have

$$\varphi(s_{D(\lambda)/D(\mu)}) = 2^{n-m} Q_{\lambda/\mu}^2.$$

(From Chapter I, §5, Example 22 we have

$$s_{D(\lambda)/D(\mu)} = (-1)^n \det\begin{pmatrix} C_\lambda & H_{\lambda,\mu} \\ -E_{\lambda,\mu} & 0 \end{pmatrix}$$

where $C_\lambda = (s_{(\lambda_i|\lambda_j - 1)})$, $H_{\lambda,\mu} = (h_{\lambda_i - \mu_j})$ and $E_{\lambda,\mu} = (e_{\lambda_j - \mu_i})$. By use of (a) it follows that

$$\varphi(s_{D(\lambda)/D(\mu)}) = 2^{n-m} \det(M_{\lambda,\mu} + B)$$

where $M_{\lambda,\mu}$ is the matrix defined in Example 9, and $B=(b_{ij})$ is given by

$$b_{ij}=\begin{cases} q_{\lambda_i}q_{\lambda_j} & \text{if } 1\leqslant i,j\leqslant m,\\ 0 & \text{otherwise.}\end{cases}$$

From Example 9 it now follows that

$$\varphi(s_{D(\lambda)/D(\mu)})=2^{n-m}Q_{\lambda/\mu}^2\det\left(1+BM_{\lambda,\mu}^{-1}\right)$$

and since rank $(B)=1$,

$$\det\left(1+BM_{\lambda,\mu}^{-1}\right)=1+\operatorname{trace}\left(BM_{\lambda,\mu}^{-1}\right)=1$$

because $M_{\lambda,\mu}^{-1}$ is skew-symmetric.)

(c) By applying φ to both sides of the identity of Chapter I, §5, Example 9(b), we obtain

$$\sum_{\lambda}2^{-l(\lambda)}Q_\lambda(x)^2=\prod_{i,j}(1+x_ix_j)/(1-x_ix_j)$$

(i.e. (8.13) with $x_i=y_i$ for each i).

11. (a) With the notation of Example 8, show that for r odd

$$p_r^\perp=\frac{1}{2}r\frac{\partial}{\partial p_r}\tag{1}$$

(Use (8.12) (ii).)

From Example 6(a) we have

$$q_n=\sum_\rho 2^{l(\rho)}z_\rho^{-1}p_\rho\tag{2}$$

summed over odd partitions ρ of n. Deduce from (1) and (2) that

$$p_r^\perp q_n=q_{n-r}\tag{3}$$

for all $n\geqslant 1$; equivalently,

$$p_r^\perp Q(t)=t^rQ(t).\tag{3'}$$

Deduce that

$$p_r^\perp Q(t_1,\dots,t_n)=p_r(t_1,\dots,t_n)Q(t_1,\dots,t_n).\tag{4}$$

(Observe that $p_r^\perp$ is a derivation, by (1) above.) More generally, for any odd partition ρ,

$$p_\rho^\perp Q(t_1,\dots,t_n)=p_\rho(t_1,\dots,t_n)Q(t_1,\dots,t_n).\tag{5}$$

Hence $p_\rho^\perp Q_\lambda$ (λ a strict partition) is equal to the coefficient of t^λ in the right-hand side of (5).

(b) Let λ be a strict partition of length l and let $r \geqslant 1$ be odd. From (4) above we have

$$p_r^\perp Q_\lambda = \sum_{i=1}^{l} Q_{\lambda - r\varepsilon_i} \tag{6}$$

where $\lambda - r\varepsilon_i = (\lambda_1, \dots, \lambda_i - r, \dots, \lambda_l)$. From Example 8(c) it follows that $Q_{\lambda - r\varepsilon_i} = 0$ unless either (i) $\lambda_i \geqslant r$ and the numbers $\lambda_1, \dots, \lambda_i - r, \dots, \lambda_l$ are all distinct, or (ii) $r = \lambda_i + \lambda_j$ for some $j > i$.

In case (i), either $\lambda_j > \lambda_i - r > \lambda_{j+1}$ for some $j < m$, or else $\lambda_m > \lambda_{i-r}$, and $Q_{\lambda - r\varepsilon_i} = (-1)^{j-i} Q_\mu$ where $\mu = (\lambda_1, \dots, \lambda_{i-1}, \lambda_{i+1}, \dots, \lambda_j, \lambda_i - r, \dots)$. The shifted diagram $S(\mu)$ of μ is obtained from $S(\lambda)$ by removing a border strip β of length r, starting in the ith row and ending in the jth row, and hence of height (Chapter I, §1) $\operatorname{ht}(\beta) = j - i$. Define $\varepsilon(\beta) = (-1)^{\operatorname{ht}(\beta)}$.

In case (ii) we have

$$Q_{\lambda - r\varepsilon_i} = (-1)^{j-i-1+\lambda_j} 2 Q_\mu \tag{7}$$

where now μ is the strict partition obtained from λ by deleting λ_i and λ_j. To express this in terms of shifted diagrams, we define a *double strip* δ to be a skew diagram formed by the union of two border ships which both end on the main diagonal. A double strip δ can be cut into two non-empty connected pieces, one piece α consisting of the diagonals in δ of length 2, and the other piece β consisting of the border strip formed by the diagonals of length 1. Define $\varepsilon(\delta) = (-1)^d$, where $d = \frac{1}{2}|\alpha| + \operatorname{ht}(\beta)$. From (6) and (7) we now have

$$p_r^\perp Q_\lambda = \sum \varepsilon(\gamma) Q_\mu \tag{8}$$

summed over strict partitions $\mu \subset \lambda$ such that $|\lambda| - |\mu| = r$ and $\gamma = S(\lambda - \mu)$ is either a border strip (case (i)) or a double strip (case (ii)).

(c) From (8) we obtain the following combinatorial rule for computing $X_\rho^\lambda(-1) = \langle Q_\lambda, p_\rho \rangle$, where λ is strict and ρ is odd:

$$X_\rho^\lambda(-1) = \sum_S \varepsilon(S)$$

summed over all sequences of strict partitions $S = (\lambda^{(0)}, \lambda^{(1)}, \dots, \lambda^{(m)})$ such that $m = l(\rho), 0 = \lambda^{(0)} \subset \lambda^{(1)} \subset \dots \subset \lambda^{(m)} = \lambda$, and such that each $S(\lambda^{(i)} - \lambda^{(i-1)}) = \gamma_i$ is a border strip or double strip of length ρ_i, and $\varepsilon(S) = \prod_{i=1}^m \varepsilon(\gamma_i)$. (Compare Chapter I, §7, Example 5.)

12. From Example 11(c) (or from §7, Example 4) it follows that if λ is a strict partition of n, then $X_{(1^n)}^\lambda(-1)$ is equal to the number g^λ of *shifted standard tableaux* of shape $S(\lambda)$, obtained by labelling the squares of $S(\lambda)$ with the numbers $1, 2, \dots, n$, with strict increase along each row and down each column.

Just as in the case of ordinary standard tableaux (Chapter I, §5, Example 2) there is a formula for g^λ in terms of hook-lengths. For each square x in $S(\lambda)$, the *hook-length* $h(x)$ is defined to be the hook-length at x in the double diagram $D(\lambda) = (\lambda_1 - 1, \lambda_2 - 1, \ldots | \lambda_1, \lambda_2, \ldots)$. Then we have

(1)
$$g^\lambda = n! \Big/ \prod_{x \in S(\lambda)} h(x).$$

This formula is equivalent to one due to Schur [S5]:

(2)
$$g^\lambda = \frac{n!}{\lambda!} \prod_{i<j} \frac{\lambda_i - \lambda_j}{\lambda_i + \lambda_j}$$

where $\lambda! = \lambda_1! \lambda_2! \ldots$. The equivalence of (1) and (2) follows from the fact that the numbers $h(x)$ for x in the ith row of $S(\lambda)$ are $\lambda_i + \lambda_{i+1}, \lambda_i + \lambda_{i+2}, \ldots$ and $\lambda_i, \lambda_i - 1, \ldots, 2, 1$ with the exception of $\lambda_i - \lambda_{i+1}, \lambda_i - \lambda_{i+2}, \ldots$.

Since the g^λ clearly satisfy the recurrence relation

$$g^\lambda = \sum_{i \geq 1} g^{\lambda^{(i)}}$$

where $\lambda^{(i)}$ is the partition obtained from λ by replacing λ_i by $\lambda_i - 1$, the proof of (2) reduces to verifying the identity

$$\sum_i \lambda_i = \sum_i \lambda_i \prod_{j \neq i} \frac{\lambda_i + \lambda_j}{\lambda_i + \lambda_j - 1} \cdot \frac{\lambda_i - \lambda_j - 1}{\lambda_i - \lambda_j}$$

which may be proved by considering the expansion into partial fractions of the function

$$(2t-1) \prod_i \frac{(t + \lambda_i)(t - \lambda_i - 1)}{(t + \lambda_i - 1)(t - \lambda_i)}.$$

13. Let A denote the skew-symmetric matrix

$$((x_i - x_j)/(x_i + x_j))_{1 \leq i,j \leq n}$$

and for each strict partition $\lambda = (\lambda_1, \ldots, \lambda_l)$ of length $l \leq n$ let B_λ denote the $n \times l$ matrix $\left(x_i^{\lambda_j}\right)$. Let

$$A_\lambda(x_1, \ldots, x_n) = \begin{pmatrix} A & B_\lambda \\ -B'_\lambda & 0 \end{pmatrix}$$

which is a skew-symmetric matrix of $n + l$ rows and columns.

Now define

$$\mathrm{Pf}_\lambda(x_1, \ldots, x_n)$$

to be the Pfaffian of $A_\lambda(x_1, \ldots, x_n)$ if $n + l$ is even, and to be the Pfaffian of $A_\lambda(x_1, \ldots, x_n, 0)$ if $n + l$ is odd. (In the latter case, the effect of introducing the extra variable $x_{n+1} = 0$ is to border A with a column of 1's and a row of -1's. Show that

$$P_\lambda(x_1, \ldots, x_n) = \mathrm{Pf}_\lambda(x_1, \ldots, x_n)/\mathrm{Pf}_0(x_1, \ldots, x_n).$$

(Multiply both sides of the formula (1) of Example 1 by

$$\mathrm{Pf}_0(x_1,\dots,x_n) = \prod_{i<j}(x_i - x_j)/(x_i + x_j).$$

This formula for P_λ may be regarded as an analogue of the definition (Chapter I, (3.1)) of the Schur function $s_\lambda(x_1,\dots,x_n)$ as a quotient of alternants.

Notes and references

As explained in the preface, we have stopped short of describing the rôle played by Schur's Q-functions in the construction of the projective representations of the symmetric groups. Apart from Schur's original paper [S5] of 1911, there are several recent accounts available: the monograph [H7] of Hoffman and Humphreys, and the papers of Józefiak [J15] and Stembridge [S28].

It has become clear in recent years that Schur's Q-functions arise naturally in other contexts: as (up to scalar factors) the characters of certain irreducible representations of the Lie superalgebra $Q(n)$ [S11]; as the cohomology classes dual to Schubert cycles in isotropic Grassmannians [J16], [P2]; as the polynomial solutions of the BKP hierarchy of partial differential equations [Y1]; and as generating functions for the zonal spherical functions (Chapter VII, §1) of the twisted Gelfand pair (S_{2n}, H_n, ζ), where H_n is the hyperoctahedral group, embedded in the symmetric group S_{2n} as the centralizer of a fixed-point free involution, as in Chapter VII, §2, and ζ is the composition of the sign character of S_n with the projection $H_n \to S_n$ [S29].

A combinatorial theory of shifted tableaux, parallel to that for ordinary (unshifted) tableaux, has been developed independently by Sagan [S1] and Worley [W4]; it includes shifted versions of the Robinson–Schensted–Knuth correspondence, Knuth's relations, jeu de taquin, etc. In particular, (8.17) (ii) (originally conjectured by R. Stanley) is due to Sagan and Worley (independently), and Stembridge's proof of (8.18) relies on their theory. Finally, we may remark that setting $t = -1$ in Chapter II, (4.11) will provide another expression for $f^\lambda_{\mu\nu} = f^\lambda_{\mu\nu}(-1)$.

Example 3(b) is due to R. Stanley.

Example 8. The operators B_n are the analogues, for Schur's Q-functions, of Bernstein's operators (Chapter I, §5, Example 29). See [H7] and [J11].

Examples 9 and 10. These results are due to Józefiak and Pragacz [J17]. Our proof of the Pfaffian formula for $Q_{\lambda/\mu}$ (Example 9) is somewhat different from theirs.

Example 13 is due to Nimmo [N2].

IV

THE CHARACTERS OF GL_n OVER A FINITE FIELD

1. The groups L and M

Let k be a finite field with q elements, $\bar{k}$ an algebraic closure of k. Let

$$F: x \mapsto x^q$$

be the Frobenius automorphism of $\bar{k}$ over k. For each $n \geqslant 1$, the set k_n of fixed points of F^n in $\bar{k}$ is the unique extension of k of degree n in $\bar{k}$.

Let M denote the multiplicative group of $\bar{k}$, and M_n the multiplicative group of k_n, so that $M_n = M^{F^n}$.

If m divides n, the norm map $N_{n,m}: M_n \to M_m$ defined by

$$N_{n,m}(x) = x^{(q^n - 1)/(q^m - 1)} = \prod_{i=0}^{d-1} F^{mi}x,$$

where $d = n/m$, is a surjective homomorphism. The groups M_n and the homomorphisms $N_{n,m}$ form an inverse system. Let

$$K = \varprojlim M_n$$

be their inverse limit, which is a profinite group. The character group of K is therefore a discrete group

$$\hat{K} = L = \varinjlim \hat{M}_n$$

where $\hat{M}_n$ is the character group of M_n. Whenever m divides n, $\hat{M}_m$ is embedded in $\hat{M}_n$ by the transpose of the norm homomorphism $N_{n,m}$.

Both groups L and M are (non-canonically) isomorphic to the group of roots of unity in $\mathbf{C}$ of order prime to $p = \text{char. } k$.

The Frobenius map $F: \xi \mapsto \xi^q$ acts on L. For each $n \geqslant 1$ let $L_n = L^{F^n}$ be the group of elements $\xi \in L$ fixed by F^n, so that $L = \bigcup_{n \geqslant 1} L_n$, and $L_m \subset L_n$ if and only if m divides n. The canonical mapping of $\hat{M}_n$ into L is an isomorphism of $\hat{M}_n$ onto L_n. By identifying $\hat{M}_n$ with L_n by means of this isomorphism, we define a pairing of L_n with M_n:

$$\langle \xi, x \rangle_n = \xi(x)$$

for $\xi \in L_n$ and $x \in M_n$. (The suffix n is necessary because these pairings (for different n) are not coherent: if m divides n and $\xi \in L_m$, $x \in M_m$ we have $\langle \xi, x \rangle_n = (\langle \xi, x \rangle_m)^{n/m}$.)

Finally, let Φ denote the set of F-orbits in M. Each such orbit is of the form $\{x, x^q, \dots, x^{q^{d-1}}\}$ where $x^{q^d} = x$, and the polynomial

$$f = \prod_{i=0}^{d-1} (t - x^{q^i})$$

belongs to $k[t]$ and is irreducible over k. Conversely, each irreducible monic polynomial $f \in k[t]$, with the single exception of the polynomial t, determines in this way an F-orbit in M. We shall therefore use the same letter f to denote the polynomial and the orbit consisting of its roots in $\bar{k}$, and the letter Φ to denote the set of these irreducible polynomials.

Remark. For each $n \geqslant 1$ let $\mu_n = (\mathbf{Z}/n\mathbf{Z})(1)(\bar{k})$ be the group of nth roots of unity in $\bar{k}$, and let

$$\hat{\mathbf{Z}}(1)(\bar{k}) = \varprojlim \mu_n$$

with respect to the homomorphisms $\mu_n \to \mu_m : x \mapsto x^{n/m}$ whenever m divides n ([D2], §2). We have

$$\hat{\mathbf{Z}}(1)(\bar{k}) = K = \hat{L},$$

the character group of L, so that $\hat{L}$ is a $\hat{\mathbf{Z}}$-module, where

$$\hat{\mathbf{Z}} = \varprojlim (\mathbf{Z}/n\mathbf{Z}) = \prod_p \mathbf{Z}_p$$

(direct product over all primes p).

Let $\hat{M}$ be the character group of M. Then we have

$$\hat{M} \cong \operatorname{Hom}_{\hat{\mathbf{Z}}}(\hat{L}, \hat{\mathbf{Z}}),$$

$$\hat{L} \cong \operatorname{Hom}_{\hat{\mathbf{Z}}}(\hat{M}, \hat{\mathbf{Z}})$$

and therefore $\hat{M} = \hat{\mathbf{Z}}(-1)(\bar{k})$.

Moreover,

$$M \cong (\mathbf{Q}/\mathbf{Z}) \otimes_{\hat{\mathbf{Z}}} \hat{L} = (\mathbf{Q}/\mathbf{Z})(1)(\bar{k}),$$

$$L \cong (\mathbf{Q}/\mathbf{Z}) \otimes_{\hat{\mathbf{Z}}} \hat{M} = (\mathbf{Q}/\mathbf{Z})(-1)(\bar{k}).$$

2. Conjugacy classes

Let G_n denote the group $GL_n(k)$ of invertible $n \times n$ matrices over the finite field k. Each $g \in G_n$ acts on the vector space k^n and hence defines a $k[t]$-module structure on k^n, such that $t.v = gv$ for $v \in k^n$. We shall denote this $k[t]$-module by V_g. Clearly, two elements $g, h \in G_n$ are conjugate in G_n if and only if V_g and V_h are isomorphic $k[t]$-modules. We may therefore write V_c in place of V_g, where c is the conjugacy class of g in

G_n. The conjugacy classes of G_n are thus in one–one correspondence with the isomorphism classes of $k[t]$-modules V such that (i) $\dim_k V = n$, (ii) $tv = 0$ implies $v = 0$.

Since $k[t]$ is a principal ideal domain, $V = V_c$ is a direct sum of cyclic modules of the form $k[t]/(f)^m$, where $m \geqslant 1$, $f \in \Phi$ and (f) is the ideal in $k[t]$ generated by the polynomial f. Hence V assigns to each $f \in \Phi$ a *partition* $\mu(f) = (\mu_1(f), \mu_2(f), \ldots)$ such that

(2.1) $$V \cong \bigoplus_{f,i} k[t]/(f)^{\mu_i(f)}.$$

In other words, V determines a partition-valued function μ on Φ. Since $\dim_k k[t]/(f)^{\mu_i(f)} = d(f)\mu_i(f)$, where $d(f)$ is the degree of f, it follows that μ must satisfy

(2.2) $$\|\mu\| = \sum_{f \in \Phi} d(f)\,|\mu(f)| = n.$$

Conversely, each function $\mu: \Phi \to \mathscr{P}$, where $\mathscr{P}$ is the set of all partitions, which satisfies (2.2) determines a $k[t]$-module V of dimension n by (2.1), and hence a conjugacy class c_μ in G_n. We shall write V_μ in place of V_{c_μ}.

For each $f \in \Phi$ let $V_{(f)}$ denote the f-primary component of $V = V_\mu$, i.e. the submodule consisting of all $v \in V$ annihilated by some power of f. In the notation of (2.1),

(2.3) $$V_{(f)} \cong \bigoplus_i k[t]/(f)^{\mu_i(f)}.$$

The $V_{(f)}$ are characteristic submodules of V, and V is their direct sum. Hence if $\mathrm{Latt}(M)$ denotes the lattice of submodules of a module M, and $\mathrm{Aut}(M)$ the automorphism group, we have

(2.4) $$\mathrm{Latt}(V) \cong \prod_{f \in \Phi} \mathrm{Latt}(V_{(f)}),$$

(2.5) $$\mathrm{Aut}(V) \cong \prod_{f \in \Phi} \mathrm{Aut}(V_{(f)}).$$

Let $k[t]_{(f)}$ denote the localization of $k[t]$ at the prime ideal (f), i.e. the ring of fractions u/v where $u, v \in k[t]$ and $v \notin (f)$. Then $k[t]_{(f)}$ is a discrete valuation ring with residue field $k_f = k[t]/(f)$, of degree $d(f)$ over k, and $V_{(f)}$ is a finite $k[t]_{(f)}$-module of type $\mu(f)$, by (2.3). Hence from Chapter II, (1.6) we have

(2.6) $$|\mathrm{Aut}(V_{(f)})| = a_{\mu(f)}(q_f)$$

where $q_f = \mathrm{Card}(k_f) = q^{d(f)}$, and for any partition λ

$$a_\lambda(q) = q^{|\lambda| + 2n(\lambda)} b_\lambda(q^{-1}).$$

For any $g \in G_n$, the automorphisms of the $k[t]$-module V_g are precisely the elements $h \in G_n$ which commute with g, i.e. $\mathrm{Aut}(V_g)$ is the *centralizer* of g in G_n. Hence, from (2.5) and (2.6), the centralizer of each $g \in c_\mu$ in G_n has order

(2.7) $$a_\mu = \prod_{f \in \Phi} a_{\mu(f)}(q_f).$$

Examples

1. For each $f \in \Phi$, say $f = t^d - \sum_{i=1}^{d} a_i t^{i-1}$, let $J(f)$ denote the 'companion matrix' for the polynomial f:

$$J(f) = \begin{pmatrix} 0 & 1 & 0 & \dots & 0 \\ 0 & 0 & 1 & \dots & 0 \\ \dots & & & & \dots \\ 0 & 0 & 0 & \dots & 1 \\ a_1 & a_2 & a_3 & \dots & a_d \end{pmatrix},$$

and for each integer $m \geqslant 1$ let

$$J_m(f) = \begin{pmatrix} J(f) & 1_d & 0 & \dots & 0 \\ 0 & J(f) & 1_d & \dots & 0 \\ \dots & & & & \dots \\ 0 & 0 & 0 & \dots & J(f) \end{pmatrix}$$

with m diagonal blocks $J(f)$. Then the Jordan canonical form for elements of the conjugacy class c_μ is the diagonal sum of the matrices $J_{\mu_i(f)}(f)$ for all $i \geqslant 1$ and $f \in \Phi$.

2. Sometimes it is more convenient to regard $\mu: \Phi \to \mathscr{P}$ as a function on M, i.e. we define $\mu(x)$ for each $x \in M$ to be the partition $\mu(f)$, where f is the F-orbit of x (or the minimal polynomial of x over k). We have then

$$\|\mu\| = \sum_{f \in \Phi} d(f)\,|\mu(f)| = \sum_{x \in M} |\mu(x)|$$

and $\mu \circ F = \mu$. With this notation, the characteristic polynomial of any element $g \in c_\mu$ is

$$\prod_{f \in \Phi} f^{|\mu(f)|} = \prod_{x \in M} (t - x)^{|\mu(x)|}$$

so that in particular

$$\mathrm{trace}(g) = \sum_{x \in M} |\mu(x)|\, x,$$

$$\det(g) = \prod_{x \in M} x^{|\mu(x)|}.$$

3. The *support* $\mathrm{Supp}(\mu)$ of $\mu: \Phi \to \mathscr{P}$ is the set of $f \in \Phi$ such that $\mu(f) \neq 0$. Let $g \in c_\mu$. Then

(i) g is unipotent $\Leftrightarrow$ $\mathrm{Supp}(\mu)$ is the polynomial $t-1$.
(ii) g is primary $\Leftrightarrow$ $\mathrm{Supp}(\mu) = \{f\}$ for some $f \in \Phi$.
(iii) g is regular $\Leftrightarrow$ $\mu(f)$ has length $\leqslant 1$, for each $f \in \Phi$.
(iv) g is semisimple $\Leftrightarrow$ $\mu(f)'$ has length $\leqslant 1$, for each $f \in \Phi$.
(v) g is regular semisimple $\Leftrightarrow$ $\mu(f) = 0$ or (1) for each $f \in \Phi$.

4. Let k_n be the unique extension of k of degree n, and let $\alpha: k_n \to k^n$ be an isomorphism of k-vector spaces. For each $x \in M_n$, multiplication by x defines an invertible k-linear map $k_n \to k_n$, hence an element $\tau(x) \in G_n$: namely $\tau(x)v = \alpha(x(\alpha^{-1}v))$ for $v \in k^n$. The mapping $\tau: M_n \to G_n$ so defined is an injective homomorphism, so that $T_n = \tau(M_n)$ is a subgroup of G_n isomorphic to M_n. If we change the isomorphism α, we replace T_n by a conjugate subgroup. Each $\tau(x) \in T_n$ is semisimple, with eigenvalues $F^i x$ $(0 \leqslant i \leqslant n-1)$. Hence if f_x is the minimal polynomial of x, the conjugacy class of $\tau(x)$ in G_n is c_μ, where $\mu(f_x) = (1^{n/d})$ and $\mu(f) = 0$ for $f \neq f_x$, and $d = d(f_x)$.

Now let $\nu = (\nu_1, \ldots, \nu_r)$ be any partition of n. Then $T_\nu = T_{\nu_1} \times \ldots \times T_{\nu_r}$ is a subgroup of $G_{\nu_1} \times \ldots \times G_{\nu_r}$, hence of G_n, and is well-defined up to conjugacy in G_n. Let N_ν be the normalizer of T_ν, and $W_\nu = N_\nu / T_\nu$. Then W_ν is isomorphic to the centralizer of an element of cycle-type ν in the symmetric group S_n. In particular, W_n is cyclic of order n, and is obtained by transporting $\mathrm{Gal}(k_n/k)$ via the isomorphism α.

The groups T_ν (up to conjugacy in G_n) are the 'maximal tori' in G_n. Every semisimple element of G_n lies in at least one such group.

5. Let $W_\nu \backslash T_\nu$ be the set of orbits of W_ν in T_ν. Then the number of conjugacy classes in G_n is equal to

$$\sum_{|\nu| = n} |W_\nu \backslash T_\nu|.$$

We may identify T_ν with $M_{\nu_1} \times \ldots \times M_{\nu_r}$, and hence an element $t \in T_\nu$ with a sequence $(x_1, \ldots, x_r)$, where $x_i \in M_{\nu_i}$ $(1 \leqslant i \leqslant r)$. Given t, we define a function $\rho = \rho(t): \Phi \to \mathscr{P}$ as follows: for each $f \in \Phi$, the parts of $\rho(f)$ are the numbers $\nu_i / d(f)$ for all i such that x_i is a root of f. We have $\|\rho(t)\| = n$, and $\rho(t)$ depends only on the W_ν-orbit of t in T_ν. Conversely, given $\mu: \Phi \to \mathscr{P}$ such that $\|\mu\| = n$, let ν be the partition of n whose parts are the non-zero numbers $d(f)|\mu(f)|$, $f \in \Phi$. Then μ determines a unique W_ν-orbit in T_ν.

3. Induction from parabolic subgroups

Let $n = n_1 + \cdots + n_r$, where the n_i are positive integers, and let $V^{(i)}$ be the subspace of $V = k^n$ spanned by the first $n_1 + \ldots + n_i$ basis vectors. Let F denote the flag

$$0 = V^{(0)} \subset V^{(1)} \subset \ldots \subset V^{(r)} = V.$$

The parabolic subgroup P of elements $g \in G_n$ which fix F consists of all matrices of the form

$$g = \begin{pmatrix} g_{11} & g_{12} & \cdots & g_{1r} \\ 0 & g_{22} & \cdots & g_{2r} \\ \cdots & & & \cdots \\ 0 & 0 & \cdots & g_{rr} \end{pmatrix}$$

where $g_{ii} \in G_{n_i}$ $(1 \leqslant i \leqslant r)$.

Let u_i be a class-function on G_{n_i} $(1 \leqslant i \leqslant r)$. Then the function u defined by

$$u(g) = \prod_{i=1}^{r} u_i(g_{ii})$$

is a class function on P. Let $u_1 \circ u_2 \circ \ldots \circ u_r$ denote the class-function on G_n obtained by inducing u from P to G_n:

$$u_1 \circ \ldots \circ u_r = \operatorname{ind}_P^{G_n}(u).$$

If each u_i is a character of G_{n_i}, then $u_1 \circ \ldots \circ u_r$ is a character of G_n.

Let $G_n = \bigcup_i t_i P$ be a left coset decomposition for P in G_n. For any $x \in G_n$ we have

$$(u_1 \circ \ldots \circ u_r)(x) = \sum_i u(t_i^{-1} x t_i)$$

with the understanding that u vanishes outside P. Now $t^{-1}xt$ belongs to P if and only if it fixes the flag F, i.e. if and only if x fixes the flag tF, or equivalently tF is a flag of *submodules* of the $k[t]$-module V_x. If we put $tV^{(i)} = W^{(i)}$ and

$$t^{-1}xt = \begin{pmatrix} h_{11} & h_{12} & \cdots & h_{1r} \\ 0 & h_{22} & \cdots & h_{2r} \\ \cdots & & & \cdots \\ 0 & 0 & \cdots & h_{rr} \end{pmatrix}$$

then the module $W^{(i)}/W^{(i-1)}$ is isomorphic to $V_{h_{ii}}$ $(1 \leqslant i \leqslant r)$. Hence

(3.1) *Let u_i be a class function on G_{n_i} $(1 \leqslant i \leqslant r)$. Then the value of the class function $u_1 \circ \ldots \circ u_r$ at a class c_μ in G_n is given by*

$$(u_1 \circ \ldots \circ u_r)(c_\mu) = \sum g^{\mu}_{\mu^{(1)} \ldots \mu^{(r)}} u_1(c_{\mu^{(1)}}) \ldots u_r(c_{\mu^{(r)}})$$

summed over all sequences $(\mu^{(1)}, \ldots, \mu^{(r)})$, where $c_{\mu^{(i)}}$ is a conjugacy class in G_{n_i} $(1 \leqslant i \leqslant r)$, and $g^{\mu}_{\mu^{(1)} \ldots \mu^{(r)}}$ is the number of sequences

$$0 = W^{(0)} \subset W^{(1)} \subset \ldots \subset W^{(r)} = V_\mu$$

of submodules of V_μ *such that* $W^{(i)}/W^{(i-1)} \cong V_{\mu^{(i)}}$ $(1 \leqslant i \leqslant r)$.

In particular, let π_μ be the characteristic function of the class c_μ, which takes the value 1 at elements of c_μ, and 0 elsewhere. Then from (3.1) we have

(3.2)
$$\pi_{\mu^{(1)}} \circ \ldots \circ \pi_{\mu^{(r)}} = \sum_\mu g^\mu_{\mu^{(1)} \ldots \mu^{(r)}} \pi_\mu .$$

Also from (2.7) it follows that

(3.3)
$$g^\mu_{\mu^{(1)} \ldots \mu^{(r)}} = \prod_{f \in \Phi} G^{\mu(f)}_{\mu^{(1)}(f) \ldots \mu^{(r)}(f)}(k[t]_{(f)})$$

where the G's on the right have the same meaning as in Chapter II.

Now let A_n denote the space of complex-valued class functions on G_n: this is a finite-dimensional complex vector space having the characteristic functions π_μ such that $\|\mu\| = n$ as a basis. It carries a hermitian scalar product:

$$\langle u, v \rangle = \frac{1}{|G_n|} \sum_{g \in G_n} u(g)\overline{v(g)}.$$

Let

$$A = \bigoplus_{n \geqslant 0} A_n$$

(with the understanding that G_0 is a one-element group, so that $A_0 = \mathbf{C}$). The induction product $u \circ v$ defined above defines a multiplication on A, and we extend the scalar product to the whole of A by requiring that the components A_n be mutually orthogonal.

From (3.2), (3.3), and Chapter II, §2 it follows that

(3.4)
$$A = H \otimes_{\mathbf{Z}} \mathbf{C}$$

where H is the tensor product (over $\mathbf{Z}$) of the Hall algebras $H(k[t]_{(f)})$ for all $f \in \Phi$. In particular, A is a commutative and associative graded ring, with identity element the characteristic function π_0 of G_0.

Next let $R_n \subset A_n$ be the $\mathbf{Z}$-module generated by the *characters* of G_n. The irreducible characters of G_n form an orthonormal basis of R_n, and $A_n = R_n \otimes_{\mathbf{Z}} \mathbf{C}$. Let

$$R = \bigoplus_{n \geqslant 0} R_n;$$

since the induction product of characters is a character, R is a *subring* of A, and we have $A = R \otimes_{\mathbf{Z}} \mathbf{C}$.

4. The characteristic map

Let $X_{i,f}$ ($i \geqslant 1$, $f \in \Phi$) be independent variables over $\mathbf{C}$. If u is any symmetric function, we denote by $u(f)$ the symmetric function $u(X_f) = u(X_{1,f}, X_{2,f}, \ldots)$. For example, $e_n(f)$ is the nth elementary symmetric function of the variables $X_{i,f}$ ($i \geqslant 1$, f fixed), and $p_n(f) = \sum_i X_{i,f}^n$.

Let

$$B = \mathbf{C}[e_n(f) : n \geqslant 1, f \in \Phi]$$

be the polynomial algebra over $\mathbf{C}$ generated by the $e_n(f)$. We grade B by assigning degree $d(f)$ to each $X_{i,f}$, so that $e_n(f)$ is homogeneous of degree $n.d(f)$.

For each partition λ and each $f \in \Phi$ let

$$\tilde{P}_\lambda(f) = q_f^{-n(\lambda)} P_\lambda(X_f; q_f^{-1}),$$

$$\tilde{Q}_\lambda(f) = q_f^{|\lambda| + n(\lambda)} Q_\lambda(X_f; q_f^{-1})$$

$$= a_\lambda(q_f) \tilde{P}_\lambda(f),$$

where P_λ, Q_λ are the symmetric functions defined in Chapter III, and $q_f = q^{d(f)}$. For each $\mu : \Phi \to \mathscr{P}$ such that $\|\mu\| < \infty$ let

$$\tilde{P}_\mu = \prod_{f \in \Phi} \tilde{P}_{\mu(f)}(f), \qquad \tilde{Q}_\mu = \prod_{f \in \Phi} \tilde{Q}_{\mu(f)}(f) = a_\mu \tilde{P}_\mu$$

(almost all terms in these products are equal to 1, because $\|\mu\|$ is finite). Clearly $\tilde{P}_\mu$ and $\tilde{Q}_\mu$ are homogeneous elements of B, of degree $\|\mu\|$. By Chapter III, (2.7), $(\tilde{P}_\mu)$ and $(\tilde{Q}_\mu)$ are $\mathbf{C}$-bases of B. We define a hermitian scalar product on B by requiring that these should be dual bases, i.e.

$$\langle \tilde{P}_\mu, \tilde{Q}_\nu \rangle = \delta_{\mu\nu}$$

for all μ, ν.

Now let ch: $A \to B$ be the $\mathbf{C}$-linear mapping defined by

$$\mathrm{ch}(\pi_\mu) = \tilde{P}_\mu$$

where as before π_μ is the characteristic function of the conjugacy class c_μ. This map ch we call the *characteristic* map, and ch(u) is the *characteristic* of $u \in A$.

(4.1) *ch*: $A \to B$ *is an isometric isomorphism of graded* $\mathbf{C}$-*algebras.*

Proof. Since the π_μ form a basis of A and the $\tilde{P}_\mu$ a basis of B, ch is a linear isomorphism, which clearly respects the gradings. It follows from

(3.2) and Chapter III, (3.6) that ch is a ring homomorphism, and from (2.7) that it is an isometry. |

From Chapter III, (4.4) we have

$$\sum_{\lambda} q^{|\lambda|} P_\lambda(x; q^{-1}) Q_\lambda(y; q^{-1}) = \prod_{i,j} (1 - x_i y_i)/(1 - q x_i y_i).$$

If we replace x_i by $X_{i,f} \otimes 1$, y_j by $1 \otimes X_{j,f}$, and q by q_f, this identity takes the form

$$\sum_{\lambda} \tilde{P}_\lambda(f) \otimes \tilde{Q}_\lambda(f) = \prod_{i,j} (1 - X_{i,f} \otimes X_{j,f})/(1 - q_f X_{i,f} \otimes X_{j,f}).$$

Hence, by taking the product over all $f \in \Phi$ on either side, we obtain

$$\text{(4.2)} \qquad \sum_{\mu} \tilde{P}_\mu \otimes \tilde{Q}_\mu = \prod_{f \in \Phi} \prod_{i,j} (1 - X_{i,f} \otimes X_{j,f})/(1 - q_f X_{i,f} \otimes X_{j,f}).$$

Now take logarithms:

$$\log\left(\sum_{\mu} \tilde{P}_\mu \otimes \tilde{Q}_\mu\right) = \sum_{f \in \Phi} \sum_{i,j} \left(\log(1 - X_{i,f} \otimes X_{j,f}) - \log(1 - q_f X_{i,f} \otimes X_{j,f})\right)$$

$$= \sum_{n \geqslant 1} \frac{1}{n} \sum_{f \in \Phi} \sum_{i,j} (q_f^n - 1) X_{i,f}^n \otimes X_{j,f}^n$$

so that

$$\text{(4.3)} \qquad \log\left(\sum_{\mu} \tilde{P}_\mu \otimes \tilde{Q}_\mu\right) = \sum_{n \geqslant 1} \frac{1}{n} \sum_{f \in \Phi} (q_f^n - 1) p_n(f) \otimes p_n(f).$$

It is convenient at this stage to modify our notation for the power sums $p_n(f)$. Let $x \in M$ and let $f \in \Phi$ be the minimal polynomial for x over k (or, equivalently, the F-orbit of x (§1)). Define

$$\tilde{p}_n(x) = \begin{cases} p_{n/d}(f) & \text{if } n \text{ is a multiple of } d = d(f), \\ 0 & \text{otherwise,} \end{cases}$$

so that $\tilde{p}_n(x) \in B_n$, and $\tilde{p}_n(x) = 0$ unless $x \in M_n$. In this notation (4.3) takes the form

$$\text{(4.4)} \qquad \log\left(\sum_{\mu} \tilde{P}_\mu \otimes \tilde{Q}_\mu\right) = \sum_{n \geqslant 1} \frac{q^n - 1}{n} \sum_{x \in M_n} \tilde{p}(x) \otimes \tilde{p}_n(x).$$

Now define, for each $\xi \in L$,

$$\tilde{p}_n(\xi) = \begin{cases} (-1)^{n-1} \sum_{x \in M_n} \langle \xi, x \rangle_n \tilde{p}_n(x) & \text{if } \xi \in L_n, \\ 0 & \text{if } \xi \notin L_n. \end{cases} \tag{4.5}$$

Since $\tilde{p}_n(x) = \tilde{p}_n(y)$ if x, y are in the same F-orbit in M, it follows that $\tilde{p}_n(\xi) = \tilde{p}_n(\eta)$ if ξ, η are in the same F-orbit in L.

Let Θ denote the set of F-orbits in L. For $\varphi \in \Theta$ we denote by $d(\varphi)$ the number of elements of φ, and for each $r \geqslant 1$ we define

$$p_r(\varphi) = \tilde{p}_{rd}(\xi)$$

where ξ is any element of φ, and $d = d(\varphi)$. We regard the $p_r(\varphi)$ as the power-sums of a set of variables $Y_{i,\varphi}$, each of degree $d(\varphi)$. We can then define other symmetric functions of the $Y_{i,\varphi}$ by means of the formulas of Chapter I, in particular Schur functions $s_\lambda(\varphi)$ by the formula (7.5) of Chapter I.

For each function $\boldsymbol{\lambda}: \Theta \to \mathscr{P}$ such that

$$\|\boldsymbol{\lambda}\| = \sum_{\varphi \in \Theta} d(\varphi) |\boldsymbol{\lambda}(\varphi)| < \infty$$

let

$$S_{\boldsymbol{\lambda}} = \prod_{\varphi \in \Theta} s_{\boldsymbol{\lambda}(\varphi)}(\varphi)$$

which is a homogeneous element of B, of degree $\|\boldsymbol{\lambda}\|$.

Now let S be the $\mathbf{Z}$-submodule of B generated by the $S_{\boldsymbol{\lambda}}$. Since the Schur functions form a $\mathbf{Z}$-basis of the ring of symmetric functions, it is clear that S is closed under multiplication and that the $S_{\boldsymbol{\lambda}}$ form a $\mathbf{Z}$-basis of S. Hence S is a graded subring of B. The equations (4.5) can be solved for the $\tilde{p}_n(x)$ in terms of the $\tilde{p}_n(\xi)$, namely

$$\tilde{p}_n(x) = (-1)^{n-1}(q^n - 1)^{-1} \sum_{\xi \in L_n} \overline{\langle \xi, x \rangle_n} \tilde{p}_n(\xi)$$

for $x \in M_n$, by orthogonality of characters of the finite group M_n. Hence we have $S \otimes_{\mathbf{Z}} \mathbf{C} = B$.

The point of setting up this machinery is that the $S_{\boldsymbol{\lambda}}$ such that $\|\boldsymbol{\lambda}\| = n$ will turn out to be the characteristics of the irreducible characters of $G_n = GL_n(k)$.

To complete this section, we shall show that the $S_{\boldsymbol{\lambda}}$ form an orthonormal basis of S.

If u is any element of B, we may write $u = \sum u_\mu \tilde{P}_\mu$ with coefficients $u_\mu \in \mathbf{C}$, and we define $\bar{u}$ to be $\sum \bar{u}_\mu \tilde{P}_\mu$. Now from Chapter I, (4.3) we have

$$\log\Big(\sum_\lambda s_\lambda(x)s_\lambda(y)\Big) = \sum_{i,j} \log(1-x_iy_j)^{-1}$$

$$= \sum_{n\geq 1} \frac{1}{n} p_n(x)p_n(y)$$

for two sets of variables x_i, y_j, and therefore

$$\log\Big(\sum_\lambda S_\lambda \otimes \bar{S}_\lambda\Big) = \sum_{n\geq 1} \frac{1}{n} \sum_{\varphi\in\Theta} p_n(\varphi)\otimes\overline{p_n(\varphi)}$$

$$= \sum_{n\geq 1} \frac{1}{n} \sum_{\xi\in L_n} \tilde{p}_n(\xi)\otimes\overline{\tilde{p}_n(\xi)}.$$

But from (4.5) we have

$$\sum_{\xi\in L_n} \tilde{p}_n(\xi)\otimes\overline{\tilde{p}_n(\xi)} = \sum_{\xi\in L_n}\sum_{x,y\in M_n} \langle \xi,x\rangle_n \overline{\langle \xi,y\rangle_n}\tilde{p}_n(x)\otimes\tilde{p}_n(y)$$

$$= (q^n-1)\sum_{x\in M_n} \tilde{p}_n(x)\otimes\tilde{p}_n(x)$$

by orthogonality of characters of the finite group M_n. It now follows from (4.4) that

$$\sum_\lambda S_\lambda \otimes \bar{S}_\lambda = \sum_\mu \tilde{P}_\mu \otimes \tilde{Q}_\mu. \tag{4.6}$$

If $S_\lambda = \sum_\mu u_{\lambda\mu}\tilde{P}_\mu = \sum_\mu v_{\lambda\mu}\tilde{Q}_\mu$, it follows from (4.6) that $\sum_\lambda u_{\lambda\mu}\bar{v}_{\lambda\nu} = \delta_{\mu\nu}$, and therefore also

$$\sum_\mu u_{\kappa\mu}\bar{v}_{\lambda\mu} = \delta_{\kappa\lambda}.$$

In other words, $\langle S_\kappa, S_\lambda\rangle = \delta_{\kappa\lambda}$, so that

(4.7) *The S_λ form an orthonormal basis of S.* |

Example

Let u be any class function on G_n. Then the value of u at elements of the class c_μ is

$$u(c_\mu) = \langle \mathrm{ch}(u), \tilde{Q}_\mu\rangle.$$

For $u(c_\mu) = a_\mu\langle u, \pi_\mu\rangle = \langle \mathrm{ch}(u), \mathrm{ch}(a_\mu\pi_\mu)\rangle = \langle \mathrm{ch}(u), \tilde{Q}_\mu\rangle$ by (2.7) and (4.1).

5. Construction of the characters

Fix a character θ of M, the multiplicative group of $\bar{k}$.

(5.1) *Let G be a finite group and let $\rho: G \to GL_n(k)$ be a modular representation of G. For each $x \in G$ let $u_i(x)$ $(1 \leqslant i \leqslant n)$ be the eigenvalues of $\rho(x)$. Let $f \in \mathbf{Z}[t_1, \ldots, t_n]^{S_n}$ be a symmetric polynomial. Then the function*

$$\chi: x \mapsto f(\theta(u_1(x)), \ldots, \theta(u_n(x)))$$

is a character *of G, i.e. an integral linear combination of characters of complex representations of G.*

We defer the proof to the Appendix. We shall apply (5.1) in the case that $G = G_n$, ρ is the identity map and f is the rth elementary symmetric function e_r $(0 \leqslant r \leqslant n)$. We shall also assume that $\theta: M \to \mathbf{C}^*$ is injective, i.e. an isomorphism of M onto the group of roots of unity in $\mathbf{C}$ of order prime to $p = \text{char. } k$. Then (5.1) asserts that

$$\chi_r: g \mapsto e_r(\theta(x_1), \ldots, \theta(x_n)),$$

where $x_i \in \bar{k}$ are the eigenvalues of $g \in G_n$, is a (complex) character of G_n.

Now the characteristic polynomial of g is (§2, Example 2)

$$\prod_{f \in \Phi} f^{|\mu(f)|} = \prod_{i=1}^{n} (t - x_i)$$

where $\mu: \Phi \to \mathscr{P}$ is the partition-valued function determined by g.

For each $f \in \Phi$, say $f = \Pi(t - y_i)$, we shall write

$$\tilde{f} = \prod (1 + y_i t)$$

and

$$\theta(\tilde{f}) = \prod (1 + \theta(y_i) t).$$

Then since

$$\sum_{r=0}^{n} \chi_r(g) t^r = \prod_{i=1}^{n} (1 + \theta(x_i) t)$$

we have

$$\sum_{r=0}^{n} \chi_r(g) t^r = \prod_{f \in \Phi} \theta(\tilde{f})^{|\mu(f)|}. \tag{5.2}$$

We need to calculate the characteristic of χ_r. Since

$$\chi_r = \sum_{\mu} \chi_r(c_\mu) \pi_\mu$$

where $\chi_r(c_\mu)$ is the value of χ_r at any element of the class c_μ, it follows from (5.2) that

(5.3) $$\sum_{r=0}^{n} \operatorname{ch}(\chi_r)t^r = \sum_{\mu} \prod_{f\in\Phi} \theta(\tilde{f})^{|\mu(f)|} \tilde{P}_\mu$$

summed over all $\mu: \Phi \to \mathscr{P}$ such that $\|\mu\| = n$.

Now from Chapter III, §3, Example 1 we have

$$\sum_{|\lambda|=m} t^{n(\lambda)} P_\lambda(x;t) = h_m(x)$$

and therefore

(5.4) $$\sum_{|\lambda|=m} \tilde{P}_\lambda(f) = h_m(f)$$

for each $f \in \Phi$. It follows that the sum on the right-hand side of (5.3) is equal to

$$\sum_{\alpha} \prod_{f\in\Phi} \theta(\tilde{f})^{\alpha(f)} h_{\alpha(f)}(f)$$

summed over all $\alpha: \Phi \to \mathbf{N}$ such that $\Sigma_f\, d(f)\alpha(f) = n$, and hence is equal to the coefficient of u^n in

$$H(t,u) = \prod_{f\in\Phi} \prod_{i\geq 1} \left(1 - \theta(\tilde{f}) X_{i,f} u^{d(f)}\right)^{-1}.$$

Hence

(5.5) $\operatorname{ch}(\chi_r)$ *is equal to the coefficient of* $t^r u^n$ *in* $H(t,u)$.

To get $H(t,u)$ into a usable form, we shall calculate its logarithm:

$$\log H(t,u) = \sum_{f\in\Phi} \sum_{i\geq 1} \log\left(1 - \theta(\tilde{f}) X_{i,f} u^{d(f)}\right)^{-1}$$

$$= \sum_{m\geq 1} \frac{1}{m} \sum_{f\in\Phi} \theta(\tilde{f})^m p_m(f) u^{md(f)}.$$

Now if x is any root of f, we have

$$\theta(\tilde{f}) = \prod_{i=0}^{d-1} (1 + t\theta(F^i x))$$

where $d = d(f)$, and therefore (since $F^d x = x$)

$$\theta(\tilde{f})^m = \prod_{i=0}^{md-1} (1 + t\theta(F^i x)).$$

Consequently,

$$\log H(t,u) = \sum_{m \geq 1} \frac{u^m}{m} \sum_{x \in M_m} \tilde{p}_m(x) \prod_{i=0}^{m-1} (1 + t\theta(F^i x)).$$

Now F^i $(0 \leq i \leq m-1)$ acting on k_m are the elements of the Galois group $\mathfrak{g}_m = \mathrm{Gal}(k_m/k)$. Hence the inner sum above may be rewritten as

$$\begin{aligned}\sum_{x \in M_m} \tilde{p}_m(x) \prod_{\gamma \in \mathfrak{g}_m} (1 + t\theta(\gamma x)) &= \sum_{I \subset \mathfrak{g}_m} t^{|I|} \sum_{x \in M_m} \tilde{p}_m(x) \prod_{\gamma \in I} \theta(\gamma x) \\ &= (-1)^{m-1} \sum_{I \subset \mathfrak{g}_m} t^{|I|} \tilde{p}_m(\theta_{m,I})\end{aligned}$$

by (4.5), where $\theta_{m,I}$ is the character $x \mapsto \prod_{\gamma \in I} \theta(\gamma x)$ of M_m, identified with its image in L_m. So we have

$$\log H(t,u) = \sum_{m \geq 1} \frac{(-1)^{m-1} u^m}{m} \sum_{I \subset \mathfrak{g}_m} t^{|I|} \tilde{p}_m(\theta_{m,I}). \tag{5.6}$$

The group $\mathfrak{g}_m$ acts by multiplication on the set of all subsets I of $\mathfrak{g}_m$. If the stabilizer of I is trivial, we shall say that I is *primitive* and that the pair (m, I) is an *index*. In that case all the translates γI $(\gamma \in \mathfrak{g}_m)$ are distinct, and it follows that the F-orbit of $\theta_{m,I}$ consists of m elements. If I is not primitive, the stabilizer of I in $\mathfrak{g}_m$ is a non-trivial subgroup $\mathfrak{h}$ of $\mathfrak{g}_m$, generated by say F^s, where s divides m. Since I is stable under multiplication by elements of $\mathfrak{h}$, it is a union of cosets of $\mathfrak{h}$ in $\mathfrak{g}_m$, say $I = \mathfrak{h}J$ where $J = I/\mathfrak{h}$ may be regarded as a subset of $\mathfrak{g}_m/\mathfrak{h} = \mathfrak{g}_s = \mathrm{Gal}(k_s/k)$. Moreover, by construction J is a primitive subset of $\mathfrak{g}_s$, i.e. (s, J) is an index, and it is clear that $\theta_{m,I} = \theta_{s,J} \circ N_{m,s}$, where $N_{m,s}: M_m \to M_s$ is the norm map (§1), so that $\theta_{m,I}$ and $\theta_{s,J}$ define the same element of L. Since (s, J) is an index, the F-orbit $\varphi_{s,J}$ of $\theta_{s,J}$ in L has s elements, and therefore $\tilde{p}_m(\theta_{m,I}) = p_{m/s}(\varphi_{s,J})$.

Hence the formula (5.6) now becomes

$$\log H(t,u) = \sum_{(s,J)} \sum_{n=1}^{\infty} \frac{(-1)^{n-1}}{n} p_n(\varphi_{s,J})(t^{|J|}u^s)^n,$$

the sum being over all indices (s, J), where two indices (s, J) and (s, J') are to be regarded as the same if J' is a translate of J, i.e. if $\varphi_{s,J} = \varphi_{s,J'}$. Now the inner sum above is by Chapter I, (2.10′) equal to

$$\log \sum_{m \geqslant 0} e_m(\varphi_{s,J})(t^{|J|}u^s)^m$$

and therefore we have proved

(5.7) *For each pair of integers r, n such that $0 \leqslant r \leqslant n$, the coefficient of $t^r u^n$ in*

$$H(t,u) = \prod_{(s,J)} \left(\sum_{m \geqslant 0} e_m(\varphi_{s,J})(t^{|J|}u^s)^m \right)$$

is the characteristic of a character of G_n. |

We shall use (5.7) to prove

(5.8) *Let φ be an F-orbit in L, let $d = d(\varphi)$ and let $n \geqslant 0$. Then $e_n(\varphi)$ is the characteristic of a character of G_{nd}.*

Proof. By induction on the pair (n, d) ordered lexicographically: $(n,d) < (n',d')$ if either $n < n'$, or $n = n'$ and $d < d'$. The result is true when $n = d = 1$, because then $\varphi = \{\xi\}$ where $\xi \in L_1$ is a character of $G_1 = M_1$, and

$$e_1(\xi) = p_1(\xi) = \sum_{x \in G_1} \xi(x) p_1(x) = \mathrm{ch}(\xi)$$

(since $\mathrm{ch}(\pi_x) = p_1(x)$ for $x \in G_1$, π_x being the characteristic function of $\{x\}$).

Let $\chi_{n,d}$ denote the coefficient of $t^n u^{nd}$ in $H(t,u)$. By (5.7) $\chi_{n,d}$ is the characteristic of a character of G_{nd}, and we have

$$\chi_{n,d} = \sum_{\upsilon} \prod_{(s,J)} e_{\upsilon(s,J)}(\varphi_{s,J})$$

summed over all $\mathbf{N}$-valued functions υ on the set of indices such that

$$\sum_{(s,J)} |J| \upsilon(s,J) = n, \qquad \sum_{(s,J)} s\upsilon(s,J) = nd.$$

Each index (s,J) that actually occurs has $\upsilon(s,J) \leqslant n$, and $s \leqslant d$ if $\upsilon(s,J) = n$, so that $(\upsilon(s,J), s) \leqslant (n,d)$, with equality if and only if $\upsilon(s,J) = n$ (so that $|J| = 1$) and $s = d$. Hence $\chi_{n,d}$ has 'leading term' $e_n(\varphi_{d,\{1\}})$, where 1 is the identity element of $\mathfrak{g}_d$, and $\varphi_{d,\{1\}}$ is the F-orbit of $\theta \mid M_d$. Now we can choose θ so that the F-orbit of $\theta \mid M_d$ is φ, and then we have

$$\chi_{n,d} = e_n(\varphi) + \dots,$$

where the terms not written are products of two or more e's, which by the inductive hypothesis are characteristics of characters of groups $G_{n'd'}$, where $(n', d') < (n, d)$. Hence $e_n(\varphi)$ is the characteristic of a character of G_{nd}. ∎

6. The irreducible characters

We can now very quickly obtain all the irreducible characters of the groups $G_n = GL_n(k)$. Since each Schur function s_λ is a polynomial in the e's with integer coefficients, it follows from (5.8) that each $s_\lambda(\varphi)$ and hence each S_λ is the characteristic of a *character* χ^λ of G_n (where $n = \|\lambda\|$).

By (4.1) and (4.7) we have $\langle \chi^\kappa, \chi^\lambda \rangle = \delta_{\kappa\lambda}$, and therefore each χ^λ such that $\|\lambda\| = n$ is, up to sign, an irreducible character of G_n. Moreover, the rank of the free **Z**-module R_n generated by the irreducible characters of G_n is equal to the number of conjugacy classes in G_n, i.e. the number of functions $\mu: \Phi \to \mathscr{P}$ such that $\|\mu\| = n$. Since the groups L, M are isomorphic, this is also the number of functions $\lambda: \Theta \to \mathscr{P}$ such that $\|\lambda\| = n$, and consequently the χ^λ form an orthonormal *basis* of R_n. Hence the characters $\pm\chi^\lambda$ exhaust all the irreducible characters of G_n.

It remains to settle the question of sign, which we shall do by computing the degree $d_\lambda = \chi^\lambda(1_n)$ of χ^λ.

Let χ^λ_μ denote the value of χ^λ at elements of the class c_μ. Then

$$\chi^\lambda = \sum_\mu \chi^\lambda_\mu \pi_\mu$$

summed over all $\mu: \Phi \to \mathscr{P}$ such that $\|\mu\| = \|\lambda\|$, and therefore

$$S_\lambda = \sum_\mu \chi^\lambda_\mu \tilde{P}_\mu \tag{6.1}$$

which shows that $\chi^\lambda_\mu = \langle S_\lambda, \tilde{Q}_\mu \rangle$, and hence by (4.7) that

$$\tilde{Q}_\mu = \sum_\lambda \chi^\lambda_\mu \bar{S}_\lambda. \tag{6.2}$$

Now suppose that c_μ is the class of the identity element $1_n \in G_n$, and let f_1 denote the polynomial $t - 1$. Then $\mu(f_1) = (1^n)$ and $\mu(f) = 0$ for $f \neq f_1$. Hence

$$\begin{aligned}\tilde{Q}_\mu = \tilde{Q}_{(1^n)}(f_1) &= q^{n(n+1)/2}\varphi_n(q^{-1}) P_{(1^n)}\left(X_{f_1}; q^{-1}\right)\\ &= \psi_n(q) e_n(f_1)\end{aligned}$$

by Chapter III, (2.8), where

$$\psi_n(q) = \prod_{i=1}^{n} (q^i - 1).$$

Hence from (6.2) we have

$$(6.3) \qquad \psi_n(q) e_n(f_1) = \sum_{\|\lambda\| = n} d_\lambda \bar{S}_\lambda.$$

Now let $\delta: B \to \mathbf{C}$ be the $\mathbf{C}$-algebra homomorphism defined by

$$(6.4) \qquad \delta(\tilde{p}_n(x)) = \begin{cases} 0 & \text{if } x \neq 1, \\ (-1)^{n-1}/(q^n - 1) & \text{if } x = 1. \end{cases}$$

From (4.6) and (4.4) it follows that

$$\begin{aligned} \log\left(\sum_\lambda \delta(S_\lambda) \bar{S}_\lambda \right) &= \log\left(\sum_\mu \delta\left(\tilde{P}_\mu\right) \tilde{Q}_\mu \right) \\ &= \sum_{n \geq 1} \frac{(-1)^{n-1}}{n} \tilde{p}_n(1) \\ &= \log \prod_i (1 + X_{i,f_1}) \end{aligned}$$

and therefore

$$\sum_{n \geq 0} e_n(f_1) = \sum_\lambda \delta(S_\lambda) \bar{S}_\lambda,$$

so that

$$e_n(f_1) = \sum_{\|\lambda\| = n} \delta(S_\lambda) \bar{S}_\lambda.$$

By comparing this formula with (6.3), we see that

$$(6.5) \qquad d_\lambda = \psi_n(q) \delta(S_\lambda).$$

To calculate the number $\delta(S_\lambda)$, observe that from (6.4) and the definition (4.5) of $\tilde{p}_n(\xi)$ we have

$$\delta(\tilde{p}_n(\xi)) = (q^n - 1)^{-1}$$

for $\xi \in L_n$, and hence for all $\varphi \in \Theta$

$$\delta(p_n(\varphi)) = (q_\varphi^n - 1)^{-1} = \sum_{i \geq 1} q_\varphi^{-in},$$

where $q_\varphi = q^{d(\varphi)}$. In other words, $\delta(p_n(\varphi))$ is the nth power sum of the numbers q_φ^{-i} ($i \geqslant 1$), and hence from Chapter I, §3, Example 2 we have

$$\delta(s_\lambda(\varphi)) = s_\lambda\left(q_\varphi^{-1}, q_\varphi^{-2}, \dots\right)$$

$$= q_\varphi^{-|\lambda|-n(\lambda)} \prod_{x\in\lambda} \left(1 - q_\varphi^{-h(x)}\right)^{-1}.$$

where $h(x)$ is the hook-length at $x \in \lambda$. Since

$$\sum_{x\in\lambda} h(x) = |\lambda| + n(\lambda) + n(\lambda')$$

(Chapter I, §1, Example 2) it follows that

$$\delta(s_\lambda(\varphi)) = q_\varphi^{n(\lambda')}\tilde{H}_\lambda(q_\varphi)^{-1} \tag{6.6}$$

where

$$\tilde{H}_\lambda(q_\varphi) = \prod_{x\in\lambda}\left(q_\varphi^{h(x)} - 1\right).$$

From (6.5) and (6.6) we obtain

$$d_\lambda = \psi_n(q) \prod_{\varphi\in\Theta} q_\varphi^{n(\lambda(\varphi)')}\tilde{H}_{\lambda(\varphi)}(q_\varphi)^{-1} \tag{6.7}$$

which is clearly positive. Hence χ^λ (and not $-\chi^\lambda$) is an irreducible character of G_n. To summarize:

(6.8) *The functions* χ^λ *defined by* $\mathrm{ch}(\chi^\lambda) = S_\lambda$, *where* $\lambda: \Theta \to \mathscr{P}$ *and* $\|\lambda\| = n$, *are the irreducible characters of the group* $G_n = GL_n(k)$. *The degree* d_λ *of* χ^λ *is given by* (6.7), *and the character table of* G_n *is the transition matrix between the bases* (S_λ) *and* $(\tilde{P}_\mu)$.

Examples

1. Let $\nu = (\nu, \dots, \nu_r)$ be a partition of n, and let θ_ν be a character of the 'maximal torus' T_ν (§2, Example 4). If we identify T_ν with $M_{\nu_1} \times \dots \times M_{\nu_r}$, θ_ν may be identified with $(\xi_1, \dots, \xi_r)$, where $\xi_i \in L_{\nu_i}$ ($1 \leqslant i \leqslant r$). Then $\prod_{i=1}^r \tilde{p}_{\nu_i}(\xi_i)$ is the characteristic of an (in general reducible) character $B(\theta_\nu)$ of G_n, which depends only on the F-orbits φ_i of ξ_i, i.e. $B(\theta_\nu)$ depends only on the W_ν-orbit of θ_ν in the character group $\hat{T}_\nu$. The distinct characters $B(\theta)$, where $\theta \in \bigcup_\nu (W_\nu \backslash \hat{T}_\nu)$, are pairwise orthogonal in A_n; and since by §2, Example 5 there are exactly as many of them as there are conjugacy classes in G_n, they form an orthogonal *basis* of the space A_n of class functions on G_n. Green [G11] calls the $B(\theta)$ the *basic* characters. $B(\theta_\nu)$ is irreducible if and only if $d(\varphi_i) = \nu_i$ ($1 \leqslant i \leqslant r$), that is to say if and only if θ_ν is a *regular* character of T_ν (i.e. its stabilizer in W_ν is trivial).

The value of $B(\theta_\nu)$ at a unipotent element of type λ is (§4, Example 1) equal to $\langle \Pi\, \tilde{p}_{\nu_i}(\xi_i), \tilde{Q}_\lambda(f_1)\rangle$, where $f_1 = t-1$. By (4.5) this is equal to

$$(-1)^{\Sigma(\nu_i-1)}\langle p_\nu(f_1), \tilde{Q}_\lambda(f_1)\rangle = (-1)^{n-l(\nu)}Q_\nu^\lambda(q)$$

where $Q_\nu^\lambda(q)$ is the Green polynomial (Chapter III, §7).

Also $B(\theta_\nu)$ vanishes at regular semisimple classes which do not intersect the torus T_ν.

2. As in §2, Example 2 it is sometimes more convenient to regard $\boldsymbol{\lambda}: \Theta \to \mathscr{P}$ as a function on L, i.e. we define $\boldsymbol{\lambda}(\xi)$ for $\xi \in L$ to be the partition $\boldsymbol{\lambda}(\varphi)$, where φ is the F-orbit of ξ. With this notation let

$$\Delta(\boldsymbol{\lambda}) = \prod_{\xi \in L} \xi^{|\boldsymbol{\lambda}(\xi)|} \in L_1.$$

Let $a \in k^* = M_1$. Then

$$(*) \qquad \chi^{\boldsymbol{\lambda}}(a.1_n) = \langle \Delta(\boldsymbol{\lambda}), a\rangle_1\, \chi^{\boldsymbol{\lambda}}(1_n).$$

This can be shown by the method used in the text to compute $\chi^{\boldsymbol{\lambda}}(1_n)$, using the homomorphism $\delta_a: B \to \mathbf{C}$ defined by $\delta_a(\tilde{p}_n(x)) = (-1)^{n-1}/(q^n-1)$ if $x = a$, and $\delta_a(\tilde{p}_n(x)) = 0$ if $x \neq a$. Then $\chi^{\boldsymbol{\lambda}}(a.1_n) = \psi_n(q)\delta_a(S_{\boldsymbol{\lambda}})$, and δ_a has the effect of replacing the φ-variables $Y_{i,\varphi}$ by $\langle \xi, a\rangle_d . q_\varphi^{-i}$, where $\xi \in \varphi$ and $d = d(\varphi)$. The formula $(*)$ now follows easily.

3. Let U_n be the set of unipotent elements in G_n. Then for every irreducible character $\chi^{\boldsymbol{\lambda}}$ of G_n we have

$$\sum_{u \in U_n} \chi^{\boldsymbol{\lambda}}(u) = (-1)^{a(\boldsymbol{\lambda})} q^{N(\boldsymbol{\lambda})} \chi^{\boldsymbol{\lambda}}(1_n)$$

where

$$a(\boldsymbol{\lambda}) = n - \Sigma_\varphi |\boldsymbol{\lambda}(\varphi)| \text{ and } N(\boldsymbol{\lambda}) = \tfrac{1}{2}n(n-1) + \Sigma_\varphi\, d(\varphi)(n(\boldsymbol{\lambda}(\varphi)) - n(\boldsymbol{\lambda}(\varphi)')).$$

Let $e_{\boldsymbol{\lambda}} = \Sigma_{u \in U_n} \chi^{\boldsymbol{\lambda}}(u)$. Then we have

$$e_{\boldsymbol{\lambda}} = |G_n| \sum_{|\rho| = n} a_\rho(q)^{-1} \langle S_{\boldsymbol{\lambda}}, \tilde{Q}_\rho(f_1)\rangle$$

where $f_1 = t-1$, and hence from (5.4)

$$e_{\boldsymbol{\lambda}} = |G_n| \langle S_{\boldsymbol{\lambda}}, h_n(f_1)\rangle$$

so that

$$q^{n(n-1)/2}\psi_n(q) h_n(f_1) = \sum_{\|\boldsymbol{\lambda}\| = n} e_{\boldsymbol{\lambda}} \bar{S}_{\boldsymbol{\lambda}}.$$

Let $\varepsilon: B \to \mathbf{C}$ be the $\mathbf{C}$-algebra homomorphism defined by $\varepsilon(\tilde{p}_n(1)) = (q^n-1)^{-1}$, $\varepsilon(\tilde{p}_n(x)) = 0$ if $x \neq 1$. Then as in the text we find that

$$h_n(f_1) = \sum_{\|\boldsymbol{\lambda}\| = n} \varepsilon(S_{\boldsymbol{\lambda}}) \bar{S}_{\boldsymbol{\lambda}}$$

so that $e_\lambda = q^{n(n-1)/2}\psi_n(q)\varepsilon(S_\lambda)$, and $\varepsilon(S_\lambda)$ can be computed in the same way as $\delta(S_\lambda)$ in the text.

4. Let ψ be a non-trivial additive character of k, and χ^λ an irreducible character of G_n. We shall compute

$$w_\lambda = \chi^\lambda(1_n)^{-1} \sum_{g \in G_n} \psi(\text{trace } g)\chi^\lambda(g)$$

(see [K14], [M2]). For this purpose we introduce the following notation: if $x \in k_n$ let $\psi_n(x) = \psi(\text{trace}_{k_n/k}(x))$, and for $\xi \in L_n$ let

$$\tau_n(\xi) = (-1)^{n-1} \sum_{x \in M_n} \langle \xi, x \rangle_n \psi_n(x).$$

If $f \in \Phi$, let $\psi(f)$ denote $\psi_d(x)$, where $d = d(f)$ and x is a root of f. If $\varphi \in \Theta$, let $\tau(\varphi) = \tau_d(\xi)$, where $d = d(\varphi)$ and $\xi \in \varphi$.

If $g \in c_\mu$, we have $\text{trace}(g) = \sum_{x \in M} |\mu(x)|\, x$ (§2, Example 2) and therefore

$$\psi(\text{trace } g) = \prod_{f \in \Phi} \psi(f)^{|\mu(f)|} = \psi_\mu, \quad \text{say.}$$

We have

$$\chi^\lambda(1_n)w_\lambda = |G_n| \sum_{\|\mu\| = n} a_\mu^{-1} \chi_\mu^\lambda \psi_\mu$$

and since $\chi_\mu^\lambda = \langle S_\lambda, \tilde{Q}_\mu \rangle$ it follows that

$$\sum_\mu \psi_\mu \tilde{P}_\mu = |G_n|^{-1} \sum_\lambda \chi^\lambda(1_n) w_\lambda \bar{S}_\lambda. \tag{1}$$

Now define a C-algebra homomorphism $\varepsilon: B \to \mathbf{C}$ by $\varepsilon(X_{i,f}) = \psi(f)q_f^{-i}$, for all $f \in \Phi$ and $i \geqslant 1$. Then from Chapter III, §2, Example 1 it follows that $\varepsilon(\tilde{Q}_\mu) = \psi_\mu$, and hence from (1) and (4.6) we obtain

$$\varepsilon(S_\lambda) = |G_n|^{-1}\, \chi^\lambda(1_n) w_\lambda. \tag{2}$$

For each $x \in M_n$, we have $\varepsilon(\tilde{p}_n(x)) = \psi_n(x)/(q^n - 1)$, and therefore

$$\varepsilon(\tilde{p}_n(\xi)) = \frac{(-1)^{n-1}}{q^n - 1} \sum_{x \in M_n} \langle \xi, x \rangle_n \psi_n(x)$$

$$= \tau_n(\xi)/(q^n - 1)$$

for $\xi \in L_n$. Now the Hasse–Davenport identity (see e.g. [W1]) states that $\tau_{nd}(\xi) = \tau_n(\xi)^d$ for all $d \geqslant 1$, and hence we have $\varepsilon(p_n(\varphi)) = \tau(\varphi)^n/(q_\varphi^n - 1)$, where $q_\varphi = q^{d(\varphi)}$, so that the effect of the homomorphism ε is to replace the φ-variables $Y_{i,\varphi}$ by $\tau(\varphi)q_\varphi^{-i}$. Consequently

$$\varepsilon(S_\lambda) = \prod_{\varphi \in \Theta} \tau(\varphi)^{|\lambda(\varphi)|} . \delta(S_\lambda)$$

where δ is the homomorphism defined in (6.4). From (2), (3), and (6.5) we obtain

$$w_\lambda = q^{n(n-1)/2} \prod_{\varphi \in \Theta} \tau(\varphi)^{|\lambda(\varphi)|}.$$

5. In this example we shall compute the sum of the degrees of the irreducible representations of G_n. Since (6.5) $d_\lambda = \psi_n(q)\delta(S_\lambda)$, we need to compute $\Sigma_{\|\lambda\|=n}\, \delta(S_\lambda)$. For this purpose we shall compute the sum of the series

$$S = \sum_\lambda \delta(S_\lambda) t^{\|\lambda\|}$$

where t is an indeterminate, and then pick out the coefficient of t^n.

We have

$$S = \prod_\varphi \sum_\lambda \delta(s_\lambda(\varphi)) t^{|\lambda| d(\varphi)},$$

and since the homomorphism δ replaces the variables $Y_{i,\varphi}$ by q_φ^{-1} $(i \geqslant 1)$, it follows from Chapter I, §5, Example 4 that

$$\sum_\lambda \delta(s_\lambda(\varphi)) t^{|\lambda| d(\varphi)} = \prod_i \left(1 - (tq^{-i})^{d(\varphi)}\right)^{-1} \prod_{i<j} \left(1 - (t^2 q^{-i-j})^{d(\varphi)}\right)^{-1}$$

and hence that

$$\text{(1)} \quad \log S = \sum_\varphi \left(\sum_i \log\left(1 - (tq^{-i})^{d(\varphi)}\right)^{-1} + \sum_{i<j} \log\left(1 - (t^2 q^{-i-j})^{d(\varphi)}\right)^{-1} \right).$$

Let d_m denote the number of F-orbits φ with cardinality $d(\varphi) = m$, for each $m \geqslant 1$; then we have

$$\text{(2)} \qquad q^m - 1 = \sum_{r|m} r d_r$$

from considering the action of F on the group L_m. We can then rewrite (1) as

$$\begin{aligned} \log S &= \sum_{m \geqslant 1} d_m \left(\sum_{i \geqslant 1} \sum_{r \geqslant 1} \frac{(tq^{-i})^{mr}}{r} + \sum_{i<j} \sum_{r \geqslant 1} \frac{(t^2 q^{-i-j})^{mr}}{r} \right) \\ &= \sum_{m \geqslant 1} \sum_{r \geqslant 1} \frac{d_m}{r} \frac{t^{mr}}{q^{mr} - 1} \left(1 + \sum_{i \geqslant 1} t^{mr} q^{-2imr} \right) \end{aligned}$$

which by use of (2) is equal to

$$\sum_{N \geqslant 1} \left(\frac{t^N}{N} + \sum_{i \geqslant 1} \frac{t^{2N} q^{-2iN}}{N} \right).$$

It follows that

$$S = \frac{1}{1-t}\prod_{i\geq 1}\frac{1}{1-t^2q^{-2i}} = (1+t)\prod_{i\geq 0}\frac{1}{1-t^2q^{-2i}}$$

$$= (1+t)\sum_{m\geq 0} t^{2m}/\varphi_m(q^{-2})$$

(Chapter I, §2, Example 4). On picking out the coefficient of t^n, and multiplying by $\psi_n(q)$, we obtain

$$\sum_{\|\lambda\|=n} d_\lambda = (q-1)q^2(q^3-1)q^4(q^5-1)\ldots$$

(n factors altogether). This number is also equal to the number of symmetric matrices in G_n, or equivalently to the number of non-degenerate symmetric bilinear forms on k^n.

6. A calculation analogous to that of Example 5 shows that

$$\sum d_\mu = (q-1)q^2(q^3-1)q^4\ldots(q^{2n}-1)$$

where the sum on the left is over μ such that $\|\mu\| = 2n$ and the partition $\mu(\varphi)'$ is *even* for each φ. This number is equal to the number of non-degenerate skew-symmetric bilinear forms on k^{2n}. In fact we have

$$\sum \chi^\mu = \operatorname{ind}_{C_n}^{G_{2n}}(1)$$

summed over μ as above, where $C_n = Sp_{2n}(k)$ is the symplectic subgroup of $G_{2n} = GL_{2n}(k)$.

7. Let N_n denote the subgroup of upper unitriangular matrices in G_n. For each sequence $b = (b_1, \ldots, b_r)$ of positive integers with sum n, define a character ψ_b of N_n by the formula

$$\psi_b((h_{ij})) = \psi\left(\sum h_{i,i+1}\right)$$

where ψ is a fixed non-degenerate additive character of k, and the sum is over the indices i other than $b_1, b_1+b_2, \ldots, b_1+\ldots+b_{r-1}$. Then the multiplicity of each irreducible character χ^λ in the induced character $\operatorname{ind}_{N_n}^{G_n}(\psi_b)$ is equal to

$$\sum_a \prod_{\varphi\in\Theta} K_{\lambda(\varphi),a(\varphi)^+}$$

where the sum is over all functions $a: \Theta \to (\mathbf{Z}^+)^r$ such that $\sum_\varphi d(\varphi)a(\varphi) = b$, and $a(\varphi)^+$ denotes the partition obtained by rearranging the sequence $a(\varphi)$. (For a proof, see [Z2].) In particular, $\operatorname{ind}_{N_n}^{G_n}(\psi_{(n)}) = \sum \chi^\nu$, summed over ν such that for each φ the partition $\nu(\varphi)$ has length ≤ 1 (this result is due to Gelfand and Graev).

Notes and references

The result of Example 4 is due to T. Kondo [K14], and the results of Example 5 to A. A. Kljačko [K11] (also to R. Gow [G10] in the case of

Example 5(a) with q odd). Example 7 was contributed by A. Zelevinsky.

Appendix: proof of (5.1)

Since f is a polynomial in the elementary symmetric functions e_r, it is enough to prove the proposition for $f = e_r$ $(1 \leqslant r \leqslant n)$. Replacing ρ by its exterior powers, it follows that we may assume that $f = e_1 = t_1 + \ldots + t_n$.

(1) Suppose first that $g = |G|$ is prime to $p = \text{char}(k)$. Then q mod. g is a unit in $\mathbf{Z}/(g)$, hence g divides $q^r - 1$ where $r = \varphi(g)$ (Euler's function). Since $u_i(x)^g = 1$ for all $x \in G$ and $1 \leqslant i \leqslant n$, it follows that all the eigenvalues $u_i(x)$ lie in M_r, and hence only the restriction of θ to M_r comes into play. Since M_r is a cyclic group it is enough to prove the proposition for a prescribed generator θ_0 of $\hat{M}_r$, for then we have $\theta \mid M_r = \theta_0^s$ say, and we can apply the proposition (supposed proved for θ_0) to the symmetric function $f(t_1^s, \ldots, t_n^s)$.

Let $\mathfrak{o}$ be the ring of integers of the cyclotomic field generated by the $(q^r - 1)$th roots of unity in $\mathbf{C}$, and let $\mathfrak{p}$ be a prime ideal of $\mathfrak{o}$ which contains p. Then $\mathfrak{o}/\mathfrak{p} \cong k_r$, and ρ is the reduction mod $\mathfrak{p}$ of a representation σ of G with coefficients in $\mathfrak{o}$. Let U be the group of $(q^r - 1)$th roots of unity in $\mathfrak{o}$. Reduction mod $\mathfrak{p}$ defines an isomorphism of U onto M_r, and we take $\theta_0 \colon M_r \to U$ to be the inverse of this isomorphism. If the eigenvalues of $\sigma(x)$ are $v_i(x)$ $(1 \leqslant i \leqslant n)$, each $v_i(x)$ is a complex gth root of unity, hence lies in U, and its reduction mod $\mathfrak{p}$ is $u_i(x)$ (for a suitable numbering of the $v_i(x)$). Hence $\chi(x) = \Sigma\, \theta_0(u_i(x)) = \Sigma\, v_i(x)$ is the trace of $\sigma(x)$.

(2) Now let G be any finite group. By a famous theorem of Brauer (see for example Huppert, *Endliche Gruppen I*, p. 586), it is enough to prove the proposition in the case that G is an elementary group, i.e. the direct product of a cyclic group $\langle x \rangle$ and an l-group, where l is a prime which does not divide the order of x. Then G is the direct product of a p-group P and a group H of order prime to p. If $y \in P$, the eigenvalues of $\rho(y)$ are p-power roots of unity in $\bar{k}$, hence are all equal to 1, and since y commutes with each $x \in H$ the matrices $\rho(x)$ and $\rho(y)$ can be simultaneously put in triangular form; hence the eigenvalues of xy are the same as those of x, whence $\chi(xy) = \chi(x)$. But $\chi \mid H$ is a character of H, by the first part of the proof, hence χ is a character of G. |

Notes and references

Our account of the character theory of $GL_n(k)$ follows Green's paper [G11] in all essential points. For another account, oriented more towards the structure theory of reductive algebraic groups, see Springer [S20]. Another approach to the characters of $GL_n(k)$, based on the theory of Hopf algebras, is developed in [Z2] and [S21].

V

THE HECKE RING OF GL_n OVER A LOCAL FIELD

1. Local fields

Let F be a locally compact topological field. We shall assume that the topology of F is not the discrete topology, since any field whatsoever is locally compact when given the discrete topology. A non-discrete locally compact field is called a *local field*.

Every local field F carries a canonical absolute value, which can be defined as follows. Let dx be a Haar measure for the additive group of F. Then for $a \in F$, $a \neq 0$, the absolute value $|a|$ is defined by $d(ax) = |a|\, dx$. Equivalently, for any measurable set E in F, the measure of aE is $|a|$ times the measure of E. To complete the definition we set $|0| = 0$. Then it can be shown that $|a + b| \leqslant |a| + |b|$ for any $a, b \in F$, and that the distance function $d(a, b) = |a - b|$ determines the topology of F.

There are now two possibilities. Either the absolute value $|a|$ satisfies the axiom of Archimedes, in which case F is connected and can be shown to be isomorphic to either $\mathbf{R}$ or $\mathbf{C}$: these are the *archimedean* local fields. The other possibility is that the absolute value $|a|$ is non-archimedean, in which case F is totally disconnected (the only connected subsets of F are single points): these are the *non-archimedean* local fields.

The classification of the non-archimedean local fields can be simply described. If F has characteristic 0, then F is a finite algebraic extension of the field $\mathbf{Q}_p$ of p-adic numbers, for some prime p. If F has characteristic > 0, then F is a field of formal power series in one variable over a finite field.

From now on, let F be a non-archimedean local field. Let $\mathfrak{o} = \{a \in F : |a| \leqslant 1\}$ and $\mathfrak{p} = \{a \in F : |a| < 1\}$. Then $\mathfrak{o}$ and $\mathfrak{p}$ are compact open subsets of F; moreover $\mathfrak{o}$ is a *subring* of F, and $\mathfrak{p}$ is an ideal in $\mathfrak{o}$. The ring $\mathfrak{o}$ is the *ring of integers* of F; it is a complete discrete valuation ring, with F as field of fractions; $\mathfrak{p}$ is the maximal ideal of $\mathfrak{o}$, and $k = \mathfrak{o}/\mathfrak{p}$ is a *finite* field (because it is both compact and discrete). Let q denote the number of elements in k, and let π be a generator of $\mathfrak{p}$. We have $|\pi| = q^{-1}$, and hence the absolute value $|a|$ of any $a \neq 0$ in F is a power (positive or negative) of q. The *normalized valuation* $v: F^* \to \mathbf{Z}$ is defined by

$$v(x) = n \Leftrightarrow x = u\pi^n$$

with u a unit in $\mathfrak{o}$ (i.e. $|u| = 1$).

2. The Hecke ring $H(G, K)$

Let F be a non-archimedean local field, and let $G = GL_n(F)$ be the group of all invertible $n \times n$ matrices over F. Also let

$$G^+ = G \cap M_n(\mathfrak{o})$$

be the subsemigroup of G consisting of all matrices $x \in G$ with entries $x_{ij} \in \mathfrak{o}$, and let

$$K = GL_n(\mathfrak{o}) = G^+ \cap (G^+)^{-1}$$

so that K consists of all $x \in G$ with entries $x_{ij} \in \mathfrak{o}$ and $\det(x)$ a unit in $\mathfrak{o}$.

We may regard G as an open subset of matrix space $M_n(F) = F^{n^2}$, whence G inherits a topology for which it is a locally compact topological group. Since $\mathfrak{o}$ is compact and open in F, it follows that G^+ is compact and open in G, and that K is a compact open subgroup of G.

A Haar measure on G is given by the formula

$$dx = \left(\prod_{i,j} dx_{ij}\right) \Big/ |\det x|^n$$

and is both left- and right-invariant. In particular, we have $dx = dx'$, where x' is the transpose of x.

Let dx henceforth denote the unique Haar measure on G for which K has measure 1.

Next, let $L(G, K)$ (resp. $L(G^+, K)$) denote the space of all complex-valued continuous functions of compact support on G (resp. G^+) which are bi-invariant with respect to K, i.e. such that

$$f(k_1 x k_2) = f(x)$$

for all $x \in G$ (resp. G^+) and k_1, $k_2 \in K$. We may and shall regard $L(G^+, K)$ as a subspace of $L(G, K)$.

We define a multiplication on $L(G, K)$ as follows: for f, $g \in L(G, K)$,

$$(f * g)(x) = \int_G f(xy^{-1}) g(y)\, dy.$$

(Since f and g are compactly supported, the integration is over a compact set.) This product is associative, and we shall see in a moment that it is commutative. Since G^+ is closed under multiplication it follows immediately from the definition that $L(G^+, K)$ is a subring of $L(G, K)$.

Each function $f \in L(G,K)$ is constant on each double coset KxK in G. These double cosets are compact, open and mutually disjoint. Since f has compact support, it follows that f takes non-zero values on only finitely many double cosets KxK, and hence can be written as a finite linear combination of their characteristic functions. Hence the characteristic functions of the double cosets of K in G form a **C**-*basis* of $L(G,K)$. The characteristic function of K is the identity element of $L(G,K)$.

If we vary the definition of the algebra $L(G,K)$ (resp. $L(G^+,K)$) by requiring the functions to take their values in **Z** instead of **C**, the resulting ring is called the *Hecke ring* of G (resp. G^+), and we denote it by $H(G,K)$ (resp. $H(G^+,K)$). Clearly we have

(2.1) $L(G,K) \cong H(G,K) \otimes_{\mathbf{Z}} \mathbf{C}, \qquad L(G^+,K) \cong H(G^+,K) \otimes_{\mathbf{Z}} \mathbf{C}.$

Our first aim is to show that the Hecke ring $H(G,K)$ is closely related to the Hall algebra $H(\mathfrak{o})$ of the discrete valuation ring $\mathfrak{o}$.

Consider a double coset KxK, where $x \in G$. By multiplying x by a suitable power of π (the generator of $\mathfrak{p}$) we can bring x into G^+. The theory of elementary divisors for matrices over a principal ideal domain now shows that by pre- and post-multiplying x by suitable elements of K we can reduce x to a diagonal matrix. Multiplying further by a diagonal matrix belonging to K will produce a diagonal matrix whose entries are powers of π, and finally conjugation by a permutation matrix will get the exponents in descending order. Hence

(2.2) *Each double coset KxK has a unique representative of the form*

$$\pi^\lambda = \operatorname{diag}(\pi^{\lambda_1}, \dots, \pi^{\lambda_n})$$

where $\lambda_1 \geqslant \lambda_2 \geqslant \dots \geqslant \lambda_n$. We have $\lambda_n \geqslant 0$ (so that λ is a partition) if and only if $x \in G^+$. |

Let c_λ denote the characteristic function of the double coset $K\pi^\lambda K$. Then from (2.2) we have

(2.3) *The c_λ* (resp. *the c_λ such that $\lambda_n \geqslant 0$*) *form a* **Z**-*basis of $H(G,K)$* (resp. *$H(G^+,K)$*). *The characteristic function c_0 of K is the identity element of $H(G,K)$ and $H(G^+,K)$.* |

We can now prove that $H(G,K)$ (and hence also $H(G^+,K)$, $L(G,K)$ and $L(G^+,K)$) is commutative:

(2.4) *Let f, $g \in H(G,K)$. Then $f * g = g * f$.*

Proof. Let $t: G \to G$ be the transposition map: $t(x) = x'$. Clearly K is stable under t, and since by (2.2) each double coset KxK contains a diagonal matrix, it follows that KxK is stable under t. Hence $f \circ t = f$ for all $f \in H(G, K)$, and therefore

$$(f * g)(x) = \int_G f(xy) g(y^{-1})\, dy$$

$$= \int_G f(y^t x^t) g\left((y^t)^{-1}\right) dy^t$$

$$= (g * f)(x^t) = (g * f)(x). \quad |$$

(2.5) $$H(G, K) = H(G^+, K)\left[c_{(1^n)}^{-1}\right].$$

Proof. Since $\pi^{(1^n)} = \pi 1_n$ is in the centre of G, an easy calculation shows that, for any λ and any integer r, we have

$$c_\lambda * c_{(r^n)} = c_{\lambda + (r^n)}$$

where $\lambda + (r^n)$ is the sequence $(\lambda_1 + r, \dots, \lambda_n + r)$. In particular it follows that $c_{(1^n)}$, the characteristic function of $K\pi^{(1^n)}K = \pi K$, is a unit in $H(G, K)$ and that its rth power is $c_{(r^n)}$, for all $r \in \mathbf{Z}$. (2.5) now follows directly. |

In view of (2.5), we may concentrate our attention on $H(G^+, K)$, which by (2.3) has a $\mathbf{Z}$-basis consisting of the characteristic functions c_λ, where λ runs through all *partitions* $(\lambda_1, \dots, \lambda_n)$ of length $\leqslant n$.

Let μ, ν be partitions of length $\leqslant n$. The product $c_\mu * c_\nu$ will be a linear combination of the c_λ. In fact

(2.6) $$c_\mu * c_\nu = \sum_\lambda g^\lambda_{\mu\nu}(q) c_\lambda$$

summed over all partitions λ of length $\leqslant n$, where $g^\lambda_{\mu\nu}(q)$ is the 'Hall polynomial' defined in Chapter II, §2.

Proof. Let $h^\lambda_{\mu\nu}$ denote the coefficient of c_λ in $c_\mu * c_\nu$. Then

$$h^\lambda_{\mu\nu} = (c_\mu * c_\nu)(\pi^\lambda) = \int_G c_\mu(\pi^\lambda y^{-1}) c_\nu(y)\, dy.$$

Since $c_\nu(y)$ vanishes for y outside $K\pi^\nu K$, the integration is over this double coset, which we shall write as a union of right cosets, say

$$K\pi^\nu K = \bigcup_j K y_j \qquad (y_j \in \pi^\nu K)$$

Likewise let

$$K\pi^\mu K = \bigcup_i K x_i, \qquad (x_i \in \pi^\mu K)$$

both unions being disjoint. Then we have

$$h^{\lambda}_{\mu\nu} = \sum_j \int_{Ky_j} c_{\mu}(\pi^{\lambda} y^{-1})\, dy$$

$$= \sum_j \int_K c_{\mu}(\pi^{\lambda} y_j^{-1} k)\, dk$$

$$= \sum_j c_{\mu}(\pi^{\lambda} y_j^{-1})$$

since K has measure 1.

Now

$$c_{\mu}(\pi^{\lambda} y_j^{-1}) \neq 0 \Leftrightarrow \pi^{\lambda} y_j^{-1} \in Kx_i \quad \text{for some } i,$$

$$\Leftrightarrow \pi^{\lambda} \in Kx_i y_i \quad \text{for some } i.$$

Hence $h^{\lambda}_{\mu\nu}$ is equal to the number of pairs (i, j) such that

$$\pi^{\lambda} = kx_i y_j$$

for some $k \in K$ (depending on i, j).

Let L denote the lattice $\mathfrak{o}^n$ in the vector space F^n. Let $\xi_1, \ldots, \xi_n$ be the standard basis of F^n, and let G act on F^n on the *right* (i.e. we think of the elements of F^n as row-vectors rather than column-vectors). Then $L\pi^{\lambda}$ is the sublattice of L generated by the vectors π^{λ_i} $(1 \leq i \leq n)$, and therefore $M = L/L\pi^{\lambda}$ is a finite $\mathfrak{o}$-module of type λ (Chapter II, §1).

Consider Lx_i. Since $x_i \in \pi^{\mu}K$ we have $Lx_i = L\pi^{\mu}k$ for some $k \in K$, hence $L/Lx_i \cong L/L\pi^{\mu}$, and is therefore a finite $\mathfrak{o}$-module of type μ. Next, consider $N = Lx_i/L\pi^{\lambda} = Lx_i/Lx_i y_j$. Since $y_j \in \pi^{\nu}K$ it follows as before that N is of type ν. Hence for each pair (i, j) such that $\pi^{\lambda} \in Kx_i y_j$ we have a submodule N of M of cotype μ and type ν. Conversely, each submodule N of M with cotype μ and type ν determines a pair (i, j) such that $\pi^{\lambda} \in Kx_i y_j$. Hence $h^{\lambda}_{\mu\nu} = g^{\lambda}_{\mu\nu}(q)$. |

From (2.6) it follows that the mapping $u_{\lambda} \mapsto c_{\lambda}$ is a homomorphism of the Hall algebra $H(\mathfrak{o})$ onto $H(G^+, K)$ whose kernel is generated by the u_{λ} such that $l(\lambda) > n$. Hence from Chapter III, (3.4) we obtain a structure theorem for $H(G^+, K)$ and $L(G^+, K)$:

(2.7) *Let* $\Lambda_n[q^{-1}]$ (resp. $\Lambda_{n,\mathbf{C}}$) *denote the ring of symmetric polynomials in* n *variables* $x_1, \ldots, x_n$ *with coefficients in* $\mathbf{Z}[q^{-1}]$ (resp. $\mathbf{C}$). *Then the* $\mathbf{Z}$*-linear mapping* θ *of* $H(G^+, K)$ *into* $\Lambda_n[q^{-1}]$ (resp. *the* $\mathbf{C}$*-linear mapping of* $L(G^+, K)$ *into* $\Lambda_{n,\mathbf{C}}$) *defined by*

$$\theta(c_{\lambda}) = q^{-n(\lambda)} P_{\lambda}(x_1, \ldots, x_n; q^{-1})$$

for all partitions λ of length $\leqslant n$, is an injective ring homomorphism (resp. *an isomorphism of* C*-algebras*). |

Remark. From (2.7) and Chapter III, (2.8) we have

$$\theta(c_{(1^n)}) = q^{-n(n-1)/2}x_1x_2\dots x_n.$$

Hence from (2.5) it follows that θ extends to an injective ring homomorphism of $H(G,K)$ into the algebra of symmetric polynomials in $x_1^{\pm 1},\dots,x_n^{\pm 1}$ with coefficients in $\mathbf{Z}[q^{-1}]$, and to an isomorphism of $L(G,K)$ onto the algebra of symmetric polynomials in $x_1^{\pm 1},\dots,x_n^{\pm 1}$ with coefficients in $\mathbf{C}$.

In the next section we shall need to know the measure of a double coset $K\pi^\lambda K$. This may be computed as follows. For $f \in L(G^+, K)$, let

$$\mu(f) = \int_G f(x)\,dx.$$

Then $\mu: L(G^+, K) \to \mathbf{C}$ is a $\mathbf{C}$-algebra homomorphism, and clearly

$$\mu(c_\lambda) = \text{measure of } K\pi^\lambda K. \tag{2.8}$$

In view of (2.7) we may write $\mu = \mu' \circ \theta$, where $\mu': \Lambda_{n,\mathbf{C}} \to \mathbf{C}$ is a $\mathbf{C}$-algebra homomorphism, hence is determined by its effect on the generators e_r $(1 \leqslant r \leqslant n)$. Since $e_r = P_{(1^r)}(x_1,\dots,x_n;q^{-1})$ (Chapter III, (2.8)) it is enough to compute the measure of $K\pi^\lambda K$ when $\lambda = (1^r)$ $(1 \leqslant r \leqslant n)$.

Now the measure of $K\pi^\lambda K$ is equal to the number of cosets Kx_i contained in $K\pi^\lambda K$. For each of these cosets we have a sublattice Lx_i of $L = \mathfrak{o}^n$ such that L/Lx_i is a finite $\mathfrak{o}$-module of type λ. Hence the measure of $K\pi^\lambda K$ is equal to the number of sublattices L' of L such that L/L' has type λ. In particular, if $\lambda = (1^r)$, each such L' will be such that $\pi(L/L') = 0$, i.e. $\pi L \subset L'$. Now $L/\pi L \cong k^n$ is an n-dimensional vector space over the finite field k, and $L'/\pi L$ is a vector subspace of codimension r. The number of such subspaces is equal to the Gaussian polynomial $\begin{bmatrix} n \\ r \end{bmatrix}(q)$ (Chapter I, §2, Example 3) and therefore

$$\mu(c_{(1^r)}) = \begin{bmatrix} n \\ r \end{bmatrix}(q) \qquad (1 \leqslant r \leqslant n).$$

From (2.7) it follows that $\mu'(e_r) = q^{r(r-1)/2}\begin{bmatrix} n \\ r \end{bmatrix}(q)$, which (*loc. cit.*) is the rth elementary symmetric function of $q^{n-1}, q^{n-2},\dots,1$. Hence μ' is the mapping which takes x_i to $q^{n-i}(1 \leqslant i \leqslant n)$. It follows therefore from (2.7) and (2.8) that the measure of $K\pi^\lambda K$ is $q^{-n(\lambda)}P_\lambda(q^{n-1}, q^{n-2},\dots,1;q^{-1})$, which by Chapter III, §2, Example 1 is equal to

$$(q^{-n(\lambda)}/v_\lambda(q^{-1}))q^{\Sigma(n-i)\lambda_i}v_n(q^{-1}).$$

Hence we have the formula

$$\text{(2.9)} \qquad \text{measure of } K\pi^\lambda K = q^{\Sigma(n-2i+1)\lambda_i} v_n(q^{-1})/v_\lambda(q^{-1}),$$

$$= q^{2\langle\lambda,\rho\rangle} v_n(q^{-1})/v_\lambda(q^{-1})$$

where $\rho = \frac{1}{2}(n-1, n-3, \ldots, 1-n)$.

3. Spherical functions

A *spherical function* on G relative to K is a complex-valued continuous function ω on G which satisfies the following conditions:
(a) ω is bi-invariant with respect to K, i.e. $\omega(k_1 x k_2) = \omega(x)$ for $x \in G$ and $k_1, k_2 \in K$;
(b) $\omega * f = \lambda_f \omega$ for each $f \in L(G, K)$, where λ_f is a complex number: in otherwords, ω is an eigenfunction of all the convolution operators defined by elements of $L(G, K)$;
(c) $\omega(1) = 1$.

From (b) and (c) the scalar λ_f is given by

$$\lambda_f = (\omega * f)(1) = \int_G f(x)\omega(x^{-1})\,dx.$$

Regarded as a function of ω, λ_f is called the *Fourier transform* of f and is written $\hat{f}(\omega)$. Regarded as a function of f for fixed ω, it is written $\hat{\omega}(f)$.
For $f, g \in L(G, K)$ we have from (b)

$$\lambda_{f*g}\omega = \omega * (f * g) = (\omega * f) * g = \lambda_f \lambda_g \omega$$

and hence $\lambda_{f*g} = \lambda_f \cdot \lambda_g$, or equivalently

$$\text{(3.1)} \qquad \hat{\omega}(f * g) = \hat{\omega}(f)\hat{\omega}(g).$$

Also it is clear that $\omega * c_0 = \omega$, where c_0 is the identity element of $L(G, K)$ (the characteristic function of K). Hence $\hat{\omega}(c_0) = 1$ and therefore (3.1) shows that

(3.2) $\hat{\omega}: L(G, K) \to \mathbf{C}$ *is a* $\mathbf{C}$*-algebra homomorphism.*

Conversely, it can be shown that all the $\mathbf{C}$-algebra homomorphisms $L(G, K) \to \mathbf{C}$ arise in this way from spherical functions, so that the set $\Omega(G, K)$ of spherical functions on G relative to K may be identified with the complex affine algebraic variety whose coordinate ring is $L(G, K)$. From the remark following (2.7), $L(G, K)$ is isomorphic to the $\mathbf{C}$-algebra $\mathbf{C}[x_1^{\pm 1}, \ldots, x_n^{\pm 1}]^{S_n}$; now $\mathbf{C}[x_1^{\pm 1}, \ldots, x_n^{\pm 1}]$ is the coordinate ring of $(\mathbf{C}^*)^n$, and therefore $\Omega(G, K)$ may be canonically identified with the nth symmetric product of the punctured affine line $\mathbf{C}^*$. Hence the spherical functions may be parametrized by n-tuples $z=(z_1,\dots,z_n)$ of non-zero complex numbers, the order of the components z_i being immaterial. However, it will be more convenient to take as parameter $s=(s_1,\dots,s_n)$, where $z_i=q^{\frac{1}{2}(n-1)-s_i}$ (so that $s_i \in \mathbf{C}$ (mod. $2\pi i\mathbf{Z}/\log q$).)

Let ω_s denote the spherical function with parameter s. We have $\omega_s=\omega_{ws}$ for all $w \in S_n$.

(3.3) *The Fourier transform of the characteristic function* c_λ *is given by*

$$\hat{c}_\lambda(\omega_s)=q^{\langle\lambda,\rho\rangle}P_\lambda(q^{-s_1},\dots,q^{-s_n};q^{-1}).$$

Proof. From the discussion above and (2.7) it is clear that $\hat{\omega}_s$ is the composition of θ with the specialization $x_i \mapsto z_i=q^{\frac{1}{2}(n-1)-s_i}$ $(1\leqslant i\leqslant n)$. The result therefore follows from (2.7). |

Now we have

$$\hat{c}_\lambda(\omega_s)=\int_G c_\lambda(x)\omega_s(x^{-1})\,dx$$

$$=\omega_s(\pi^{-\lambda})\times \text{measure of } K\pi^\lambda K.$$

From (3.3) and (2.9) it follows that

$$\omega_s(\pi^{-\lambda})=\frac{q^{-\langle\lambda,\rho\rangle}}{v_n(q^{-1})}v_\lambda(q^{-1})P_\lambda(q^{-s_1},\dots,q^{-s_n};q^{-1})$$

and therefore, from the definition (Chapter III, §2) of the symmetric functions P_λ, we have (always under the assumption that $\lambda_1\geqslant\dots\geqslant\lambda_n$)

$$\omega_s(\pi^{-\lambda})=\frac{q^{-\langle\lambda,\rho\rangle}}{v_n(q^{-1})}\sum_{w\in S_n} w\left(q^{-\langle\lambda,s\rangle}\prod_{i<j}\frac{q^{-s_i}-q^{-(s_j+1)}}{q^{-s_i}-q^{-s_j}}\right). \tag{3.4}$$

Since ω_s is by definition constant on each double coset of K in G, the formula (3.4) gives the value of the spherical function ω_s at each element of G.

In particular, when $s=\rho$, we have $\omega_\rho(\pi^{-\lambda})=1$ for all λ, so that ω_ρ is the constant function 1 (the trivial spherical function).

From (3.4) it follows easily that

$$\omega_s(x^{-1})=\omega_{-s}(x) \tag{3.5}$$

for all $x\in G$.

Example

ω_s is bounded $\Leftrightarrow$ Re(s) lies in the convex hull of the set $S_n\rho=\{w\rho; w\in S_n\}$. Let $\sigma=\text{Re}(s)$, i.e. $\sigma_i=\text{Re}(s_i)$ $(1\leqslant i\leqslant n)$. We may assume, since $\omega_{ws}=\omega_s$ for all

$w \in S_n$, that $\sigma_1 \geqslant \ldots \geqslant \sigma_n$. Then the leading term in (3.4) is $q^{-\langle \lambda, s+\rho \rangle}$, with absolute value $q^{-\langle \lambda, \sigma+\rho \rangle}$. Hence ω_s is bounded if and only if $q^{-\langle \lambda, \sigma+\rho \rangle} \leqslant 1$ for all λ such that $\lambda_1 \geqslant \ldots \geqslant \lambda_n$, that is to say if and only if $\langle \lambda, \sigma+\rho \rangle \geqslant 0$ for all such λ, and it is easy to see that this condition is equivalent to $\sigma \in \mathrm{Conv}\,(S_n\, \rho)$.

4. Hecke series and zeta functions for $GL_n(F)$

For each integer $m \geqslant 0$ let

$$G_m^+ = \{x \in G^+ : v(\det x) = m\}$$

where v is the normalized valuation on F (§1). The set G_m^+ is the disjoint union of the double cosets $K\pi^\lambda K$, where λ runs through all partitions of m of length $\leqslant n$. In particular, $G_0^+ = K$. Let τ_m denote the characteristic function of G_m^+, so that

$$\tau_m = \sum_{|\lambda| = m} c_\lambda.$$

The *Hecke series* of (G, K) is by definition the formal power series

$$\tau(X) = \sum_{m=0}^{\infty} \tau_m X^m \in H(G^+, K)[[X]]. \tag{4.1}$$

The results we have obtained enable us to calculate $\tau(X)$ with no difficulty. For by (2.7) we have

$$\begin{aligned}\theta(\tau_m) &= \sum_{|\lambda| = m} q^{-n(\lambda)} P_\lambda(x_1, \ldots, x_n; q^{-1}) \\ &= h_m(x_1, \ldots, x_m)\end{aligned}$$

by Chapter III, §3, Example 1. Hence

$$\sum_{m=0}^{\infty} \theta(\tau_m) X^m = \prod_{i=1}^{n} (1 - x_i X)^{-1}. \tag{4.2}$$

Since

$$\begin{aligned}\prod_{i=1}^{n} (1 - x_i X) &= \sum_{r=0}^{n} (-1)^r e_r X^r \\ &= \sum_{r=0}^{n} (-1)^r P_{(1^r)}(x_1, \ldots, x_n; q^{-1}) X^r \\ &= \sum_{r=0}^{n} (-1)^r q^{r(r-1)/2} \theta(c_{(1^r)}) X^r\end{aligned}$$

by (2.7) again, it follows from (4.2) that

$$\tau(X) = \left(\sum_{r=0}^{n} (-1)^r q^{r(r-1)/2} c_{(1^r)} X^r \right)^{-1}, \tag{4.3}$$

a formula first obtained by Tamagawa [T2].

Now let ω be a spherical function on G relative to K, and let s be a complex variable. The *zeta function* $\zeta(s, \omega)$ is defined by

$$\zeta(s, \omega) = \int_G \varphi(x) \|x\|^s \omega(x^{-1})\, dx \tag{4.4}$$

where as usual dx is Haar measure on G, normalized so that $\int_K dx = 1$; φ is the characteristic function of G^+, and $\|x\| = |\det(x)| = q^{-v(\det x)}$. For the moment we shall ignore questions of convergence and treat s (or rather q^{-s}) as an indeterminate. We have

$$\begin{aligned}\zeta(s, \omega) &= \sum_{m=0}^{\infty} q^{-ms} \int_G \tau_m(x)\, \omega(x^{-1})\, dx \\ &= \sum_{m=0}^{\infty} \hat{\omega}(\tau_m) q^{-ms} \\ &= \left(\sum_{r=0}^{n} (-1)^r q^{r(r-1)/2} \hat{\omega}(c_{(1^r)}) q^{-rs} \right)^{-1}\end{aligned}$$

by (4.3).

If $\omega = \omega_s$, it follows from (4.2) that

$$\zeta(s, \omega_s) = \prod_{i=1}^{n} (1 - q^{\frac{1}{2}(n-1) - s_i - s})^{-1}. \tag{4.5}$$

These formal calculations will be valid, and the integral (4.4) will converge, provided that $|q^{\frac{1}{2}(n-1) - s_i - s}| < 1$ for $1 \leqslant i \leqslant n$, i.e. provided that s lies in the half-plane $\mathrm{Re}(s) > \sigma_0$, where

$$\sigma_0 = \max_{1 \leqslant i \leqslant n} \left(\frac{n-1}{2} - \mathrm{Re}(s_i) \right).$$

In this half-plane the integral (4.4) defines a meromorphic function of s, given by (4.5), which is the analytic continuation of $\zeta(s, \omega)$ to the whole s-plane.

In particular, for the trivial spherical function $\omega_\rho = 1$ we have $s_i = \frac{1}{2}(n+1) - i$, and therefore

$$\zeta(s, 1) = \prod_{i=1}^{n} (1 - q^{i-1-s})^{-1}. \tag{4.6}$$

5. Hecke series and zeta functions for $GSp_{2n}(F)$

In this section we shall show how knowledge of the spherical functions for $GL_n(F)$ enables us to compute explicitly the Hecke series for the group of symplectic similitudes $GSp_{2n}(F)$, thereby completing a calculation started by Satake in [S3], to which we refer for proofs.

Let i denote the $n \times n$ matrix with 1's along the reverse diagonal and zeros elsewhere, and let

$$j = \begin{pmatrix} 0 & i \\ -i & 0 \end{pmatrix}.$$

The group of symplectic similitudes $G = GSp_{2n}(F)$ is the group of matrices $x \in GL_{2n}(F)$ such that xjx' is a scalar multiple of j, say

$$xjx' = \mu(x)j$$

where $\mu(x) \in F^*$. Let

$$K = GSp_{2n}(\mathfrak{o}) = G \cap GL_{2n}(\mathfrak{o}),$$

$$G^+ = G \cap M_{2n}(\mathfrak{o}).$$

Then G is a locally compact group, K is a maximal compact subgroup of G and is open in G, and G^+ is a subsemigroup of G. The Hecke rings $H(G,K)$, $H(G^+,K)$ and the C-algebras $L(G,K) \cong H(G,K) \otimes_{\mathbf{Z}} \mathbf{C}$, $L(G^+,K) \cong H(G^+,K) \otimes_{\mathbf{Z}} \mathbf{C}$ are defined as in §2. The characteristic functions of the double cosets $KxK \subset G$ form a Z-basis of $H(G,K)$. Also $H(G,K)$ is commutative, for the same reason as in the case of $GL_n(F)$ (2.4).

Spherical functions ω on G relative to K are defined as in §3, and are in one-one correspondence with the C-algebra homomorphisms $\hat{\omega}: L(G,K) \to \mathbf{C}$. The spherical functions are parametrized by vectors $s = (s_0, s_1, \ldots, s_n) \in \mathbf{C}^{n+1}$ ([S3], Chapter III).

Define $\varepsilon_i: \mathbf{C}^{n+1} \to \mathbf{C}^{n+1}$ $(1 \leqslant i \leqslant n)$ by

$$\varepsilon_i(s_0, s_1, \ldots, s_n) = (s_0 + s_i, s_1, \ldots, s_{i-1}, -s_i, s_{i+1}, \ldots, s_n).$$

The ε_i generate a group E of order 2^n. Also the symmetric group S_n acts by permuting $s_1, \ldots, s_n$. Let $W \subset GL_{n+1}(\mathbf{C})$ be the group generated by E and S_n, which is the semidirect product $E \times S_n$. Then we have [S3]

$$\omega_s = \omega_{ws}$$

for all $w \in W$, where ω_s is the spherical function on $G = GSp_{2n}(F)$ with parameter s.

Now let

$$G_m^+ = \{x \in G^+ : v(\mu(x)) = m\}$$

for all $n \geqslant 0$, and let τ_m denote the characteristic function of G_m^+. The Hecke series $\tau(X)$ and the series $\hat{\tau}(s, X)$ are defined as in §4:

$$\tau(X) = \sum_{m=0}^{\infty} \tau_m X^m,$$

$$\hat{\tau}(s, X) = \sum_{m=0}^{\infty} \hat{\tau}_m(\omega_s) X^m.$$

Now from [S3], Appendix 1 we have the following expression for $\hat{\tau}_m(\omega_s)$:

$$\hat{\tau}_m(\omega_s) = \sum_{\lambda} q^{-\langle\lambda, \rho\rangle} \hat{c}_\lambda(\omega_{s'}) q^{m(N-s_0)} \tag{5.1}$$

where $s = s_0, s_1, \ldots, s_n)$, $s' = (s_1, \ldots, s_n)$, $N = \frac{1}{4}n(n+1)$; $\hat{c}_\lambda(\omega_{s'})$ has the same meaning as in §3, and the summation is over all partitions λ such that $\lambda_1 \leqslant m$. From (3.3) and (5.1) it follows that

$$\hat{\tau}_m(\omega_s) = \sum_{\lambda} P_\lambda(q^{-s_1}, \ldots, q^{-s_n}; q^{-1}) q^{m(N-s_0)}$$

and hence that

$$\hat{\tau}(s, X) = \sum_{m, \lambda} P_\lambda(q^{-s_1}, \ldots, q^{-s_n}; q^{-1})(q^{N-s_0}X)^m \tag{5.2}$$

summed over $m \geqslant \lambda_1 \geqslant \ldots \geqslant \lambda_n \geqslant 0$.

Now the sum on the right-hand side of (5.2) is one that we have calculated in Chapter III, §5, Example 5. Hence we obtain the following expression for $\hat{\tau}(s, X)$ as a sum of partial fractions:

$$\hat{\tau}(s, X) = \sum_{w \in E} w\big(\Phi(q^{-s_1}, \ldots, q^{-s_n}; q^{-1})(1 - q^{N-s_0}X)^{-1}\big) \tag{5.3}$$

where

$$\Phi(x_1, \ldots, x_n; t) = \prod_{i=1}^{n} (1 - x_i)^{-1} \prod_{i<j} (1 - tx_i x_j)(1 - x_i x_j)^{-1}.$$

From (5.3) it follows that $\hat{\tau}(s, X)$ is a rational function $f(X_0)/g(X_0)$ where $X_0 = q^{N-s_0}X$, with denominator

$$g(X_0) = \prod_J (1 - q^{-s_J} X_0),$$

the product being over all subsets J of $\{1, 2, \ldots, n\}$, and $s_J = \sum_{i \in J} s_i$. Also it is not difficult to show that the degree of the numerator $f(X_0)$ is $2^n - 2$.

Finally, we can attach to the spherical function ω_s a zeta function as in §4: we define as before

$$\zeta(s, \omega_s) = \int_G \varphi(x) \|x\|^s \omega_s(x^{-1})\, dx$$

where s is a complex variable, φ is the characteristic function of G^+, and $\|x\| = |\mu(x)|$. Then we have

$$\begin{aligned}\zeta(s, \omega_s) &= \sum_{m=0}^{\infty} \hat{\tau}_m(\omega_s) q^{-ms} \\ &= \hat{\tau}(s, q^{-s})\end{aligned}$$

and therefore from (5.3) we obtain the following formula for this zeta function:

$$\zeta(s, \omega_s) = \sum_{w \in E} w\Big(\Phi(q^{-s_1}, \ldots, q^{-s_n}; q^{-1})(1 - q^{N-s_0-s})^{-1}\Big). \tag{5.4}$$

Notes and references

For background on algebraic groups over local fields, their Hecke rings and spherical functions, we refer to [M1] and [S3] and the references given there.

We have derived the formula (3.4) for the spherical function by exploiting our knowledge of the Hall algebra. It is possible to obtain (3.4) directly by computing the spherical function from an integral formula. This is done in [M1], Chapter IV in much greater generality, and in the case of GL_n provides a more natural (though less elementary) proof of the structure theorem (Chapter III, (3.4)) for the Hall algebra.

The formula (3.4) was also obtained by Luks [L14].

VI

SYMMETRIC FUNCTIONS WITH TWO PARAMETERS

1. Introduction

The Schur functions s_λ (Chapter I, §3) are characterized by the following two properties:

(a) The transition matrix $M(s,m)$ that expresses the Schur functions in terms of the monomial symmetric functions is strictly upper unitriangular (Chapter I, (6.5)), that is to say s_λ is of the form

$$s_\lambda = m_\lambda + \sum_{\mu<\lambda} K_{\lambda\mu} m_\mu$$

for suitable coefficients $K_{\lambda\mu}$ (the Kostka numbers).

(b) The s_λ are pairwise orthogonal relative to the scalar product of Chapter I, §4, which may be defined by

(1.1) $$\langle p_\lambda, p_\mu \rangle = \delta_{\lambda\mu} z_\lambda.$$

Again, the Hall–Littlewood functions $P_\lambda(x;t)$ (Chapter III, §1) are characterized by the same two properties (a) and (b), except that in (b) the scalar product is now that of Chapter III, §4, defined by

(1.2) $$\langle p_\lambda, p_\mu \rangle = \delta_{\lambda\mu} z_\lambda(t) = \delta_{\lambda\mu} z_\lambda \prod_{i=1}^{l(\lambda)} (1-t^{\lambda_i})^{-1}.$$

Another example is provided by the zonal polynomials (Chapter VII). Up to a scalar factor they are characterized by the same properties (a) and (b), but with a different choice of scalar product, namely

(1.3) $$\langle p_\lambda, p_\mu \rangle = 2^{l(\lambda)} z_\lambda \delta_{\lambda\mu}.$$

More generally, Jack's symmetric functions (§10 below) are characterized by (a) and (b), the scalar product now being defined by

(1.4) $$\langle p_\lambda, p_\mu \rangle = \alpha^{l(\lambda)} z_\lambda \delta_{\lambda\mu}$$

where α is a positive real number. They reduce to the zonal polynomials when $\alpha = 2$, and to the Schur functions when $\alpha = 1$.

In this Chapter we shall study a class of symmetric functions $P_\lambda(x;q,t)$ depending rationally on two parameters q and t, which include all the examples above as particular cases. They will be characterized by the same two properties (a) and (b), relative now to the scalar product defined by

$$(1.5) \qquad \langle p_\lambda, p_\mu\rangle = \langle p_\lambda, p_\mu\rangle_{q,t} = \delta_{\lambda\mu} z_\lambda \prod_{i=1}^{l(\lambda)} \frac{1-q^{\lambda_i}}{1-t^{\lambda_i}}.$$

When $q=t$, they reduce to the Schur functions s_λ, and when $q=0$ to the Hall–Littlewood functions $P_\lambda(x;t)$. To obtain Jack's symmetric functions, we set $q=t^\alpha$ and let $t\to 1$, so that $(1-q^r)/(1-t^r)\to\alpha$ for each $r\geqslant 1$, and hence in the limit the scalar product (1.5) becomes that given by (1.4).

Example

Let F be a field of characteristic zero, and let $\langle\ ,\ \rangle$ be a non-degenerate symmetric F-bilinear form on Λ_F with values in F, such that the homogeneous components Λ_F^n of Λ_F are pairwise orthogonal. If $(u_i),(v_i)$ are dual bases of Λ_F^n (i.e. $\langle u_i, v_j\rangle=\delta_{ij}$) then

$$T_n(x|y)=\sum_i u_i(x)v_i(y)$$

is independent of the choice of dual bases (so that in particular $T_n(x|y)=T_n(y|x)$). Let

$$T(x|y)=\sum_{n\geqslant 0} T_n(x|y)$$

(the 'metric tensor').

Let $\Delta:\Lambda_F\to\Lambda_F\otimes\Lambda_F$ be the diagonal map (or comultipliction) defined by $\Delta f=f(x,y)$ (Chapter I, §5, Example 25).

A family (a_λ) of elements of a commutative monoid, indexed by partitions λ, will be called *multiplicative* if $a_\lambda a_\mu=a_{\lambda\cup\mu}$ for each pair of partitions λ,μ. Equivalently, $a_\lambda=a_{\lambda_1}a_{\lambda_2}\dots$ for each partition $\lambda=(\lambda_1,\lambda_2,\dots)$, where $a_r=a_{(r)}$ for each integer $r\geqslant 1$, and $a_0=1$. For example, the bases (e_λ), (h_λ) and (p_λ) of Λ_F are multiplicative.

Show that the following conditions on the scalar product $\langle\ ,\ \rangle$ are equivalent:

(a) Δ is the adjoint of multiplication, i.e.

$$\langle\Delta f, g\otimes h\rangle=\langle f, gh\rangle$$

for all $f,g,h\in\Lambda_F$.

(b) There exists a multiplicative family (v_λ) in F^* such that

$$\langle p_\lambda, p_\mu\rangle=\delta_{\lambda\mu}z_\lambda v_\lambda$$

for all partitions λ,μ.

(c) There exists a character (i.e. F-algebra homomorphism) $\chi: \Lambda_F \to F$ such that $\chi(p_n) \neq 0$ for all $n \geqslant 1$ and

$$\langle f, g \rangle = \chi(f * g)$$

for all $f, g \in \Lambda_F$, where $f * g$ is the internal product defined in Chapter I, §7.

(d) The basis of Λ_F dual to (m_λ) is multiplicative.

(e) The metric tensor T satisfies

$$T(x|y, z) = T(x|y)T(x|z).$$

(Let $p_\lambda^* = z_\lambda^{-1} p_\lambda$ for all partitions λ. Then

$$\Delta p_\lambda^* = \sum p_\alpha^* \otimes p_\beta^*$$

summed over pairs of partitions (α, β) such that $\alpha \cup \beta = \lambda$. Hence (a) implies that

$$\langle p_\lambda^*, p_\rho p_\sigma \rangle = \sum \langle p_\alpha^*, p_\rho \rangle \langle p_\beta^*, p_\sigma \rangle \tag{1}$$

summed over (α, β) such that $|\alpha| = |\rho|$, $|\beta| = |\sigma|$ and $\alpha \cup \beta = \lambda$.

By iterating (1) we obtain

$$\langle p_\lambda^*, p_\mu \rangle = \sum \langle p_{\alpha^{(1)}}^*, p_{\mu_1} \rangle \langle p_{\alpha^{(2)}}^*, p_{\mu_2} \rangle \ldots \tag{2}$$

summed over all sequences $(\alpha^{(1)}, \alpha^{(2)}, \ldots)$ of partitions such that $|\alpha^{(i)}| = \mu_i$ for each $i \geqslant 1$, and $\bigcup \alpha^{(i)} = \lambda$. The sum on the right-hand side of (2) will be zero unless the partition λ is a refinement of μ (Chapter I, § 6), hence $\langle p_\lambda, p_\mu \rangle = 0$ unless $\lambda \leqslant \mu$, and therefore also (by symmetry) unless $\mu \leqslant \lambda$. Thus $\langle p_\lambda, p_\mu \rangle = 0$ unless $\lambda = \mu$, in which case (2) gives

$$\langle p_\lambda^*, p_\lambda \rangle = \prod_{i \geqslant 1} \langle p_{\lambda_i}^*, p_{\lambda_i} \rangle.$$

Hence if we define $v_n = \langle p_n^*, p_n \rangle$ for each $n \geqslant 1$ (and $v_0 = 1$) we have

$$\langle p_\lambda, p_\mu \rangle = \delta_{\lambda\mu} z_\lambda v_\lambda$$

where $v_\lambda = v_{\lambda_1} v_{\lambda_2} \ldots$. Thus (a) implies (b).

Next, since (Chapter I, (7.12)) $p_\lambda * p_\mu = \delta_{\lambda\mu} z_\lambda p_\lambda$, it follows easily that (b) and (c) are equivalent, the character χ being defined by $\chi(p_n) = v_n$ for all $n \geqslant 1$.

To show next that (b) implies (d), observe that it follows from (b) that the bases (p_λ) and $(z_\lambda^{-1} v_\lambda^{-1} p_\lambda)$ of Λ_F are dual to each other, so that

$$\begin{aligned} T(x|y) &= \sum_\lambda z_\lambda^{-1} v_\lambda^{-1} p_\lambda(x) p_\lambda(y) \\ &= \exp\left(\sum_{n \geqslant 1} \frac{v_n^{-1}}{n} p_n(x) p_n(y) \right). \end{aligned}$$

Let

$$\exp\left(\sum_{n\geq 1}\frac{v_n^{-1}}{n}p_n t^n\right)=\sum_{n\geq 0}q_n t^n,$$

then we have

$$\begin{aligned}T(x|y)&=\prod_j\left(\sum_{n\geq 0}q_n(x)y_j^n\right)\\&=\sum_\lambda q_\lambda(x)m_\lambda(y)\end{aligned}$$

where $q_\lambda=q_{\lambda_1}q_{\lambda_2}\ldots$. This shows that the basis of Λ_F dual to (m_λ) is (q_λ), which is multiplicative. Conversely, if (q_λ) is multiplicative then we have

$$T(x|y)=\prod_j T(x|y_j)$$

and so (d) implies (e).

Finally, to show that (e) implies (a), let $(u_\lambda),(v_\lambda)$ be dual bases of Λ_F, indexed by partitions, and let

$$v_\lambda(y,z)=\sum_{\mu,\nu}a_{\mu\nu}^\lambda v_\mu(y)v_\nu(z) \tag{3}$$

Then

$$\begin{aligned}T(x|y,z)&=\sum_\lambda u_\lambda(x)v_\lambda(y,z)\\&=\sum_{\lambda,\mu,\nu}a_{\mu\nu}^\lambda u_\lambda(x)v_\mu(y)v_\nu(z)\end{aligned} \tag{4}$$

and on the other hand

$$T(x|y)T(x|z)=\sum_{\mu,\nu}u_\mu(x)u_\nu(x)v_\mu(y)v_\nu(z). \tag{5}$$

Since the right-hand sides of (4) and (5) are equal, we conclude that

$$u_\mu u_\nu=\sum_\lambda a_{\mu\nu}^\lambda u_\lambda. \tag{6}$$

From (3) and (6) it follows that

$$\langle\Delta v_\lambda,u_\mu\otimes u_\nu\rangle=a_{\mu\nu}^\lambda=\langle v_\lambda,u_\mu u_\nu\rangle$$

which completes the proof.)

The scalar products considered in the text all satisfy these conditions: we have

$$\chi(p_n)=1,(1-t^n)^{-1},2,\alpha,(1-q^n)/(1-t^n)\qquad(n\geq 1)$$

corresponding to (1.1), (1.2), (1.3), (1.4), (1.5) respectively.

Notes and references

An earlier account of the symmetric functions $P_\lambda(x;q,t)$ appeared in [M6]. For some indications of the historical background, see the notes and references at the end of this Chapter.

2. Orthogonality

Let q, t be independent indeterminates, let $F = \mathbb{Q}(q,t)$ be the field of rational functions in q and t, and let $\Lambda_F = \Lambda \otimes_{\mathbb{Z}} F$ denote the F-algebra of symmetric functions with coefficients in F. If we define

$$z_\lambda(q,t) = z_\lambda \prod_{i=1}^{l(\lambda)} \frac{1-q^{\lambda_i}}{1-t^{\lambda_i}}, \tag{2.1}$$

the scalar product (1.5) takes the form

$$\langle p_\lambda, p_\mu \rangle = \langle p_\lambda, p_\mu \rangle_{q,t} = \delta_{\lambda\mu} z_\lambda(q,t). \tag{2.2}$$

On each homogeneous component Λ_F^n of Λ_F, this scalar product differs only by a scalar factor from that defined by the parameters q^{-1}, t^{-1}. For we have

$$z_\lambda(q^{-1}, t^{-1}) = (q^{-1}t)^{|\lambda|} z_\lambda(q,t)$$

and hence

$$\langle f, g \rangle_{q^{-1},t^{-1}} = (q^{-1}t)^n \langle f, g \rangle_{q,t} \tag{2.3}$$

for $f, g \in \Lambda_F^n$.

If a is an indeterminate we denote by $(a;q)_\infty$ the infinite product

$$(a;q)_\infty = \prod_{r=0}^{\infty} (1 - aq^r) \tag{2.4}$$

regarded as a formal power series in a and q.

Let now $x = (x_1, x_2, \ldots)$ and $y = (y_1, y_2, \ldots)$ be two sequences of independent indeterminates, and define

$$\Pi(x,y;q,t) = \prod_{i,j} \frac{(tx_iy_j;q)_\infty}{(x_iy_j;q)_\infty}. \tag{2.5}$$

Then we have

$$\Pi(x,y;q,t) = \sum_\lambda z_\lambda(q,t)^{-1} p_\lambda(x) p_\lambda(y). \tag{2.6}$$

Proof. We calculate

$$\begin{aligned}
\log \Pi(x,y;q,t) &= \sum_{i,j} \sum_{r \geq 0} \left(\log(1 - q^r t x_i y_j) - \log(1 - q^r x_i y_j) \right) \\
&= \sum_{i,j} \sum_{r \geq 0} \sum_{n \geq 1} \frac{1}{n}(1 - t^n)(q^r x_i y_j)^n \\
&= \sum_{i,j} \sum_{n \geq 1} \frac{1}{n} \frac{1-t^n}{1-q^n} (x_i y_j)^n \\
&= \sum_{n \geq 1} \frac{1}{n} \frac{1-t^n}{1-q^n} p_n(x) p_n(y)
\end{aligned}$$

from which it follows that

$$\Pi(x,y;q,t) = \prod_{n \geq 1} \exp\left(\frac{1}{n} \frac{1-t^n}{1-q^n} p_n(x) p_n(y) \right)$$

which as in Chapter III, (4.1) is seen to be equal to the right-hand side of (2.6). ∎

(2.7) *For each $n \geq 0$, let $(u_\lambda), (v_\lambda)$ be F-bases of Λ^n_F, indexed by the partitions of n. Then the following conditions are equivalent:*

(a) $$\langle u_\lambda, v_\mu \rangle = \delta_{\lambda\mu} \text{ for all } \lambda, \mu;$$

(b) $$\sum_\lambda u_\lambda(x) v_\lambda(y) = \Pi(x,y;q,t).$$

Proof. The proof is almost the same as that of Chapter I, (4.6). Let $p^*_\lambda = z_\lambda(q,t)^{-1} p_\lambda$, so that $\langle p^*_\lambda, p_\mu \rangle = \delta_{\lambda\mu}$. If

$$u_\lambda = \sum_\rho a_{\lambda\rho} p^*_\rho, \qquad v_\mu = \sum_\sigma b_{\mu\sigma} p_\sigma$$

then we have

$$\langle u_\lambda, v_\mu \rangle = \sum_\rho a_{\lambda\rho} b_{\mu\rho}$$

so that (a) is equivalent to

(a′) $$\sum_\rho a_{\lambda\rho} b_{\mu\rho} = \delta_{\lambda\mu}.$$

On the other hand, by virtue of (2.6), the condition (b) is equivalent to

$$\sum_\lambda u_\lambda(x) u_\lambda(y) = \sum_\rho p^*_\rho(x) p_\rho(y)$$

and hence to

(b′) $$\sum_\lambda a_{\lambda\rho} b_{\lambda\sigma} = \delta_{\rho\sigma}.$$

Since (a′) and (b′) are equivalent, so are (a) and (b). ∎

Now let $g_n(x; q, t)$ denote the coefficient of y^n in the power-series expansion of the infinite product

(2.8) $$\prod_{i \geq 1} \frac{(tx_i y; q)_\infty}{(x_i y; q)_\infty} = \sum_{n \geq 0} g_n(x; q, t) y^n,$$

and for any partition $\lambda = (\lambda_1, \lambda_2, \ldots)$ define

$$g_\lambda(x; q, t) = \prod_{i \geq 1} g_{\lambda_i}(x; q, t).$$

Then we have

(2.9) $$g_n(x; q, t) = \sum_{|\lambda| = n} z_\lambda(q, t)^{-1} p_\lambda(x)$$

by setting $y_2 = y_3 = \ldots = 0$ in (2.6), and hence

$$\Pi(x, y; q, t) = \prod_j \left(\sum_{n \geq 0} g_n(x; q, t) y_j^n \right)$$

(2.10) $$= \sum_\lambda g_\lambda(x; q, t) m_\lambda(y).$$

It follows now from (2.7) that

(2.11) $$\langle g_\lambda(x; q, t), m_\mu(x) \rangle = \delta_{\lambda\mu}$$

so that the g_λ form a basis of Λ_F dual to the basis (m_λ). Hence

(2.12) *The* g_n $(n \geq 1)$ *are algebraically independent over* F, *and* $\Lambda_F = F[g_1, g_2, \ldots]$. ∎

Next we have

(2.13) *Let* $E: \Lambda_F \to \Lambda_F$ *be an* F*-linear operator. Then the following conditions on* E *are equivalent*:
(a) E *is self-adjoint, i.e.* $\langle Ef, g \rangle = \langle f, Eg \rangle$ *for all* $f, g \in \Lambda_F$;
(b) $E_x \Pi(x, y; q, t) = E_y \Pi(x, y; q, t)$, *where the suffix* x *(resp.* y*) indicates operation on the* x *(resp.* y*) variables.*

Proof. For any two partitions λ, μ let

$$e_{\lambda\mu} = \langle Em_\lambda, m_\mu \rangle.$$

Then (a) is equivalent to

(c) $$e_{\lambda\mu} = e_{\mu\lambda}.$$

Next, from (2.11) we have

$$Em_\lambda = \sum_\mu e_{\lambda\mu} g_\mu$$

and since by (2.10)

$$\Pi(x,y;q,t) = \sum_\lambda m_\lambda(x) g_\lambda(y) = \sum_\lambda m_\lambda(y) g_\lambda(x)$$

it follows that (b) is equivalent to

$$\sum_{\lambda,\mu} e_{\lambda\mu} g_\mu(x) g_\lambda(y) = \sum_{\lambda,\mu} e_{\lambda\mu} g_\mu(y) g_\lambda(x)$$

and hence to (c). |

For $u, v \in F$ such that $v \neq \pm 1$, let $\omega_{u,v}$ denote the F-algebra endomorphism of Λ_F defined by

(2.14) $$\omega_{u,v}(p_r) = (-1)^{r-1} \frac{1-u^r}{1-v^r} p_r$$

for all $r \geqslant 1$, so that

(2.14′) $$\omega_{u,v}(p_\lambda) = \varepsilon_\lambda p_\lambda \prod_{i=1}^{l(\lambda)} \frac{1-u^{\lambda_i}}{1-v^{\lambda_i}}$$

for any partition λ, where $\varepsilon_\lambda = (-1)^{|\lambda| - l(\lambda)}$. Clearly

$$\omega_{v,u} = \omega_{u,v}^{-1}$$

(if $u \neq \pm 1$), and $\omega_{u,u}$ is the involution ω of Chapter I, §2.

From (2.9) and Chapter I, (2.14′) it follows that

(2.15) $$\omega_{q,t}(g_n(x;q,t)) = e_n(x),$$

the nth elementary symmetric function.

The endomorphism $\omega_{u,v}$ is self-adjoint:

(2.16) $$\langle \omega_{u,v} f, g \rangle = \langle f, \omega_{u,v} g \rangle$$

for all $f, g \in \Lambda_F$. By linearity it is enough to verify this for $f = p_\lambda$ and $g = p_\mu$, where it is immediate from the definitions.

Also we have

(2.17) $$\langle \omega_{t,q} f, g \rangle_{q,t} = \langle \omega f, g \rangle$$

for $f, g \in \Lambda_F$, where the scalar product on the right is that of Chapter I. Again we may take $f = p_\lambda$ and $g = p_\mu$, where the result becomes obvious. From (2.17) it follows, for example, that (s_λ) and $(\omega_{t,q} s_{\lambda'})$ are dual bases of Λ_F.

Finally, it follows from (2.6) and Chapter I, (4.1′) that

(2.18) $$\omega_{q,t} \Pi(x, y; q, t) = \prod_{i,j} (1 + x_i y_j)$$

where $\omega_{q,t}$ acts on the symmetric functions in the x variables.

So far we have worked in the algebra Λ_F of symmetric functions in infinitely many variables $x_1, x_2, \ldots$, with coefficients in F. However, in the next section we shall need to work temporarily in $\Lambda_{n,F} = F[x_1, \ldots, x_n]^{S_n}$, and to adapt the scalar product (1.5) to $\Lambda_{n,F}$. The original definition (1.5) no longer makes sense in the context of $\Lambda_{n,F}$, because when $x = (x_1, \ldots, x_n)$ the power sum products $p_\lambda(x)$ (for all partitions λ) are no longer linearly independent. We therefore proceed as follows.

(2.19) *The $g_\lambda(x; q, t)$ such that $l(\lambda) \leqslant n$ form an F-basis of $\Lambda_{n,F}$.*

Proof. Suppose first that $q = t = 0$. Then g_λ becomes h_λ in the notation of Chapter I, §2, and from Chapter I, §6 we have

(1) $$h_\mu(x) = \sum_\lambda K_{\lambda\mu} s_\lambda(x).$$

Now in the ring Λ_n the Schur function $s_\lambda(x)$ is zero if $l(\lambda) > n$, and the $s_\lambda(x)$ with $l(\lambda) \leqslant n$ from a $\mathbf{Z}$-basis of Λ_n. Since the matrix $(K_{\lambda\mu})$ is unitriangular, so is its principal submatrix $(K_{\lambda\mu})_{l(\lambda), l(\mu) \leqslant n}$. Hence it follows from (1) that the $h_\lambda(x)$ such that $l(\lambda) \leqslant n$ form a $\mathbf{Z}$-basis of Λ_n.

Now let q, t be arbitrary. Since the monomial symmetric functions $m_\mu(x)$ with $l(\mu) \leqslant n$ form a $\mathbf{Z}$-basis of Λ_n, we may write

(2) $$g_\lambda(x; q.t) = \sum_\mu b_{\lambda\mu}(q, t) m_\mu(x)$$

with coefficients $b_{\lambda\mu}(q, t) \in F$. From above, the matrix $B(q, t) = (b_{\lambda\mu}(q, t))_{l(\lambda), l(\mu) \leqslant n}$ is well-defined and nonsingular when $q = t = 0$. Hence its determinant, which is a rational function of q and t, cannot vanish identically. In other words, the matrix $B(q, t)$ is invertible over F, and hence (2) shows that the $g_\lambda(x; q, t)$ with $l(\lambda) \leqslant n$ form an F-basis of $\Lambda_{n,F}$, as required. |

We now define a (symmetric) scalar product on $\Lambda_{n,F}$ by requiring that the bases $(g_\lambda)_{l(\lambda)\leq n}$ and $(m_\lambda)_{l(\lambda)\leq n}$ be dual to each other:

(2.20)
$$\langle g_\lambda, m_\mu\rangle = \delta_{\lambda\mu}$$

whenever λ, μ are partitions of length $\leq n$. With the obvious modifications, (2.7) remains valid for this scalar product, since by virtue of (2.8) we have

(2.21)
$$\Pi(x,y;q,t) = \sum_{l(\lambda)\leq n} g_\lambda(x;q,t)m_\lambda(y)$$

when $x = (x_1,\dots,x_n)$ and $y = (y_1,\dots,y_n)$.

Examples

1. For each integer $m \geq 0$ let

$$\begin{aligned}(a;q)_m &= \prod_{r=0}^{m-1}(1-aq^r)\\ &= (a;q)_\infty/(aq^m;q)_\infty\end{aligned}$$

and for any partition $\mu = (\mu_1, \mu_2, \dots)$ let

$$(a;q)_\mu = \prod_{i\geq 1}(a;q)_{\mu_i}.$$

Then we have

$$g_n(x;q,t) = \sum_{|\mu|=n}\frac{(t;q)_\mu}{(q;q)_\mu}m_\mu(x).$$

This follows from the identity

$$\frac{(ty;q)_\infty}{(y;q)_\infty} = \sum_{m\geq 0}\frac{(t;q)_m}{(q;q)_m}y^m$$

which is a particular case of Chapter I, §2, Example 5 (with a, b, t there replaced by $1, t, y$ respectively).

Since $(q;q)_n/(q;q)_\mu$ is a polynomial in q for all partitions μ of n (because it is a product of Gaussian polynomials), it follows that $(q;q)_n g_n(x;q,t)$ is a linear combination of the monomial symmetric functions with coefficients in $\mathbb{Z}[q,t]$, i.e. it lies in the subring $\Lambda[q,t]$ of Λ_F.

2. Show that

$$\omega_{q,t}g_r(x;0;t^{-1}) = (-t)^{-r}g_r(x;0,q).$$

3. The operators D_n^r

In this section we shall work with a finite set of variables $x=(x_1,\dots,x_n)$, and we shall define F-linear operators

$$D_n^r ; \Lambda_{n,F} \to \Lambda_{n,F}$$

for each integer r such that $0 \leq r \leq n$. It will appear in §4 that the symmetric functions $P_\lambda(x;q,t)$ are simultaneous eigenfunctions of these operators.

For each $u \in F$ and $1 \leq i \leq n$ we define the 'shift operator' T_{u,x_i} by

$$(T_{u,x_i}f)(x_1,\dots,x_n) = f(x_1,\dots,ux_i,\dots,x_n) \tag{3.1}$$

for any polynomial $f \in F[x_1,\dots,x_n]$. Next, let X be another indeterminate, and define

$$D_n(X;q,t) = a_\delta(x)^{-1} \sum_{w \in S_n} \varepsilon(w) x^{w\delta} \prod_{i=1}^{n} \left(1 + X t^{(w\delta)_i} T_{q,x_i}\right) \tag{3.2}$$

where as usual $\delta = (n-1, n-2, \dots, 1, 0)$, $a_\delta(x)$ is the Vandermonde determinant (Chapter I, §3), $\varepsilon(w) = \pm 1$ is the sign of $w \in S_n$, and $(w\delta)_i$ is the ith component of $w\delta$.

For $r = 0,1,\dots,n$ let D_n^r denote the coefficient of X^r in $D_n(X;q,t)$:

$$D_n(X;q,t) = \sum_{r=0}^{n} D_n^r X^r. \tag{3.3}$$

We have $D_n^0 = 1$, and

$$D_n^1 = \sum_{i=1}^{n} A_i(x;t) T_{q,x_i} \tag{3.4}$$

where

$$A_i(x;t) = a_\delta(x)^{-1} \sum_{w \in S_n} \varepsilon(w) t^{(w\delta)_i} x^{w\delta}$$

$$= a_\delta(x)^{-1} T_{t,x_i} a_\delta(x)$$

so that

$$A_i(x;t) = \prod_{j \neq i} \frac{tx_i - x_j}{x_i - x_j}. \tag{3.5}$$

More generally, for any $r = 0,1,\dots,n$ we have

$$D_n^r = \sum_{I} A_I(x;t) \prod_{i \in I} T_{q,x_i} \tag{3.4$_r$}$$

summed over all r-element subsets I of $\{1,2,\dots,n\}$, where

$$A_I(x;t) = a_\delta(x)^{-1}\Big(\prod_{i\in I} T_{t,x_i}\Big)a_\delta(x)$$

$$(3.5)_r \qquad = t^{r(r-1)/2}\prod_{\substack{i\in I\\ j\notin I}} \frac{tx_i - x_j}{x_i - x_j}.$$

Each of the operators D_n^r maps symmetric polynomials to symmetric polynomials, and hence is an F-linear endomorphism of $\Lambda_{n,F}$. To establish this fact we shall compute $D_n(X;q,t)m_\lambda(x)$, where $\lambda = (\lambda_1,\dots,\lambda_n)$ is any partition of length $\leqslant n$. First of all, for any $\nu \in \mathbf{N}^n$ we have from (3.2)

$$D_n(X;q,t)x^\nu = \sum_{w_1\in S_n} \varepsilon(w_1)\prod_{i=1}^n (1 + Xt^{(w_1\delta)_i}q^{\nu_i})x^{w_1\delta+\nu},$$

since $T_{q,x_i}(x^\nu) = q^{\nu_i}x^\nu$. Now replace ν by $w_2\lambda$, where $w_2 \in S_n$, and sum over w_2. We shall obtain

$$|S_n^\lambda|D_n(X;q,t)m_\lambda(x)$$

$$= a_\delta(x)^{-1}\sum_{w_1,w_2} \varepsilon(w_1)\prod_{i=1}^n (1 + Xt^{(w_1\delta)_i}q^{(w_2\lambda)_i})x^{w_1\delta+w_2\lambda},$$

where S_n^λ is the subgroup of S_n that fixes λ.

In the sum on the right, let us put $w_2 = w_1 w$. Then it becomes

$$a_\delta(x)^{-1}\sum_{w,w_1} \varepsilon(w_1)\prod_{i=1}^n (1 + Xt^{n-i}q^{(w\lambda)_i})x^{w_1(w\lambda+\delta)}$$

$$= \sum_{w\in S_n} \prod_{i=1}^n (1 + Xt^{n-i}q^{(w\lambda)_i})s_{w\lambda}(x),$$

so that finally we have

$$(3.6) \qquad D_n(X;q,t)m_\lambda(x) = \sum_\alpha \prod_{i=1}^n (1 + Xt^{n-i}q^{\alpha_i})s_\alpha(x)$$

summed over all derangements $\alpha \in \mathbf{N}^n$ of λ.

From this formula we see that each operator D_n^r is a degree-preserving F-linear endomorphism of $\Lambda_{n,F}$. Moreover, since each Schur function s_α that occurs on the right-hand side of (3.6) is either zero or else is equal to $\pm s_\mu$ for some partition $\mu < \lambda$ (unless $\alpha = \lambda$), and since the transition matrix $M(s,m)$ is strictly upper unitriangular (Chapter I, (6.5)) we conclude from (3.6) that

(3.7) $$D_n(X;q,t)m_\lambda = \sum_{\mu \leqslant \lambda} a_{\lambda\mu}(X;q,t)m_\mu$$

with coefficients $a_{\lambda\mu} \in \mathbf{Z}[X,q,t]$, and in particular

(3.8) $$a_{\lambda\lambda}(X;q,t) = \prod_{i=1}^{n}(1 + Xt^{n-i}q^{\lambda_i}).$$

Thus the matrix of each operator D_n^r, relative to the basis of $\Lambda_{n,F}$ formed by the monomial symmetric functions, is strictly upper triangular, and its eigenvalues are the coefficients of X^r in the polynomials (3.8). In particular, we have

(3.9) $$D_n^1 m_\lambda = \sum_{\mu \leqslant \lambda} c_{\lambda\mu}(q,t)m_\mu$$

with coefficients $c_{\lambda\mu} \in \mathbf{Z}[q,t]$, and the eigenvalues of D_n^1 are

(3.10) $$c_{\lambda\lambda}(q,t) = \sum_{i=1}^{n} q^{\lambda_i}t^{n-i},$$

which are visibly all distinct.

The second basic fact about the operators D_n^r is

(3.11) *Each operator D_n^r is self-adjoint for the scalar product* (2.20), *that is to say*

$$\langle D_n^r f, g\rangle_n = \langle f, D_n^r g\rangle_n$$

for all $f, g \in \Lambda_{n,F}$ and $0 \leqslant r \leqslant n$.

It follows from (2.13) that (3.11) is equivalent to the identity

(3.12) $$D_n(X;q,t)_x \Pi(x,y;q,t) = D_n(X;q,t)_y \Pi(x,y;q,t)$$

where $\Pi(x,y;q,t)$ is the product (2.5) (with i,j running from 1 to n) and the suffix x (resp. y) indicates operation on the x (resp. y) variables.

To prove (3.12) we observe that from the definition (2.5) of $\Pi = \Pi(x,y;q,t)$ we have

$$\Pi^{-1}T_{q,x_i}\Pi = \prod_{j=1}^{n} \frac{1 - x_i y_j}{1 - tx_i y_j}$$

which is independent of q. It follows now from (3.2) that

$$\Pi^{-1}D(X;q,t)_x\Pi$$

is independent of q. Hence to prove (3.12), or equivalently (3.11), we may assume that $q = t$.

Now for any polynomial $f(x_1, \dots, x_n)$ and $w \in S_n$ we have

$$x^{w\delta} t^{(w\delta)_i} T_{t,x_i} f = T_{t,x_i}(x^{w\delta} f)$$

and therefore

$$x^{w\delta} \prod_{i=1}^{n} \left(1 + X t^{(w\delta)_i} T_{t,x_i}\right) f = \prod_{i=1}^{n} (1 + X T_{t,x_i})(x^{w\delta} f),$$

so that

$$D_n(X; t, t) f = a_\delta^{-1} \prod_{i=1}^{n} (1 + X T_{t,x_i})(a_\delta f).$$

In particular, therefore,

$$\begin{aligned} D_n(X; t, t) s_\lambda &= a_\delta^{-1} \prod_{i=1}^{n} (1 + X T_{t,x_i}) a_{\lambda+\delta} \\ &= \prod_{i=1}^{n} (1 + X t^{\lambda_i + n - i}) s_\lambda \end{aligned}$$

for any partition λ of length $\leqslant n$.

Now the Schur functions s_λ such that $l(\lambda) \leqslant n$ form an orthonormal basis of $\Lambda_{n,F}$ relative to the scalar product (2.20) when $q = t$, by virtue of the Cauchy formula (Chapter I, (4.3)). Hence we have $\langle D_n(X; t, t) s_\lambda, s_\mu \rangle_n = 0$ if $\lambda \neq \mu$, and therefore

$$\langle D_n(X; t, t) s_\lambda, s_\mu \rangle_n = \langle s_\lambda, D_n(X; t, t) s_\mu \rangle_n$$

for all λ, μ of length $\leqslant n$. This shows that each operator D_n^r is self-adjoint when $q = t$, and completes the proof of (3.11). |

Examples

1. (a) The coefficient of m_μ in $D_n(X; q, t) m_\lambda$ is explicitly

$$a_{\lambda\mu}(X; q, t) = \sum \varepsilon(w) K_{\pi\mu} \prod_{i=1}^{n} (1 + X q^{\alpha_i} t^{n-i})$$

summed over all triples (w, α, π) where $w \in S_n$, $\alpha \in \mathbf{N}^n$ is a derangement of λ, and π is a partition, such that $\alpha + \delta = w(\pi + \delta)$; and $K_{\pi\mu}$ is a Kostka number (Chapter I, §6).

(b) Suppose that $\lambda_1 = \mu_1$. Deduce from (a) that

$$a_{\lambda\mu}(X; q, t) = (1 + X q^{\lambda_1} t^{n-1}) a_{\lambda^* \mu^*}(X; q, t)$$

where $\lambda^* = (\lambda_2, \lambda_3, \dots)$, $\mu^* = (\mu_2, \mu_3, \dots)$.

2. (a) Show that, if $r \geq 0$ and $x = (x_1, \dots, x_n)$,

$$(t-1)\sum_{i=1}^{n} A_i(x;t)x_i^r = t^n g_r(x;0,t^{-1}) - \delta_{0r}.$$

(Express the product

$$\prod_{i=1}^{n} \frac{tX - x_i}{X - x_i}$$

as a sum of partial fractions.)

(b) Let $\Pi = \Pi(x, y; q, t)$, where $x = (x_1, \dots, x_n)$ and $y = (y_1, \dots, y_n)$, and let $\Pi_0 = \omega_{q,t}\Pi = \Pi(1 + x_i y_j)$ (2.18). Then

$$\Pi^{-1}T_{q,x_i}\Pi = \sum_{r \geq 0} g_r(y;0,t^{-1})t^r x_i^r,$$

$$\Pi_0^{-1}T_{q,x_i}\Pi_0 = \sum_{r \geq 0} (-1)^r g_r(y;0,q)x_i^r.$$

(c) Let $\tilde{E} = \tilde{E}_{q,t} = t^{-n}(1 + (t-1)D_n^1)$. Deduce from (a) and (b) that

$$\Pi^{-1}\tilde{E}\Pi = \sum_{r \geq 0} g_r(x;0,t^{-1})g_r(y;0,t^{-1})t^r,$$

$$\Pi_0^{-1}\tilde{E}\Pi_0 = \sum_{r \geq 0} (-1)^r g_r(x;0,t^{-1})g_r(y;0,q),$$

where in each case $\tilde{E}$ acts on symmetric functions in the x variables.

(d) Deduce from (c) and §2, Example 2 that

$$\omega_{q,t}\left(\Pi^{-1}\tilde{E}_{q,t}\Pi\right) = \Pi_0^{-1}\tilde{E}_{t^{-1},q^{-1}}\Pi_0$$

where $\omega_{q,t}$ acts on the x-variables, and hence that

$$\omega_{q,t}\tilde{E}_{q,t} = \tilde{E}_{t^{-1},q^{-1}}\,\omega_{q,t}.$$

3. Let α be a positive real number, X an indeterminate and let

$$D_n(X;\alpha) = a_\delta(x)^{-1} \sum_{w \in S_n} \varepsilon(w) x^{w\delta} \prod_{i=1}^{n} (X + (w\delta)_i + \alpha x_i D_i)$$

in the notation of (3.2), where $D_i = \partial/\partial x_i$. We have

$$D_n(X;\alpha) = \sum_{r=0}^{n} X^{n-r} D_n^r$$

say, where the D_n^r are linear differential operators on functions of $x_1, \dots, x_n$. In particular, when $\alpha = 1$, we have

$$D_n(X;1)f = a_\delta^{-1} \prod_{i=1}^{n} (X + x_i D_i)(a_\delta f)$$

for a function $f = f(x_1, \ldots, x_n)$.

(a) Show that if λ is a partition of length $\leqslant n$, then

$$D_n(X; \alpha) m_\lambda(x) = \sum_\beta \prod_{i=1}^n (X + n - i + \alpha\beta_i) s_\beta(x)$$

summed over all derangements $\beta \in \mathrm{N}^n$ of λ.

(b) Show that each operator D_n^r is self-adjoint with respect to the scalar product defined by the metric tensor (§1, Example) $T(x|y) = \prod_{i,j}(1 - x_i y_j)^{-\alpha}$. (The proofs of (a) and (b) are analogous to those of (3.6) and (3.11).)

(c) Let $Y = (t-1)X - 1$. Then

$$\frac{Y + t^{(w\delta)_i} T_{q, x_i}}{t - 1} = X + \frac{1 - t^{(w\delta)_i}}{1 - t} + t^{(w\delta)_i} \frac{1 - T_{q, x_i}}{1 - q} \cdot \frac{1 - q}{1 - t}.$$

If $(q, t) \to (1, 1)$ in such a way that $(1 - q)/(1 - t) \to \alpha$ (for example, if $q = t^\alpha$ and $t \to 1$), then this operator tends to $X + (w\delta)_i + \alpha x_i D_i$. Hence $D_n(X; \alpha)$ is the limit of $(t-1)^{-n} Y^n D_n(Y^{-1}; q, t)$.

(d) If f is a homogeneous polynomial of degree r, show that

$$D_n^0 f = f,$$

$$D_n^1 f = (\alpha r + \tfrac{1}{2} n(n-1)) f,$$

$$D_n^2 f = (-\alpha^2 U_n - \alpha V_n + c_n) f,$$

where

$$U_n = \tfrac{1}{2} \sum_{i=1}^n x_i^2 D_i^2, \qquad V_n = \sum_{i \neq j} (x_i - x_j)^{-1} x_i^2 D_i,$$

and

$$c_n = \tfrac{1}{2} \alpha^2 r(r-1) + \tfrac{1}{2} \alpha rn(n-1) + \tfrac{1}{24} n(n-1)(n-2)(3n-1).$$

(e) The Laplace–Beltrami operator $\square_n^\alpha$ is defined by

$$\square_n^\alpha f = (\alpha U_n + V_n - (n-1)r) f$$

for f homogeneous of degree r, as in (d) above. Thus

$$D_n^2 f = (-\alpha \, \square_n^\alpha + c_n') f,$$

where $c_n' = c_n - \alpha(n-1)r$.

Let E_n^α be the operator defined by

$$E_n^\alpha f = a_\delta^{-1/\alpha} U_n (a_\delta^{1/\alpha} f) - (a_\delta^{-1/\alpha} U_n a_\delta^{1/\alpha}) f.$$

Show that

$$E_n^\alpha = U_n + \alpha^{-1} V_n.$$

Notes and references

Example 3. The differential operators D_n^r (depending on α) were introduced by Sekiguchi [S9] and later by Debiard [D1]. See also [M5].

4. The symmetric functions $P_\lambda(x;q,t)$

The operators $D_n^r: \Lambda_{n,F} \to \Lambda_{n,F}$ defined in §3 are (for fixed r and varying n) not compatible with the restriction homomorphisms $\rho_{m,n}: \Lambda_{m,F} \to \Lambda_{n,F}$ of Chapter I, §2. However, at least when $r = 1$ (which will be sufficient for our purposes) it is easy to modify them. Let E_n be the operator on $\Lambda_{n,F}$ defined by

(4.1) $$E_n = t^{-n} D_n^1 - \sum_{i=1}^{n} t^{-i}.$$

If $\lambda = (\lambda_1, \dots, \lambda_n)$ is a partition of length $\leqslant n$, we have

$$m_\lambda = \sum_\alpha s_\alpha$$

in Λ_n, where the sum is over all distinct derangements $\alpha \in \mathbf{N}^n$ of λ, as one sees by multiplying $m_\lambda = \sum x^\alpha$ by $a_\delta = \sum \varepsilon(w) x^{w\delta}$ and rearranging (Chapter I, § 6, Example 11). Hence it follows from (3.6) that in $\Lambda_{n,F}$ we have

(4.2) $$E_n m_\lambda = \sum_\alpha \left(\sum_{i=1}^{n} (q^{\alpha_i} - 1) t^{-i} \right) s_\alpha$$

summed over derangements α of λ as above.

This formula shows that

$$\rho_{n,n-1} \circ E_n = E_{n-1} \circ \rho_{n,n-1}$$

where $\rho_{n,n-1}: \Lambda_{n,F} \to \Lambda_{n-1,F}$ is the homomorphism defined by setting $x_n = 0$ (observe that $s_\alpha(x_1, \dots, x_{n-1}, 0) = 0$ if $\alpha_n \neq 0$). Hence we have a well-defined degree-preserving operator

(4.3) $$E = E_{q,t} = \varprojlim E_n : \Lambda_F \to \Lambda_F$$

such that for each partition λ

(4.4) $$E m_\lambda = \sum_{\mu \leqslant \lambda} e_{\lambda\mu} m_\mu$$

with coefficients $e_{\lambda\mu} \in \mathbf{Z}[q, t^{-1}]$, and in particular

(4.5) $$e_{\lambda\lambda} = \sum_{i \geqslant 1} (q^{\lambda_i} - 1) t^{-i}.$$

Since D_n^1 is self-adjoint for the scalar product $\langle f, g\rangle_n$ (3.11), so is E_n, and hence

(4.6) *E is self-adjoint for the scalar product* (1.5) *on* Λ_F. ∎

We come now to the main result of this section.

(4.7) *For each partition λ there is a unique symmetric function $P_\lambda = P_\lambda(x; q, t) \in \Lambda_F$ such that*

(a)
$$P_\lambda = \sum_{\mu \leqslant \lambda} u_{\lambda\mu} m_\mu$$

where $u_{\lambda\mu} \in F$ and $u_{\lambda\lambda} = 1$,

(b)
$$\langle P_\lambda, P_\mu\rangle = 0 \text{ if } \lambda \neq \mu.$$

Proof. We shall construct the P_λ as eigenfunctions of the operator E. If P_λ satisfies (a) above and

(c)
$$EP_\lambda = e_{\lambda\lambda} P_\lambda$$

then by (4.4) we have

$$e_{\lambda\lambda} u_{\lambda\nu} = \sum_{\nu \leqslant \mu \leqslant \lambda} u_{\lambda\mu} e_{\mu\nu}$$

for all pairs of partitions ν, λ such that $\nu < \lambda$, or equivalently

(1)
$$(e_{\lambda\lambda} - e_{\nu\nu}) u_{\lambda\nu} = \sum_{\nu < \mu \leqslant \lambda} u_{\lambda\mu} e_{\mu\nu}.$$

Since by (4.5) we have $e_{\lambda\lambda} \neq e_{\nu\nu}$ if $\nu \neq \lambda$ (i.e. the eigenvalues of the operator E are all distinct), this equation determines $u_{\lambda\nu}$ uniquely in terms of the $u_{\lambda\mu}$ such that $\nu < \mu \leqslant \lambda$. Hence symmetric functions P_λ exist satisfying the conditions (a) and (b). But then we have by (4.6)

$$\begin{aligned} e_{\lambda\lambda}\langle P_\lambda, P_\mu\rangle &= \langle EP_\lambda, P_\mu\rangle \\ &= \langle P_\lambda, EP_\mu\rangle = e_{\mu\mu}\langle P_\lambda, P_\mu\rangle \end{aligned}$$

and since $e_{\lambda\lambda} \neq e_{\mu\mu}$ if $\lambda \neq \mu$, it follows that the P_λ satisfy condition (b).

Finally, to show that the P_λ are uniquely determined by (a) and (b), let λ be any partition and assume that P_μ is determined for all $\mu < \lambda$. Then by (a), P_λ must be of the form

$$P_\lambda = m_\lambda + \sum_{\mu < \lambda} v_{\lambda\mu} P_\mu;$$

taking the scalar product of each side with P_μ, we obtain

$$v_{\lambda\mu} = -\langle m_\lambda, P_\mu\rangle / \langle P_\mu, P_\mu\rangle$$

and hence P_λ is uniquely determined. (Note that $\langle P_\mu, P_\mu\rangle \neq 0$, because when q and t are taken to be real numbers between 0 and 1 the scalar product (1.5) is positive definite.) |

Remark. In fact, conditions (a) and (b) of (4.7) *overdetermine* the symmetric functions P_λ. For if we arrange the partitions of each positive integer n in lexicographical order (so that (1^n) comes first and (n) comes last), we can use the Gram–Schmidt orthogonalization process to derive a unique basis (P_λ) of Λ_F^n satisfying (a′) and (b), where

(a′) $P_\lambda = m_\lambda +$ a linear combination of the m_μ for μ preceding λ in lexicographical order.

If we replace the lexicographical ordering by some other total ordering compatible with the natural partial ordering $\lambda \geqslant \mu$, and apply Gram–Schmidt as before, we shall obtain the same basis (P_λ): this is the content of (4.7).

From (4.7) (a) we have in particular

(4.8) $$P_{(1^r)} = e_r$$

(since $m_{(1^r)} = e_r$). Also

(4.9) $$P_{(r)} = \frac{(q;q)_r}{(t;q)_r} g_r.$$

For by (2.11), g_r is orthogonal to m_μ for all partitions $\mu \neq (r)$, hence to all P_μ except for $\mu = (r)$. It follows that g_r must be a scalar multiple of $P_{(r)}$, and the scalar factor is given by §2, Example 1.

Next we have

(4.10) $$P_\lambda(x_1, \dots, x_n; q, t) = 0 \quad \text{if } n < l(\lambda).$$

Proof. If $\mu \leqslant \lambda$, then $\mu' \geqslant \lambda'$ (Chapter I, (1.11)), so that

$$l(\mu) = \mu'_1 \geqslant \lambda'_1 = l(\lambda) > n,$$

and hence $m_\mu(x_1, \dots, x_n) = 0$ for each $\mu \leqslant \lambda$. |

Let

(4.11) $$b_\lambda = b_\lambda(q,t) = \langle P_\lambda, P_\lambda\rangle^{-1}$$

and

(4.12) $$Q_\lambda = b_\lambda P_\lambda,$$

so that we have

$$\langle P_\lambda, Q_\mu \rangle = \delta_{\lambda\mu}$$

for all λ, μ, i.e. the bases $(P_\lambda), (Q_\lambda)$ of Λ_F are duals of each other for the scalar product (1.5). Hence by (2.7) we have

(4.13) $$\sum_\lambda P_\lambda(x;q,t) Q_\lambda(y;q,t) = \Pi(x,y;q,t).$$

We shall next consider some particular cases.

(4.14) (i) When $q = t$ we have

$$P_\lambda(x;t,t) = Q_\lambda(x;t,t) = s_\lambda(x)$$

(Chapter I, (4.8) and (6.5)).

(ii) When $q = 0$ we have

$$P_\lambda(x;0,t) = P_\lambda(x;t),$$

$$Q_\lambda(x;0,t) = Q_\lambda(x;t),$$

the Hall–Littlewood symmetric functions studied in Chapter III. This follows from Chapter III, (2.6) and (4.9).

(iii) Let $q = t^\alpha$ and let $t \to 1$. The resulting symmetric functions are Jack's symmetric functions, which we shall consider in §10 below.

(iv) From (2.3) and (4.7) it follows that

$$P_\lambda(x;q^{-1},t^{-1}) = P_\lambda(x;q,t),$$

$$Q_\lambda(x;q^{-1},t^{-1}) = (qt^{-1})^{|\lambda|} Q_\lambda(x;q,t).$$

(v) When $t = 1$, we have

$$P_\lambda(x;q,1) = m_\lambda(x).$$

For it follows from (3.2) that $D_n(X;q,1) = \prod_{i=1}^n (1 + XT_{q,x_i})$, so that $D_n(X;q,1) m_\lambda = \prod(1 + Xq^{\lambda_i}) m_\lambda$. Hence the m_λ are the eigenfunctions of D_n^1, and hence of E_n, when $t = 1$, which proves the assertion.

(vi) When $q = 1$, we have

$$P_\lambda(x,1,t) = e_{\lambda'}(x)$$

for all t. We defer the proof of this statement to §5.

(4.15) *If $x = (x_1, \dots, x_n)$ and $l(\lambda) \leqslant n$, we have*

$$D_n(X;q,t) P_\lambda(x;q,t) = \prod_{i=1}^n (1 + Xq^{\lambda_i} t^{n-i}) P_\lambda(x;q,t).$$

Proof. As μ runs through the partitions of length $\leqslant n$, the m_μ form a basis of $\Lambda_{n,F}$; hence so do the P_μ, by (4.7) (a), and therefore also the Q_μ. Hence $D_n(X;q,t)P_\lambda$ is of the form

$$D_n(X;q,t)P_\lambda = \sum_\mu v_{\lambda\mu}(X;q,t)Q_\mu.$$

By (4.7) and (3.7), the matrix $(v_{\lambda\mu})$ is strictly upper triangular, and by (3.11) it is symmetric, hence *diagonal*. It follows that $D_n(X;q,t)P_\lambda$ is a scalar multiple of P_λ, and by (3.7), (3.8) the scalar multiple is

$$\prod_{i=1}^{n}(1+Xq^{\lambda_i}t^{n-i}). \quad |$$

(4.16) *The operators of $D_n^r(0\leqslant r\leqslant n)$ on $\Lambda_{n,F}$ commute with each other.*

For by (4.15) they are simultaneously diagonalized by the basis (P_λ). |

(4.17) *Let λ be a partition of length n. Then*

$$P_\lambda(x_1,\dots,x_n;q,t) = x_1\dots x_n P_\mu(x_1,\dots,x_n;q,t)$$

where $\mu=(\lambda_1-1,\dots,\lambda_n-1)$.

Proof. We have

$$T_{q,x_i}(x_1\dots x_n P_\mu) = qx_1\dots x_n T_{q,x_i}(P_\mu)$$

and therefore

$$D_n^1(x_1\dots x_n P_\mu) = qx_1\dots x_n D_n^1(P_\mu)$$

$$= q\left(\sum_{i=1}^{n} q^{\mu_i}t^{n-i}\right)x_1\dots x_n P_\mu$$

by (4.15). Hence both P_λ and $x_1\dots x_n P_\mu$ are eigenfunctions of the operator D_n^1 for the eigenvalue $\sum q^{\lambda_i}t^{n-i}$. Since the eigenvalues of D_n^1 are all distinct, it follows that P_λ and $x_1\dots x_n P_\mu$ can differ by at most a scalar factor; but they both have leading term x^λ, and so they are equal. |

Examples

1. For any partition λ, the coefficient of $x_1^{\lambda_1}$ in $P_\lambda(x;q,t)$ is equal to $P_{\lambda^*}(x^*;q,t)$, where $\lambda^*=(\lambda_2,\lambda_3,\dots)$ and $x^*=(x_2,x_3,\dots)$. (With the notation of (4.7), the coefficient of $x_1^{\lambda_1}$ in P_λ is equal to

$$\sum_\nu u_{\lambda\nu}m_{\nu^*}(x^*)$$

summed over partitions $\nu \leq \lambda$ such that $\nu_1 = \lambda_1$, where $\nu^* = (\nu_2, \nu_3, \dots)$. Hence it has to be shown that $u_{\lambda\nu} = u_{\lambda^*\nu^*}$ for all such partitions. The coefficients $u_{\lambda\nu}$ are determined recursively by the equations (1) in the proof of (4.7). From §3, Example 1(b) it follows that $e_{\lambda^*\mu^*} = te_{\lambda\mu}$ if $\mu < \lambda$, and $e_{\lambda^*\lambda^*} - e_{\nu^*\nu^*} = t(e_{\lambda\lambda} - e_{\nu\nu})$. Hence $u_{\lambda\nu} = u_{\lambda^*\nu^*}$, as required.)

2. From §3, Example 3(a) it follows that

$$D_n(X;\alpha)m_\lambda(x) = \sum_{\mu \leq \lambda} c_{\lambda\mu}(X;\alpha)m_\mu(x)$$

with coefficients $c_{\lambda\mu} \in \mathbf{Z}[X,\alpha]$ and

$$c_{\lambda\lambda} = \prod_{i=1}^{n} (X + \alpha\lambda_i + n - i).$$

Since $\alpha > 0$, the sequence $(\alpha\lambda_i + n - i)_{1 \leq i \leq n}$ is strictly decreasing, and therefore $c_{\lambda\lambda} \neq c_{\mu\mu}$ if $\lambda \neq \mu$.

(a) Let C denote the matrix $(c_{\lambda\mu})$. Show that there is a unique strictly upper unitriangular matrix U such that UCU^{-1} is a diagonal matrix. For each partition λ of length $\leq n$ define

(1) $$P_\lambda^\alpha(x) = \sum_{\mu \leq \lambda} u_{\lambda\mu} m_\mu(x)$$

where $x = (x_1, \dots, x_n)$. Show that

(2) $$D_n(X;\alpha)P_\lambda^\alpha = c_{\lambda\lambda}(X;\alpha)P_\lambda^\alpha$$

and deduce from §3, Example 3(b) that

(3) $$\langle P_\lambda^\alpha, P_\mu^\alpha \rangle_\alpha = 0$$

if $\lambda \neq \mu$.

(b) The symmetric polynomials $P_\lambda^\alpha(x)$ so defined are characterized by the properties (1) and (3). Hence the coefficients $u_{\lambda\mu}$ can be computed recursively by the Gram–Schmidt process. Deduce that the $u_{\lambda\mu}$ are rational functions of α, independent of X and of n. The symmetric functions P_λ^α so defined are *Jack's symmetric functions* (§10 below).

(c) The differential operators D_n^r defined in §3, Example 3 commute with each other. (By (2) above, they are simultaneously diagonalized by the P_λ^α.)

3. Let $\Box_n^\alpha$ be the Laplace–Beltrami operator (§3, Example 3(e)) acting on symmetric polynomials in $x_1, \dots, x_n$.

(a) If $\rho = \rho_{n,n-1}: \Lambda_n \to \Lambda_{n-1}$ is the restriction homomorphism, show that for $f \in \Lambda_n^r$ we have

$$\rho(U_n f) = U_{n-1}(\rho f),$$

$$\rho(V_n f) = (V_{n-1} + r)(\rho f)$$

and hence that $\rho \circ \Box_n^\alpha = \Box_{n-1}^\alpha \circ \rho$. It follows that

$$\Box^\alpha = \lim_{\leftarrow} \Box_n^\alpha$$

is a well-defined operator on Λ_F, where $F = \mathbf{Q}(\alpha)$.

(b) Show that

$$\square^\alpha P_\lambda^\alpha = e_\lambda(\alpha) P_\lambda^\alpha$$

where $e_\lambda(\alpha) = n(\lambda')\alpha - n(\lambda)$.

If $\lambda > \mu$ we have $n(\lambda) < n(\mu)$ and $n(\lambda') > n(\mu')$, hence $e_\lambda(\alpha) \neq e_\mu(\alpha)$.

(c) Show that, if λ is a partition of length n,

$$\square^\alpha m_\lambda = e_\lambda(\alpha) m_\lambda + \sum_{1 \leqslant i < j \leqslant n} (\lambda_i - \lambda_j) \sum_{r \geqslant 1} m_{(\lambda - r\varepsilon_i + r\varepsilon_j)^+}$$

where $(\lambda - r\varepsilon_i + r\varepsilon_j)^+$ is the partition obtained by rearranging the sequence $(\lambda_1, \dots, \lambda_i - r, \dots, \lambda_j + r, \dots, \lambda_n)$, and in the inner sum r runs from 1 to $[\frac{1}{2}(\lambda_i - \lambda_j)]$.

(d) Let $u_{\lambda\mu}$ denote the coefficient of m_μ in P_λ^α, as in Example 2 above. Deduce from (b) and (c) that if $\mu < \lambda$ we have

$$u_{\lambda\mu} = (e_\lambda(\alpha) - e_\mu(\alpha))^{-1} \sum (\mu_i - \mu_j + 2r) u_{\lambda, (\mu + r\varepsilon_i - r\varepsilon_j)^+}$$

summed over (i, j, r) satisfying $1 \leqslant i < j \leqslant l(\mu)$ and $r \geqslant 1$, such that $(\mu + r\varepsilon_i - r\varepsilon_j)^+$ is a partition $< \lambda$.

The coefficients $u_{\lambda\mu}$ may be computed recursively from this formula. In particular, it shows that $u_{\lambda\mu}$ is a rational function of α whose numerator and denominator are polynomials in α with positive integral coefficients.

5. Duality

The effect of the involution ω (Chapter I, §2) on the Schur functions is given by

$$\omega s_\lambda = s_{\lambda'}$$

(Chapter I, (3.8)). This result is generalized in the following proposition, in which $\omega_{q,t}$ is the automorphism of Λ_F defined in (2.14):-

(5.1) *We have*

$$\omega_{q,t} P_\lambda(x; q, t) = Q_{\lambda'}(x; t, q),$$

$$\omega_{q,t} Q_\lambda(x; q, t) = P_{\lambda'}(x; t, q).$$

Since $\omega_{t,q} = \omega_{q,t}^{-1}$, these two assertions are equivalent.

Clearly (5.1) is equivalent to

$$\langle \omega_{q,t} P_{\lambda'}(q, t), P_\mu(t, q) \rangle_{t,q} = \delta_{\lambda\mu}$$

and hence by (2.17) to

(5.1′) $$\langle \omega P_{\lambda'}(q, t), P_\mu(t, q) \rangle = \delta_{\lambda\mu}$$

in which the scalar product is that of Chapter I. Next, let

$$A(q,t) = M(P(q,t), s),$$

the transition matrix from the P's to the Schur functions. Then (5.1′) is equivalent to

$$JA(q,t)JA(t,q)' = 1 \tag{5.1″}$$

where $J = (\delta_{\lambda'\mu})$ is the conjugation matrix and $A(t,q)'$ is the transpose of $A(t,q)$.

To prove (5.1″) we introduce the matrix of scalar products

$$U(q,t) = (\langle s_\lambda, s_\mu \rangle_{q,t})$$

which has the following properties:

(5.2) (a) $U(q,t) = U(t,q)^{-1}$,
(b) $U(q,t) = JU(q,t)J$,
(c) $D(q,t) = A(q,t)U(q,t)A(q,t)'$ *is a diagonal matrix.*

Proof. Let $X = (\chi_\rho^\lambda)$ be the matrix of characters of S_n (Chapter I, §7). Then the (ρ, μ) entry of $XU(q,t)$ is

$$\sum_\lambda \chi_\rho^\lambda \langle s_\lambda, s_\mu \rangle_{q,t} = \langle p_\rho, s_\mu \rangle_{q,t}$$

$$= \chi_\rho^\mu \prod \frac{1-q^{\rho_i}}{1-t^{\rho_i}}$$

from which it follows that $XU(q,t)X^{-1}$ is a diagonal matrix whose inverse is $XU(t,q)X^{-1}$. This proves (a), and since $XJ = \varepsilon X$, where ε is a diagonal matrix of ± 1's, it also proves (b). Finally, (c) expresses the orthogonality of the $P_\lambda(q,t)$. ∎

Now the matrix $A(q,t)$ is strictly upper unitriangular, and hence

$$B = JA(q,t)JA(t,q)'$$

is strictly lower unitriangular. We now compute

$$C = D(t,q)B^{-1} = D(t,q)A(t,q)'^{-1}JA(q,t)^{-1}J$$

$$= A(t,q)U(t,q)JA(q,t)^{-1}J \qquad \text{by (5.2)(c)}$$

$$= A(t,q)JU(q,t)^{-1}A(q,t)^{-1}J \qquad \text{by (5.2)(a), (b)}$$

$$= A(t,q)JA(q,t)'D(q,t)^{-1}J \qquad \text{by (5.2)(c)}$$

$$= B'JD(q,t)^{-1}J.$$

So the matrix C is lower triangular (because B^{-1} is) and upper triangular (because B' is), hence diagonal. It follows that $B = 1$, which proves (5.1″) and hence also (5.1). |

In particular, we may take $q = 1$ in (5.1′). Then $P_\mu(t,1) = m_\mu$ by (4.14)(v), and hence (5.1′) shows that $\omega P_{\lambda'}(1,t) = h_\lambda$, or equivalently $P_{\lambda'}(1,t) = e_\lambda$, as stated in (4.14)(vi).

From (5.1) and (4.12) it also follows that

$$b_\lambda(q,t)b_{\lambda'}(t,q) = 1, \tag{5.3}$$

and by applying $\omega_{q,t}$ to both sides of (4.13) that

$$\begin{aligned}\prod_{i,j}(1 + x_i y_j) &= \sum_\lambda P_\lambda(x;q,t)P_{\lambda'}(x;t,q) \\ &= \sum_\lambda Q_\lambda(x;q,t)Q_{\lambda'}(x;t,q).\end{aligned} \tag{5.4}$$

Finally we have

$$Q_{(n)}(x;q,t) = g_n(x;q,t) \tag{5.5}$$

by (2.13), (4.8), and (5.1).

Examples

1. If λ is any partition let

$$f_\lambda(q,t) = (1-t)\sum_{i \geq 1}(q^{\lambda_i} - 1)t^{i-1}.$$

(a) Show that

$$f_\lambda(q,t) = f_{\lambda'}(t,q)$$

where λ' is the conjugate partition.

(b) The eigenvalues of the operator $(t-1)E$ (4.3) are $f_\lambda(q,t^{-1})$.

2. We shall use Example 1 above and §3, Example 2(d) to deduce another proof of the duality theorem (5.1). The operator $\tilde{E}$ of §3, Example 2 is equal to $1 + (t-1)E$, and hence

$$\begin{aligned}\tilde{E}_{t^{-1},q^{-1}}\omega_{q,t}P_\lambda(q,t) &= \omega_{q,t}\tilde{E}_{q,t}P_\lambda(q,t) \\ &= (1 + f_\lambda(q,t^{-1}))\,\omega_{q,t}P_\lambda(q,t) \\ &= (1 + f_{\lambda'}(t^{-1},q))\,\omega_{q,t}P_\lambda(q,t).\end{aligned}$$

It follows that $\omega_{q,t}P_\lambda(q,t)$ is an eigenfunction of $\tilde{E}_{t^{-1},q^{-1}}$ with eigenvalue $1+f_{\lambda'}(t^{-1},q)$, and hence must be a scalar multiple of $P_{\lambda'}(t^{-1},q^{-1})=P_{\lambda'}(t,q)$. To complete the proof of (5.1) it remains to show that

$$\langle \omega_{q,t}P_\lambda(q,t), P_{\lambda'}(t,q)\rangle_{t,q}=1$$

or equivalently (2.17) that

$$\langle \omega P_\lambda(q,t), P_{\lambda'}(t,q)\rangle = 1 \tag{1}$$

with respect to the scalar product of Chapter I. Finally, prove (1) by expressing $P_\lambda(q,t)$ and $P_{\lambda'}(q,t)$ as linear combinations of Schur functions, and using the fact that $\omega s_\mu = s_{\mu'}$.

3. Let α be a positive real number, let $F=\mathbb{Q}(\alpha)$ and let ω_α be the F-algebra automorphism of Λ_F defined by

$$\omega_\alpha p_r = (-1)^{r-1}\alpha^r p_r$$

for each $r \geqslant 1$.

(a) Show that

$$\omega_\alpha \square^\alpha \omega_\alpha^{-1} = -\alpha \square^{\alpha^{-1}}$$

where $\square^\alpha$ is the operator defined in §4, Example 3(a). (Verify that

$$U_n p_\lambda = n(\lambda')p_\lambda + S,$$

$$V_n p_\lambda = ((n-1)|\lambda| - n(\lambda'))p_\lambda + S',$$

where U_n, V_n are the operators defined in §3, Example 3(d) and S (resp. S') is a linear combination of power-sum products p_μ such that $|\mu|=|\lambda|$ and $l(\mu)=l(\lambda)-1$ (resp. $l(\mu)=l(\lambda)+1$). Hence show that

$$\left(\omega_\alpha \square^\alpha + \alpha \square^{\alpha^{-1}} \omega_\alpha\right)p_\lambda = 0$$

for all partitions λ.)

(b) Let $Q_\lambda^\alpha = \omega_\alpha^{-1}P_{\lambda'}^{\alpha^{-1}}$ (§4, Example 2(b)). Deduce from (a) that Q_λ^α is an eigenfunction of $\square^\alpha$ with eigenvalue $e_\lambda(\alpha)$ (§4, Example 3).

(c) Show that (Q_λ^α) is the basis dual to (P_λ^α) for the scalar product (1.4). (Let $a_{\lambda\mu}=\langle P_\lambda^\alpha, Q_\mu^\alpha\rangle$. By expressing P_λ^α and $P_{\mu'}^{\alpha^{-1}}$ as linear combinations of Schur functions, show that the matrix $(a_{\lambda\mu})$ is strictly upper unitriangular, so that $a_{\lambda\lambda}=1$ and $a_{\lambda\mu}=0$ unless $\lambda \geqslant \mu$. Finally, if $\lambda > \mu$ use the self-adjointness of $\square^\alpha$ and the fact that $e_\lambda(\alpha) \neq e_\mu(\alpha)$ (§4, Example 3) to conclude that $a_{\lambda\mu}=0$.)

Notes and references

The proof of the duality theorem (5.1) in the text was suggested by A. Garsia. See [G3].

6. Pieri formulas

For any partition μ and positive integer r we have (Chapter I, (5.16))

$$s_\mu s_{(r)} = \sum_\lambda s_\lambda$$

summed over all partitions $\lambda \supset \mu$ such that $\lambda - \mu$ is a horizontal r-strip, and dually

$$s_\mu s_{(1^r)} = \sum_\lambda s_\lambda$$

summed over partitions $\lambda \supset \mu$ such that $\lambda - \mu$ is a vertical r-strip. In this section we shall generalize these facts to the symmetric functions $P_\lambda(x; q, t)$.

For each partition μ of length $\leqslant n$, define a homomorphism (or specialization)

$$u_\mu : F[x_1, \dots, x_n] \to F$$

by $u_\mu(x_i) = q^{\mu_i} t^{n-i} (1 \leqslant i \leqslant n)$. We extend u_μ to those elements of the field $F(x_1, \dots, x_n)$ for which the specialized denominator does not vanish. In particular,

$$u_0(f) = f(t^{n-1}, t^{n-2}, \dots, t, 1)$$

for any polynomial $f \in F[x_1, \dots, x_n]$.

In this notation (4.15) takes the form

$$D_n^r P_\lambda = u_\lambda(e_r) P_\lambda \tag{6.1}$$

for $0 \leqslant r \leqslant n$ and λ of length $\leqslant n$. Hence by $(3.4)_r$ we have

$$u_\lambda(e_r) P_\lambda = \sum_I A_I \left(\prod_{i \in I} T_{q, x_i} \right) P_\lambda \tag{6.2}$$

summed over all r-element subsets I of $\{1, 2, \dots, n\}$, where

$$\begin{aligned} A_I &= a_\delta^{-1} \left(\prod_{i \in I} T_{t, x_i} \right) a_\delta \\ &= \prod_{1 \leqslant i < j \leqslant n} \frac{x_i t^{\theta_i} - x_j t^{\theta_j}}{x_i - x_j} \end{aligned}$$

where $\theta_i = 1$ if $i \in I$, and $\theta_i = 0$ otherwise. Hence for a partition μ of length $\leqslant n$ we have

$$u_\mu(A_I) = \prod_{i<j} \frac{q^{\mu_i} t^{n-i+\theta_i} - q^{\mu_j} t^{n-j+\theta_j}}{q^{\mu_i} t^{n-i} - q^{\mu_j} t^{n-j}}.$$

(6.3) $u_\mu(A_I) \neq 0$ *if and only if* $\mu + \theta = (\mu_1 + \theta_1, \ldots, \mu_n + \theta_n)$ *is a partition.*

Proof. Suppose that $u_\mu(A_I) = 0$. Then there exist $i < j$ such that $\mu_i = \mu_j$ and $i - j = \theta_i - \theta_j$. Since $|\theta_i - \theta_j| \leqslant 1$, it follows that $j = i + 1$, $\theta_i = 0$, and $\theta_j = 1$, so that $\mu_i = \mu_{i+1}$ and $\mu_i + \theta_i < \mu_{i+1} + \theta_{i+1}$, which means that $\mu + \theta$ is not a partition.

Conversely, if $\mu + \theta$ is not a partition there exists $i \leqslant n - 1$ such that $\mu_i = \mu_{i+1}$, $\theta_i = 0$, and $\theta_{i+1} = 1$, whence $u_\mu(A_I) = 0$. ∎

Now let μ, ν be partitions of length $\leqslant n$ such that $\nu \supset \mu$ and $\theta = \nu - \mu$ is a vertical strip, and define

$$(6.4) \qquad B_{\nu/\mu} = t^{n(\nu) - n(\mu)} \prod_{1 \leqslant i < j \leqslant n} \frac{1 - q^{\mu_i - \mu_j} t^{j-i+\theta_i - \theta_j}}{1 - q^{\mu_i - \mu_j} t^{j-i}},$$

so that $B_{\nu/\mu} = u_\mu(A_I)$ where $I = \{i: \theta_i = 1\}$. Then by applying u_μ to (6.2) we obtain

$$(6.5) \qquad u_\lambda(e_r) u_\mu(P_\lambda) = \sum_\nu B_{\nu/\mu} u_\nu(P_\lambda)$$

where by virtue of (6.3) the sum is over partitions $\nu \supset \mu$ of length $\leqslant n$ such that $\nu - \mu$ is a vertical r-strip.

Observe next that $u_0(P_\lambda)$ is not identically zero, because when $q = t$ we have $u_0(P_\lambda) = u_0(s_\lambda) = s_\lambda(t^{n-1}, t^{n-2}, \ldots, 1)$ which is not zero (Chapter I, §3, Example 1). Let us temporarily write

$$\tilde{P}_\lambda = P_\lambda / u_0(P_\lambda).$$

We shall now prove simultaneously

(6.6) (Symmetry) *For all partitions* λ, μ *of length* $\leqslant n$ *we have*

$$u_\lambda(\tilde{P}_\mu) = u_\mu(\tilde{P}_\lambda).$$

(6.7) (Pieri formula) *For all partitions* σ *of length* $\leqslant n$ *and all positive integers* r *we have*

$$\tilde{P}_\sigma e_r = \sum_\nu B_{\nu/\sigma} \tilde{P}_\nu$$

summed over partitions $\nu \supset \sigma$ *of length* $\leqslant n$ *such that* $\nu - \sigma$ *is a vertical* r*-strip.*

Proof. We shall proceed by induction. First of all, (6.6) is true for all λ when $\mu = 0$, because $u_\lambda(\tilde{P}_0) = u_\lambda(1) = 1$ and $u_0(\tilde{P}_\lambda) = 1$. So let μ be a non-zero partition of length $\leqslant n$, and assume as inductive hypothesis that

(6.6) is true for all λ and for all π such that either $|\pi|<|\mu|$, or $|\pi|=|\mu|$ and $\pi<\mu$.

Let $r\geqslant 1$ be such that $\mu_r>\mu_{r+1}$ and let σ be the partition defined by $\sigma_i=\mu_i-1$ for $1\leqslant i\leqslant r$, and $\sigma_i=\mu_i$ for $i>r$. We shall show that (6.7) is true for this σ.

Consider the product $\tilde{P}_\sigma e_r$. The leading monomial in $\tilde{P}_\sigma$ is x^σ, and the leading monomial in e_r is $x_1\dots x_r$. Hence the leading monomial in $\tilde{P}_\sigma e_r$ is x^μ, and therefore we have

(1)
$$e_r\tilde{P}_\sigma=\sum_{\nu\leqslant\mu}b_\nu\tilde{P}_\nu$$

for suitable coefficients b_ν. Next, from (6.5) we have, for any partition π,

(2)
$$u_\pi(e_r)u_\sigma(\tilde{P}_\pi)=\sum_\nu B_{\nu/\sigma}u_\nu(\tilde{P}_\pi)$$

summed over partitions $\nu\supset\sigma$ such that $\nu-\sigma$ is a vertical r-strip.

Suppose that $|\pi|=|\mu|$ and $\pi\leqslant\mu$. By the inductive hypothesis we have

$$u_\sigma(\tilde{P}_\pi)=u_\pi(\tilde{P}_\sigma)$$

because $|\sigma|<|\mu|$, and

$$u_\nu(\tilde{P}_\pi)=u_\pi(\tilde{P}_\nu)$$

for all $\nu\leqslant\mu$ (this is assured by the inductive hypothesis if either $\pi<\mu$ or $\nu<\mu$, and the only other possibility is $\pi=\nu=\mu$). Hence (2) now takes the form

(3)
$$u_\pi(e_r\tilde{P}_\sigma)=\sum_\nu B_{\nu/\sigma}u_\pi(\tilde{P}_\nu).$$

On the other hand, by applying u_π to both sides of (1) we obtain

(4)
$$u_\pi(e_r\tilde{P}_\sigma)=\sum_{\nu\leqslant\mu}b_\nu u_\pi(\tilde{P}_\nu).$$

At this point we require

(6.8)
$$\det\left(u_\pi(\tilde{P}_\nu)\right)_{\pi,\nu\leqslant\mu}\neq 0.$$

Assuming this for the moment, it follows from comparison of (3) and (4) that $b_\nu=B_{\nu/\sigma}$ if $\nu-\sigma$ is a vertical r-strip, and that $b_\nu=0$ otherwise, so that (1) becomes

(5)
$$e_r\tilde{P}_\sigma=\sum_\nu B_{\nu/\sigma}\tilde{P}_\nu$$

which is (6.7).

Now let λ be any partition of length $\leqslant n$, and apply u_λ to both sides of (5):

(6) $$u_\lambda(e_r)u_\lambda(\tilde{P}_\sigma) = \sum_\nu B_{\nu/\sigma}u_\lambda(\tilde{P}_\nu).$$

By the inductive hypothesis we have $u_\lambda(\tilde{P}_\sigma) = u_\sigma(\tilde{P}_\lambda)$ since $|\sigma| < |\mu|$, and $u_\lambda(\tilde{P}_\nu) = u_\nu(\tilde{P}_\lambda)$ if $\nu \neq \mu$. Hence (6) takes the form

(7) $$u_\lambda(e_r)u_\sigma(\tilde{P}_\lambda) = B_{\mu/\sigma}u_\lambda(\tilde{P}_\mu) + \sum_{\nu<\mu} B_{\nu/\sigma}u_\nu(\tilde{P}_\lambda).$$

On the other hand we have from (6.5)

(8) $$u_\lambda(e_r)u_\sigma(\tilde{P}_\lambda) = B_{\mu/\sigma}u_\mu(\tilde{P}_\lambda) + \sum_{\nu<\mu} B_{\nu/\sigma}u_\nu(\tilde{P}_\lambda).$$

Since $B_{\mu/\sigma} \neq 0$, comparison of (7) and (8) shows that $u_\lambda(\tilde{P}_\mu) = u_\mu(\tilde{P}_\lambda)$, and completes the induction step.

Thus it remains only to prove (6.8). In the determinant we may replace $\tilde{P}_\nu$ by P_ν and then by m_ν, since the transition matrix $M(P,m)$ is unitriangular. So it is enough to show that

$$\det(u_\pi(m_\nu))_{\pi,\nu\leqslant\mu} \neq 0.$$

Regarded as a polynomial in t, the dominant term in $u_\pi(m_\nu)$ is $u_\pi(x^\nu) = q^{\langle\nu,\pi\rangle}t^{\langle\nu,\delta\rangle}$. Hence it is enough to show that

$$\det(q^{\langle\nu,\pi\rangle})_{\nu,\pi\leqslant\mu} \neq 0.$$

which is a particular case of Example 1 below. |

We can restate (6.7) in the form

(6.7′) $$P_\mu e_r = \sum_\lambda \psi'_{\lambda/\mu}P_\lambda$$

summed over partitions $\lambda \supset \mu$ (of length $\leqslant n$) such that $\lambda - \mu$ is a vertical r-strip, and

(6.9) $$\psi'_{\lambda/\mu} = B_{\lambda/\mu}u_0(P_\mu)/u_0(P_\lambda).$$

When $\lambda = \mu + (1^r)$, consideration of the coefficient of x^λ on either side of (6.7′) shows that $\psi'_{\lambda/\mu} = 1$, and hence

(6.10) $$u_0(P_\lambda) = B_{\lambda/\mu}u_0(P_\mu)$$

when $\lambda = \mu + (1^r)$.

Since $B_{\lambda/\mu}$ is given explicitly by (6.4), we can use (6.10) to compute $u_0(P_\lambda)$ and then (6.9) to compute the coefficients $\psi'_{\lambda/\mu}$ in the Pieri formula

(6.7′), thus making it fully explicit. To state the result, we define

$$\Delta^+ = \Delta_n^+(q,t) = \prod_{1\leqslant i<j\leqslant n} \frac{(x_i x_j^{-1};q)_\infty}{(tx_i x_j^{-1};q)_\infty}.$$

Then we have, for any partition λ of length $\leqslant n$,

(6.11)
$$\begin{aligned} u_0(P_\lambda) &= P_\lambda(1,t,\dots,t^{n-1};q,t) \\ &= t^{n(\lambda)} u_\lambda(\Delta^+)/u_0(\Delta^+). \end{aligned}$$

Proof. Since (6.11) is true when $\lambda = 0$ (both sides being equal to 1) it is enough by (6.10) to verify that

(1)
$$u_\lambda(\Delta^+)/u_\mu(\Delta^+) = t^{n(\mu)-n(\lambda)} B_{\lambda/\mu}$$

when $\mu = (\lambda_1 - 1, \lambda_2 - 1, \dots)$ is the partition obtained from λ by removing the first column. Both sides of (1) are products of terms of the form $(1 - q^a t^b)^{\pm 1}$, and to verify (1) and similar identities we shall encounter later it is helpful to switch to additive notation.

Thus, if P is any (finite or infinite) product of the form

$$P = \prod_{a,b\geqslant 0} (1 - q^a t^b)^{n_{ab}}$$

with exponents $n_{ab} \in \mathbf{Z}$ we define

(6.12)
$$L(P) = \sum_{a,b\geqslant 0} n_{ab} q^a t^b \in \mathbf{Z}[[q,t]].$$

The mapping L is injective, so that if P and Q are two such products and $L(P) = L(Q)$, we may conclude that $P = Q$.

For example, if $P = (q^a t^b; q)_\infty$ we have

$$L(P) = q^a t^b (1 + q + q^2 + \dots) = \frac{q^a t^b}{1-q}$$

Hence

$$L\big(u_\lambda(\Delta^+)/u_\mu(\Delta^+)\big) = \sum_{1\leqslant i<j\leqslant n} \big(u_\lambda(x_i x_j^{-1}) - u_\mu(x_i x_j^{-1})\big)\frac{1-t}{1-q},$$

and since

$$\begin{aligned} u_\lambda(x_i x_j^{-1}) - u_\mu(x_i x_j^{-1}) &= (q^{\lambda_i - \lambda_j} - q^{\mu_i - \mu_j}) t^{j-i} \\ &= \begin{cases} q^{\mu_i}(q-1)t^{j-i} & \text{if } i \leqslant r < j, \\ 0 & \text{otherwise,} \end{cases} \end{aligned}$$

where $r = l(\lambda)$, it follows that

$$L\big(u_\lambda(\Delta^+)/u_\mu(\Delta^+)\big) = \sum_{i=1}^{r} \sum_{j=r+1}^{n} q^{\mu_i} t^{j-i}(t-1). \tag{2}$$

On the other hand, from (6.4) we have

$$L\big(t^{n(\mu)-n(\lambda)} B_{\lambda/\mu}\big) = \sum_{1 \leqslant i < j \leqslant n} q^{\mu_i - \mu_j} t^{j-i}(t^{\theta_i - \theta_j} - 1) \tag{3}$$

where $\theta_i = 1$ if $1 \leqslant i \leqslant r$, and $\theta_i = 0$ if $i > r$. The right-hand sides of (2) and (3) are visibly equal, and therefore (1) is established. |

Next, from (6.9) and (6.11) we have

$$L(\psi'_{\lambda/\mu}) = \sum q^{\mu_i - \mu_j} t^{j-i}(t^{\theta_i - \theta_j} - 1) - \sum q^{\mu_i - \mu_j} t^{j-i}(q^{\theta_i - \theta_j} - 1)\frac{1-t}{1-q}$$

$$= \sum q^{\mu_i - \mu_j} t^{j-i}\left((t^{\theta_i - \theta_j} - 1) - (q^{\theta_i - \theta_j} - 1)\frac{1-t}{1-q}\right),$$

all sums being over pairs (i, j) such that $1 \leqslant i < j \leqslant n$. The pairs (i, j) such that $\theta_i = \theta_j$ give a zero contribution, and so do the pairs (i, j) such that $\theta_i = 1$ and $\theta_j = 0$. When $\theta_i = 0$ and $\theta_j = 1$, the expression in braces above is equal to

$$(t^{-1} - 1) - (q^{-1} - 1)(1-t)/(1-q) = t^{-1} + q^{-1}t - 1 - q^{-1},$$

and hence we have

$$\psi'_{\lambda/\mu} = \prod \frac{(1 - q^{\mu_i - \mu_j} t^{j-i-1})(1 - q^{\lambda_i - \lambda_j} t^{j-i+1})}{(1 - q^{\mu_i - \mu_j} t^{j-i})(1 - q^{\lambda_i - \lambda_j} t^{j-i})} \tag{6.13}$$

where the product is taken over all pairs (i, j) *such that* $i < j$ *and* $\lambda_i = \mu_i$, $\lambda_j = \mu_j + 1$.

When $q = t$, the formula (6.11) reduces to the formula

$$s_\lambda(1, t, \ldots, t^{n-1}) = t^{n(\lambda)} \prod_{1 \leqslant i < j \leqslant n} \frac{1 - t^{\lambda_i - \lambda_j - i + j}}{1 - t^{j-i}} \tag{1}$$

of Chapter I, §3, Example 1. Now (*loc. cit.*) this can also be written in the form

$$s_\lambda(1, t, \ldots, t^{n-1}) = t^{n(\lambda)} \prod_{s \in \lambda} \frac{1 - t^{n+c(s)}}{1 - t^{h(s)}} \tag{2}$$

where $c(s)$ is the content and $h(s)$ the hook-length of the square $s \in \lambda$. In the first version (1), the dependence on n appears only in the range $1 \leqslant i < j \leqslant n$ of the product, whereas in the second version (2) it appears as an exponent in each factor.

There is a corresponding reformulation of (6.11) that reduces to (2) when $q = t$. For each square $s = (i, j)$ in the diagram of a partition λ, let

$$(6.14) \qquad \begin{aligned} a(s) &= a_\lambda(s) = \lambda_i - j, & a'(s) &= j - 1, \\ l(s) &= l_\lambda(s) = \lambda'_j - i, & l'(s) &= i - 1, \end{aligned}$$

so that $l'(s)$, $l(s)$, $a(s)$, and $a'(s)$ are respectively the numbers of squares in the diagram of λ to the north, south, east, and west of the square s. The numbers $a(s)$ and $a'(s)$ may be called respectively the *arm-length* and the *arm-colength* of s, and $l(s)$, $l'(s)$ the *leg-length* and the *leg-colength*. The *hook length* $h(s)$ is thus $a(s) + l(s) + 1$, and the *content* $c(s)$ is $a'(s) - l'(s)$.

With this notation established, we have

$$(6.11') \qquad P_\lambda(1, t, \dots, t^{n-1}; q, t) = t^{n(\lambda)} \prod_{s \in \lambda} \frac{1 - q^{a'(s)} t^{n - l'(s)}}{1 - q^{a(s)} t^{l(s)+1}}.$$

Proof. By use of the operator L (6.12) we reduce to showing that

$$(6.15) \qquad \sum_{s \in \lambda} \left(q^{a'(s)} t^{n - l'(s)} - q^{a(s)} t^{l(s)+1}\right) = \frac{1-t}{1-q} \sum_{1 \leqslant i < j \leqslant n} (q^{\lambda_i - \lambda_j} - 1) t^{j-i}$$

for a partition λ of length $\leqslant n$. First we have

$$\sum_{s \in \lambda} q^{a'(s)} t^{n - l'(s)} = \sum_{i=1}^{r} (1 + q + \dots + q^{\lambda_i - 1}) t^{n-i+1}$$

where $r = l(\lambda)$, so that

$$(1) \qquad (1-q) \sum_{s \in \lambda} q^{a'(s)} t^{n - l'(s)} = \sum_{i=1}^{n} (1 - q^{\lambda_i}) t^{n+1-i}.$$

Next consider the sum $\sum_{s \in \lambda} q^{a(s)} t^{l(s)+1}$. The contribution to this sum from the squares in the ith row of λ is

$$\begin{aligned} &\sum_{j=i}^{r} (q^{\lambda_i - \lambda_j} + \dots + q^{\lambda_i - \lambda_{j+1} + 1}) t^{j-i+1} \\ &\qquad = (1-q)^{-1} \sum_{j=i}^{r} (q^{\lambda_i - \lambda_j} - q^{\lambda_i - \lambda_{j+1}}) t^{j-i+1}. \end{aligned}$$

Summing for $i=1,2,\dots,r$ we obtain

(2)

$$(1-q)\sum_{s\in\lambda} q^{a(s)}t^{l(s)+1}=\sum_{i=1}^{n}(t-q^{\lambda_i}t^{n+1-i})-(1-t)\sum_{1\leqslant i<j\leqslant n} q^{\lambda_i-\lambda_j}t^{j-i}$$

and (6.15) follows from (1) and (2) without difficulty. |

Now let u be an indeterminate, and define a homomorphism

$$\varepsilon_{u,t}:\Lambda_F\to F$$

by

(6.16)
$$\varepsilon_{u,t}(p_r)=\frac{1-u^r}{1-t^r}$$

for each integer $r\geqslant 1$. In particular, if u is replaced by t^n we have

$$\varepsilon_{t^n,t}(p_r)=\frac{1-t^{nr}}{1-t^r}=p_r(1,t,\dots,t^{n-1})$$

and hence for any $f\in\Lambda_F$

$$\varepsilon_{t^n,t}(f)=f(1,t,\dots,t^{n-1}),$$

previously denoted by $u_0(f)$. We now have

(6.17)
$$\varepsilon_{u,t}P_\lambda=\prod_{s\in\lambda}\frac{t^{l'(s)}-q^{a'(s)}u}{1-q^{a(s)}t^{l(s)+1}}.$$

Proof. By (6.11′), this is true when $u=t^n$ for any $n\geqslant l(\lambda)$, because $\sum_{s\in\lambda}l'(s)=n(\lambda)$. Both sides of (6.17) are polynomials in u with coefficients in F, and agree for infinitely many values of u, hence are identically equal. |

Next, recall ((4.11), (4.12)) that

$$Q_\lambda(q,t)=b_\lambda(q,t)P_\lambda(q,t)$$

where

$$b_\lambda(q,t)=\langle P_\lambda,P_\lambda\rangle^{-1}.$$

We shall now compute $b_\lambda(q,t)$. For this purpose we require

(6.18) *Let $f\in\Lambda_F$ be homogeneous of degree r. Then*

$$\varepsilon_{u,t}\omega_{t,q}(f)=(-q)^{-r}\varepsilon_{u,q^{-1}}(f).$$

Proof. Since both $\varepsilon_{u,t}$ and $\omega_{t,q}$ are algebra homomorphisms it is enough to verify this when $f = p_r$, in which case it follows directly from the definitions (6.16), (2.14). ∎

We now calculate

$$\begin{aligned}\varepsilon_{u,t} P_\lambda(q,t) &= \varepsilon_{u,t}\,\omega_{t,q} Q_{\lambda'}(t,q) \\ &= (-q)^{-|\lambda|} \varepsilon_{u,q^{-1}} Q_{\lambda'}(t,q) \\ &= (-q)^{-|\lambda|} b_{\lambda'}(t,q)\, \varepsilon_{u,q^{-1}} P_{\lambda'}(t^{-1},q^{-1})\end{aligned}$$

by (5.1), (6.18), and (4.14) (iv). From this it follows that

$$b_{\lambda'}(t,q) = \frac{(-q)^{|\lambda|}\varepsilon_{u,t} P_\lambda(q,t)}{\varepsilon_{u,q^{-1}} P_{\lambda'}(t^{-1},q^{-1})}$$

in which, by (6.17),

$$\begin{aligned}\varepsilon_{u,q^{-1}} P_{\lambda'}(t^{-1},q^{-1}) &= \prod_{s\in\lambda'} \frac{q^{-l'(s)} - t^{-a'(s)}u}{1 - t^{-a(s)}q^{-l(s)-1}} \\ &= (-q)^{|\lambda|} \prod_{s\in\lambda} \frac{t^{l'(s)} - q^{a'(s)}u}{1 - q^{a(s)+1}t^{l(s)}}\end{aligned}$$

(since $\Sigma_{s\in\lambda} a(s) = \Sigma_{s\in\lambda} a'(s)$ and $\Sigma_{s\in\lambda} l(s) = \Sigma_{s\in\lambda} l'(s)$). Hence we obtain

$$b_{\lambda'}(t,q) = \prod_{s\in\lambda} \frac{1 - q^{a(s)+1}t^{l(s)}}{1 - q^{a(s)}t^{l(s)+1}}$$

or, on replacing (λ,q,t) by (λ',t,q),

(6.19) $$b_\lambda(q,t) = \prod_{s\in\lambda} \frac{1 - q^{a(s)}t^{l(s)+1}}{1 - q^{a(s)+1}t^{l(s)}}.$$

Remark. When $q = t$ we obtain $b_\lambda(t,t) = 1$, in agreement with $\langle s_\lambda, s_\lambda\rangle = 1$. When $q = 0$, the denominator in the product (6.19) is 1, and the only $s \in \lambda$ that contribute to the numerator are those for which $a(s) = 0$, i.e. the end squares of each now. Thus we obtain

$$b_\lambda(0,t) = \prod_{i\geq 1} \prod_{j=1}^{m_i(\lambda)} (1 - t^j)$$

in agreement with Chapter III, (2.12).

Let us now return to the Pieri formula (6.7′), in which the coefficients were computed explicitly in (6.13). We shall now give an alternative expression for $\psi'_{\lambda/\mu}$. For this purpose we define, for each partition λ and each square s,

$$(6.20)\qquad b_\lambda(s)=b_\lambda(s;q,t)=\begin{cases}\dfrac{1-q^{a_\lambda(s)}t^{l_\lambda(s)+1}}{1-q^{a_\lambda(s)+1}t^{l_\lambda(s)}} & \text{if } s\in\lambda,\\ 1 & \text{otherwise.}\end{cases}$$

If $s=(i,j)\in\lambda$, let $s'=(j,i)\in\lambda'$. Then

$$(6.21)\qquad a_\lambda(s)=l_{\lambda'}(s'),\qquad l_\lambda(s)=a_{\lambda'}(s'),$$

so that

$$(6.22)\qquad b_\lambda(s;q,t)=b_{\lambda'}(s';t,q)^{-1}.$$

Furthermore, if λ and μ are partitions such that $\lambda\supset\mu$, let $C_{\lambda/\mu}$ (resp. $R_{\lambda/\mu}$) denote the union of the columns (resp. rows) that intersect $\lambda-\mu$. Then we have

$$(6.23)\qquad \psi'_{\lambda/\mu}=\prod_{s\in C_{\lambda/\mu}-R_{\lambda/\mu}}\frac{b_\lambda(s)}{b_\mu(s)}.$$

Proof. From (6.13) we have

$$L(\psi'_{\lambda/\mu})=(t-q)\sum q^{\mu_i-\mu_j-1}(t^{j-i}-t^{j-i-1})$$

summed over all pairs (i,j) such that $i<j$, $\lambda_i=\mu_i$ and $\lambda_j=\mu_j+1$. To each such pair we associate the square $s=(i,\lambda_j)\in C_{\lambda/\mu}-R_{\lambda/\mu}$. The contribution to $L(\psi'_{\lambda/\mu})$ from the pairs (i,j) such that $(i,\lambda_j)=s$ is easily seen to be equal to

$$(t-q)q^{a_\lambda(s)}(t^{l_\lambda(s)}-t^{l_\mu(s)})$$

and hence $L(\psi'_{\lambda/\mu})$ is the sum of these expressions, for all $s\in\mu\cap(C_{\lambda/\mu}-R_{\lambda/\mu})$. But this is precisely the image under L of the product (6.23). |

By applying duality to (6.7′) we shall obtain $Q_\mu g_r$ as a linear combination of the Q_λ. Altogether there are four 'Pieri formulas':

(6.24) *Let μ be a partition and r a positive integer. Then*

(i) $P_\mu g_r=\sum_\lambda \varphi_{\lambda/\mu}P_\lambda$,

(ii) $Q_\mu g_r=\sum_\lambda \psi_{\lambda/\mu}Q_\lambda$,

(iii) $Q_\mu e_r=\sum_\lambda \varphi'_{\lambda/\mu}Q_\lambda$,

(iv) $P_\mu e_r=\sum_\lambda \psi'_{\lambda/\mu}P_\lambda$.

In (i) *and* (ii) (*resp.* (iii) *and* (iv)) *the sum is over partitions* λ *such that* $\lambda - \mu$ *is a horizontal* (*resp. vertical*) *r-strip, and the coefficients are given by*

(i)
$$\varphi_{\lambda/\mu} = \prod_{s \in C_{\lambda/\mu}} \frac{b_\lambda(s)}{b_\mu(s)},$$

(ii)
$$\psi_{\lambda/\mu} = \prod_{s \in R_{\lambda/\mu} - C_{\lambda/\mu}} \frac{b_\mu(s)}{b_\lambda(s)},$$

(iii)
$$\varphi'_{\lambda/\mu} = \prod_{s \in R_{\lambda/\mu}} \frac{b_\mu(s)}{b_\lambda(s)},$$

(iv)
$$\psi'_{\lambda/\mu} = \prod_{s \in C_{\lambda/\mu} - R_{\lambda/\mu}} \frac{b_\lambda(s)}{b_\mu(s)}.$$

Proof. Equation (iv) is a restatement of (6.7′) and (6.23). By applying duality (5.1) to (iv) we obtain (ii), with coefficients

$$\psi_{\lambda/\mu}(q,t) = \psi'_{\lambda'/\mu'}(t,q)$$

which by virtue of (6.21) and (6.22) shows that $\psi_{\lambda/\mu}$ is given by (ii) above. Finally the coefficients in (i) and (iii) are

$$\varphi_{\lambda/\mu} = b_\lambda b_\mu^{-1} \psi_{\lambda/\mu}, \qquad \varphi'_{\lambda/\mu} = b_\lambda^{-1} b_\mu \psi'_{\lambda/\mu}$$

and (6.19) shows that they are given by the formulas above. |

Remarks. 1. When $q = 0$, P_λ (resp. Q_λ) is the Hall–Littlewood symmetric function denoted by the same symbol in Chapter III, and (6.24) (i), (ii), and (iv) reduce respectively to the formulas (5.7), (5.7′), and (3.2) of Chapter III.

2. We have

$$\varphi'_{\lambda/\mu}(q,t) = \varphi_{\lambda'/\mu'}(t,q), \qquad \psi'_{\lambda/\mu}(q,t) = \psi_{\lambda'/\mu'}(t,q).$$

Examples

1. Let $v_1, \dots, v_n$ be distinct vectors in a real vector space V equipped with a positive definite scalar product $\langle u, v \rangle$. Then

$$\sum \langle v_i - v_{\sigma(i)}, v_i - v_{\sigma(i)} \rangle > 0$$

for any permutation $\sigma \neq 1$ in S_n, and hence

$$\sum \langle v_i, v_i \rangle > \sum \langle v_i, v_{\sigma(i)} \rangle.$$

Deduce that $\det(q^{\langle v_i, v_j \rangle})$ is not identically zero as a function of the real variable q.

2. (a) Let λ, μ be partitions such that $\lambda \supset \mu$ and $\lambda - \mu$ is a horizontal strip. Then

$$L(\varphi_{\lambda/\mu}) = \frac{t-q}{1-q} \sum (q^{\lambda_i - \lambda_j} - q^{\lambda_i - \mu_j} - q^{\mu_i - \lambda_{j+1}} + q^{\mu_i - \mu_{j+1}}) t^{j-i}$$

summed over $1 \leqslant i \leqslant j \leqslant l(\lambda)$, and hence

$$\varphi_{\lambda/\mu} = \prod_{1 \leqslant i \leqslant j \leqslant l(\lambda)} \frac{f(q^{\lambda_i - \lambda_j} t^{j-i}) f(q^{\mu_i - \mu_{j+1}} t^{j-i})}{f(q^{\lambda_i - \mu_j} t^{j-i}) f(q^{\mu_i - \lambda_{j+1}} t^{j-i})}$$

where $f(u) = (tu; q)_\infty / (qu; q)_\infty$.

(b) With λ, μ as in (a) we have

$$L(\psi_{\lambda/\mu}) = \frac{t-q}{1-q} \sum (q^{\mu_i - \mu_j} - q^{\lambda_i - \mu_j} - q^{\mu_i - \lambda_{j+1}} + q^{\lambda_i - \lambda_{j+1}}) t^{j-i}$$

summed over $1 \leqslant i \leqslant j \leqslant l(\mu)$, and hence

$$\psi_{\lambda/\mu} = \prod_{1 \leqslant i \leqslant j \leqslant l(\mu)} \frac{f(q^{\mu_i - \mu_j} t^{j-i}) f(q^{\lambda_i - \lambda_{j+1}} t^{j-i})}{f(q^{\lambda_i - \mu_j} t^{j-i}) f(q^{\mu_i - \lambda_{j+1}} t^{j-i})}.$$

(c) Let λ, μ be partitions such that $\lambda \supset \mu$ and $\lambda - \mu$ is a vertical strip. Then

$$L(\varphi'_{\lambda/\mu}) = (t-q) \sum q^{\mu_i - \mu_j} (t^{j-i} - t^{j-i-1})$$

summed over all pairs (i, j) such that $1 \leqslant i < j < \infty$ and $\lambda_i > \mu_i$, $\lambda_j = \mu_j$. Hence

$$\varphi'_{\lambda/\mu} = \prod \frac{(1 - q^{\lambda_i - \lambda_j} t^{j-i-1})(1 - q^{\mu_i - \mu_j} t^{j-i+1})}{(1 - q^{\lambda_i - \lambda_j} t^{j-i})(1 - q^{\mu_i - \mu_j} t^{j-i})}$$

the product being taken over the same set of pairs (i, j).

3. Let $\lambda = (\lambda_1, \lambda_2, \dots)$ be a partition, thought of as an infinite sequence, and let

$$G_\lambda(q, t) = (1-t)^2 \sum_{1 \leqslant i < j < \infty} (1 - q^{\lambda_i - \lambda_j}) t^{j-i-1}.$$

Then

$$\sum_{s \in \lambda} q^{a(s)} t^{l(s)} = \frac{G_\lambda(q, t)}{(1-q)(1-t)}$$

and hence $G_\lambda(q, t) = G_{\lambda'}(t, q)$.

4. Let λ be a partition of length $\leqslant n$ and let

$$v_\lambda(q, t) = \prod_{1 \leqslant i < j \leqslant n} (q^{\lambda_i - \lambda_j} t^{j-i}; q)_k$$

$$v'_\lambda(q, t) = \prod_{1 \leqslant i < j \leqslant n} (q^{\lambda_i - \lambda_j + 1} t^{j-i-1}; q)_k$$

when $t = q^k$. Then

$$\varepsilon_{u,t}(P_\lambda) = t^{n(\lambda)} v_\lambda(q,t) \prod_{i=1}^{n} \frac{(ut^{1-i};q)_{\lambda_i}}{(t;q)_{\lambda_i+k(n-i)}}$$

and

$$\varepsilon_{u,t}(Q_\lambda) = t^{n(\lambda)} v'_\lambda(q,t) \prod_{i=1}^{n} \frac{(ut^{1-i};q)_{\lambda_i}}{(q;q)_{\lambda_i+k(n-i)}}.$$

5. Show that

$$\varepsilon_{t^n,t}(P_\lambda(q,t)) = t^{n(\lambda)} \prod_{1 \leqslant i<j \leqslant n} \frac{(t^{j-i+1};q)_{\lambda_i-\lambda_j}}{(t^{j-i};q)_{\lambda_i-\lambda_j}}$$

and

$$\varepsilon_{qt^{n-1},t}(Q_\lambda(q,t)) = t^{n(\lambda)} \prod_{1 \leqslant i<j \leqslant n} \frac{(qt^{j-i};q)_{\lambda_i-\lambda_j}}{(qt^{j-i-1};q)_{\lambda_i-\lambda_j}},$$

if λ is a partition of length $\leqslant n$.

Notes and references

Proposition (6.6), together with the proof of (6.6) and (6.7) given in the text, is due to T. Koornwinder [K15].

7. The skew functions $P_{\lambda/\mu}, Q_{\lambda/\mu}$

For any three partitions λ, μ, ν let

(7.1) $$f^\lambda_{\mu\nu} = f^\lambda_{\mu\nu}(q,t) = \langle Q_\lambda, P_\mu P_\nu \rangle \in F.$$

Equivalently, $f^\lambda_{\mu\nu}$ is the coefficient of P_λ in the product $P_\mu P_\nu$:

(7.1′) $$P_\mu P_\nu = \sum_\lambda f^\lambda_{\mu\nu} P_\lambda.$$

In particular:

(7.2) (i) *$f^\lambda_{\mu\nu}(t,t)$ is the Littlewood–Richardson coefficient $c^\lambda_{\mu\nu} = \langle s_\lambda, s_\mu s_\nu \rangle$* (Chapter I, §9).
(ii) *$f^\lambda_{\mu\nu}(0,t)$ is the Hall polynomial $f^\lambda_{\mu\nu}(t)$* (Chapter III, §3).
(iii) *$f^\lambda_{\mu\nu}(q,1)$ is the coefficient of x^λ in $m_\mu m_\nu$.*
(iv) *$f^\lambda_{\mu\nu}(1,t) = 1$ if $\lambda = \mu + \nu$, and is zero otherwise.*
(v) *$f^\lambda_{\mu\nu}(q,t) = f^\lambda_{\mu\nu}(q^{-1},t^{-1})$.*

These assertions are all consequences of (4.14). As to (iv), when $q = 1$ we have $P_\lambda = e_{\lambda'}$, hence $P_\mu P_\nu = e_{\mu'} e_{\nu'} = e_{\mu' \cup \nu'} = e_{(\mu+\nu)'}$.

By applying $\omega_{q,t}$ to each side of (7.1′) and using duality (5.1) we obtain

$$Q_{\mu'}(t,q)Q_{\nu'}(t,q)=\sum_{\lambda} f^{\lambda}_{\mu\nu}(q,t)Q_{\lambda'}(t,q)$$

or equivalently

$$Q_{\mu}(q,t)Q_{\nu}(q,t)=\sum_{\lambda} f^{\lambda'}_{\mu'\nu'}(t,q)Q_{\lambda}(q,t).$$

Comparison of this relation with (7.1′) yields

(7.3) $$f^{\lambda}_{\mu\nu}(q,t)=f^{\lambda'}_{\mu'\nu'}(t,q)b_{\lambda}(q,t)/b_{\mu}(q,t)b_{\nu}(q,t)$$

by (4.12).

Clearly, $f^{\lambda}_{\mu\nu}$ vanishes unless $|\lambda|=|\mu|+|\nu|$, for reasons of degree. Moreover

(7.4) $f^{\lambda}_{\mu\nu}=0$ *unless* $\lambda\supset\mu$ *and* $\lambda\supset\nu$.

Proof. For each partition μ, let I_{μ} denote the subspace of Λ_F spanned by the P_{λ} such that $\lambda\supset\mu$. The Pieri formula (6.24)(i) shows that $g_r I_{\mu}\subset I_{\mu}$ for all $r\geqslant 1$. Since by (2.12) the g_r generate Λ_F as an F-algebra, it follows that I_{μ} is an *ideal* in Λ_F. Hence $P_{\mu}P_{\nu}\in I_{\mu}\cap I_{\nu}$, which proves (7.4). ∎

Now let λ, μ be partitions and define $Q_{\lambda/\mu}\in\Lambda_F$ by

(7.5) $$Q_{\lambda/\mu}=\sum_{\nu} f^{\lambda}_{\mu\nu}Q_{\nu}$$

so that

(7.6) $$\langle Q_{\lambda/\mu},P_{\nu}\rangle=\langle Q_{\lambda},P_{\mu}P_{\nu}\rangle$$

(and hence, by linearity, $\langle Q_{\lambda/\mu},f\rangle=\langle Q_{\lambda},P_{\mu}f\rangle$ for all $f\in\Lambda_F$. From (7.4) it follows that

(7.7) (i) $Q_{\lambda/\mu}=0$ *unless* $\lambda\supset\mu$;
(ii) *If* $\lambda\supset\mu$, $Q_{\lambda/\mu}$ *is homogeneous of degree* $|\lambda|-|\mu|$. ∎

Likewise we define $P_{\lambda/\mu}$ by interchanging the P's and Q's in (7.6):

(7.6′) $$\langle P_{\lambda/\mu},Q_{\nu}\rangle=\langle P_{\lambda},Q_{\mu}Q_{\nu}\rangle.$$

Since $Q_{\lambda}=b_{\lambda}P_{\lambda}$ it follows that

(7.8) $$Q_{\lambda/\mu}=b_{\lambda}b_{\mu}^{-1}P_{\lambda/\mu}.$$

From (7.5) we have

$$\begin{aligned}\sum_{\lambda} Q_{\lambda/\mu}(x)P_\lambda(y) &= \sum_{\lambda,\nu} f^\lambda_{\mu\nu} Q_\nu(x)P_\lambda(y)\\ &= \sum_\nu Q_\nu(x)P_\mu(y)P_\nu(y)\\ &= P_\mu(y)\Pi(x,y)\end{aligned}$$

by (4.13), where $\Pi(x,y) = \Pi(x,y;q,t)$. Consequently

$$\begin{aligned}\sum_{\lambda,\mu} Q_{\lambda/\mu}(x)P_\lambda(y)Q_\mu(z) &= \sum_\mu P_\mu(y)Q_\mu(z)\Pi(x,y)\\ &= \Pi(x,y)\Pi(y,z)\end{aligned}$$

which by (4.13) again is equal to

$$\sum_\lambda Q_\lambda(x,z)P_\lambda(y).$$

It follows from this that

$$Q_\lambda(x,z) = \sum_\mu Q_{\lambda/\mu}(x)Q_\mu(z) \tag{7.9}$$

where the sum on the right is over partitions $\mu \subset \lambda$, by (7.7). Likewise we have

$$P_\lambda(x,z) = \sum_\mu P_{\lambda/\mu}(x)P_\mu(z). \tag{7.9'}$$

All of this parallels exactly the developments of Chapter I, §5 (for the Schur functions) and Chapter III, §5 (for the Hall–Littlewood functions). Just as in those cases, the formulas (7.9) and (7.9′) enable us to express the symmetric functions $P_{\lambda/\mu}$ and $Q_{\lambda/\mu}$ explicitly as sums of monomials. Since (m_ν) and (g_ν) are dual bases of Λ_F (2.11), it follows that

$$\begin{aligned}Q_{\lambda/\mu} &= \sum_\nu \langle Q_{\lambda/\mu}, g_\nu\rangle m_\nu\\ &= \sum_\nu \langle Q_\lambda, P_\mu g_\nu\rangle m_\nu. \end{aligned}\tag{7.10}$$

Now we can use (6.24) (i) to express $P_\mu g_\nu$ as a linear combination of the P_λ. Let T be a (column-strict) tableau of shape $\lambda - \mu$ and weight ν, thought of as a sequence of partitions $(\lambda^{(0)},\dots,\lambda^{(r)})$ such that $\mu=\lambda^{(0)}\subset\lambda^{(1)}\subset\dots\subset\lambda^{(r)}=\lambda$ and such that each $\lambda^{(i)}-\lambda^{(i-1)}$ is a horizontal strip. Let

$$\varphi_T(q,t)=\prod_{i=1}^{r}\varphi_{\lambda^{(i)}/\lambda^{(i-1)}}(q,t), \tag{7.11}$$

then iteration of (6.24) (i) shows that

$$P_\mu g_\nu=\sum_\lambda\Big(\sum_T\varphi_T\Big)P_\lambda \tag{7.12}$$

summed over all partitions $\lambda\supset\mu$ such that $|\lambda-\mu|=\nu$, the inner sum being over all tableaux T of shape $\lambda-\mu$ and weight ν.

From (7.10) and (7.12) it follows that

$$Q_{\lambda/\mu}=\sum_T\varphi_T(q,t)x^T \tag{7.13}$$

summed over all tableaux T of shape $\lambda-\mu$, where (as in Chapter I, (5.13)) x^T is the monomial defined by the tableau T.

Likewise, if we define

$$\psi_T(q,t)=\prod_{i=1}^{r}\psi_{\lambda^{(i)}/\lambda^{(i-1)}}(q,t) \tag{7.11'}$$

for a tableau T as above, we have

$$P_{\lambda/\mu}=\sum_T\psi_T(q,t)x^T \tag{7.13'}$$

summed over tableaux T of shape $\lambda-\mu$, as before.

Remarks. 1. In the case where there is only one variable x, (7.13) gives

$$Q_{\lambda/\mu}(x;q,t)=\varphi_{\lambda/\mu}(q,t)x^{|\lambda-\mu|} \tag{7.14}$$

if $\lambda-\mu$ is a horizontal strip, and $Q_{\lambda/\mu}(x;q,t)=0$ otherwise. Likewise, from (7.13'), we have

$$P_{\lambda/\mu}(x;q,t)=\psi_{\lambda/\mu}(q,t)x^{|\lambda-\mu|} \tag{7.14'}$$

if $\lambda-\mu$ is a horizontal strip, and $P_{\lambda/\mu}(x;q,t)=0$ otherwise.

2. If $x=(x_1,\dots,x_n)$, it follows from (7.13) that

(7.15) $Q_{\lambda/\mu}(x_1,\dots,x_n;q,t)=0$ *unless* $0\leqslant\lambda'_i-\mu'_i\leqslant n$ *for each* $i\geqslant 1$.

For if $\lambda'_i-\mu'_i>n$ for some i, there is no tableau of shape $\lambda-\mu$ formed with the symbols $1,2,\dots,n$.

3. Duality (5.1) extends to the skew functions $P_{\lambda/\mu}$ and $Q_{\lambda/\mu}$: namely

$$\omega_{q,t}P_{\lambda/\mu}(q,t)=Q_{\lambda'/\mu'}(t,q), \tag{7.16}$$

$$\omega_{q,t}Q_{\lambda/\mu}(q,t)=P_{\lambda'/\mu'}(t,q). \tag{7.16'}$$

For we have

$$\omega_{q,t}Q_{\lambda/\mu}(q,t)=\sum_\nu f^\lambda_{\mu\nu}(q,t)P_{\nu'}(t,q)$$

by (7.5) and (5.1), and by (7.3) this is equal to

$$\sum_\nu f^{\lambda'}_{\mu'\nu'}(t,q)\frac{b_{\mu'}(t,q)}{b_{\lambda'}(t,q)}Q_{\nu'}(t,q)$$

$$=\frac{b_{\mu'}(t,q)}{b_{\lambda'}(t,q)}Q_{\lambda'/\mu'}(t,q)=P_{\lambda'/\mu'}(t,q).$$

This proves (7.16′), and (7.16) follows by replacing (q,t,λ,μ) by (t,q,λ',μ'). |

4. The explicit calculation of the structure constants $f^\lambda_{\mu\nu}(q,t)$ for arbitrary λ,μ,ν remains an open problem. (The cases where μ or ν is a single row or a single column are covered by (6.24).) For example, it is not known to the author whether the vanishing of the Littlewood–Richardson coefficient $c^\lambda_{\mu\nu}$ implies the vanishing of $f^\lambda_{\mu\nu}(q,t)$.

Examples

1. In this and the following examples we shall sketch an alternative approach to the results of §6. We shall not, therefore, make use of any result in §6 (with the exception of the elementary fact (6.18) in Example 2).

Let μ be a partition of length n, and let r be a positive integer. Then $Q_\mu g_r$ is a linear combination of the Q_λ, say

$$Q_\mu(x;q,t)g_r(x;q,t)=\sum_\lambda \psi_{\lambda/\mu}(q,t)Q_\lambda(x;q,t) \tag{1}$$

with certain (as yet undetermined) coefficients $\psi_{\lambda/\mu}\in F$. By duality (5.1) we obtain

$$P_{\mu'}(x;t,q)e_r(x)=\sum_\lambda \psi_{\lambda/\mu}(q,t)P_{\lambda'}(x;t,q). \tag{2}$$

(a) Suppose that $\varphi_{\lambda/\mu}\neq 0$. By considering the leading monomials in (1) and (2), show that $\lambda\leqslant\mu+(r)$ and that $\lambda'\leqslant\mu'+(1^r)$, and hence that $l(\lambda)=n$ or $n+1$.

(b) Suppose first that $l(\lambda) = n$, and let $x = (x_1, \dots, x_n)$. Use §4, Example 1 to show that

(3) $$\psi_{\lambda/\mu} = \psi_{\lambda^*/\mu^*} b_{\lambda^*} b_\mu / b_\lambda b_{\mu^*}.$$

where $\lambda^* = (\lambda_1 - 1, \lambda_2 - 1, \dots)$ and $\mu^* = (\mu_1 - 1, \mu_2 - 1, \dots)$.

(c) Suppose next that $l(\lambda) = n + 1$, and consider the coefficient of x_1^{n+1} on either side of (2). Use §4, Example 2 to show that

(4) $$\psi_{\lambda/\mu} = \psi_{\lambda^*/\mu^*}.$$

(d) Deduce from (b) and (c) that $\psi_{\lambda/\mu} = 0$ unless $\lambda \supset \mu$ and $\lambda - \mu$ is a horizontal strip.

2. In this example we shall indicate an alternative proof of the specialization formula (6.17). Let

$$\Phi_\lambda(u; q, t) = \varepsilon_{u,t} P_\lambda(x; q, t)$$

which is a polynomial in u with coefficients in F, of degree $\leqslant |\lambda|$.

The developments of §7 use from §6 only the result of Example 1(d), as the reader may easily verify. In particular, we have from (7.9′) and (7.14′)

(1) $$P_\lambda(x; q, t) = \sum_\mu \psi_{\lambda/\mu}(q, t) x^{|\lambda - \mu|} P_\mu(x^*; q, t)$$

where $x^* = (x_2, x_3, \dots)$.

(a) By setting $x = (1, t, \dots, t^{n-1})$ in (1), where n is any positive integer, deduce that the relation

(2) $$\Phi_\lambda(u; q, t) = \sum_\mu \psi_{\lambda/\mu}(q, t) t^{|\mu|} \Phi_\mu(t^{-1}u; q, t)$$

is true for $u = t, t^2, \dots$ and hence identically.

(b) Use duality (5.1) together with (6.18) to show that

(3) $$\Phi_\lambda(u; q, t) = (-q)^{-|\lambda|} b_{\lambda'}(t, q) \Phi_{\lambda'}(u; t^{-1}, q^{-1}).$$

(c) The polynomial $\Phi_\lambda(u; q, t)$ vanishes for $u = 1, t, \dots, t^{l(\lambda)-1}$ by (4.10). Deduce from (3) that $\Phi_\lambda(u; q, t)$ is divisible in $F[u]$ by $\prod_{j=1}^{\lambda_1} (q^{j-1}u - 1)$.

(d) In each partition μ on the right-hand side of (2) we have $\lambda_1 \geqslant \mu_1 \geqslant \lambda_2 \geqslant \dots$, since $\lambda - \mu$ is a horizontal strip (Example 1(d)). Hence by (c) above each summand on the right of (2) is divisible by the product $\prod_{j=1}^{\lambda_2} (q^{j-1}u - t)$, and therefore $\Phi_\lambda(u; q, t)$ is also divisible by this product. By repeating this argument, conclude that Φ_λ is divisible in $F[u]$ by

$$\prod_{i=1}^{l(\lambda)} \prod_{j=1}^{\lambda_i} (q^{j-1}u - t^{i-1}) = \prod_{s \in \lambda} (q^{a'(s)}u - t^{l'(s)}).$$

(e) Since the degree in u of Φ_λ is at most $|\lambda|$, it follows from (d) that

(4) $$\Phi_\lambda(u; q, t) = v_\lambda(q, t) \prod_{s \in \lambda} (q^{a'(s)}u - t^{l'(s)})$$

for some $v_\lambda(q,t) \in F$. If $l(\lambda) = n$ it follows from §4, Example 1 that

$$\Phi_\lambda(t^n;q,t) = t^{n(n-1)/2}\Phi_{\lambda^*}(t^n;q,t), \tag{5}$$

where $\lambda^* = (\lambda_1 - 1, \ldots, \lambda_n - 1)$. Deduce from (4) and (5) that

$$v_\lambda(q,t) = v_{\lambda^*}(q,t)\prod_{i=1}^{n}(q^{\lambda_i - 1}t^{n-i+1} - 1)^{-1}$$

and hence that

$$v_\lambda(q,t) = \prod_{s\in\lambda}(q^{a(s)}t^{l(s)+1} - 1)^{-1}. \tag{6}$$

3. From (3), (4), and (6) of Example 2 we can now obtain the formula (6.19) for $b_\lambda(q,t)$. Then, knowing b_λ, we can derive the formula (6.24)(ii) for $\psi_{\lambda/\mu}$ by induction from the relations (3) and (4) of Example 1.

4. For each partition λ, let

$$b_\lambda^{\mathrm{el}} = \prod_{\substack{s\in\lambda \\ l(s)\ \mathrm{even}}} b_\lambda(s), \qquad b_\lambda^{\mathrm{oa}} = \prod_{\substack{s\in\lambda \\ a(s)\ \mathrm{odd}}} b_\lambda(s)$$

(the superscripts el and oa stand for 'even legs' and 'odd arms' respectively). Then we have

(i) $$\sum_\nu b_\nu^{\mathrm{el}}P_\nu = \prod_{i<j}\frac{(tx_ix_j;q)_\infty}{(x_ix_j;q)_\infty},$$

(ii) $$\sum_\lambda b_\lambda^{\mathrm{el}}P_\lambda = \prod_i\frac{(tx_i;q)_\infty}{(x_i;q)_\infty}\prod_{i<j}\frac{(tx_ix_j;q)_\infty}{(x_ix_j;q)_\infty},$$

(iii) $$\sum_\mu b_\mu^{\mathrm{oa}}P_\mu = \prod_i\frac{(qtx_i^2;q^2)_\infty}{(x_i^2;q^2)_\infty}\prod_{i<j}\frac{(tx_ix_j;q)_\infty}{(x_ix_j;q)_\infty},$$

(iv) $$\sum_\lambda b_\lambda^{\mathrm{oa}}P_\lambda = \prod_i\frac{(qtx_i^2;q^2)_\infty}{(1-x_i)(q^2x_i^2;q^2)_\infty}\prod_{i<j}\frac{(tx_ix_j;q)_\infty}{(x_ix_j;q)_\infty}.$$

In these sums, λ runs through all partitions, μ through all even partitions (i.e. with all parts even), and ν through all partitions such that ν' is even (so that all the columns of ν are of even length).

We shall prove (i) first, and then deduce the other three identities from (i). It is enough to prove (i) for a finite set of variables $x_1, \ldots, x_n$. By induction on n, it is enough to show that if $\Phi(x)$ denotes the left-hand side of (i), then

$$\Phi(x_1,\ldots,x_n,y) = \Phi(x_1,\ldots,x_n)\prod_{i=1}^{n}\frac{(tx_iy;q)_\infty}{(x_iy;q)_\infty}. \tag{1}$$

Now the left-hand side of (1) is by (7.9′) and (7.14′) equal to

$$\sum_{\lambda,\mu} b_\lambda^{\mathrm{el}}\psi_{\lambda/\mu}P_\mu(x_1,\ldots,x_n)y^{|\lambda-\mu|} \tag{2}$$

summed over pairs of partitions λ, μ such that λ' is even, $\lambda \supset \mu$ and $\lambda - \mu$ is a horizontal strip. Again, the right-hand side of (1) is by (2.8) and (6.24) equal to

$$\sum_{\nu, r} b_\nu^{\mathrm{el}} P_\nu(x_1, \dots, x_n) g_r(x_1, \dots, x_n) y^r$$

(3)
$$= \sum_{\mu, \nu} b_\nu^{\mathrm{el}} \varphi_{\mu/\nu} P_\mu(x_1, \dots, x_n) y^{|\mu - \nu|}$$

summed over pairs of partitions μ, ν such that ν' is even, $\mu \supset \nu$ and $\mu - \nu$ is a horizontal strip.

For a given partition μ of length $\leqslant n$ both the partitions λ and ν are uniquely determined: we have $\lambda_i' = \nu_i' = \mu_i'$ if μ_i' is even, and $\lambda_i' = \mu_i' + 1, \nu_i' = \mu_i' - 1$ if μ_i' is odd. Hence $|\lambda - \mu| = |\mu - \nu|$, and a column contains a square belonging to $\lambda - \mu$ if and only if it contains a square belonging to $\mu - \nu$: the condition in either case is that it should be a column of odd length in μ.

To prove that (2) and (3) are equal therefore reduces to showing that

(4)
$$b_\lambda^{\mathrm{el}} \psi_{\lambda/\mu} = b_\nu^{\mathrm{el}} \varphi_{\mu/\nu}.$$

For this purpose it is convenient to introduce the following notation: if λ is any partition and S is any set of squares, then $b_\lambda(S)$ denotes the product of the $b_\lambda(s)$ for $s \in S$. With this notation, if C (resp. $\overline{C}$) denotes the union of the columns of odd (resp. even) length in μ, we have $\varphi_{\mu/\nu} = b_\mu(C)/b_\nu(C)$ and $\psi_{\lambda/\mu} = b_\mu(\overline{C})/b_\lambda(\overline{C})$ by (6.24). Let R_1 (resp. R_2) denote the union of the odd (resp. even) numbered rows, and let $C_i = C \cap R_i, \overline{C}_i = \overline{C} \cap R_i$. Then $b_\lambda^{\mathrm{el}} = b_\lambda(R_2)$ and $b_\nu^{\mathrm{el}} = b_\nu(R_2)$, since all the columns of λ and ν have even length. We may therefore rewrite (4) as

$$b_\lambda(R_2) b_\mu(\overline{C}) b_\nu(C) = b_\lambda(\overline{C}) b_\mu(C) b_\nu(R_2)$$

or, since $C = C_1 \cup C_2, \overline{C} = \overline{C}_1 \cup \overline{C}_2, R_2 = C_2 \cup \overline{C}_2$ (the unions in each case being disjoint) as

(5)
$$b_\lambda(C_2) b_\mu(\overline{C}_1) b_\mu(\overline{C}_2) b_\nu(C_1) = b_\lambda(\overline{C}_1) b_\mu(C_1) b_\mu(C_2) b_\nu(\overline{C}_2).$$

For each square $s \in \overline{C} \cap \lambda$ we have $l_\lambda(s) = l_\mu(s) = l_\nu(s)$; for each square $s \in R_1 \cap \lambda$ we have $a_\lambda(s) = a_\mu(s)$, and for each $s \in R_2 \cap \mu$ we have $a_\mu(s) = a_\nu(s)$. Hence we have

(6)
$$b_\lambda(\overline{C}_1) = b_\mu(\overline{C}_1), \qquad b_\mu(\overline{C}_2) = b_\nu(\overline{C}_2).$$

Furthermore, for each square $s \in C_2 \cap \lambda$ let $t \in C_1$ be the square immediately above s. Then we have $a_\lambda(s) = a_\mu(t)$ and $l_\lambda(s) = l_\mu(t)$; also $a_\mu(s) = a_\nu(t)$ and $l_\mu(s) = l_\nu(t)$, and therefore

(7)
$$b_\lambda(C_2) = b_\mu(C_1), \qquad b_\mu(C_2) = b_\nu(C_1).$$

These relations (6) and (7) together imply (5), and complete the proof of (i).

Next consider (ii). By (2.8) and (i), the right-hand side of (ii) is equal to

(8)
$$\sum_{\nu, r} b_\nu^{\mathrm{el}} P_\nu g_r = \sum_{\mu, \nu} b_\nu^{\mathrm{el}} \varphi_{\mu/\nu} P_\mu$$

summed over partitions $\mu \supset \nu$ such that ν' is even and $\mu - \nu$ is a horizontal strip. From the proof of (i) above we have

$$\begin{aligned} b_\nu^{\mathrm{el}} \varphi_{\mu/\nu} &= b_\nu(R_2) b_\mu(C)/b_\nu(C) \\ &= b_\nu(\overline{C}_2) b_\mu(C_1) b_\mu(C_2)/b_\nu(C_1) \\ &= b_\mu(C_1) b_\mu(\overline{C}_2) = b_\mu^{\mathrm{el}} \end{aligned}$$

by (6) and (7). Hence (8) is equal to $\Sigma_\mu b_\mu^{\mathrm{el}} P_\mu$ summed over all partitions μ, which proves (ii).

Finally, (iii) and (iv) are obtained from (i) and (ii) respectively by duality, i.e. by operating on both sides with $\omega_{q,t}$ and applying (5.1).

When $q = 0$, so that P_λ becomes the Hall–Littlewood symmetric function, the identities (i)–(iv) above reduce to Examples 3, 4, 2, 1 of Chapter III, § 5, respectively.

5. For partitions λ, μ, π such that $\lambda \supset \mu$ and $|\lambda| = |\mu| + |\pi|$, let

$$A_{\lambda/\mu,\pi} = \sum_T \varphi_T$$

summed over all (column-strict) tableaux T of shape $\lambda - \mu$ and weight π, so that by (7.13) we have

(1) $$Q_{\lambda/\mu} = \sum_\pi A_{\lambda/\mu,\pi} m_\pi$$

and hence

(2) $$A_{\lambda/\mu,\pi} = \langle Q_{\lambda/\mu}, g_\pi \rangle$$

from which it follows that

(3) $$g_\pi = \sum_\nu A_{\nu\pi} P_\nu .$$

From (2) and (3) we have

$$\begin{aligned} A_{\lambda/\mu,\pi} &= \sum_\nu A_{\nu\pi} \langle Q_{\lambda/\mu}, P_\nu \rangle \\ &= \sum_\nu A_{\nu\pi} \langle Q_\lambda, P_\mu P_\nu \rangle \\ &= \sum_\nu A_{\nu\pi} f_{\mu\nu}^\lambda . \end{aligned}$$

The matrix $(A_{\nu\pi})$ is strictly lower triangular and invertible; if its inverse is $(B_{\nu\pi})$ we have

$$f_{\mu\nu}^\lambda = \sum_\pi A_{\lambda/\mu,\pi} B_{\pi\nu}$$

and hence the structure constants $f_{\mu\nu}^\lambda$ are in principle computable in terms of the φ_T, which are given explicitly by (7.11) and (6.24).

6. Let λ, μ be partitions. Then

(a) $\sum_{\rho} P_{\rho/\lambda}(x;q,t)Q_{\rho/\mu}(y;q,t) = \prod(x,y;q,t)\sum_{\sigma} P_{\mu/\sigma}(x;q,t)Q_{\lambda/\sigma}(y;q,t).$

The proof is the same as that of Chapter I, §5, Example 26.

(b) By applying $\omega_{q,t}$ to the symmetric functions of the x variables in (a) we obtain

$$\sum_{\rho} Q_{\rho'/\lambda'}(x;t,q)Q_{\rho/\mu}(y;q,t) = \Pi_0(x,y)\sum_{\sigma} Q_{\mu'/\sigma'}(x;t,q)Q_{\lambda/\sigma}(y;q,t),$$

where $\Pi_0(x,y) = \prod_{i,j}(1+x_iy_j)$.

(c) Likewise

$$\sum_{\rho} P_{\rho/\lambda}(x;q,t)P_{\rho'/\mu'}(y;t,q) = \Pi_0(x,y)\sum_{\sigma} P_{\mu/\sigma}(x;q,t)P_{\lambda'/\sigma'}(y;t,q).$$

Notes and references

Examples 1–3. The results of these Examples, in the context of Jack's symmetric functions, are due to R. Stanley [S25].

Example 4(i). This identity, again in the context of Jack's symmetric functions, is due to K. Kadell.

8. Integral forms

For each partition λ we define

(8.1) $$c_\lambda(q,t) = \prod_{s\in\lambda}(1-q^{a(s)}t^{l(s)+1}),$$

(8.1') $$c'_\lambda(q,t) = \prod_{s\in\lambda}(1-q^{a(s)+1}t^{l(s)}),$$

so that

(8.2) $$c'_\lambda(q,t) = c_{\lambda'}(t,q)$$

and by (6.19)

$$b_\lambda(q,t) = c_\lambda(q,t)/c'_\lambda(q,t).$$

Now let

(8.3) $$J_\lambda = J_\lambda(x;q,t) = c_\lambda(q,t)P_\lambda(x;q,t) = c'_\lambda(q,t)Q_\lambda(x;q,t).$$

The symmetric functions J_λ are in some sense 'integral forms' of the P_λ (or Q_λ). It seems likely that when they are expressed as linear combinations of the monomial symmetric functions, the coefficients are polynomials, not just rational functions, in q and t. We shall make a more precise conjecture later in this section.

(8.4) *Remarks.*

(i) When $q = t$ we have

$$c_\lambda(t,t) = c'_\lambda(t,t) = \prod_{s\in\lambda}(1 - t^{h(s)}),$$

the hook polynomial of λ, denoted by $H_\lambda(t)$ in Chapter I, §3, Example 2. Hence $J_\lambda = H_\lambda(t)s_\lambda$ when $q = t$.

(ii) When $q = 0$ we have $c'_\lambda(0,t) = 1$ and $c_\lambda(0,t) = b_\lambda(t)$, so that $J_\lambda(x;0,t)$ is the Hall–Littlewood function $Q_\lambda(x;t)$.

(iii) When $q = 1$,

$$c_\lambda(1,t) = \prod_{s\in\lambda}(1 - t^{l(s)+1}) = (t;t)_{\lambda'}$$

in the notation of §2, Example 1, and hence by (4.14) (vi)

$$J_\lambda(x;1,t) = (t;t)_{\lambda'} e_{\lambda'}(x).$$

(iv) When $t = 0$, $J_\lambda(x;q,0) = P_\lambda(x;q,0)$.

(v) When $t = 1$ we have $c_\lambda(q,1) = 0$ if $\lambda \neq 0$, so that $J_\lambda = 0$.

(vi) Let $q = t^\alpha$ and let $t \to 1$. The symmetric functions

$$J_\lambda^{(\alpha)}(x) = \lim_{t\to 1}\frac{J_\lambda(x;t^\alpha,t)}{(1-t)^{|\lambda|}}$$

are the integral forms of Jack's symmetric functions (see §10 below).

We have

$$\begin{aligned} c_\lambda(q^{-1},t^{-1}) &= \prod_{s\in\lambda}(1 - q^{-a(s)}t^{-l(s)-1}) \\ &= (-1)^{|\lambda|}q^{-n(\lambda')}t^{-n(\lambda)-|\lambda|}c_\lambda(q,t), \end{aligned}$$

since $\Sigma_{s\in\lambda}\, a(s) = n(\lambda')$ and $\Sigma_{s\in\lambda}\, l(s) = n(\lambda)$, and hence

(8.5) $$J_\lambda(x;q^{-1},t^{-1}) = (-1)^{|\lambda|}q^{-n(\lambda')}t^{-n(\lambda)-|\lambda|}J_\lambda(x;q,t)$$

by virtue of (4.14) (iv).

Next, duality (5.1) now takes the form

(8.6) $$\omega_{q,t}J_\lambda(q,t) = J_{\lambda'}(t,q)$$

and we have

(8.7) $$\langle J_\lambda, J_\lambda\rangle_{q,t} = c_\lambda(q,t)c'_\lambda(q,t) = \langle J_{\lambda'}, J_{\lambda'}\rangle_{t,q}.$$

The specialization formula (6.17) gives

(8.8)
$$\begin{aligned}\varepsilon_{u,t}J_\lambda(x;q,t) &= \prod_{s\in\lambda}(t^{l'(s)}-q^{a'(s)}u)\\ &= \prod_{(i,j)\in\lambda}(t^{i-1}-q^{j-1}u).\end{aligned}$$

As in Chapter III, (4.5) let $S_\lambda(x;t)$ denote the Schur functions associated with the product $\Pi(1-tx_i)/(1-x_i)$, so that (Chapter I, §7)

(8.9)
$$S_\lambda(x;t) = \sum_\rho z_\rho^{-1}\chi_\rho^\lambda p_\rho(x;t)$$

where

(8.10)
$$p_\rho(x;t) = p_\rho(x)\prod_{i=1}^{l(\rho)}(1-t^{\rho_i}).$$

The $S_\lambda(x;t)$ form an F-basis of Λ_F, and hence we may express the $J_\mu(x;q,t)$ in terms of them, say

(8.11)
$$J_\mu(x;q,t) = \sum_\lambda K_{\lambda\mu}(q,t)S_\lambda(x;t).$$

When $q=0$ we have $J_\mu(x;q,t) = Q_\mu(x;t)$ (8.4 (ii)), hence

(8.12)
$$K_{\lambda\mu}(0,t) = K_{\lambda\mu}(t)$$

where the $K_{\lambda\mu}(t)$ are the polynomials defined in Chapter III, §6. In particular,

(8.13)
$$K_{\lambda\mu}(0,0) = \delta_{\lambda\mu}, \qquad K_{\lambda\mu}(0,1) = K_{\lambda\mu},$$

where the $K_{\lambda\mu}$ are the Kostka numbers defined in Chapter I, §6, so that $K_{\lambda\mu}$ is the number of (column-strict) tableaux of shape λ and weight μ.

From (8.9) and (8.10) it follows that $S_\lambda(x;t^{-1}) = (-t)^{-|\lambda|}S_{\lambda'}(x;t)$ and hence by (8.5) that

(8.14)
$$K_{\lambda\mu}(q,t) = q^{n(\mu')}t^{n(\mu)}K_{\lambda'\mu}(q^{-1},t^{-1}).$$

Again, it follows from (8.9) that

$$\begin{aligned}\omega_{q,t}S_\lambda(x;t) &= \sum_\rho z_\rho^{-1}\chi_\rho^\lambda\varepsilon_\rho p_\rho(x;q)\\ &= S_{\lambda'}(x;q)\end{aligned}$$

and hence by (8.6) that

(8.15)
$$K_{\lambda\mu}(q,t) = K_{\lambda'\mu'}(t,q).$$

For each integer $n \geqslant 0$, let $K_n(q,t)$ denote the matrix $(K_{\lambda\mu}(q,t))$ where λ and μ run through the partitions of n. Unlike the matrices $K_n(t) = K_n(0,t)$ of Chapter III, the matrices $K_n(q,t)$ (for $n > 1$) are not upper triangular; indeed, it follows from (8.15) that $K_{\lambda\mu}(q,0) = K_{\lambda'\mu'}(q)$, so that $K_n(q,0)$ is lower triangular.

Another special case in

(8.16) $$K_{\lambda\mu}(1,1) = \chi^{\lambda}_{(1^n)} = n!/h(\lambda)$$

(where λ, μ are partitions of n).

Proof. From (8.9) and (8.11) we have

$$J_\mu(x;q,t) = \sum_\lambda K_{\lambda\mu}(q,t) \sum_\rho z_\rho^{-1} \chi_\rho^\lambda p_\rho(x;t).$$

On setting $q = t$ we obtain from (8.4)(i)

$$H_\mu(t) s_\mu(x) = \sum_{\lambda,\rho} z_\rho^{-1} \chi_\rho^\lambda K_{\lambda\mu}(t,t) p_\rho(x;t).$$

But also (Chapter I, §7)

$$s_\mu(x) = \sum_\rho z_\rho^{-1} \chi_\rho^\mu p_\rho(x)$$

and therefore

$$\sum_\lambda K_{\lambda\mu}(t,t) \chi_\rho^\lambda = \chi_\rho^\mu H_\mu(t) / \prod (1 - t^{\rho_i})$$

so that we obtain

(8.17) $$K_{\lambda\mu}(t,t) = \sum_\rho z_\rho^{-1} \chi_\rho^\lambda \chi_\rho^\mu H_\mu(t) / \prod (1 - t^{\rho_i})$$

by orthogonality of the characters of the symmetric group S_n. Now let $t \to 1$, and we obtain

$$K_{\lambda\mu}(1,1) = \chi^\lambda_{(1^n)}$$

since the only term that survives on the right-hand side of (8.17) is that corresponding to the partition $\rho = (1^n)$. ∎

The matrices $K_n(q,t)$ have been computed for $n \leqslant 8$. The results suggest the conjecture

(8.18?) *$K_{\lambda\mu}(q,t)$ is a polynomial in q and t with positive integral coefficients.*

Some further partial evidence in favour of this conjecture is contained in the Examples at the end of this section. In particular, we know from the theorem of Lascoux and Schützenberger (Chapter III, (6.5)) that $K_{\lambda\mu}(0,t)$ is a polynomial in t with positive integral coefficients. By (8.15) the same is true of $K_{\lambda\mu}(q,0)$.

Since $S_\lambda(x;t)$ is a linear combination of the monomial symmetric functions $m_\lambda(x)$ with coefficients in $\mathbf{Z}[t]$ (see the table of transition matrices in Chapter III, §6), the conjecture (8.18?) would imply that the J_λ are linear combinations of the m_μ with coefficients in $\mathbf{Z}[q,t]$.

On the assumption that (8.18?) is true, the number of monomials $q^a t^b$ in $K_{\lambda\mu}(q,t)$ would by (8.16) be equal to the number of standard tableaux of shape λ. One may therefore ask whether there is a combinatorial rule that attaches to each pair (T,μ), where T is a standard tableau containing n symbols and μ is a partition of n, a monomial $q^{a(T,\mu)}t^{b(T,\mu)}$, so that $K_{\lambda\mu}(q,t)$ is the sum of these monomials as T runs through the standard tableaux of shape λ.

Finally, we shall introduce generalizations of the polynomials $X_\rho^\lambda(t)$ of Chapter III, §7. For each pair of partitions λ,ρ we define $X_\rho^\lambda(q,t)\in F$ by

$$J_\lambda(x;q,t)=\sum_\rho z_\rho^{-1}X_\rho^\lambda(q,t)p_\rho(x;t) \tag{8.19}$$

(so that $X_\rho^\lambda(q,t)=0$ unless $|\lambda|=|\rho|$). By (8.4) (ii) and Chapter III, (7.5) we have

$$X_\rho^\lambda(0,t)=X_\rho^\lambda(t).$$

From (8.9) and (8.11) it follows that

$$J_\lambda(x;q,t)=\sum_\mu K_{\mu\lambda}(q,t)\sum_\rho z_\rho^{-1}\chi_\rho^\mu p_\rho(x;t)$$

and hence that

$$X_\rho^\lambda(q,t)=\sum_\mu \chi_\rho^\mu K_{\mu\lambda}(q,t). \tag{8.20}$$

By orthogonality of the characters of the symmetric group, this relation is equivalent to

$$K_{\mu\lambda}(q,t)=\sum_\rho z_\rho^{-1}\chi_\rho^\mu X_\rho^\lambda(q,t). \tag{8.20'}$$

The conjecture (8.18?), together with (8.20), would imply that $X_\rho^\lambda(q,t) \in \mathbf{Z}[q,t]$, since the χ_ρ^μ are integers.

From (8.14) and (8.15) we deduce that

$$X_\rho^\lambda(q^{-1},t^{-1}) = \varepsilon_\rho q^{-n(\lambda')}t^{-n(\lambda)}X_\rho^\lambda(q,t), \tag{8.21}$$

$$X_\rho^\lambda(t,q) = \varepsilon_\rho X_\rho^{\lambda'}(q,t). \tag{8.22}$$

The $X_\rho^\lambda(q,t)$ satisfy orthogonality relations that generalize those of Chapter III, §7. Namely, from (2.6), (4.13), and the definition (8.3) of J_λ we have

$$\sum_\rho z_\rho(q,t)^{-1} p_\rho(x) p_\rho(y) = \sum_\lambda c_\lambda(q,t)^{-1} c'_\lambda(q,t)^{-1} J_\lambda(x;q,t) J_\lambda(y;q,t).$$

If we now substitute (8.19) in the right-hand side of this relation, and compare the coefficients of $p_\rho(x)p_\sigma(y)$ on either side, we shall obtain

$$\sum_{|\lambda|=n} c_\lambda(q,t)^{-1} c_{\lambda'}(q,t)^{-1} X_\rho^\lambda(q,t) X_\sigma^\lambda(q,t) = \delta_{\rho\sigma}\zeta_\rho(q,t) \tag{8.23}$$

where

$$\zeta_\rho(q,t) = z_\rho \prod_{i=1}^{l(\rho)} (1-q^{\rho_i})^{-1}(1-t^{\rho_i})^{-1}. \tag{8.24}$$

An equivalent statement is

$$\sum_{|\rho|=n} \zeta_\rho(q,t)^{-1} X_\rho^\lambda(q,t) X_\rho^\mu(q,t) = \delta_{\lambda\mu} c_\lambda(q,t) c'_\lambda(q,t). \tag{8.25}$$

Finally, it follows from (8.16) and (8.20) that

$$X_\rho^\lambda(1,1) = \sum_\mu \chi_\rho^\mu \chi_{(1^n)}^\mu$$

and therefore by orthogonality of the χ_ρ^μ (Chapter I, §7)

$$X_\rho^\lambda(1,1) = \begin{cases} n! & \text{if } \rho = (1^n), \\ 0 & \text{otherwise}. \end{cases} \tag{8.26}$$

Remark (8.27). Let $V = \mathbb{Z}[S_n]$ be the regular representation of S_n, and for each partition λ of n let M_λ be an irreducible S_n-module with character χ^λ, so that in particular $M_{(1^n)}$ is the one-dimensional module affording the sign character ε. Then the conjecture (8.18?), together with (8.26), would imply that for each partition μ of n there is a bigrading of the regular representation

$$V = \bigoplus_{h,k} V_\mu^{h,k}$$

where $0 \leqslant h \leqslant n(\mu)$ and $0 \leqslant k \leqslant n(\mu')$, such that for each λ the multiplicity of M_λ in $V_\mu^{h,k}$ is equal to the coefficient of $t^h q^k$ in $K_{\lambda\mu}(q,t)$. By (8.14) and (8.15) these bigradings would satisfy

$$V_\mu^{h,k} \otimes M_{(1^n)} \cong V_\mu^{h',k} \cong V_{\mu'}^{k,h},$$

where $h' = n(\mu) - h$ and $k' = n(\mu') - k$.

Recently Garsia and Haiman [G4] have put forward a conjecture in this direction. Let $A = \mathbb{Z}[x_1, \dots, x_n, y_1, \dots, y_n]$, where the x's and y's are $2n$ independent indeterminates. The polynomial ring A is bigraded: $A = \bigoplus_{h,k \geqslant 0} A^{h,k}$, where $A^{h,k}$ consists of the polynomials $f \in A$ that are homogeneous of degree h (resp. k) in the x's (resp. y's). The symmetric group S_n acts diagonally on A:

$$wf(x_1, \dots, x_n, y_1, \dots, y_n) = f(x_{w(1)}, \dots, x_{w(n)}, y_{w(1)}, \dots, y_{w(n)})$$

for $w \in S_n$ and $f \in A$, and this action respects the bigrading.

Now let μ be a partition of n and let $(p_1, q_1), \dots, (p_n, q_n)$ denote the set of pairs $\{(i-1, j-1) : (i,j) \in \mu\}$ arranged in lexicographical order. We have $\sum p_i = n(\mu)$ and $\sum q_i = n(\mu')$. Let

$$\Delta_\mu(x,y) = \det(x_j^{p_i} y_j^{q_i})_{1 \leqslant i,j \leqslant n} \in A^{n(\mu), n(\mu')}.$$

Clearly, $w\Delta_\mu = \varepsilon(w)\Delta_\mu$ for $w \in S_n$. Let $H_\mu \subset A$ be the linear span of all the partial derivatives of $\Delta_\mu(x,y)$ of all orders with respect to the x's and y's. Then H_μ is stable under the action of S_n, and moreover

$$H_\mu = \bigoplus_{h,k} H_\mu^{h,k}$$

where $H_\mu^{h,k} = H_\mu \cap A^{h,k}$ is S_n-stable. Let $\varphi_\mu^{h,k}$ denote the character of the S_n-module $H_\mu^{h,k}$, and for each partition λ of n let

$$C_{\lambda\mu}(q,t) = \sum_{h,k} \langle \chi^\lambda, \varphi_\mu^{h,k} \rangle t^h q^k$$

where as in Chapter I, §7 the χ^λ are the irreducible characters of S_n. Then Garsia and Haiman conjecture (*loc. cit.*) that

$$C_{\lambda\mu}(q,t) = K_{\lambda\mu}(q,t^{-1})t^{n(\mu)}.$$

In particular they have verified that this is so for all pairs (λ, μ) of partitions of $n \leqslant 6$, and in other special cases.

The matrices $K(q,t)'$, $n \leqslant 6$

	1
1	1

	2	1^2
2	1	q
1^2	t	1

	3	21	1^3
3	1	$q+q^2$	q^3
21	t	$1+qt$	q
1^3	t^3	$t+t^2$	1

	4	31	2^2	21^2	1^4
4	1	$q+q^2+q^3$	q^2+q^4	$q^3+q^4+q^5$	q^6
31	t	$1+qt+q^2t$	$q+q^2t$	$q+q^2+q^3t$	q^3
2^2	t^2	$t+qt+qt^2$	$1+q^2t^2$	$q+qt+q^2t$	q^2
21^2	t^3	$t+t^2+qt^3$	$t+qt^2$	$1+qt+qt^2$	q
1^4	t^6	$t^3+t^4+t^5$	t^2+t^4	$t+t^2+t^3$	1

	5	41	32
5	1	$q+q^2+q^3+q^4$	$q^2+q^3+q^4+q^5+q^6$
41	t	$1+qt+q^2t+q^3t$	$q+q^2+q^2t+q^3t+q^4t$
32	t^2	$t+qt+qt^2+q^2t^2$	$1+qt+q^2t+q^2t^2+q^3t^2$
31^2	t^3	$t+t^2+qt^3+q^2t^3$	$t+qt+qt^2+q^2t^2+q^2t^3$
2^21	t^4	$t^2+t^3+qt^3+qt^4$	$t+t^2+qt^2+qt^3+q^2t^4$
21^3	t^6	$t^3+t^4+t^5+qt^6$	$t^2+t^3+t^4+qt^4+qt^5$
1^5	t^{10}	$t^6+t^7+t^8+t^9$	$t^4+t^5+t^6+t^7+t^8$

	31^2	2^21	21^3	1^5
5	$q^3+q^4+2q^5+q^6+q^7$	$q^4+q^5+q^6+q^7+q^8$	$q^6+q^7+q^8+q^9$	q^{10}
41	$q+q^2+q^3+q^3t+q^4t+q^5t$	$q^2+q^3+q^4+q^4t+q^5t$	$q^3+q^4+q^5+q^6t$	q^6
32	$q+qt+2q^2t+q^3t+q^3t^2$	$q+q^2+q^2t+q^3t+q^4t^2$	$q^2+q^3+q^3t+q^4t$	q^4
31^2	$1+qt+q^2t+qt^2+q^2t^2+q^3t^3$	$q+qt+q^2t+q^2t^2+q^3t^2$	$q+q^2+q^3t+q^3t^2$	q^3
2^21	$t+qt+2qt^2+qt^3+q^2t^3$	$1+qt+qt^2+q^2t^2+q^2t^3$	$q+qt+q^2t+q^2t^2$	q^2
21^3	$t+t^2+t^3+qt^3+qt^4+qt^5$	$t+t^2+qt^2+qt^3+qt^4$	$1+qt+qt^2+qt^3$	q
1^5	$t^3+t^4+2t^5+t^6+t^7$	$t^2+t^3+t^4+t^5+t^6$	$t+t^2+t^3+t^4$	1

	6	51	42
6	1	$q + q^2 + q^3 + q^4 + q^5$	$q^2 + q^3 + 2q^4 + q^5 + 2q^6 + q^7 + q^8$
51	t	$1 + qt + q^2t + q^3t + q^4t$	$q + q^2 + q^3 + q^2t + q^3t + 2q^4t + q^5t + q^6t$
42	t^2	$t + qt + qt^2 + q^2t^2 + q^3t^2$	$1 + qt + 2q^2t + q^3t + q^2t^2 + q^3t^2 + 2q^4t^2$
41^2	t^3	$t + t^2 + qt^3 + q^2t^3 + q^3t^3$	$t + qt + q^2t + qt^2 + q^2t^2 + q^3t^2 + q^2t^3 + q^3t^3 + q^4t^3$
3^2	t^3	$t^2 + qt^2 + q^2t^2 + qt^3 + q^2t^3$	$t + qt + q^2t + qt^2 + q^2t^2 + q^3t^2 + q^2t^3 + q^3t^3 + q^4t^3$
321	t^4	$t^2 + t^3 + qt^3 + qt^4 + q^2t^4$	$t + t^2 + 2qt^2 + qt^3 + 2q^2t^3 + q^2t^4 + q^3t^4$
31^3	t^6	$t^3 + t^4 + t^5 + qt^6 + q^2t^6$	$t^2 + t^3 + qt^3 + t^4 + qt^4 + q^2t^4 + qt^5 + q^2t^5 + q^2t^6$
2^3	t^6	$t^4 + qt^4 + t^5 + qt^5 + qt^6$	$t^2 + t^3 + qt^3 + t^4 + qt^4 + q^2t^4 + qt^5 + q^2t^5 + q^2t^6$
2^21^2	t^7	$t^4 + t^5 + t^6 + qt^6 + qt^7$	$2t^3 + t^4 + qt^4 + t^5 + 2qt^5 + qt^6 + q^2t^7$
21^4	t^{10}	$t^6 + t^7 + t^8 + t^9 + qt^{10}$	$t^4 + t^5 + 2t^6 + t^7 + qt^7 + t^8 + qt^8 + qt^9$
1^6	t^{15}	$t^{10} + t^{11} + t^{12} + t^{13} + t^{14}$	$t^7 + t^8 + 2t^9 + t^{10} + 2t^{11} + t^{12} + t^{13}$

	41^2	3^2
6	$q^3 + q^4 + 2q^5 + 2q^6 + 2q^7 + q^8 + q^9$	$q^3 + q^5 + q^6 + q^7 + q^9$
51	$q + q^2 + q^3 + q^4 + q^3t + q^4t + 2q^5t + q^6t + q^7t$	$q^2 + q^4 + q^3t + q^5t + q^6t$
42	$q + qt + 2q^2t + 2q^3t + q^4t + q^3t^2 + q^4t^2 + q^5t^2$	$q + q^2t + q^3t + q^3t^2 + q^5t^2$
41^2	$1 + qt + q^2t + q^3t + qt^2 + q^2t^2 + q^3t^2 + q^3t^3 + q^4t^3 + q^5t^3$	$qt + q^2t + q^2t^2 + q^4t^2 + q^3t^3$
3^2	$qt + q^2t + q^3t + qt^2 + 2q^2t^2 + 2q^3t^2 + q^4t^2 + q^3t^3$	$1 + q^2t^2 + q^3t^2 + q^4t^2 + q^3t^3$
321	$t + qt + 2qt^2 + q^2t^2 + qt^3 + 2q^2t^3 + q^3t^3 + q^3t^4$	$t + qt^2 + q^2t^2 + q^2t^3 + q^3t^4$
31^3	$t + t^2 + t^3 + qt^3 + q^2t^3 + qt^4 + q^2t^4 + qt^5 + q^2t^5 + q^3t^6$	$qt^2 + t^3 + qt^4 + q^2t^4 + q^2t^5$
2^3	$qt^2 + t^3 + 2qt^3 + q^2t^3 + 2qt^4 + q^2t^4 + qt^5 + q^2t^5$	$qt^2 + t^3 + qt^3 + qt^4 + q^3t^6$
2^21^2	$t^2 + t^3 + qt^3 + t^4 + 2qt^4 + 2qt^5 + qt^6 + q^2t^6$	$t^2 + t^4 + qt^4 + qt^5 + q^2t^6$
21^4	$t^3 + t^4 + 2t^5 + t^6 + qt^6 + t^7 + qt^7 + qt^8 + qt^9$	$t^4 + t^5 + qt^6 + t^7 + qt^8$
1^6	$t^6 + t^7 + 2t^8 + 2t^9 + 2t^{10} + t^{11} + t^{12}$	$t^6 + t^8 + t^9 + t^{10} + t^{12}$

	321
6	$q^4 + 2q^5 + 2q^6 + 3q^7 + 3q^8 + 2q^9 + 2q^{10} + q^{11}$
51	$q^2 + 2q^3 + 2q^4 + 2q^5 + q^6 + q^4t + 2q^5t + 2q^6t + 2q^7t + q^8t$
42	$q + 2q^2 + q^3 + q^2t + 3q^3t + 3q^4t + q^5t + q^4t^2 + 2q^5t^2 + q^6t^2$
41^2	$q + q^2 + qt + 2q^2t + 2q^3t + q^4t + q^2t^2 + 2q^3t^2 + 2q^4t^2 + q^5t^2 + q^4t^3 + q^5t^3$
3^2	$q + q^2 + qt + 2q^2t + 2q^3t + q^4t + q^2t^2 + 2q^3t^2 + 2q^4t^2 + q^5t^2 + q^4t^3 + q^5t^3$
321	$1 + 3qt + q^2t + qt^2 + 4q^2t^2 + q^3t^2 + q^2t^3 + 3q^3t^3 + q^4t^4$
31^3	$t + qt + t^2 + 2qt^2 + q^2t^2 + 2qt^3 + 2q^2t^3 + qt^4 + 2q^2t^4 + q^3t^4 + q^2t^5 + q^3t^5$
2^3	$t + qt + t^2 + 2qt^2 + q^2t^2 + 2qt^3 + 2q^2t^3 + qt^4 + 2q^2t^4 + q^3t^4 + q^2t^5 + q^3t^5$
2^21^2	$t + 2t^2 + qt^2 + t^3 + 3qt^3 + 3qt^4 + q^2t^4 + qt^5 + 2q^2t^5 + q^2t^6$
21^4	$t^2 + 2t^3 + 2t^4 + qt^4 + 2t^5 + 2qt^5 + t^6 + 2qt^6 + 2qt^7 + qt^8$
1^6	$t^4 + 2t^5 + 2t^6 + 3t^7 + 3t^8 + 2t^9 + 2t^{10} + t^{11}$

	31^3	2^3
6	$q^6+q^7+2q^8+2q^9+2q^{10}+q^{11}+q^{12}$	$q^6+q^8+q^9+q^{10}+q^{12}$
51	$q^3+q^4+2q^5+q^6+q^7+q^6t+q^7t+q^8t+q^9t$	$q^4+q^5+q^7+q^6t+q^8t$
42	$q^2+q^3+q^4+q^3t+2q^4t+2q^5t+q^6t+q^6t^2$	$q^2+q^4+q^4t+q^5t+q^6t^2$
41^2	$q+q^2+q^3+q^3t+q^4t+q^5t+q^3t^2+q^4t^2+q^5t^2+q^6t^3$	$q^3+q^2t+q^4t+q^4t^2+q^5t^2$
3^2	$q^3+q^2t+2q^3t+2q^4t+q^5t+q^3t^2+q^4t^2+q^5t^2$	$q^3+q^2t+q^3t+q^4t+q^6t^3$
321	$q+qt+2q^2t+q^3t+q^2t^2+2q^3t^2+q^3t^3+q^4t^3$	$q+q^2t+q^2t^2+q^3t^2+q^4t^3$
31^3	$1+qt+q^2t+qt^2+q^2t^2+qt^3+q^2t^3+q^3t^3+q^3t^4+q^3t^5$	$qt+qt^2+q^2t^2+q^3t^3+q^2t^4$
2^3	$qt+q^2t+qt^2+2q^2t^2+qt^3+2q^2t^3+q^3t^3+q^2t^4$	$1+q^2t^2+q^2t^3+q^3t^3+q^2t^4$
2^21^2	$t+qt+2qt^2+2qt^3+q^2t^3+qt^4+q^2t^4+q^2t^5$	$t+qt^2+qt^3+q^2t^3+q^2t^5$
21^4	$t+t^2+t^3+qt^3+t^4+qt^4+2qt^5+qt^6+qt^7$	$t^2+qt^3+t^4+qt^5+qt^6$
1^6	$t^3+t^4+2t^5+2t^6+2t^7+t^8+t^9$	$t^3+t^5+t^6+t^7+t^9$

	2^21^2	21^4	1^6
6	$q^7+q^8+2q^9+q^{10}+2q^{11}+q^{12}+q^{13}$	$q^{10}+q^{11}+q^{12}+q^{13}+q^{14}$	q^{15}
51	$q^4+q^5+2q^6+q^7+q^8+q^7t+q^8t+q^9t$	$q^6+q^7+q^8+q^9+q^{10}t$	q^{10}
42	$2q^3+q^4+q^5+q^4t+2q^5t+q^6t+q^7t^2$	$q^4+q^5+q^6+q^6t+q^7t$	q^7
41^2	$q^2+q^3+q^4+q^3t+q^4t+q^5t+q^4t^2+q^5t^2+q^6t^2$	$q^3+q^4+q^5+q^6t+q^6t^2$	q^6
3^2	$q^2+q^3+q^4+q^3t+q^4t+q^5t+q^4t^2+q^5t^2+q^6t^2$	$q^4+q^5+q^4t+q^5t+q^6t$	q^6
321	$q+q^2+2q^2t+q^3t+2q^3t^2+q^4t^2+q^4t^3$	$q^2+q^3+q^3t+q^4t+q^4t^2$	q^4
31^3	$q+qt+q^2t+qt^2+q^2t^2+q^3t^2+q^2t^3+q^3t^3+q^3t^4$	$q+q^2+q^3t+q^3t^2+q^3t^3$	q^3
2^3	$q+qt+q^2t+qt^2+q^2t^2+q^3t^2+q^2t^3+q^3t^3+q^3t^4$	$q^2+q^2t+q^3t+q^2t^2+q^3t^2$	q^3
2^21^2	$1+qt+2qt^2+q^2t^2+qt^3+q^2t^3+2q^2t^4$	$q+qt+q^2t+q^2t^2+q^2t^3$	q^2
21^4	$t+t^2+qt^2+t^3+qt^3+2qt^4+qt^5+qt^6$	$1+qt+qt^2+qt^3+qt^4$	q
1^6	$t^2+t^3+2t^4+t^5+2t^6+t^7+t^8$	$t+t^2+t^3+t^4+t^5$	1

Examples

1. We have

$$\sum_{n\geq 0} g_n(x;q,t) = \prod_{i,j} \frac{1-tx_iq^{j-1}}{1-x_iq^{j-1}}$$

$$= \sum_\lambda s_\lambda(1,q,q^2,\dots)S_\lambda(x;t)$$

$$= \sum_\lambda q^{n(\lambda)}H_\lambda(q)^{-1}S_\lambda(x;t)$$

by Chapter I, (4.3) and Chapter I, §3, Example 2, where $H_\lambda(q) = c_\lambda(q,q)$ is the hook-length polynomial. Hence

$$J_{(n)}(x;q,t) = (q;q)_n g_n(x;q,t)$$
$$= \sum_{|\lambda|=n} \frac{q^{n(\lambda)}(q;q)_n}{H_\lambda(q)} S_\lambda(x;t)$$

and therefore

$$K_{\lambda(n)}(q,t) = q^{n(\lambda)}(q;q)_n / H_\lambda(q)$$
$$= \sum_T q^{r(T)}$$

(Chapter I, §5, Example 14); the sum is over all standard tableaux T of shape λ, and $r(T)$ is the sum of the positive integers k such that $k+1$ lies in a lower row than k in T.

By duality (8.15) it follows that

$$K_{\lambda(1^n)}(q,t) = t^{n(\lambda')}(t;t)_n / H_\lambda(t).$$

2. Let $\lambda = (r+1, 1^s) = (r|s)$ in Frobenius notation. For each partition μ of $n = r+s+1$, $K_{\lambda\mu}(q,t)$ is the coefficient of u^s in the polynomial

$$\prod_{(i,j)} (t^{i-1} + q^{j-1}u)$$

where the product is over all $(i,j) \in \mu$ with the exception of $(i,j) = (1,1)$.

This can be proved by applying the specialization $\varepsilon_{-u,t}$ to both sides of (8.11). By (8.8) we have

$$\varepsilon_{-u,t} J_\mu(x;q,t) = \prod_{(i,j)\in\mu} (t^{i-1} + q^{j-1}u)$$

and

$$\varepsilon_{-u,t} S_\lambda(x;t) = \sum_\rho z_\rho^{-1} \chi_\rho^\lambda \prod_{i=1}^{l(\rho)} (1 - (-u)^{\rho_i})$$

which is the Schur function s_λ corresponding to the series $\sum h_n y^n = (1+uy)/(1-y)$, so that $s_\lambda = 0$ unless λ is a hook, and $s_\lambda = (1+u)u^s$ if $\lambda = (r|s)$. It follows that

$$(*) \qquad \prod_{(i,j)\in\mu} (t^{i-1} + q^{j-1}u) = \sum_{r+s=n-1} K_{(r|s)\mu}(q,t)(1+u)u^s,$$

which gives the result stated.

3. Consider the functions $K_{\lambda\mu}(q,t)$ when $q = t$. We have then $J_\mu = H_\mu(t)s_\mu(x)$, so that

$$H_\mu(t)s_\mu(x) = \sum_\lambda K_{\lambda\mu}(t,t)S_\lambda(x;t).$$

Since $(s_\lambda(x))$ and $(S_\lambda(x;t))$ are dual bases for the scalar product $\langle u,v\rangle_{0,t}$, it follows that

$$K_{\lambda\mu}(t,t) = H_\mu(t)\langle s_\lambda, s_\mu\rangle_{0,t}.$$

Now

$$\begin{aligned}\langle s_\lambda, s_\mu\rangle_{0,t} &= \sum_\rho z_\rho^{-1}\chi_\rho^\lambda\chi_\rho^\mu\prod_{i=1}^{l(\rho)}(1-t^{\rho_i})^{-1}\\ &= (s_\lambda * s_\mu)(1,t,t^2,\dots)\end{aligned}$$

where $s_\lambda * s_\mu$ is the internal product defined in Chapter I, §7.

If $\chi_\rho^\mu \neq 0$, the polynomial $\prod(1-t^{\rho_i})$ divides $H_\mu(t)$ (Stanley [S24]), from which it follows that $K_{\lambda\mu}(t,t)\in\mathbf{Q}[t]$. But also $s_\lambda * s_\mu$ is of the form $\sum_\nu a_\nu s_\nu$ with coefficients $a_\nu \in \mathbf{N}$, so that

$$\langle s_\lambda, s_\mu\rangle_{0,t} = \sum_\nu a_\nu t^{n(\nu)}H_\nu(t)^{-1}\in \mathbf{Z}[[t]].$$

Hence $K_{\lambda\mu}(t,t)\in\mathbf{Q}[t]\cap\mathbf{Z}[[t]] = \mathbf{Z}[t]$.

Also

$$X_\rho^\lambda(t,t) = \chi_\rho^\lambda H_\lambda(t)/\prod(1-t^{\rho_i})\in\mathbf{Z}[t].$$

4. Let λ be a partition. For each square s in the positive quadrant define

$$c_\lambda(s) = 1 - q^{a_\lambda(s)}t^{l_\lambda(s)+1},$$

$$c'_\lambda(s) = 1 - q^{a_\lambda(s)+1}t^{l_\lambda(s)}$$

if $s\in\lambda$, and $c_\lambda(s) = c'_\lambda(s) = 1$ otherwise. If S is any set of squares, let

$$c_\lambda^S = \prod_{s\in S} c'_\lambda(s)\,.\prod_{s\notin S} c_\lambda(s).$$

With this notation we have by (6.24)

$$J_\mu J_{(r)} = \sum_\lambda \eta_{\lambda/\mu}J_\lambda$$

summed over partitions $\lambda\supset\mu$ such that $\lambda-\mu$ is a horizontal strip of length r, where

$$\eta_{\lambda/\mu}(q,t) = (q;q)_r c_\mu^S(q,t)/c_\lambda^S(q,t),$$

and S is the union of the columns that contain squares $s\in\lambda-\mu$.

In particular

$$(1-q^rt^s)J_{(r)}J_{(1^s)} = (1-t^s)J_{(r|s-1)} + (1-q^r)J_{(r-1|s)}$$

if $r,s\geqslant 1$.

5. Let $\mu = (21^{n-2})$ and let λ be a partition of n. Then

$$K_{\lambda\mu}(q,t) = K_{\lambda\mu}(0,t) + qt^{n(\mu)}K_{\lambda'\mu}(0,t^{-1}).$$

(From the last equation of Example 4 above, when $r=1$ we have

$$(1-t^n)J_\mu=(1-qt^{n-1})J_{(1)}J_{(1^{n-1})}-(1-q)J_{(1^n)}.$$

By replacing (q,t) by $(0,t)$ and then by $(0,t^{-1})$, and eliminating e_1e_{n-1} and e_n from the three resulting equations, we obtain

$$J_\mu(q,t)=J_\mu(0,t)+(-1)^n qt^{1+n(n-1)/2}J_\mu(0,t^{-1}),$$

from which the result follows by picking out the coefficient of $S_\lambda(x;t)$ on either side, and bearing in mind that $S_\lambda(x;t^{-1})=(-t)^{-|\lambda|}S_{\lambda'}(x;t)$.)

6. We have

$$\det K_n(q,t)=\prod_{|\lambda|=n}F_\lambda(q,t)$$

where

$$F_\lambda(q,t)=\prod_{\substack{s\in\lambda\\ a(s)>0}}(1-q^{a(s)}t^{l(s)+1}).$$

For

$$K(q,t)'=M(J(q,t),S(t))$$

$$=M(J(q,t),P(q,t))M(P(q,t),m)M(S(t),m)^{-1}$$

where $M(J,P)$ is diagonal, with determinant $\prod_\lambda c_\lambda(q,t)$; $M(P,m)$ is unitriangular, hence has determinant 1; and (Chapter III, §6) $M(S,m)$ has determinant $\prod_\lambda b_\lambda(t)=\prod_\lambda c_\lambda(0,t)$. It follows that

$$\det K_n(q,t)=\prod_{|\lambda|=n}c_\lambda(q,t)/c_\lambda(0,t)$$

which is equivalent to the result stated.

7. When $q=1$ we have $P_\mu=e_{\mu'}$, so that

$$J_\mu(x;1,t)=(t;t)_{\mu'}e_{\mu'}(x). \tag{1}$$

Moreover, it follows from Chapter I, §3, Example 2 and Chapter I, (4.3′) that

$$e_r(x)=\sum_{|\lambda|=r}S_\lambda(x;t)t^{n(\lambda')}/H_\lambda(t). \tag{2}$$

Let us write

$$u_\lambda(t)=t^{n(\lambda')}(t;t)_r/H_\lambda(t) \tag{3}$$

for λ a partition of r. By Chapter I, §5, Example 14 we have

$$u_\lambda(t)=\sum_T t^{p(T)} \tag{4}$$

summed over the standard tableaux T of shape λ, where $p(T)$ is the sum of the entries k in T such that $k+1$ lies in a column to the right of k. In particular, therefore, $u_\lambda(t)$ is a polynomial in t with positive integral coefficients.

From (1), (2), and (3) we have

$$J_\mu(x;1,t) = \prod_{i\geq 1}\left(\sum u_{\lambda^i}(t) S_{\lambda^i}(x;t)\right)$$

where the inner sum is over partitions λ^i of μ_i'. It follows that

$$K_{\lambda\mu}(1,t) = \sum c^{\lambda}_{\lambda^1\lambda^2\ldots} u_{\lambda^1}(t) u_{\lambda^2}(t)\ldots \tag{5}$$

(where the c's are Littlewood–Richardson coefficients), and hence that $K_{\lambda\mu}(1,t)$ is a polynomial in t with positive integral coefficients. Dually, by (8.15), $K_{\lambda\mu}(q,1)$ is a polynomial in q with positive integral coefficients.

A closer examination of the equation (5), in the context of the 'algebra of tableaux' [S7] shows that

$$K_{\lambda\mu}(1,t) = \sum_T t^{p(T,\mu)} \tag{6}$$

summed over all standard tableaux T of shape λ, where

$$p(T,\mu) = \sum_{i\geq 1} p(\rho_i T)$$

and $\rho_i T$ is the restriction of T to the ith segment of $[1,n]$, of length μ_i', determined by the partition $\mu' = (\mu_1', \mu_2', \ldots)$. An equivalent formulation is

$$K_{\lambda\mu}(1,t) = \sum_T t^{c(T,\mu)} \tag{7}$$

summed as before over the standard tableaux T of shape λ, where

$$c(T,\mu) = \sum_{i\geq 1} c(\rho_i T)$$

and $c(\rho_i T)$ is the *charge* (Chapter III, (6.5)) of the skew standard tableau $\rho_i T$.

It can also be shown that if $w \in S_n$ is a transposition,

$$\frac{\partial K_{\lambda\mu}}{\partial t}(1,1) = \tfrac{1}{2}(\chi^\lambda(1) + \chi^\lambda(w)) n(\mu),$$

$$\frac{\partial K_{\lambda\mu}}{\partial q}(1,1) = \tfrac{1}{2}(\chi^\lambda(1) - \chi^\lambda(w)) n(\mu).$$

8. (a) Since

$$J_{(n)}(x;q,t) = (q;q)_n \sum_{|\rho|=n} z_\rho^{-1} p_\rho(x) \prod_{i=1}^{l(\rho)} \frac{1-t^{\rho_i}}{1-q^{\rho_i}},$$

it follows that

$$X_\rho^{(n)}(q,t) = (q;q)_n \prod (1-q^{\rho_i})^{-1}$$

for all partitions ρ of n. Dually,

$$X_\rho^{(1^n)}(q,t) = \varepsilon_\rho(t;t)_n \prod (1-t^{\rho_i})^{-1}.$$

(b) Another special case is

$$X_{(n)}^\lambda(q,t) = \prod_{(i,j)} (t^{i-1} - q^{j-1})$$

where the product is over all $(i,j) \in \lambda$ with the exception of $(1,1)$. For by (8.20) we have

$$X_{(n)}^\lambda(q,t) = \sum_\mu \chi_{(n)}^\mu K_{\lambda\mu}(q,t)$$

and $\chi_{(n)}^\mu$ is zero unless μ is a hook $(r \mid s)$, in which case it is equal to $(-1)^s$. The result now follows by setting $u = -1$ in the formula $(*)$ of Example 2.

(c) We have

$$X_{(1^n)}^\lambda(q,t) = \frac{c'_\lambda(q,t)}{(1-t)^n} \sum_T \varphi_T(q,t)$$

summed over the standard tableaux T of shape λ.

9. $$J_\lambda(1,t,t^2,\ldots;q,t) = t^{n(\lambda)}.$$

(Set $u = t^n$ in (8.8), and then let $n \to \infty$.)

10. For each partition λ, let $\Sigma_\lambda(x;q)$ denote the Schur function s_λ in the variables $x_i q^{j-1} (i,j \geqslant 1)$. Then $(\Sigma_\lambda(x;q))$ is the basis of Λ_F dual to the basis $(S_\lambda(x;t))$, and hence

$$K_{\lambda\mu}(q,t) = \langle \Sigma_\lambda(x;q), J_\mu(x;q,t) \rangle.$$

We can now define

$$K_{\lambda/\nu,\mu}(q,t) = \langle \Sigma_{\lambda/\nu}(x;q), J_\mu(x;q,t) \rangle$$

where $\lambda \supset \nu$ and $|\lambda| = |\mu| + |\nu|$ (otherwise it is zero). Since

$$\Sigma_{\lambda/\nu} = \sum_\pi c_{\nu\pi}^\lambda \Sigma_\pi$$

where the c's are Littlewood–Richardson coefficients, hence are integers $\geqslant 0$, it follows that

$$K_{\lambda/\nu,\mu}(q,t) = \sum_\pi c_{\nu\pi}^\lambda K_{\pi\mu}(q,t)$$

which will be a polynomial with positive integer coefficients if (8.18?) is true.

Since $\chi^{\lambda/\nu} = \Sigma_\pi c_{\nu\pi}^\lambda \chi^\pi$, it follows from (8.16) that $K_{\lambda/\nu,\mu}(1,1) = \chi^{\lambda/\nu}(1)$, the number of standard tableaux of shape $\lambda - \nu$.

We have

$$K_{\lambda/\nu,\mu}(t,q) = K_{\lambda'/\nu',\mu'}(q,t),$$

$$K_{\lambda/\nu,\mu}(q^{-1},t^{-1}) = q^{-n(\mu')}t^{-n(\mu)}K_{\lambda'/\nu',\mu}(q,t).$$

11. In this example we shall assume the truth of (8.18?).
(a) Let

$$J_\lambda(x;q,t) = \sum_{\mu \leqslant \lambda} v_{\lambda\mu}(q,t) m_\mu(x)$$

Then $(1-t)^{-l(\mu)} v_{\lambda\mu}(q,t) \in \mathbf{Z}[q,t]$.

We have already observed that (8.18?) implies that $v_{\lambda\mu} \in \mathbf{Z}[q,t]$. On the other hand, we have

$$J_\lambda(x;q,t) = \sum_\mu K_{\mu\lambda}(q,t) S_\mu(x;t)$$

$$= \sum_\mu K_{\mu\lambda}(q,t) \sum_\rho z_\rho^{-1} \chi_\rho^\mu p_\rho \prod_{i=1}^{l(\rho)} (1-t^{\rho_i}),$$

and

$$p_\rho = \sum_{\sigma \geqslant \rho} L_{\rho\sigma} m_\sigma$$

with coefficients $L_{\rho\sigma} \in \mathbf{N}$. Hence

$$v_{\lambda\sigma}(q,t) = \sum_\mu K_{\mu\lambda}(q,t) \sum_{\rho \leqslant \sigma} z_\rho^{-1} \chi_\rho^\mu L_{\rho\sigma} \prod_{i=1}^{l(\rho)} (1-t^{\rho_i}),$$

and since $\rho \leqslant \sigma$ implies that $l(\rho) \geqslant l(\sigma)$, every term in the inner sum is divisible by $(1-t)^{l(\sigma)}$.

(b) For any three partitions λ, μ, ν we have

$$\langle J_\lambda, J_\mu J_\nu \rangle \in \mathbf{Z}[q,t]$$

—again on the assumption that (8.18?) is true. For this it is enough to show that $\langle S_\lambda, S_\mu S_\nu \rangle \in \mathbf{Z}[q,t]$, and hence it is enough to show that $\langle S_\lambda, S_\mu \rangle \in \mathbf{Z}[q,t]$. But

$$\langle S_\lambda, S_\mu \rangle = \sum_\rho z_\rho^{-1} \chi_\rho^\lambda \chi_\rho^\mu \prod_{i=1}^{l(\rho)} (1-q^{\rho_i})(1-t^{\rho_i})$$

$$= \varphi(s_\lambda * s_\mu)$$

where $s_\lambda * s_\mu$ is the internal product defined in Chapter I, §7, and φ is the specialization defined by $\varphi(p_r) = (1-q^r)(1-t^r)$ for $r \geqslant 1$. We have then

$$\sum_{r \geqslant 0} \varphi(h_r) u^r = \frac{(1-qu)(1-tu)}{(1-u)(1-qtu)}$$

so that $\varphi(h_r) \in \mathbf{Z}[q,t]$ for all $r \geq 0$, and hence successively $\varphi(s_\lambda)$ for all λ and $\varphi(s_\lambda * s_\mu)$ for all λ, μ, lie in $\mathbf{Z}[q,t]$.

Notes and references

For other cases in which $K_{\lambda\mu}(q,t)$ can be shown to be a polynomial, see [S32].

9. Another scalar product

In this section we shall work with a finite number of variables $x = (x_1, \dots, x_n)$. We shall also assume, for simplicity of exposition, that $t = q^k$ where k is a non-negative integer (see, however, the remarks at the end of the section).

Let $L_n = F[x_1^{\pm 1}, \dots, x_n^{\pm 1}]$ be the F-algebra of Laurent polynomials in $x_1, \dots, x_n$ (i.e. polynomials in the x_i and x_i^{-1}). If $f = f(x_1, \dots, x_n) \in L_n$, let $\bar{f} = f(x_1^{-1}, \dots, x_n^{-1})$ and let $[f]_1$ denote the constant term in f. Also recall from §3 that $T_{q,x_i} f(x_1, \dots, x_n) = f(x_1, \dots, qx_i, \dots, x_n)$.

(9.1) *Let $f, g \in L_n$. Then*

$$\left[(T_{q,x_i} f)\bar{g}\right]_1 = \left[(T_{q,x_i} g)\bar{f}\right]_1.$$

Proof. Since both sides are linear in each of f and g, we may assume that f and g are monomials, and then the result is obvious. |

Now let

$$\Delta = \Delta(x;q,t) = \prod_{i \neq j} \left(x_i x_j^{-1}; q\right)_\infty \Big/ \left(t x_i x_j^{-1}; q\right)_\infty$$

$$= \prod_{i \neq j} \prod_{r=0}^{k-1} \left(1 - q^r x_i x_j^{-1}\right) \tag{9.2}$$

since $t = q^k$. Clearly $\Delta \in L_n$, and is symmetric in $x_1, \dots, x_n$. We define a scalar product on L_n as follows:

$$\langle f, g \rangle' = \langle f, g \rangle'_n = \frac{1}{n!} [f \bar{g} \Delta]_1. \tag{9.3}$$

(9.4) *Let $f, g \in \Lambda_{n,F}$. Then*

$$\langle D_n^1 f, g \rangle' = \langle f, D_n^1 g \rangle'.$$

where D_n^1 is the operator defined by (3.4).

Proof. Let

$$\Delta_+ = \prod_{i<j} (x_i x_j^{-1}; q)_\infty / (tx_i x_j^{-1}; q)_\infty$$

so that $\Delta = \Delta_+ \overline{\Delta}_+$. Then we have

$$\Delta_+^{-1} T_{q,x_1} \Delta_+ = \prod_{j=2}^{n} \frac{1 - tx_1 x_j^{-1}}{1 - x_1 x_j^{-1}} = A_1(x;t)$$

in the notation of §3, so that

$$\big(A_1(x;t) T_{q,x_1} f\big)\Delta_+ = T_{q,x_1}(f\Delta_+)$$

for all $f \in \Lambda_{n,F}$, and therefore

$$\big[A_1(x;t)(T_{q,x_1} f)\bar{g}\Delta\big]_1 = \big[T_{q,x_1}(f\Delta_+)\overline{g\Delta_+}\big]_1$$

which by (9.1) is symmetrical in f and g. By interchanging x_1 and x_i, it follows that

$$\big[A_i(x;t)(T_{q,x_i} f)\bar{g}\Delta\big]_1$$

is symmetrical in f and g for each $i = 1, 2, \dots, n$. Hence $[(D_n^1 f)\bar{g}\Delta]_1$ is symmetrical in f and g, which establishes (9.4). |

(9.5) *The polynomials* $P_\lambda(x;q,t)$, *where* $x = (x_1, \dots, x_n)$ *and* $l(\lambda) \leqslant n$, *are pairwise orthogonal for the scalar product* (9.3).

Proof. Since $D_n^1 P_\lambda = c_{\lambda\lambda} P_\lambda$ by (4.15), where

$$c_{\lambda\lambda} = \sum_{i=1}^{n} q^{\lambda_i} t^{n-i},$$

it follows from (9.4) that

$$c_{\lambda\lambda}\langle P_\lambda, P_\mu\rangle' = c_{\mu\mu}\langle P_\lambda, P_\mu\rangle'.$$

Since $c_{\lambda\lambda} \neq c_{\mu\mu}$ if $\lambda \neq \mu$, the result follows. |

Remarks. 1. It follows from (9.5) that each of the operators D_n^r defined in §3 is self-adjoint for the scalar product (9.3). For the P_λ are simultaneous eigenfunctions of these operators, by (4.15). Alternatively, this result can be established by direct computation, using the expression $(3.4)_r$ for D_n^r.

2. When $q = t$ (but not otherwise) the two scalar products $\langle f, g\rangle_n$ (2.20)

and $\langle f,g\rangle'_n$ coincide. For $\Delta(x;t,t)=\prod_{i\neq j}(1-x_ix_j^{-1})=a_\delta\bar{a}_\delta$, where $a_\delta=\prod_{i<j}(x_i-x_j)$ (Chapter I, §3), and therefore

$$\langle s_\lambda,s_\mu\rangle'_n=\frac{1}{n!}\left[s_\lambda\bar{s}_\mu a_\delta\bar{a}_\delta\right]_1$$

$$=\frac{1}{n!}[a_{\lambda+\delta}\bar{a}_{\mu+\delta}]_1=\delta_{\lambda\mu}.$$

On the other hand, when $q=t$ the scalar product $\langle f,g\rangle_n$ is that of Chapter I, so that $\langle s_\lambda,s_\mu\rangle_n=\delta_{\lambda\mu}$ by Chapter I, (4.8).

It remains to calculate the scalar product $\langle P_\lambda,P_\lambda\rangle'$. One form of the answer is, with the notation of (6.14)

(9.6) $$\langle P_\lambda,Q_\lambda\rangle'_n=c_n\prod_{s\in\lambda}\frac{1-q^{a'(s)}t^{n-l'(s)}}{1-q^{a'(s)+1}t^{n-l'(s)-1}}$$

where $l(\lambda)\leqslant n$ and

(9.7) $$c_n=\langle 1,1\rangle'_n=\frac{1}{n!}[\Delta]_1$$

Proof. We shall prove (9.6) by induction on $|\lambda|$. Let λ,μ be partitions of length $\leqslant n$ such that $\lambda\supset\mu$ and $\lambda-\mu$ consists of a single square s. Also let $\nu=\mu+(1^n)$, so that $\nu\supset\lambda$ and $\nu-\lambda$ is a vertical strip of length $n-1$. Since $e_n=x_1\dots x_n$ we have $e_n\bar{e}_n=1$ and hence from the definition (9.3) of the scalar product

$$\langle P_\lambda,e_1Q_\mu\rangle'=\langle\bar{e}_1e_nP_\lambda,e_nQ_\mu\rangle'.$$

But $\bar{e}_1e_n=e_{n-1}$, and $e_nQ_\mu=b_\mu e_nP_\mu=b_\mu P_\nu$ (4.17). Hence

(1) $$\langle P_\lambda,e_1Q_\mu\rangle'=b_\mu b_\nu^{-1}\langle e_{n-1}P_\lambda,Q_\nu\rangle'.$$

On the other hand, by (6.24) and (9.5) we have

(2) $$\langle P_\lambda,e_1Q_\mu\rangle'=\varphi'_{\lambda/\mu}\langle P_\lambda,Q_\lambda\rangle',$$

(3) $$\langle e_{n-1}P_\lambda,Q_\nu\rangle'=\psi'_{\nu/\lambda}\langle P_\nu,Q_\nu\rangle'=\psi'_{\nu/\lambda}b_\nu\langle P_\nu,P_\nu\rangle'$$

and

(4) $$\langle P_\nu,P_\nu\rangle'=\langle e_nP_\mu,e_nP_\mu\rangle'=\langle P_\mu,P_\mu\rangle'=b_\mu^{-1}\langle P_\mu,Q_\mu\rangle'.$$

From (1)–(4) we obtain

(5) $$\varphi'_{\lambda/\mu}\langle P_\lambda,Q_\lambda\rangle'=\psi'_{\nu/\lambda}\langle P_\mu,Q_\mu\rangle'.$$

Now from (6.24) we have

$$\psi'_{\nu/\lambda} = \prod_{\sigma \in R} b_\nu(\sigma)/b_\lambda(\sigma),$$

$$\varphi'_{\lambda/\mu} = \prod_{\sigma \in R} b_\mu(\sigma)/b_\lambda(\sigma)$$

where R is the row containing the unique square $s \in \lambda - \mu$. Hence

$$\psi'_{\nu/\lambda}/\varphi'_{\lambda/\mu} = \prod_{\sigma \in R} b_\nu(\sigma)/b_\mu(\sigma).$$

Since $\nu = \mu + (1^n)$ it follows that if $\sigma \in R \cap \mu$ and τ is the square immediately to the right of σ, we have $a_\nu(\tau) = a_\mu(\sigma)$ and $l_\nu(\tau) = l_\mu(\sigma)$ and therefore $b_\nu(\tau) = b_\mu(\sigma)$. Hence

(6) $$\psi'_{\nu/\lambda}/\varphi'_{\lambda/\mu} = b_\nu(\sigma_1)$$

where σ_1 is the leftmost square in the row R. For this square we have $a_\nu(\sigma_1) = a'(s)$ and $l_\nu(\sigma_1) = n - 1 - l'(s)$. Hence it follows from (5) and (6) that

$$\frac{\langle P_\lambda, Q_\lambda \rangle'}{\langle P_\mu, Q_\mu \rangle'} = \frac{1 - q^{a'(s)} t^{n - l'(s)}}{1 - q^{a'(s)+1} t^{n - l'(s) - 1}},$$

and the proof of (9.6) is complete. |

Another (equivalent) formula for the scalar product $\langle P_\lambda, P_\lambda \rangle'_n$ is given in Example 1 below.

We shall now renormalize the scalar product (9.3), and define

(9.8) $$\langle f, g \rangle''_n = c_n^{-1} \langle f, g \rangle'_n$$

for $f, g \in \Lambda_{n,F}$, so that $\langle 1, 1 \rangle''_n = 1$. The original scalar product (1.5) is then the limit of the scalar product (9.8) as $n \to \infty$. Precisely, we have

(9.9) *Let* $f, g \in \Lambda_F$ *and let* $\rho_n \colon \Lambda_F \to \Lambda_{F,n}$ *be the canonical homomorphism* (Chapter I, §2). *Then*

$$\langle \rho_n f, \rho_n g \rangle''_n \to \langle f, g \rangle$$

as $n \to \infty$, *where the scalar product on the right is that defined by* (1.5).

Proof. It is enough to verify this when $f = P_\lambda$ and $g = Q_\mu$. If $\lambda \neq \mu$, both scalar products are zero, by (9.5) and (4.7). If on the other hand $\lambda = \mu$, it follows from (9.6) that $\langle P_\lambda, Q_\lambda \rangle''_n \to 1$ as $n \to \infty$. |

Remark. We have assumed throughout this section that $t = q^k$ where k is a non-negative integer. This restriction is not essential, and may be avoided as follows. Assume that q is a complex number such that $|q| < 1$, so that the infinite product $(z; q)_\infty$ converges for all $z \in \mathbf{C}$, and define

$$\langle f, g \rangle_n' = \frac{1}{n!} \int_T f(z) \overline{g(z)} \Delta(z; q, t)\, dz \tag{9.10}$$

where $T = \{z = (z_1, \ldots, z_n) \in \mathbf{C}^n : |z_i| = 1,\ 1 \leqslant i \leqslant n\}$ is the n-dimensional torus, and dz is normalized Haar measure on T. The integrand in (9.10) is a continuous function on T, provided that $|t| < 1$.

When $t = q^k, k \in \mathbf{N}$, this definition agrees with (9.3), since for a Laurent polynomial $f \in L_n$ the integral of f over T is equal to the constant term of f.

Examples

1. (a) Let $x = (x_1, \ldots, x_n)$ and let

$$\Delta'(x; q, t) = \Delta(x; q, t) \prod_{i<j} \frac{1 - tx_i x_j^{-1}}{1 - x_i x_j^{-1}}$$

$$= \prod_{1 \leqslant i < j \leqslant n} \prod_{r=1}^{k} (1 - q^r x_i x_j^{-1})(1 - q^{r-1} x_i^{-1} x_j)$$

if $t = q^k$. The constant term in Δ' is [Z1] equal to

$$\prod_{i=2}^{n} \begin{bmatrix} ik \\ k \end{bmatrix}$$

where the square brackets denote q-binomial coefficients (Chapter I, §2, Example 3).

(b) From Chapter III, (1.3) we have

$$\sum_{w \in S_n} w \left(\prod_{i<j} \frac{1 - tx_i x_j^{-1}}{1 - x_i x_j^{-1}} \right) = \prod_{i=2}^{n} \frac{1 - t^i}{1 - t}.$$

(c) Deduce from (a) and (b) that the constant c_n of (9.7) is given by

$$c_n = \prod_{i=2}^{n} \begin{bmatrix} ik - 1 \\ k - 1 \end{bmatrix}$$

and hence that

$$L(c_n) = \frac{(1-t)(1-qt^{-1})}{1-q} \sum_{i<j} t^{j-i}$$

where L is the operator (6.12).

(d) From (c) and (9.6) we have

$$L(\langle P_\lambda, P_\lambda\rangle'_n) = L(c_n) + (1-qt^{-1}) \sum_{s\in\lambda} (q^{a'(s)}t^{n-l'(s)} - q^{a(s)}t^{l(s)+1})$$

which by (6.15) is equal to

$$\frac{(1-t)(1-qt^{-1})}{1-q} \sum_{i<j} q^{\lambda_i-\lambda_j}t^{j-i}.$$

Hence

$$\langle P_\lambda, P_\lambda\rangle'_n = \prod_{i<j} \frac{(q^{\lambda_i-\lambda_j}t^{j-i};q)_\infty (q^{\lambda_i-\lambda_j+1}t^{j-i};q)_\infty}{(q^{\lambda_i-\lambda_j}t^{j-i+1};q)_\infty (q^{\lambda_i-\lambda_j+1}t^{j-i-1};q)_\infty}$$

$$= \prod_{i<j} \prod_{r=1}^{k-1} \frac{1-q^{\lambda_i-\lambda_j+r}t^{j-i}}{1-q^{\lambda_i-\lambda_j-r}t^{j-i}}$$

if $t=q^k$.

2. (a) Let $0<q<1$ and let f be a function defined on the closed interval $[0,1]$. The *q-integral* of f is defined to be

$$\int_0^1 f(x)\,d_qx = (1-q)\sum_{r=0}^\infty q^r f(q^r)$$

for all f such that the series on the right converges. Thus it is the limit as $n\to\infty$ of the Riemann sums of f corresponding to the subdivisions of $[0,1]$ at the points $q, q^2, \dots, q^n$ (provided that $xf(x)\to 0$ as $x\to 0$). More generally, if $f(x) = f(x_1,\dots,x_n)$ is a function of n variables defined on the unit cube $C_n = [0,1]^n$, the q-integral of f is defined to be

$$\int_{C_n} f(x)\,d_qx = (1-q)^n \sum_{\alpha\in\mathbf{N}^n} q^{|\alpha|} f(q^\alpha)$$

where $\alpha = (\alpha_1,\dots,\alpha_n), |\alpha| = \alpha_1+\dots+\alpha_n$ and $f(q^\alpha) = f(q^{\alpha_1},\dots,q^{\alpha_n})$.

(b) Let r, s be positive integers (or, more generally, positive real numbers). Then

$$\int_0^1 x^{r-1}(qx;q)_{s-1}\,d_qx = \Gamma_q(r)\Gamma_q(s)/\Gamma_q(r+s) \tag{1}$$

$$= B_q(r,s)$$

where $\Gamma_q(r) = (q;q)_{r-1}/(1-q)^{r-1}$ is the q-gamma function. The formula (1) is a q-analogue of Euler's beta integral, which is the limiting case as $q\to 1$. (It is equivalent to Chapter I, §2, Example 5, which may be rewritten in the form

$$\sum_{m\geq 0} a^m \frac{(q^{m+1};q)_\infty}{(a^{-1}bq^m;q)_\infty} = \frac{(b;q)_\infty(q;q)_\infty}{(a;q)_\infty(a^{-1}b;q)_\infty}.$$

If we set $a=q^r$ and $b=q^{r+s}$, then this sum, multiplied by $1-q$, is just the q-integral (1).)

3. Let $x=(x_1,\dots,x_n)$, let $t=q^k$ and define

$$\Delta^*(x;q,t)=\prod_{1\leqslant i<j\leqslant n}\prod_{r=0}^{k-1}(x_i-q^r x_j)(x_i-q^{-r}x_j)$$

(1) $$=(-1)^A q^B (x_1\dots x_n)^{k(n-1)}\Delta(x;q,t)$$

where $A=\frac{1}{2}kn(n-1)$ and $B=\frac{1}{4}k(k-1)n(n-1)$. In this example we shall evaluate the multiple q-integral

$$I_\lambda=\frac{1}{n!}\int_{C_n}P_\lambda(x;q,t)\Delta^*_{r,s}(x;q,t)\,d_q x$$

where

$$\Delta^*_{r,s}(x;q,t)=\Delta^*(x;q,t)\prod_{i=1}^n x_i^{r-1}(qx_i;q)_{s-1},$$

and λ is a partition of length $\leqslant n$. (When $\lambda=0$, I_0 is a q-analogue of Selberg's integral.)

(a) For this purpose we shall expand $P_\lambda\Delta^*$ as a sum of monomials, say

(2) $$P_\lambda\Delta^*=\sum_\beta c_{\lambda\beta}x^\beta$$

summed over $\beta\in\mathbf{N}^n$ such that $|\beta|=|\lambda|+kn(n-1)$. Then we have

$$n!I_\lambda=\sum_\beta c_{\lambda\beta}\int_{C_n}\prod_{i=1}^n x_i^{r+\beta_i-1}(qx_i;q)_{s-1}\,d_q x_i$$

$$=\sum_\beta c_{\lambda\beta}\prod_{i=1}^n\frac{\Gamma_q(r+\beta_i)\Gamma_q(s)}{\Gamma_q(r+s+\beta_i)}$$

by Example 2(b), so that

(3) $$n!I_\lambda=B_q(r,s)^n\sum_\beta c_{\lambda\beta}(q^r;q)_\beta/(q^{r+s};q)_\beta$$

where (as in §2, Example 1) $(a;q)_\beta$ means $\prod(a;q)_{\beta_i}$.

(b) To evaluate the sum in (3) we shall apply the specialization $\varepsilon_{u,t}$ (§6) to the y-variables in the Cauchy formula (4.13). In this way we obtain

(4) $$\sum_\mu\varepsilon_{u,t}(Q_\mu)P_\mu(x)=\prod_{i=1}^n\frac{(ux_i;q)_\infty}{(x_i;q)_\infty}.$$

Let $\mu=(\mu_1,\dots,\mu_n)$ be the partition defined by $\mu_i=\lambda_i+(n-1)k+a$ $(1\leqslant i\leqslant n)$, where a is a positive integer to be determined later. From (4) we have

(5) $$\varepsilon_{u,t}(Q_\mu)\langle P_\mu,P_\mu\rangle'=\langle\prod_{i=1}^n\frac{(ux_i;q)_\infty}{(x_i;q)_\infty},P_\mu\rangle'.$$

By (4.17) the right-hand side of (5) is the constant term in

$$\frac{1}{n!}P_\lambda(x;q,t)(x_1\dots x_n)^{(n-1)k+a}\Delta(x;q,t)\prod_{i=1}^{n}\frac{(ux_i^{-1};q)_\infty}{(x_i^{-1};q)_\infty}$$

which by (1) and (2) is equal to

$$\frac{1}{n!}(-1)^A q^B(x_1\dots x_n)^a\sum_\beta c_{\lambda\beta}x^\beta\prod_{i=1}^{n}\frac{(ux_i^{-1};q)_\infty}{(x_i^{-1};q)_\infty}.$$

Now the constant term in $x_i^{a+\beta_i}(ux_i^{-1};q)_\infty/(x_i^{-1};q)_\infty$ is the coefficient of $x_i^{a+\beta_i}$ in $(ux_i;q)_\infty/(x_i;q)_\infty$, which is (Chapter I, §2, Example 5)

$$\frac{(u;q)_{a+\beta_i}}{(q;q)_{a+\beta_i}}=\frac{(u;q)_a}{(q;q)_a}\frac{(uq^a;q)_{\beta_i}}{(q^{a+1};q)_{\beta_i}}.$$

Hence from (5) we have

$$\varepsilon_{u,t}(Q_\mu)\langle P_\mu,P_\mu\rangle'=\frac{1}{n!}(-1)^A q^B\frac{(u;q)_a^n}{(q;q)_a^n}\sum_\beta c_{\lambda\beta}\frac{(uq^a;q)_\beta}{(q^{a+1};q)_\beta}. \tag{6}$$

(c) The left-hand side of (6) can be evaluated by use of Example 1 above and §6, Example 4. We thus obtain

$$\frac{1}{n!}\sum_\beta c_{\lambda\beta}\frac{(uq^a;q)_\beta}{(q^{a+1};q)_\beta}=(-1)^A q^{-B+kn(\mu)}v_\lambda(q,t)\prod_{i=1}^{n}\frac{(q;q)_a(uq^{k(1-i)};q)_{\mu_i}}{(q;q)_{\mu_i+k(n-i)}(u;q)_a}$$

where (§6, Example 4)

$$v_\lambda(q,t)=\prod_{1\leqslant i<j\leqslant n}(q^{\lambda_i-\lambda_j}t^{j-i};q)_k.$$

Since

$$(uq^{k(1-i)};q)_{\mu_i}=(uq^{k(1-i)};q)_{k(i-1)}(u;q)_a(uq^a;q)_{\lambda_i+k(n-i)}$$

and

$$(uq^{k(1-i)};q)_{k(i-1)}=(-u)^{k(i-1)}q^{-C_i}(qu^{-1};q)_{k(i-1)},$$

where $C_i=\frac{1}{2}(i-1)k((i-1)k+1)$, it follows that

$$\frac{1}{n!}\sum_\beta c_{\lambda\beta}\frac{(uq^a;q)_\beta}{(q^{a+1};q)_\beta}=u^A q^E v_\lambda(q,t)\prod_{i=1}^{n}\frac{(qu^{-1};q)_{k(i-1)}(uq^a;q)_{\lambda_i+k(n-i)}}{(q^{a+1};q)_{\lambda_i+k(2n-i-1)}} \tag{7}$$

where the exponent of q is

$$E = -B + kn(\mu) - \sum_{i=1}^{n} C_i$$

$$= k\left(n(\lambda) + a\binom{n}{2}\right) + 2k^2\binom{n}{3}.$$

(d) Finally, if we now set $a = r+s-1$ and $u = q^{1-s}$ in (7), we shall obtain from (3) and (7) the desired result:

$$I_\lambda = q^F \prod_{i=1}^{n} \frac{\Gamma_q(\lambda_i + r + k(n-i))\Gamma_q(s + k(i-1))}{\Gamma_q(\lambda_i + r + s + k(2n-i-1))}$$

$$\times \prod_{1 \leqslant i < j \leqslant n} \frac{\Gamma_q(\lambda_i - \lambda_j + k(j-i+1))}{\Gamma_q(\lambda_i - \lambda_j + k(j-i))}$$

where $F = k(n(\lambda) + \frac{1}{2}rn(n-1)) + \frac{1}{3}k^2 n(n-1)(n-2)$.

(e) Deduce from (d) and §6, Example 4 that

$$I_\lambda / I_0 = \varepsilon_{u,t}(P_\lambda)\varepsilon_{v,t}(P_\lambda)/\varepsilon_{w,t}(P_\lambda),$$

where $u = q^r t^{n-1}$, $v = t^n$, and $w = q^{r+s}t^{2n-2}$.

Notes and references

The relationship (9.9) between the two scalar products was first remarked by Kadell, in the context of Jack's symmetric functions. Likewise, Example 3 is a q-analogue of an integral formula due to Kadell.

In the definition (9.2) of Δ, and in the scalar product formula of Example 1, the structure of the root system of type A_{n-1} is clearly visible. In fact, this aspect of the theory generalizes to other root systems: see the last section of [M6].

10. Jack's symmetric functions

In this section we shall summarize the main properties of Jack's symmetric functions. As we have already observed in §1, they are obtained from the preceding theory by means of the specialization

$$q = t^\alpha, \qquad t \to 1.$$

Here q and t are to be thought of as real variables, and α as a positive real number.

More generally, let

$$(q,t) \to_\alpha (1,1) \tag{10.1}$$

mean that $(q,t)\to(1,1)$ in such a way that $(1-q)/(1-t)\to\alpha$. (For example, $q=t^\alpha$ as above, or again $q=1-\alpha(1-t)$.) Then for any real numbers a,b,c,d (such that $c\alpha+d\neq 0$) we have

$$\frac{1-q^a t^b}{1-q^c t^d}\to\frac{a\alpha+b}{c\alpha+d} \tag{10.2}$$

as $(q,t)\to_\alpha(1,1)$.

(10.3) *We have*

$$(tx;q)_\infty/(x;q)_\infty\to(1-x)^{-1/\alpha}$$

as $(q,t)\to_\alpha(1,1)$.

This statement, and others of the same sort that will occur later, is to be interpreted in the sense of termwise convergence of formal power series. The coefficient of x^n in $(tx;q)_\infty/(x;q)_\infty$ is (Chapter I, §2, Example 5)

$$\frac{(t;q)_n}{(q;q)_n}=\prod_{r=0}^{n-1}\frac{1-q^r t}{1-q^{r+1}}$$

and by (10.2) this tends to the limit

$$\prod_{r=0}^{n-1}\frac{r\alpha+1}{(r+1)\alpha}=(-1)^n\binom{-1/\alpha}{n}$$

which is the coefficient of x^n in $(1-x)^{-1/\alpha}$. |

Again, from (10.2) we have

$$z_\lambda(q,t)=z_\lambda\prod_{i=1}^{l(\lambda)}\frac{1-q^{\lambda_i}}{1-t^{\lambda_i}}\to z_\lambda\alpha^{l(\lambda)}$$

as $(q,t)\to_\alpha(1,1)$, for all partitions λ. Hence the scalar product $\langle f,g\rangle_{q,t}$ defined by (1.5) becomes in the limit the scalar product $\langle f,g\rangle_\alpha$ on Λ_F (where now $F=\mathbf{Q}(\alpha)$) defined by (1.4):

$$\langle p_\lambda,p_\mu\rangle_\alpha=\delta_{\lambda\mu}\alpha^{l(\lambda)}z_\lambda.$$

By (10.3), the product $\Pi(x,y;q,t)$ defined in (2.5) is replaced by

$$\Pi(x,y;\alpha)=\prod_{i,j}(1-x_iy_j)^{-1/\alpha}.$$

In place of (2.7) we have

(10.4) *For each $n \geqslant 0$, let (u_λ) and (v_λ) be F-bases of Λ^n_F, indexed by the partitions of n. Then the following conditions are equivalent:*

(a) $\langle u_\lambda, v_\mu\rangle_\alpha = \delta_{\lambda\mu}$ *for all* λ, μ;

(b) $\Sigma_\lambda u_\lambda(x) v_\lambda(y) = \Pi(x;y;\alpha)$

The specialization of $g_n(x;q,t)$ (2.8) is $g_n^{(\alpha)}(x)$, defined by the generating function

$$\sum_{n \geqslant 0} g_n^{(\alpha)}(x) y^n = \prod_i (1 - x_i y)^{-1/\alpha}.$$

As in §2, we define

$$g_\lambda^{(\alpha)}(x) = \prod_{i \geqslant 1} g_{\lambda_i}^{(\alpha)}(x)$$

for any partition $\lambda = (\lambda_1, \lambda_2, \dots)$. By specializing (2.9) and (2.11) we obtain

(10.5) $$g_n^{(\alpha)} = \sum_{|\lambda| = n} z_\lambda^{-1} \alpha^{-l(\lambda)} p_\lambda$$

and

$$\langle g_\lambda^{(\alpha)}, m_\mu\rangle_\alpha = \delta_{\lambda\mu}$$

so that the $g_\lambda^{(\alpha)}$ form a basis of Λ_F dual to the basis (m_λ). Hence the $g_n^{(\alpha)}, n \geqslant 1$, are algebraically independent over $F = \mathbf{Q}(\alpha)$, and $\Lambda_F = F[g_1^{(\alpha)}, g_2^{(\alpha)}, \dots]$.

For each real number β let ω_β denote the F-algebra endomorphism of Λ_F defined by

(10.6) $$\omega_\beta(p_r) = (-1)^{r-1} \beta p_r$$

for $r \geqslant 1$. We have $\omega_\beta^{-1} = \omega_{\beta^{-1}}$ if $\beta \neq 0$, and ω_1 is the involution ω of Chapter I. From Chapter I, (2.14)′ we have

(10.7) $$\omega_\alpha(g_n^{(\alpha)}) = e_n$$

and it follows directly from the definitions (1.4) and (10.6) that ω_β is self-adjoint, i.e.

(10.8) $$\langle \omega_\beta f, g\rangle_\alpha = \langle f, \omega_\beta g\rangle_\alpha$$

for all $f, g \in \Lambda_F$. Moreover we have

(10.9) $$\langle \omega_{\alpha^{-1}} f, g\rangle_\alpha = \langle \omega f, g\rangle_1.$$

Consider next the behaviour of $P_\lambda(x;q,t)$ as $(q,t) \to_\alpha (1,1)$. From (7.13′), the coefficient of $m_\mu(x)$ in $P_\lambda(x;q,t)$ is

$$u_{\lambda\mu}(q,t) = \sum_T \psi_T(q,t)$$

summed over tableaux T of shape λ and weight μ, where ψ_T is given explicitly by (7.11′), (6.24), and (6.20). These formulas show that $\psi_T(q,t)$ is a product of terms of the form (10.2), and hence has a well-defined limit $\psi_T^{(\alpha)}$ as $(q,t) \to_\alpha (1,1)$.

Explicitly, the limit of $b_\lambda(s;q,t)$ (6.20) is

(10.10)
$$b_\lambda^{(\alpha)}(s) = \frac{\alpha a_\lambda(s) + l_\lambda(s) + 1}{\alpha a_\lambda(s) + l_\lambda(s) + \alpha},$$

and the limit of $\psi_{\lambda/\mu}(q,t)$ (6.24) is

(10.11)
$$\psi_{\lambda/\mu}^{(\alpha)} = \prod_{s \in R_{\lambda/\mu} - C_{\lambda/\mu}} b_\mu^{(\alpha)}(s)/b_\lambda^{(\alpha)}(s)$$

where (*loc. cit.*) $C_{\lambda/\mu}$ (resp. $R_{\lambda/\mu}$) is the union of the columns (resp. rows) that intersect $\lambda - \mu$. Hence (7.11′)

(10.12)
$$\psi_T^{(\alpha)} = \prod_{i=1}^{r} \psi_{\lambda^{(i)}/\lambda^{(i-1)}}^{(\alpha)}$$

where $0 = \lambda^{(0)} \subset \lambda^{(1)} \subset \ldots \subset \lambda^{(r)} = \lambda$ is the sequence of partitions determined by the tableau T (so that each skew diagram $\lambda^{(i)} - \lambda^{(i-1)}$ is a horizontal strip).

Hence the limit of $u_{\lambda\mu}(q,t)$ as $(q,t) \to_\alpha (1,1)$ is

$$u_{\lambda\mu}^{(\alpha)} = \sum_T \psi_T^{(\alpha)}$$

summed as above over tableaux T of shape λ and weight μ, and therefore

(10.13)
$$P_\lambda^{(\alpha)} = m_\lambda + \sum_{\mu < \lambda} u_{\lambda\mu}^{(\alpha)} m_\mu$$

is the limit of $P_\lambda(q,t)$ as $(q,t) \to_\alpha (1,1)$. We have

(10.14)
$$\langle P_\lambda^{(\alpha)}, P_\mu^{(\alpha)} \rangle_\alpha = 0$$

whenever $\lambda \neq \mu$.

The $P_\lambda^{(\alpha)}$ are *Jack's symmetric functions*. They are characterized by the properties (10.13), (10.14). From (10.10)–(10.12) it is clear that $u_{\lambda\mu}^{(\alpha)}$ is a sum of products of the form $(a\alpha + b)/(c\alpha + d)$ where a, b, c, d are non-negative integers. Also $P_\lambda^{(1)}$ is the Schur function s_λ, so that $u_{\lambda\mu}^{(1)}$ is the Kostka number $K_{\lambda\mu}$, hence is positive whenever $\lambda \geqslant \mu$ (Chapter I, §7, Example 9). Hence

(10.15) *We have $u_{\lambda\mu}^{(\alpha)} > 0$ whenever $\lambda \geqslant \mu$ and $\alpha > 0$.* |

Another proof of the existence of the $P_\lambda^{(\alpha)}$ is indicated in §4, Example 2. See also §4, Example 3(d).

Let $(Q_\lambda^{(\alpha)})$ be the basis of Λ_F dual to the basis $(P_\lambda^{(\alpha)})$, so that

$$Q_\lambda^{(\alpha)} = b_\lambda^{(\alpha)} P_\lambda^{(\alpha)}$$

where $b_\lambda^{(\alpha)} = \langle P_\lambda^{(\alpha)}, P_\lambda^{(\alpha)} \rangle_\alpha^{-1} \in F$, so that from (6.19) and (10.2) we have

$$\begin{aligned} b_\lambda^{(\alpha)} &= \prod_{s \in \lambda} b_\lambda^{(\alpha)}(s) \\ &= \prod_{s \in \lambda} \frac{\alpha a(s) + l(s) + 1}{\alpha a(s) + l(s) + \alpha}. \end{aligned} \tag{10.16}$$

In particular we have

$$P_{(1^r)}^{(\alpha)} = e_r, \qquad Q_{(r)}^{(\alpha)} = g_r^{(\alpha)}.$$

Also it follows from (4.14) that the $P_\lambda^{(\alpha)}$ are well-defined at $\alpha = 0$ and $\alpha = \infty$, and that

$$P_\lambda^{(0)} = e_{\lambda'}, \quad P_\lambda^{(1)} = Q_\lambda^{(1)} = s_\lambda, \quad P_\lambda^{(\infty)} = m_\lambda.$$

The duality theorem (5.1) becomes

$$\omega_\alpha P_\lambda^{(\alpha)} = Q_{\lambda'}^{(\alpha^{-1})}. \tag{10.17}$$

Next, let $f_{\mu\nu}^\lambda(\alpha)$† denote the coefficient of $P_\lambda^{(\alpha)}$ in the product $P_\mu^{(\alpha)} P_\nu^{(\alpha)}$, so that

$$f_{\mu\nu}^\lambda(\alpha) = \langle Q_\lambda^{(\alpha)}, P_\mu^{(\alpha)} P_\nu^{(\alpha)} \rangle_\alpha.$$

This is a rational function of α that remains finite at $\alpha = 0$ and $\alpha = \infty$. By (7.4) it vanishes identically unless $|\lambda| = |\mu| + |\nu|$ and $\lambda \supset \mu$, $\lambda \supset \nu$.

The skew functions $P_{\lambda/\mu}^{(\alpha)}, Q_{\lambda/\mu}^{(\alpha)}$ are defined as in §7 by

$$Q_{\lambda/\mu}^{(\alpha)} = \sum_\nu f_{\mu\nu}^\lambda(\alpha) Q_\nu^{(\alpha)} = \left(b_\lambda^{(\alpha)} / b_\mu^{(\alpha)}\right) P_{\lambda/\mu}^{(\alpha)}.$$

They are zero unless $\lambda \supset \mu$, in which case they are homogeneous of degree $|\lambda| - |\mu|$. For a finite number of variables $x = (x_1, \dots, x_n)$ we have by (7.15)

$$Q_{\lambda/\mu}^{(\alpha)}(x_1, \dots, x_n) = 0 \tag{10.18}$$

unless $0 \leqslant \lambda_i' - \mu_i' \leqslant n$ *for each* $i \geqslant 1$.

† There is a conflict here with the notation $f_{\mu\nu}^\lambda(t)$ of Chapter III, §3, but it should cause no confusion.

Duality (10.17) generalizes to give

(10.19) $$\omega_\alpha P^{(\alpha)}_{\lambda/\mu} = Q^{(\alpha^{-1})}_{\lambda'/\mu'}.$$

Next, let X be an indeterminate and let

$$\varepsilon_X : \Lambda_F \to F[X]$$

be the F-algebra homomorphism defined by $\varepsilon_X(p_r) = X$ for all $r \geqslant 1$, so that $\varepsilon_x(p_\lambda) = X^{l(\lambda)}$. If X is replaced by a positive integer n, then $\varepsilon_X(f) = f(1,\dots,1)$ (with n 1's) for $f \in \Lambda_F$. The specialization theorem (6.17) then gives

(10.20) $$\varepsilon_X P^{(\alpha)}_\lambda = \prod_{s\in\lambda} \frac{X + \alpha a'(s) - l'(s)}{\alpha a(s) + l(s) + 1}.$$

Let

(10.21) $$c_\lambda(\alpha) = \prod_{s\in\lambda} (\alpha a(s) + l(s) + 1),$$
$$c'_\lambda(\alpha) = \prod_{s\in\lambda} (\alpha a(s) + l(s) + \alpha)$$

be respectively the numerator and denominator of $b^{(\alpha)}_\lambda$ (10.16). We have

$$c'_\lambda(\alpha) = \alpha^{|\lambda|} c_{\lambda'}(\alpha^{-1}).$$

Now define

(10.22) $$J^{(\alpha)}_\lambda = c_\lambda(\alpha) P^{(\alpha)}_\lambda = c'_\lambda(\alpha) Q^{(\alpha)}_\lambda.$$

By comparison with (8.1) and (8.3) it is clear that

(10.23) *$J^{(\alpha)}_\lambda(x)$ is the limit, as $(q,t) \to_\alpha (1,1)$, of*

$$(1-t)^{-|\lambda|} J_\lambda(x;q,t).$$

For the $J^{(\alpha)}_\lambda$, duality (10.17) and specialization (10.20) take the forms

(10.24) $$\omega_\alpha J^{(\alpha)}_\lambda = \alpha^{|\lambda|} J^{(\alpha^{-1})}_{\lambda'},$$

(10.25) $$\varepsilon_X J^{(\alpha)}_\lambda = \prod_{s\in\lambda} (X + a'(s)\alpha - l'(s)).$$

For each partition λ, let $\tilde{m}_\lambda = u_\lambda m_\lambda$ denote the 'augmented' monomial symmetric function corresponding to λ, where (as in Chapter I, §6, Example 10)

$$u_\lambda = \prod_{i\geqslant 1} m_i(\lambda)!.$$

The transition matrices $M(J^{(\alpha)}, \tilde{m})_n$ have been computed up to weight $n = 7$, and suggest the conjecture

(10.26?) *The entries in* $M(J^{(\alpha)}, \tilde{m})$ *are polynomials in* α *with positive integral coefficients.*

Next, let $\theta_\rho^\lambda(\alpha)$ denote the coefficient of p_ρ in $J_\lambda^{(\alpha)}$:

(10.27) $$J_\lambda^{(\alpha)} = \sum_\rho \theta_\rho^\lambda(\alpha) p_\rho.$$

From (8.19) and (10.23) it follows that

(10.28) $$\theta_\rho^\lambda(\alpha) = \lim_{(q,t)\to_\alpha(1,1)} \frac{X_\rho^\lambda(q,t)}{(1-t)^n z_\rho} \prod_{i=1}^{l(\rho)} (1-t^{\rho_i}).$$

In particular,

$$\theta_{(1^n)}^\lambda(\alpha) = \frac{1}{n!} X_{(1^n)}^\lambda(1,1)$$

and hence by (8.26) (or also from (10.25))

(10.29) $$\theta_{(1^n)}^\lambda(\alpha) = 1$$

for all partitions λ of n.

From (10.24) it follows that

(10.30) $$\theta_\rho^\lambda(\alpha) = \varepsilon_\rho \alpha^{n-l(\rho)} \theta_\rho^{\lambda'}(\alpha^{-1})$$

if λ, ρ are partitions of n.

The $\theta_\rho^\lambda(\alpha)$ satisfy the orthogonality relations

(10.31) $$\sum_\rho z_\rho \alpha^{l(\rho)} \theta_\rho^\lambda(\alpha) \theta_\rho^\mu(\alpha) = \delta_{\lambda\mu} c_\lambda(\alpha) c'_\lambda(\alpha),$$

(10.32) $$\sum_\lambda c_\lambda(\alpha)^{-1} c'_\lambda(\alpha)^{-1} \theta_\rho^\lambda(\alpha) \theta_\sigma^\lambda(\alpha) = \delta_{\rho\sigma} z_\rho^{-1} \alpha^{-l(\rho)}.$$

For (10.31) follows from (10.26) and the orthogonality of the $J_\lambda^{(\alpha)}$, and (10.32) is equivalent to (10.31).

(10.33) *Remark.* Since the augmented monomial symmetric functions $\tilde{m}_\lambda$ form a **Z**-basis of the subring $\Pi = \mathbf{Z}[p_1, p_2, \dots]$ of Λ, it would follow from the conjecture (10.26?) that $\theta_\rho^\lambda(\alpha) \in \mathbf{Z}[\alpha]$ for all λ, ρ, and hence that $\langle J_\lambda^{(\alpha)}, J_\mu^{(\alpha)} J_\nu^{(\alpha)} \rangle_\alpha \in \mathbf{Z}[\alpha]$ for all λ, μ, ν. Stanley [S25] makes the apparently stronger conjecture that this scalar product should be a polynomial in α with non-negative integer coefficients.

Finally, the function Δ defined in (9.2) is replaced by

(10.34) $$\Delta(x;\alpha)=\prod_{i\neq j}\left(1-x_ix_j^{-1}\right)^{1/\alpha}$$

where $x=(x_1,\ldots,x_n)$. As in §9 we define a new scalar product on $\Lambda_{n,F}$ by

(10.35) $$\langle f,g\rangle'_n=\frac{1}{n!}\int_T f(z)\overline{g(z)}\Delta(z;\alpha)\,dz$$

in the notation of (9.10). From (9.5) we have

(10.36) *The polynomials* $P_\lambda^{(\alpha)}(x_1,\ldots,x_n)$, *where* $l(\lambda)\leqslant n$, *are pairwise orthogonal for the scalar product* (10.35). |

Also from (9.6) we have, on passing to the limit as $(q,t)\to_\alpha(1,1)$,

(10.37) $$\langle P_\lambda^{(\alpha)},Q_\lambda^{(\alpha)}\rangle'_n=c_n\prod_{s\in\lambda}\frac{n+a'(s)\alpha-l'(s)}{n+(a'(s)+1)\alpha-l'(s)-1}$$

where

$$c_n=\frac{1}{n!}\int_T\Delta(z;\alpha)\,dz.$$

Equivalently, by §9, Example 1(d),

(10.38) $$\langle P_\lambda^{(\alpha)},P_\lambda^{(\alpha)}\rangle'_n=\prod_{1\leqslant i<j\leqslant n}\frac{\Gamma(\xi_i-\xi_j+k)\Gamma(\xi_i-\xi_j-k+1)}{\Gamma(\xi_i-\xi_j)\Gamma(\xi_i-\xi_j+1)}$$

where $k=\alpha^{-1}$ and $\xi_i=\lambda_i+k(n-i)$, $1\leqslant i\leqslant n$.

As in §9, the formula (10.37) shows that the scalar product (1.4) is the limit as $n\to\infty$ of the scalar product $c_n^{-1}\langle f,g\rangle'_n=\langle f,g\rangle'_n/\langle 1,1\rangle'_n$.

Examples

1. (a) Since $J_{(n)}^{(\alpha)}=n!\alpha^n g_n^{(\alpha)}$, it follows from (10.5) that

$$\theta_\rho^{(n)}(\alpha)=\frac{n!}{z_\rho}\alpha^{n-l(\rho)}$$

and hence by (10.30) that also

$$\theta_\rho^{(1^n)}(\alpha)=\varepsilon_\rho n!/z_\rho.$$

(b) Deduce from (10.25) that

$$\theta_{(n)}^\lambda(\alpha)=\prod(\alpha(j-1)-(i-1))$$

where the product is over the squares $(i,j) \in \lambda$ with the exception of $(1,1)$, and that

$$\theta^{\lambda}_{(21^{n-2})}(\alpha) = n(\lambda')\alpha - n(\lambda).$$

2. (a) Suppose that $f(q,t) \in \mathbf{Z}[q,t]$ is such that for some positive integer m the limit

$$L = \lim_{(q,t)\to_\alpha(1,1)} \frac{f(q,t)}{(1-t)^m}$$

exists. Then $L \in \mathbf{Z}[\alpha]$. (We may write f in the form

$$f(q,t) = \sum_{r,s\geqslant 0} a_{rs}(1-q)^r(1-t)^s$$

with coefficients $a_{rs} \in \mathbf{Z}$, from which it follows that $a_{rs} = 0$ if $r+s<m$ and that $L = \sum_{r+s=m} a_{rs}\alpha^r$.)

(b) From (8.19) and (10.27) we have

$$\theta^{\lambda}_{\rho}(\alpha) = u_{\lambda}^{-1} \lim_{(q,t)\to_\alpha(1,1)} \frac{X^{\lambda}_{\rho}(q,t)}{(1-t)^{n-l(\rho)}}$$

if λ and ρ are partitions of n, where $u_\lambda = \prod m_i(\lambda)!$ Hence the conjecture (8.18?), which implies that $X^{\lambda}_{\rho}(q,t) \in \mathbf{Z}[q,t]$, also implies that $\theta^{\lambda}_{\rho}(\alpha) \in \mathbf{Q}[\alpha]$.

(c) Let

$$J^{(\alpha)}_{\lambda} = \sum_{\mu} v_{\lambda\mu}(\alpha) m_\mu$$

so that, in the notation of §8, Example 11(a) we have

$$v_{\lambda\mu}(\alpha) = \lim_{(q,t)\to_\alpha(1,1)} \frac{v_{\lambda\mu}(q,t)}{(1-t)^n}.$$

Since (8.18?) implies that $v_{\lambda\mu}(q,t) \in \mathbf{Z}[q,t]$, it also implies that $v_{\lambda\mu}(\alpha) \in \mathbf{Z}[\alpha]$.

(d) If λ, μ, ν are partitions such that $|\lambda| = |\mu| + |\nu| = n$, we have

$$\langle J^{(\alpha)}_{\lambda}, J^{(\alpha)}_{\mu} J^{(\alpha)}_{\nu} \rangle = \lim_{(q,t)\to_\alpha(1,1)} \frac{\langle J_\lambda, J_\mu J_\nu \rangle_{q,t}}{(1-t)^{2n}},$$

Since (8.18?) implies that $\langle J_\lambda, J_\mu J_\nu \rangle_{q,t} \in \mathbf{Z}[q,t]$ (§8, Example 11 (b)), it also implies that $\langle J^{(\alpha)}_{\lambda}, J^{(\alpha)}_{\mu} J^{(\alpha)}_{\nu} \rangle \in \mathbf{Z}[\alpha]$.

3. (a) The coefficient of $\tilde{m}_\mu$ in $J^{(\alpha)}_{(n)}$ is

$$c_\mu \prod_{s\in\mu} (a_\mu(s)\alpha + 1)$$

where $a_\mu(s)$ is the arm-length at $s \in \mu$, and c_μ is the number of decompositions of a set of n elements into disjoint subsets containing $\mu_1, \mu_2, \ldots$ elements, as in Chapter I, §2, Example 11.

(b) The coefficient of $\tilde{m}_{(1^n)} = n!e_n$ in $J_\lambda^{(\alpha)}$ is equal to 1, and the coefficient of $\tilde{m}_{(21^{n-2})}$ in J_λ^α is

$$\alpha n(\lambda') - n(\lambda) + \tfrac{1}{2}n(n-1).$$

4. (a) From §8, Example 4 it follows that

$$(\alpha r + s) J_{(r)}^{(\alpha)} J_{(1^s)}^{(\alpha)} = s J_{(r|s-1)}^{(\alpha)} + r\alpha J_{(r-1|s)}^{(\alpha)}.$$

Deduce that

$$J_{(r|s)}^{(\alpha)} = D(1,2,\ldots,s,-r\alpha,-(r-1)\alpha,\ldots,-\alpha)$$

where in general $D(a_1, a_2, \ldots, a_n)$ denotes the determinant

$$\begin{vmatrix} p_1 & a_1 & 0 & \cdots & 0 & 0 \\ p_2 & p_1 & a_2 & \cdots & 0 & 0 \\ \vdots & \vdots & \vdots & & \vdots & \vdots \\ p_n & p_{n-1} & p_{n-2} & \cdots & p_1 & a_n \\ p_{n+1} & p_n & p_{n-1} & \cdots & p_2 & p_1 \end{vmatrix}.$$

(b) Another formula for $J_{(r|s)}^{(\alpha)}$ is the following, due to P. Hanlon. Fix a standard tableau T of hook shape $(r|s)$ and let R, C denote respectively the row and column stabilizers of T (so that $R \cong S_{r+1}$ and $C \cong S_{s+1}$). Then

$$J_{(r|s)}^{(\alpha)} = \sum \varepsilon(v) \alpha^{r+1-c(u)} \psi(uv)$$

summed over $(u,v) \in R \times C$, where $c(u)$ is the number of cycles in u, and ψ is as in Chapter I, §7.

5. If $\lambda = (2^r 1^s)$ is a partition with two columns, the coefficient of $\tilde{m}_\mu$ in $J_\lambda^{(\alpha)}$, where $\mu = (2^{r-i} 1^{s+2i})$ and $i \geqslant 0$ is

$$\binom{r}{i}(\alpha + s + i + 1)\ldots(\alpha + s + r).$$

6. Let $\Delta: \Lambda_F \to \Lambda_F$ be the derivation defined by

$$\Delta p_r = r p_{r+1}$$

for all $r \geqslant 1$. Show that

$$J_{(n)}^{(\alpha)} = (p_1 + \alpha\Delta)^n(1)$$

and that

$$J_{(1^n)}^{(\alpha)} = (p_1 - \Delta)^n(1).$$

7. As $(q,t) \to_\alpha (1,1)$, the q-integrals of §9, Examples 2 and 3 become ordinary (Lebesgue) integrals. It follows that (with $k = 1/\alpha$) the value of the integral

$$I_\lambda = \frac{1}{n!} \int_{C_n} P_\lambda^{(1/k)}(x) a_\delta(x)^{2k} \prod_{i=1}^n x_i^{r-1}(1-x_i)^{s-1}\, dx,$$

where $C_n = [0,1]^n$, $x = (x_1, \dots, x_n)$, and $dx = dx_1 \dots dx_n$, is

$$I_\lambda = v_\lambda(k) \prod_{i=1}^{n} \frac{\Gamma(\lambda_i + r + k(n-i))\Gamma(s + k(n-i))}{\Gamma(\lambda_i + r + s + k(2n - i - 1))}$$

where

$$v_\lambda(k) = \prod_{1 \leqslant i < j \leqslant n} \frac{\Gamma(\lambda_i - \lambda_j + k(j - i + 1))}{\Gamma(\lambda_i - \lambda_j + k(j-i))}.$$

(a) By making the change of variables $y_i = ux_i$ and then letting $u \to \infty$, deduce that

$$\frac{1}{n!} \int_0^\infty \dots \int_0^\infty P_\lambda^{(1/k)}(y) a_\delta(y)^{2k} (y_1 \dots y_n)^{r-1} e^{-y_1 - \dots - y_n} \, dy$$

$$= v_\lambda(k) \prod_{i=1}^{n} \Gamma(\lambda_i + r + k(n-i)).$$

(b) By making the change of variables $y_i = tx_i$ in (a), multiplying by $t^s e^{-t}$ and integrating from 0 to ∞ with respect to t, show that

$$\frac{1}{n!} \int_0^\infty \dots \int_0^\infty P_\lambda^{(1/k)}(x) a_\delta(x)^{2k} (x_1 \dots x_n)^{r-1} \left(1 + \sum x_i\right)^{-a-s} dx$$

$$= \frac{\Gamma(s)}{\Gamma(a+s)} v_\lambda(k) \prod_{i=1}^{n} \Gamma(\lambda_i + r + k(n-i))$$

where $a = |\lambda| + nr + n(n-1)k$.

(c) In the integral of (b) above put $x_i = y_i/(1 - \Sigma y_j)$ $(1 \leqslant i \leqslant n)$, so that $1 + \Sigma x_i = (1 - \Sigma y_j)^{-1} = u$ say. We have

$$\partial x_i / \partial y_j = \delta_{ij} u + y_i u^2$$

and therefore

$$\det(\partial x_i / \partial y_j) = u^2 \det(\delta_{ij} + y_i u)$$
$$= u^n \left(1 + u \sum y_i\right) = u^{n+1}.$$

Deduce from (b) that

$$\frac{1}{n!} \int_{D_n} P_\lambda^{(1/k)}(y) a_\delta(y)^{2k} (y_1 \dots y_n)^{r-1} \left(1 - \sum y_i\right)^{s-1} dy$$

$$= \frac{\Gamma(s)}{\Gamma(a+s)} v_\lambda(k) \prod_{i=1}^{n} \Gamma(\lambda_i + r + k(n-i))$$

where $a = |\lambda| + nr + n(n-1)k$ as before, and D_n is the simplex in $\mathbf{R}^n$ bounded by the hyperplanes $x_i = 0$ $(1 \leqslant i \leqslant n)$ and $x_1 + \dots + x_n = 1$.

8. Let

$$J_{\lambda,\mu}(k) = \frac{1}{n!}\int_{C_n} P_\lambda^{(1/k)}(x)P_\mu^{(1/k)}(x)a_\delta(x)^{2k}\prod_{i=1}^n x_i^{r-1}(1-x_i)^{k-1}\,dx,$$

where (as in Example 7) $C_n = [0,1]^n$, $x = (x_1,\dots,x_n)$, $dx = dx_1\dots dx_n$ is Lebesgue measure, and λ, μ are partitions of length $\leqslant n$. Then

$$J_{\lambda,\mu}(k) = v_\lambda(k)v_\mu(k)(\Gamma(k))^n \Big/ \prod_{i,j=1}^n (\lambda_i+\mu_j+r+(2n-i-j)k)_k$$

where v_λ, v_μ are as in Example 7, and $(x)_k = \Gamma(x+k)/\Gamma(x)$. (See [K6] for a proof.)

Notes and references

Jack's symmetric functions were first defined by the statistician Henry Jack in 1969 ([J1], [J2]). He showed that when $\alpha = 1$ they reduce to the Schur functions, and conjectured that when $\alpha = 2$ they should give the zonal polynomials (Chapter VII). Later, H. O. Foulkes [F7] raised the question of finding a combinatorial interpretation of Jack's symmetric functions. About ten years ago the author began to investigate their properties, taking as a starting point the differential operators of Sekiguchi [S9] (see §3, Example 3), and showed in particular [M5] that they satisfy (10.13), (10.14), and (10.17) (duality). Shortly afterwards R. Stanley [S25] advanced the subject further, and established the scalar product formula (10.16), the Pieri formula, the explicit expression as a sum over tableaux, and the specialization theorem (10.20). (Stanley worked with $J_\lambda^{(\alpha)}$ rather than $P_\lambda^{(\alpha)}$.)

At about the same time, K. Kadell was investigating generalizations of Selberg's integral [S10] and its extension by Aomoto [A6]: in this way he was led to introduce as integrands a family of symmetric polynomials, depending on a parameter k and indexed by partitions, which he called Selberg polynomials, and which turned out to be precisely Jack's symmetric functions, with parameter $\alpha = 1/k$ [K5]. (The integral formula generalizing Selberg's integral is that given in Example 7 above.) In a different direction, R. Askey [A7] had proposed a q-analogue of Selberg's integral, and Kadell ([K3], conjecture 8) was led to conjecture the existence of 'q-Selberg polynomials' which would feature in a q-analogue of the integral formula above. In the event it turned out that these q-Selberg polynomials are our $P_\lambda(x;q,t)$, and the q-integral formula is that of §9, Example 3.

For another method of construction of the $P_\lambda(x;q,t)$, see B. Srinivasan [S22].

VII

ZONAL POLYNOMIALS

1. Gelfand pairs and zonal spherical functions

Let G be a finite group and let $A = \mathbb{C}[G]$ be the complex group algebra of G. If $f \in A$, say $f = \Sigma_{x \in G} f(x)x$, we may identify f with the function $x \mapsto f(x)$ on G, and hence A with the space of all complex-valued functions on G. From this point of view, the multiplication in A is convolution:

$$(fg)(x) = \sum_{yz = x} f(y)g(z)$$

and G acts on A by the rule $(xf)(y) = f(x^{-1}y)$. The centre of A consists of the functions f on G such that $f(xy) = f(yx)$ for all x, $y \in G$, and has the irreducible characters of G as a basis. The scalar product on A is

$$\langle f, g \rangle_G = \frac{1}{|G|} \sum_{x \in G} f(x)\overline{g(x)}.$$

Let K be a subgroup of G and let

$$e = e_K = \frac{1}{|K|} \sum_{k \in K} k,$$

so that $e^2 = e$. As a function on G, we have $e(x) = |K|^{-1}$ if $x \in K$, and $e(x) = 0$ otherwise. If $x_1, \dots, x_r$ are representatives of the left cosets xK of K in G, the elements $x_i e$ are a basis of the left A-module Ae, and the corresponding representation 1_K^G of G is that obtained by inducing the trivial one-dimensional representation of K, or equivalently it is the permutation representation of G on $G/K = \{x_i K : 1 \leqslant i \leqslant r\}$. Since e is idempotent, the endomorphism ring $\mathrm{End}_A(Ae)$ is anti-isomorphic to eAe, the anti-isomorphism being $\varphi \mapsto \varphi(e)$. Now eAe as a subalgebra of A consists of the functions f on G such that $f(kxk') = f(x)$ for all $x \in G$ and k, $k' \in K$, that is to say the functions constant on each double coset KxK in G. We shall denote this algebra by $C(G, K)$.

An A-module V (or equivalently a representation of G) is *multiplicity-free* if it is a direct sum of inequivalent irreducible A-modules, that is to say if the intertwining number $\langle V, U \rangle = \dim \mathrm{Hom}_A(V, U)$ is 0 or 1 for each irreducible A-module U. By Schur's lemma, V is multiplicity-free if and only if the algebra $\mathrm{End}_A(V)$ is commutative. Hence

(1.1) *For a subgroup K of G, the following conditions are equivalent:*
(a) *the induced representation 1_K^G is multiplicity-free;*
(b) *the algebra $C(G, K)$ is commutative.* |

If these equivalent conditions are satisfied, the pair (G, K) is called a *Gelfand pair*.

(1.2) *Suppose that $KxK = Kx^{-1}K$ for all $x \in G$. Then (G, K) is a Gelfand pair.*

Proof. We have $f(x) = f(x^{-1})$ for all $f \in C(G, K)$ and $x \in G$. Hence if f, $g \in C(G, K)$,

$$(fg)(x) = \sum_{y \in G} f(y)g(y^{-1}x) = \sum_{y \in G} g(x^{-1}y)f(y^{-1})$$

$$= (gf)(x^{-1}) = (gf)(x)$$

which shows that $C(G, K)$ is commutative. |

Assume from now on that (G, K) is a Gelfand pair. Then with the notation used above, Ae is a direct sum of non-isomorphic irreducible A-modules, say

$$Ae = \bigoplus_{i=1}^{s} V_i.$$

Let χ_i be the character of V_i and let

$$\omega_i = \bar{\chi}_i e = e\bar{\chi}_i \qquad (1 \leqslant i \leqslant s)$$

where $\bar{\chi}_i$ is the complex conjugate of χ_i. Thus

$$(1.3) \qquad \omega_i(x) = \frac{1}{|K|} \sum_{k \in K} \bar{\chi}_i(xk) = \frac{1}{|K|} \sum_{k \in K} \chi_i(x^{-1}k).$$

The functions ω_i are the *zonal spherical functions* of the Gelfand pair (G, K).

Let $d_i = \dim V_i = \chi_i(1)$.

(1.4) (i) *$\omega_i(1) = 1$ and $\omega_i(x^{-1}) = \overline{\omega_i(x)}$ for all $x \in G$.*
(ii) *$\omega_i \in C(G, K) \cap V_i$.*
(iii) *$\omega_i \omega_j = \delta_{ij} c_i \omega_i$, where $c_i = |G|/d_i$.*
(iv) *$\langle \omega_i, \omega_j \rangle_G = \delta_{ij} d_i^{-1}$.*
(v) *$(\omega_1, \ldots, \omega_s)$ is an orthogonal basis of $C(G, K)$.*
(vi) *The characters of the algebra $C(G, K)$ are $f \mapsto (\omega_i f)(1)$, for $1 \leqslant i \leqslant s$.*

Proof. (i) From the definition (1.3) we have

$$\omega_i(1) = \langle \chi_i | K, 1_K \rangle_K = \langle \chi_i, 1_K^G \rangle_G$$

by Frobenius reciprocity. Hence $\omega_i(1)$ is the multiplicity of V_i in the A-module Ae, hence is equal to 1. The relation $\omega_i(x^{-1}) = \overline{\omega_i(x)}$ follows from (1.3).

(ii) It is clear from (1.3) that $\omega_i(kx) = \omega_i(xk) = \omega_i(x)$ for $x \in G$ and $k \in K$, so that $\omega_i \in C(G, K)$. On the other hand, the V_i-isotypic component of A is $A\bar{\chi}_i$, hence the V_i-isotypic component of Ae is $Ae\bar{\chi}_i = A\omega_i$. Thus $A\omega_i = V_i$ and consequently $\omega_i \in V_i$.

(iii) Since $\chi_i \chi_j = \delta_{ij} c_i \chi_i$ it follows that

$$\omega_i \omega_j = \bar{\chi}_i e \bar{\chi}_j e = \bar{\chi}_i \bar{\chi}_j e^2 = \bar{\chi}_i \bar{\chi}_j e = \delta_{ij} c_i \bar{\chi}_i e = \delta_{ij} c_i \omega_i.$$

(iv) Since $\overline{\omega_j(x)} = \omega_j(x^{-1})$ we have

$$\langle \omega_i, \omega_j \rangle_G = \frac{1}{|G|} \omega_i \omega_j(1) = \frac{1}{|G|} \delta_{ij} c_i \omega_i(1) = \delta_{ij} d_i^{-1}$$

by (iii) and (i).

(v) It follows from (ii) that the ω_i are linearly independent elements of $C(G, K)$. On the other hand, Schur's lemma shows that the dimension of $C(G, K) \cong \mathrm{End}_A(Ae)$ is equal to s. Hence the ω_i form a basis of $C(G, K)$, and they are pairwise orthogonal by (iv).

(vi) Let θ be a character of $C(G, K)$. From (iii) we have $\theta(\omega_i)\theta(\omega_j) = 0$ if $i \neq j$, and $\theta(\omega_i)^2 = c_i \theta(\omega_i)$. Hence for some i we have $\theta(\omega_j) = \delta_{ij} c_i = (\omega_i \omega_j)(1)$, and hence by linearity $\theta(f) = (\omega_i f)(1)$ for all $f \in C(G, K)$. ∎

Let V be an irreducible G-module (i.e. A-module) with character χ, and let V^K be the subspace of vectors in V fixed by K. The dimension of V^K is the multiplicity of the trivial one-dimensional K-module in V regarded as a K-module, so that

$$\dim V^K = \langle \chi | K, 1_K \rangle_K = \langle \chi, 1_K^G \rangle_G$$

by Frobenius reciprocity. Since 1_K^G is multiplicity-free it follows that, with the previous notation,

$$\dim V^K = \begin{cases} 1 & \text{if } V \cong V_i \text{ for some } i, \\ 0 & \text{otherwise.} \end{cases} \tag{1.5}$$

Moreover, if $V = V_i$, then ω_i is the unique element of the one-dimensional space V_i^K such that $\omega_i(1) = 1$, by (1.4)(i) and (ii). The zonal spherical functions ω_i are therefore distinguished generators of the irreducible components V_i of the induced representation $Ae = 1_K^G$.

Now let V be an irreducible G-module such that $V^K \neq 0$. Then the zonal spherical function ω corresponding to V may be constructed as follows:

(1.6) *Let $\langle u, v \rangle$ be a positive-definite G-invariant Hermitian scalar product on V, and let $v \in V^K$ be a unit vector (i.e. $\langle v, v \rangle = 1$). Then*

$$\omega(x) = \langle v, xv \rangle$$

for all $x \in G$.

Proof. Choose an orthonormal basis $v_1, \ldots, v_n$ of V such that $v_1 = v$. For each $x \in G$, xv_j is a linear combination of the v_i, say

$$xv_j = \sum_{i=1}^{n} R_{ij}(x) v_i,$$

and $x \mapsto (R_{ij}(x))$ is an irreducible matrix representation of G afforded by V. Moreover $\langle v_i, xv_j \rangle = \overline{R_{ij}(x)}$, so that in particular $\langle v, xv \rangle = \overline{R_{11}(x)}$.

Now the space U spanned by the $\overline{R}_{ij}$ is the V-isotypic component of the G-module A, and eUe is one-dimensional, generated by ω. Since v is fixed by K it follows that $\overline{R_{11}(xk)} = \overline{R_{11}(x)}$ and that

$$\overline{R_{11}(kx)} = \langle v, kxv \rangle = \langle k^{-1}v, xv \rangle = \overline{R_{11}(x)}$$

for all $k \in K$ and $x \in G$. Hence $\overline{R}_{11} \in eUe$ and is therefore a scalar multiple of ω. But $\overline{R_{11}(1)} = \langle v, v \rangle = 1 = \omega(1)$, hence $\overline{R}_{11} = \omega$. |

A function $f \in A$ is said to be of *positive type* if

$$\sum_{x, y \in G} f(x^{-1}y) g(x) \overline{g(y)}$$

is real and $\geqslant 0$ for all functions $g \in A$.

(1.7) *Let $f \in C(G, K)$. Then f is of positive type if and only if*

$$f = \sum_{i=1}^{s} a_i \omega_i$$

with real coefficients $a_i \geqslant 0$.

Proof. If ω is one of the ω_i then with the notation of (1.6) we have $\omega(x^{-1}y) = \langle v, x^{-1}yv \rangle = \langle xv, yv \rangle$ and therefore

$$\sum_{x, y \in G} \omega(x^{-1}y) g(x) \overline{g(y)} = \langle u, u \rangle \geqslant 0$$

where $u = \sum_{x \in G} g(x) xv$. Hence each ω_i is of positive type and therefore so also are all non-negative linear combinations of the ω_i.

Conversely, if $f \in C(G, K)$ is of positive type, we may write

$$f = \sum_{i=1}^{s} a_i \omega_i$$

with coefficients $a_i \in \mathbf{C}$. Taking $g = \omega_i$ in the definition, $(\omega_i f \omega_i)(1)$ is real and $\geqslant 0$. But by (1.4) (iii), $\omega_i f \omega_i = a_i \omega_i^3 = a_i c_i^2 \omega_i$, so that $(\omega_i f \omega_i)(1) = a_i c_i^2$. Hence a_i is real and $\geqslant 0$. |

(1.8) *Let (G, K) and (G', K') be Gelfand pairs such that G' is a subgroup of G and K' is a subgroup of K, and let ω be a zonal spherical function of (G, K). Then the restriction of ω to G' is a non-negative linear combination of the zonal spherical functions of (G', K').*

Proof. By (1.7), $\omega | G'$ is of positive type and belongs to $C(G', K')$. Hence the result follows from (1.7). |

The zonal spherical functions of a Gelfand pair (G, K) may be characterized in other ways:

(1.9) *Let ω be a complex-valued function on G. Then the following conditions are equivalent:*

(a) *ω is a zonal spherical function of (G, K).*

(b) (i) $\omega \in C(G, K)$; (ii) $\omega(1) = 1$; (iii) *for all $f \in C(G, K)$, $f\omega = \lambda_f \omega$ for some $\lambda_f \in \mathbf{C}$ (depending on f).*

(c) *$\omega \neq 0$ and*

$$\omega(x)\omega(y) = \frac{1}{|K|} \sum_{k \in K} \omega(xky)$$

for all $x, y \in G$.

Proof. (a) $\Rightarrow$ (b). Conditions (i) and (ii) follow from (1.4). As to (iii), if $f \in C(G, K)$, say $f = \sum a_j \omega_j$, then $f\omega_i = \sum a_j \omega_j \omega_i$ is a scalar multiple of ω_i by (1.4)(iii).

(b) $\Rightarrow$ (c). Let

$$\omega_y(x) = \frac{1}{|K|} \sum_{k \in K} \omega(xky).$$

From (i) it follows that $\omega_y \in C(G, K)$. Next, for each $k \in K$ and $f \in C(G, K)$ we have

$$\begin{aligned} \lambda_f \omega(y) = \lambda_f \omega(ky) &= (f\omega)(ky) \\ &= \sum_{x \in G} f(x^{-1}) \omega(xky) \end{aligned}$$

and hence, on averaging over K,

$$\lambda_f \omega(y) = (f\omega_y)(1) = |G| \langle \omega_y, f^* \rangle_G, \tag{1}$$

where $f^*(x) = \overline{f(x^{-1})}$. On the other hand,

$$\lambda_f = \lambda_f \omega(1) = (f\omega)(1) = |G| \langle \omega, f^* \rangle_G. \tag{2}$$

From (1) and (2) it follows that

$$\langle \omega_y - \omega(y)\omega, f^* \rangle_G = 0$$

for all $f \in C(G,K)$, from which we conclude that $\omega_y = \omega(y)\omega$, i.e.

$$\frac{1}{|K|} \sum_{k \in K} \omega(xky) = \omega(y)\omega(x)$$

for all x, $y \in G$.

(c) $\Rightarrow$ (a). If f, $g \in C(G,K)$ we have

$$\begin{aligned}(\omega fg)(1) &= \sum_{x,y \in G} \omega(xy) f(y^{-1}) g(x^{-1}) \\ &= \sum_{x,y \in G} \omega(xky) f(y^{-1}) g(x^{-1})\end{aligned}$$

for all $k \in K$, and hence on averaging over K we obtain

$$\begin{aligned}(\omega fg)(1) &= \sum_{x,y \in G} \omega(x)\omega(y) f(y^{-1}) g(x^{-1}) \\ &= (\omega f)(1)(\omega g)(1).\end{aligned}$$

Since $\omega \neq 0$, this shows that $f \mapsto (\omega f)(1)$ is a character of $C(G,K)$, and hence ω is a zonal spherical function of (G,K) by (1.4)(vi). |

Let $K_1, K_2, \dots, K_s$ be the double cosets of K in G, and let $f_i (1 \leqslant i \leqslant s)$ be the functions on G defined by

$$f_i(x) = \begin{cases} 1/|K| & \text{if } x \in K_i, \\ 0 & \text{otherwise.} \end{cases}$$

They form a basis of $C(G,K)$. Hence $f_i f_j$ is a linear combination of the f_k, say

$$f_i f_j = \sum_k a_{ijk} f_k.$$

If $x \in K_k$, then

$$a_{ijk} = |K| f_i f_j(x)$$
$$= |K| \sum_{y \in G} f_i(y) f_j(y^{-1}x).$$

Now $f_i(y) f_j(y^{-1}x)$ is zero unless $y \in K_i \cap xK_j^{-1}$, so that

$$a_{ijk} = |K_i \cap xK_j^{-1}|/|K|.$$

Since both K_i and xK_j^{-1} are unions of cosets gK of K in G, it follows that *the structure constants a_{ijk} are integers $\geqslant 0$.* We shall use this to prove

(1.10) *Let ω be a zonal spherical function of (G, K), and let $x \in G$. Then $|KxK/K| \omega(x)$ is an algebraic integer.*

Proof. Let $\hat{\omega}$ be the character of $C(G, K)$ corresponding to ω, so that $\hat{\omega}(f) = (f\omega)(1)$ by (1.4)(vi). Then we have

$$\hat{\omega}(f_i)\hat{\omega}(f_j) = \sum_k a_{ijk}\hat{\omega}(f_k)$$

from which it follows that the set of s linear equations (in s unknowns $x_1, \ldots, x_s$)

$$\hat{\omega}(f_i)x_j = \sum_k a_{ijk}x_k$$

has a non-trivial solution $x_j = \hat{\omega}(f_j)$. Hence

$$\det\left(\hat{\omega}(f_i)\delta_{jk} - a_{ijk}\right)_{j,k} = 0.$$

On expanding the determinant, we see that $\hat{\omega}(f_i)$ is an algebraic integer, since the a_{ijk} are integers. But if $x \in K_i$ we have

$$\hat{\omega}(f_i) = (f_i\omega)(1) = |KxK/K| \omega(x^{-1}). \quad |$$

Examples

1. If ω is a zonal spherical function of (G, K), then $|\omega(x)| \leqslant 1$ for all $x \in G$. (Use (1.6) and the Cauchy–Schwarz inequality.)

2. The constant function equal to 1 on G is always a zonal spherical function of (G, K). (Use (1.5).)

3. If ω is a zonal spherical function of (G, K), then so also is its complex conjugate $\overline{\omega}$. (If $\omega = e\overline{\chi}$ we have $\langle \chi, 1_K^G \rangle_G = 1$, hence also $\langle \overline{\chi}, 1_K^G \rangle_G = 1$.)

In the notation of (1.4), suppose that all the characters χ_i are real-valued. Then each ω_i is real-valued and hence $\omega_i(x^{-1}) = \omega_i(x)$ for $1 \leqslant i \leqslant s$ and all $x \in G$. It follows that $f(x^{-1}) = f(x)$ for all $f \in C(G,K)$, and hence that $KxK = Kx^{-1}K$ for all $x \in G$. This is a partial converse of (1.2).

4. Let (G,K) be a Gelfand pair, let $\Omega = \{\omega_1, \dots, \omega_s\}$ be the set of zonal spherical functions of (G,K), and let $C(\Omega)$ denote the algebra of all complex-valued functions on Ω, with pointwise multiplication: $(fg)(\omega_i) = f(\omega_i)g(\omega_i)$. If $f \in C(G,K)$, the *spherical transform* (or *Fourier transform*) of f is the function Sf on Ω defined by

$$(Sf)(\omega_i) = (f\omega_i)(1) = \sum_{x \in G} f(x)\overline{\omega_i(x)}.$$

Show that S is an algebra isomorphism of $C(G,K)$ onto $C(\Omega)$, with inverse S^{-1} given by

$$S^{-1}\varphi = \frac{1}{|G|}\sum_i \varphi(\omega_i)d_i\omega_i \tag{1}$$

for $\varphi \in C(\Omega)$, in the notation of (1.4). (From (1.4)(iv) and (v) it follows that

$$f = \sum_i \langle f, \omega_i \rangle_G \, d_i \omega_i = \frac{1}{|G|}\sum_i (Sf)(\omega_i) d_i \omega_i$$

for $f \in C(G,K)$, whence S is a linear isomorphism whose inverse S^{-1} is given by the formula (1). That $S(fg) = (Sf)(Sg)$ follows from (1.4)(vi).)

5. If (G,K) and (G',K') are Gelfand pairs with zonal spherical functions $\omega_i (1 \leqslant i \leqslant s)$, $\omega'_j (1 \leqslant j \leqslant t)$ respectively, then $(G \times G', K \times K')$ is a Gelfand pair with zonal spherical functions $\omega_i \times \omega'_j$.

6. (a) Let G be a finite group and let σ be an automorphism of G such that $\sigma^2 = 1$. Let $K = \{x \in G : \sigma x = x\}$ and let $P = \{x \in G ; \sigma x = x^{-1}\}$; K is a subgroup of G, but P is not, in general. If $G = KP$, show that (G,K) is a Gelfand pair. (Let $x \in G$, say $x = kp$ where $k \in K$, $p \in P$. Then $\sigma x = kp^{-1} = kx^{-1}k$, so that $\sigma(KxK) = Kx^{-1}K$ and hence $f(\sigma x) = f(x^{-1})$ for all $f \in C(G,K)$. Hence for f, $g \in C(G,K)$ we have

$$(fg)(x) = \sum_{y \in G} f(xy^{-1})g(y) = \sum_{y \in G} f(\sigma(yx^{-1}))g(\sigma(y^{-1}))$$

$$= (gf)(\sigma x^{-1}) = gf(x)$$

which shows that $C(G,K)$ is commutative.)

(b) Suppose that $G = KA$ where K, A are subgroups of G, $K \cap A = \{1\}$, and the subgroup A is abelian, normal, and of odd order. Define $\sigma : G \to G$ by $\sigma(ka) = ka^{-1}$, where $k \in K$ and $a \in A$. Clearly σ^2 is the identity. Moreover, σ is a homomorphism, because if $x = k_1a_1$ and $y = k_2a_2$ are any elements of G, where k_1, $k_2 \in K$ and a_1, $a_2 \in A$, then $xy = k_1a_1k_2a_2 = ka$ where $k = k_1k_2 \in K$ and $a = (k_2^{-1}a_1k_2)a_2 = a_2(k_2^{-1}a_1k_2)$ (because A is abelian); and hence $\sigma(xy) = ka^{-1} = k_1k_2(k_2^{-1}a_1^{-1}k_2)a_2^{-1} = (k_1a_1^{-1})(k_2a_2^{-1}) = \sigma(x)\sigma(y)$. Since A has odd order, the subgroup of G fixed by σ is precisely K; also $A \subset P = \{x \in G: \sigma x = x^{-1}\}$, so that $G = KP$. Hence (G, K) is a Gelfand pair, by (a) above.

(c) As an example of (b), let V be a finite-dimensional vector space over a finite field of odd characteristic, and let K_0 be any subgroup of $GL(V)$. Since K_0 acts on V we can form the semidirect product $G = V \rtimes K_0$, whose elements are pairs $(v, k) \in V \times K_0$, with multiplication defined by $(v, k)(v', k') = (v + kv', kk')$. The elements $(v, 1)$ form a normal subgroup of G isomorphic to V, hence abelian and of odd order, and the elements $(0, k)$ form a subgroup K of G isomorphic to K_0. By (b) above, (G, K) is a Gelfand pair.

7. In the situation of Example 6(b), every $x \in G$ can be uniquely written in the form $x = k(x)a(x)$, where $k(x) \in K$ and $a(x) \in A$, and we have $KxK = Ka(x)K$. The group K acts on A by conjugation, hence also on the character group $\hat{A}$ of A. The double cosets of K in G are in one–one correspondence with the K-orbits in A, the double coset KxK corresponding to the K-orbit of $a(x)$.

For each character $\alpha \in \hat{A}$ let ω_α be the function on G defined by

$$\omega_\alpha(x) = \frac{1}{|K|} \sum_{k \in K} \alpha(ka(x)k^{-1}). \tag{1}$$

Show that ω_α is a zonal spherical function of (G, K). (Verify that ω_α satisfies the functional equation of (1.9)(c).)

It is clear from (1) that $\omega_\alpha = \omega_\beta$ if $\alpha, \beta \in \hat{A}$ are in the same K-orbit. If S is a set of representatives of the K-orbits in $\hat{A}$, the ω_α, $\alpha \in S$ are linearly independent (because the characters $\alpha \in \hat{A}$ are linearly independent), and hence are all the zonal spherical functions of (G, K).

8. If $f, g \in C(G, K)$ let $f \cdot g$ denote the pointwise product of f and $g: (f \cdot g)(x) = f(x)g(x)$. Show that the coefficients a_{ij}^k defined by

$$\omega_i \cdot \omega_j = \sum_k a_{ij}^k \omega_k$$

(where the ω_i are the zonal spherical functions of (G, K)) are real and $\geqslant 0$. (We may identify G with its image in $G \times G$ under the diagonal map $x \mapsto (x, x)$. By Example 4, the zonal spherical functions of $(G \times G, K \times K)$ are $\omega_i \times \omega_j$, and $\omega_i \cdot \omega_j$ is the restriction of $\omega_i \times \omega_j$ to G. Hence the result follows from (1.8).)

9. Let G be any finite group and let χ_i $(1 \leqslant i \leqslant r)$ be the irreducible characters of G. Let $G^* = \{(x, x): x \in G\}$ be the diagonal subgroup of $G \times G$. The irreducible characters of $G \times G$ are $\chi_i \times \chi_j: (x, y) \mapsto \chi_i(x)\chi_j(y)$ $(1 \leqslant i, j \leqslant r)$ and we have

$$\begin{aligned}\langle \chi_i \times \chi_j, 1_{G^*}^{G \times G} \rangle_{G \times G} &= \langle (\chi_i \times \chi_j) | G^*, 1_{G^*} \rangle_{G^*} \\ &= \frac{1}{|G|} \sum_{x \in G} \chi_i(x)\chi_j(x) = \langle \chi_i, \bar{\chi}_j \rangle_G\end{aligned}$$

which is equal to 1 if $\chi_i = \bar{\chi}_j$, and is zero otherwise. Hence the induced representation $1_{G^*}^{G \times G}$ is multiplicity-free, with character

$$\sum_{i=1}^{r} \bar{\chi}_i \times \chi_i.$$

Consequently $(G \times G, G^*)$ is a Gelfand pair, and the zonal spherical functions are ω_i $(1 \leqslant i \leqslant r)$, where

$$\begin{aligned}\omega_i(x,y) &= \frac{1}{|G|} \sum_{g \in G} \bar{\chi}_i(gx)\chi_i(gy) \\ &= \frac{1}{|G|} \sum_{g \in G} \chi_i(x^{-1}g^{-1})\chi_i(gy) \\ &= \frac{1}{|G|} \chi_i^2(x^{-1}y) = d_i^{-1}\chi_i(x^{-1}y)\end{aligned}$$

where $d_i = \chi_i(1)$. Hence $\omega_i|(1 \times G) = d_i^{-1}\chi_i$.

The functional equation (1.9)(c) for zonal spherical functions in this case becomes the functional equation for characters:

$$\chi_i(x)\chi_i(y) = \frac{\chi_i(1)}{|G|} \sum_{g \in G} \chi_i(g^{-1}xgy)$$

for all $x,\ y \in G$.

10. Let G be a finite group, A the complex group algebra of G; also let K be a subgroup of G and φ a linear character of K (i.e. a homomorphism of K into $\mathbf{C}^*$). Let

$$e_\varphi = \frac{1}{|K|} \sum_{k \in K} \varphi(k^{-1})k$$

so that $e_\varphi^2 = e_\varphi$ and the A-module Ae_φ affords the induced representation $\varphi^G = \operatorname{ind}_K^G \varphi$. Let $C^\varphi(G,K) = e_\varphi A e_\varphi$, which is anti-isomorphic to the endomorphism algebra $\operatorname{End}_A(Ae_\varphi)$. As an algebra of functions on G, $C^\varphi(G,K)$ consists of the functions $f \in A$ that satisfy $f(kx) = f(xk) = \varphi(k^{-1})f(x)$ for all $k \in K$ and $x \in G$. The representation φ^G is multiplicity-free if and only if $C^\varphi(G,K)$ is commutative. If these equivalent conditions are satisfied, as we shall assume from now on, the triple (G, K, φ) is called (illogically) a *twisted Gelfand pair*.

The A-module Ae_φ is a direct sum of inequivalent irreducible A-modules, say

$$Ae_\varphi = \bigoplus_{i=1}^{s} V_i.$$

Let χ_i be the character of V_i and let

$$\omega_i^\varphi = \bar{\chi}_i e_\varphi = e_\varphi \bar{\chi}_i$$

or explicitly

$$\omega_i^\varphi(x) = \frac{1}{|K|} \sum_{k \in K} \chi_i(x^{-1}k)\varphi(k^{-1}).$$

The functions $\omega_i^\varphi (1 \leqslant i \leqslant s)$ are the *φ-spherical functions* of the triple (G, K, φ). All the reults in the text, appropriately modified, remain valid in this more general context.

(a) Show that (1.4)(i)–(vi) remain valid if ω_i is replaced by ω_i^φ and $C(G,K)$ by $C^\varphi(G,K)$ throughout.

(b) If V is an irreducible G-module, let

$$V^\varphi = \{v \in V : kv = \varphi(k)v \text{ for all } k \in K\}.$$

Then $\dim V^\varphi = 1$ if $V \cong V_i$ for some i, and $\dim V^\varphi = 0$ otherwise. The φ-spherical function ω_i^φ is the unique element of V_i^φ such that $\omega_i^\varphi(1) = 1$.

(c) If V is an irreducible G-module such that $V^\varphi \neq 0$, let $\langle u, v\rangle$ be a positive-definite G-invariant scalar product on V, as in (1.6), and let $v \in V^\varphi$ be a unit vector. Then the φ-spherical function corresponding to V is given by $\omega^\varphi(x) = \langle v, xv\rangle$ for all $x \in G$.

(d) The functions $f \in C^\varphi(G,K)$ that are of positive type are the non-negative linear combinations of the ω_i^φ.

(e) Show that the following conditions are equivalent, for a complex-valued function ω on G:

(α) ω is a φ-spherical function for (G, K, φ).

(β) (i) $\omega \in C^\varphi(G,K)$; (ii) $\omega(1) = 1$; (iii) for all $f \in C^\varphi(G,K)$, we have $f\omega = \lambda_f \omega$ for some $\lambda_f \in \mathbf{C}$.

(γ) $\omega \neq 0$ and

$$\omega(x)\omega(y) = \frac{1}{|K|} \sum_{k \in K} \omega(xky)\varphi(k)$$

for all $x, y \in G$.

11. In continuation of Example 10, let ε be a linear character of G. Then the mapping $\theta: A \to A$ defined by $\theta(x) = \varepsilon(x)x$ for $x \in G$ (or, in terms of functions, by $(\theta f)(x) = \varepsilon(x)f(x)$) is an automorphism of A and induces an isomorphism of $C^{\varepsilon\varphi}(G,K)$ onto $C^\varphi(G,K)$, where $\varepsilon\varphi$ is the character $k \mapsto \varepsilon(k)\varphi(k)$ of K. Hence if (G, K, φ) is a twisted Gelfand pair so also is $(G, K, \varepsilon\varphi)$, and the spherical functions of $(G, K, \varepsilon\varphi)$ are the images under θ^{-1} of the spherical functions of (G, K, φ).

In particular, if (G, K) is a Gelfand pair, and φ is a linear character of K which extends to a linear character of G, then (G, K, φ) is a twisted Gelfand pair.

12. Let G be a finite group, K a subgroup of G. Let φ be the character of a (not necessarily irreducible) representation of K, and let $\varphi^G = \operatorname{ind}_K^G \varphi$. We regard φ as a function on G, vanishing outside K. Let χ be an irreducible character of G, and let

$$\omega(x) = \frac{1}{|K|} \sum_{k \in K} \chi(x^{-1}k)\varphi(k^{-1}) = \frac{1}{|K|} \chi\varphi(x^{-1}).$$

(a) We have

$$\omega(1) = \langle \chi|K, \varphi\rangle_K = \langle \chi, \varphi^G\rangle_G.$$

(b) $\overline{\omega(x)} = \omega(x^{-1})$ for all $x \in G$.

(c) $\omega^2 = c\omega$ for some positive scalar c (because χ, φ are positive scalar multiples of commuting idempotents).

It follows that

$$\langle \omega, \omega \rangle_G = \frac{1}{|G|} \omega^2(1) = \frac{c}{|G|} \omega(1)$$

and hence that $\omega \neq 0$ if and only if χ occurs in the induced representation φ^G.

13. Let m, n be positive integers such that $m \leqslant n$, and let $G = S_{m+n}$, $K = S_m \times S_n$. Then (Chapter I, §7)

$$\mathrm{ch}(1_K^G) = h_m h_n = \sum_{i=0}^{m} s_{(m+n-i,i)}$$

which shows that 1_K^G is multiplicity free, with character

$$\sum_{i=0}^{m} \chi^{(m+n-i,i)}.$$

Consequently (G, K) is a Gelfand pair, $\dim C(G,K) = m+1$, and the zonal spherical functions are

$$\omega_i = \chi^{(m+n-i,i)} e_K \qquad (0 \leqslant i \leqslant m).$$

Let $E = \{1, 2, \ldots, m\}$, $F = \{1, 2, \ldots, m+n\}$. Under the mapping $w \mapsto wE$, where $w \in G = S_{m+n}$, the cosets wK of K in G are in one-one correspondence with the m-element subsets of F. Let

$$\varphi(w) = |E - wE| = m - |E \cap wE|$$

for $w \in G$. The function φ lies in $C(G,K)$ and takes the values $0, 1, \ldots, m$, and the double cosets of K in G are the fibres $K_r = \varphi^{-1}(r)$ of φ. In particular, $K_0 = K$, and $|K_r|/|K| = \binom{m}{r}\binom{n}{r}$ for $0 \leqslant r \leqslant m$. If $\omega_i(r)$ denotes the value of ω_i at elements of K_r, the orthogonality relation (1.4)(iv) takes the form

$$\sum_{r=0}^{m} \binom{m}{r}\binom{n}{r} \omega_i(r)\omega_j(r) = \frac{h(m+n-i,i)}{m!n!} \delta_{ij},$$

where $h(m+n-i,i)$ is the product of the hook-lengths of the partition $(m+n-i,i)$. The numbers $\omega_i(r)$ can be calculated explicitly (see e.g. [B1], p. 218): if

$$Q_i(x) = \sum_{s \geqslant 0} (-1)^s q_{i,s} \binom{x}{s},$$

where

$$q_{i,s} = \binom{i}{s}\binom{m+n+1-i}{s} \Big/ \binom{m}{s}\binom{n}{s}$$

so that $Q_i(x)$ is a polynomial of degree i, then $\omega_i(r) = Q_i(r)$ for $0 \leqslant i, r \leqslant m$.

14. Let V be a vector space of dimension $m+n$ over the field $\mathbf{F}_q$ of q elements, and let U be a subspace of V of dimension m, where $m \leqslant n$. Let $G = GL(V) \cong GL_{m+n}(\mathbf{F}_q)$, and let $K = \{g \in G: gU = U\}$, which is a maximal parabolic subgroup of G. The induced representation 1_K^G is the permutation representation of G on the Grassmannian $G/K = G_{m,n}$, whose points are the m-dimensional subspaces of V.

For each $g \in G$ let

$$\varphi(g) = m - \dim(U \cap gU)$$

in analogy with Example 10. Then $\varphi \in C(G,K)$ and takes the values $0, 1, \ldots, m$, and the double cosets of K in G are the fibres $K_r = \varphi^{-1}(r)$ of φ: in particular, $K_0 = K$. Since $\varphi(g) = \varphi(g^{-1})$ it follows that $K_r = K_r^{-1}$ for $0 \leqslant r \leqslant m$ and hence (G,K) is a Gelfand pair, by (1.2).

Let $k_r = |K_r|/|K|$, so that k_r is the number of m-dimensional subspaces U' of V such that $\dim(U \cap U') = m - r$. Then we have

$$k_r = q^{r^2} \begin{bmatrix} m \\ r \end{bmatrix} \begin{bmatrix} n \\ r \end{bmatrix}$$

where $\begin{bmatrix} m \\ r \end{bmatrix}$ and $\begin{bmatrix} n \\ r \end{bmatrix}$ are q-binomial coefficients (Chapter I, §2, Example 3). (Let $E = U \cap U'$ and $F = U + U'$, so that E is an $(m-r)$-dimensional subspace of U, and F/U is an r-dimensional subspace of V/U. The number of choices for E is $\begin{bmatrix} m \\ m-r \end{bmatrix} = \begin{bmatrix} m \\ r \end{bmatrix}$, and the number of choices for F is $\begin{bmatrix} n \\ r \end{bmatrix}$. Moreover, U'/E is a complement of U/E in F/E; since $\dim(U'/E) = \dim(U/E) = r$, it follows that for each choice of E and F there are q^{r^2} possibilities for U'.)

From Chapter IV it follows that

$$\operatorname{ch}(1_K^G) = h_m(1) h_n(1) = \sum_{i=0}^{m} s_{(m+n-i,i)}(1)$$

and hence the character of 1_K^G is $\sum_{i=0}^m \chi_i$, where χ_i is the character of G whose characteristic is $s_{(m+n-i,i)}(1)$. Hence the zonal spherical functions of (G,K) are $\omega_i = \chi_i e_K$ $(0 \leqslant i \leqslant m)$. If $\omega_i(r)$ denotes the value of ω_i at elements of K_r, the orthogonality relations (1.4)(iv) take the form

$$\sum_{r=0}^{m} q^{r^2} \begin{bmatrix} m \\ r \end{bmatrix} \begin{bmatrix} n \\ r \end{bmatrix} \omega_i(r) \omega_j(r) = d_i^{-1} \begin{bmatrix} m+n \\ m \end{bmatrix} \delta_{ij},$$

where d_i is the degree of χ_i, which by Chapter IV, (6.7) is

$$d_i = q^i \psi_{m+n}(q) / \prod_{x \in \lambda} (q^{h(x)} - 1)$$

where λ is the partition $(m+n-i, i)$, and $\psi_r(q) = \prod_{i=1}^{r} (q^i - 1)$.

Notes and references

The theory of Gelfand pairs and zonal spherical functions was originally developed in the context of the representation theory of locally compact groups, and in particular Lie groups (see, for example, [H6], Chapter IV). The prototype example is that in which $G = SO(3)$ and $K = SO(2)$, so that G/K may be identified with the 2-sphere, and the zonal spherical functions are the 'spherical harmonics' or Legendre polynomials (*loc. cit.*, p. 404).

2. The Gelfand pair (S_{2n}, H_n)

Let $t \in S_{2n}$ be the product of the transpositions (12), (34), ..., $(2n-1, 2n)$, and let H_n be the centralizer of t in S_{2n}. The group H_n is the *hyperoctahedral group* of degree n, of order $2^n n!$, and isomorphic to the wreath product of S_2 with S_n.

To each permutation $w \in S_{2n}$ we attach a graph $\Gamma(w)$ with vertices $1, 2, \ldots, 2n$ and edges ε_i, $w\varepsilon_i$ $(1 \leqslant i \leqslant n)$, where ε_i joins $2i-1$ and $2i$. If we think of the edges ε_i as red and the others as blue, then each vertex of the graph lies on one red edge and one blue edge, and hence the components of $\Gamma(w)$ are cycles of even lengths $2\rho_1, 2\rho_2, \ldots$, where $\rho_1 \geqslant \rho_2 \geqslant \ldots$. Thus w gives rise to a partition $\rho = (\rho_1, \rho_2, \ldots)$ of n, called the *coset-type* of w. This terminology is justified by the following proposition:

(2.1) (i) *Two permutations* w, $w_1 \in S_{2n}$ *have the same coset-type if and only if* $w_1 \in H_n w H_n$.
(ii) w, w^{-1} *have the same coset-type.*

Proof. (i) The permutations w, w_1 have the same coset-type if and only if their graphs are isomorphic, that is to say if and only if there is a permutation $h \in S_{2n}$ that preserves edge-colours and maps $\Gamma(w)$ onto $\Gamma(w_1)$. Since h permutes the red edges ε_i, it follows that $h \in H_n$; and since the blue edges $w_1\varepsilon_i$ of $\Gamma(w_1)$ are a permutation of the blue edges $hw\varepsilon_i$ of $h\Gamma(w)$, the ε_i are a permutation of the $w_1^{-1}hw\varepsilon_i$, whence $w_1^{-1}hw \in H_n$ and consequently $w_1 \in H_n w H_n$.
(ii) Since $w\Gamma(w^{-1}) = \Gamma(w)$, the graphs $\Gamma(w)$ and $\Gamma(w^{-1})$ are isomorphic, hence determine the same partition of n. |

From (2.1)(ii) and (1.2) it follows that

(2.2) (S_{2n}, H_n) *is a Gelfand pair.* |

From (2.1)(i), the double cosets H_nwH_n of $H = H_n$ in S_{2n} are indexed by the partitions of n. Let

$$H_\rho = HwH$$

if $w \in S_{2n}$ has coset-type ρ. Then we have

(2.3) $$|H_\rho| = |H_n|^2/z_{2\rho} = |H_n|^2/2^{l(\rho)}z_\rho.$$

Proof. If w has coset-type ρ then

$$|H_\rho| = |HwH| = |HwHw^{-1}| = |H|\cdot|wHw^{-1}|/|H \cap wHw^{-1}|$$

so that

(1) $$|H_\rho| = |H|^2/|H \cap wHw^{-1}|.$$

Now $H \cap wHw^{-1}$ is the group of colour-preserving automorphisms of the graph $\Gamma(w)$. On a given cycle in $\Gamma(w)$ of length $2r$, it acts as the dihedral group of order $2r$, from which it follows that

$$|H \cap wHw^{-1}| = z_{2\rho}$$

which together with (1) completes the proof.

From (2.2), the induced representation $1_{H_n}^{S_{2n}}$ of S_{2n} on the cosets of H_n is multiplicity-free. In fact

(2.4) *We have*

$$1_{H_n}^{S_{2n}} \cong \bigoplus_{|\lambda|=n} M_{2\lambda}$$

where $M_{2\lambda}$ is an irreducible S_{2n}-module corresponding to the partition $2\lambda = (2\lambda_1, 2\lambda_2, \ldots)$.

Proof. With the notation of Chapter I we have

$$\begin{aligned}\operatorname{ch}\left(1_{H_n}^{S_{2n}}\right) &= c(H_n) = c(S_2 \sim S_n) = c(S_n) \circ c(S_2)\\ &= h_n \circ h_2 = \sum_{|\lambda|=n} s_{2\lambda}\end{aligned}$$

by use of Chapter I, §7, Example 4 and §8, Examples 5 and 6(a). |

From (2.4) and (1.5) it follows that

(2.5) *Let μ be a partition of $2n$ and let M_μ be an irreducible S_{2n}-module corresponding to μ. Then* $\dim M_\mu^{H_n} = 1$ *if μ is even* (i.e. $\mu = 2\lambda$), *and $M_\mu^{H_n} = 0$ otherwise.* |

From (2.4), the zonal spherical functions of the Gelfand pair (S_{2n}, H_n) are indexed by the partitions of n: they are

$$\omega^\lambda = \chi^{2\lambda}\tilde{e}_n$$

where $\chi^{2\lambda}$ is the character of $M_{2\lambda}$, and

$$\tilde{e}_n = \frac{1}{|H_n|}\sum_{h\in H_n} h.$$

Explicitly, since $\chi^{2\lambda}$ is real-valued, we have

$$\omega^\lambda(s) = \frac{1}{|H_n|}\sum_{h\in H_n}\chi^{2\lambda}(sh) \tag{2.6}$$

for $s \in S_{2n}$. Since $\chi^{2\lambda}$ is integer-valued, $\omega^\lambda(s)$ is a rational number for all partitions λ of n and all $s \in S_{2n}$.

For each integer $n \geqslant 0$, let

$$C_n = C(S_{2n}, H_n)$$

be the space of complex-valued functions on S_{2n} that are constant on each double coset H_ρ, and let

$$C = \bigoplus_{n\geqslant 0} C_n.$$

We define a bilinear multiplication in C as follows: if $f \in C_m$ and $g \in C_n$, then $f \times g$ is a function on $S_{2m} \times S_{2n} \subset S_{2(m+n)}$ (where S_{2m} permutes the first $2m$ symbols and S_{2n} the last $2n$ symbols), and we define

$$f * g = \tilde{e}_{m+n}(f \times g)\tilde{e}_{m+n}. \tag{2.7}$$

Thus $f * g$ is obtained from $f \times g$ by averaging over H_{m+n} on either side, and lies in C_{m+n}.

For each partition ρ of n let

$$\varphi_\rho = \tilde{e}_n w_\rho \tilde{e}_n$$

where $w_\rho \in H_\rho$. Thus $\varphi_\rho(w) = |H_\rho|^{-1}$ if $w \in H_\rho$, and $\varphi_\rho(w) = 0$ otherwise. Clearly the φ_ρ such that $|\rho| = n$ form a basis of C_n, and we have

$$\varphi_\rho * \varphi_\sigma = \varphi_{\rho\cup\sigma} \tag{2.8}$$

for any two partitions ρ, σ.

Proof. Let $|\rho| = m$, $|\sigma| = n$, and let $u \in H_\rho$, $v \in H_\sigma$. Then $\Gamma(u \times v)$ is the disjoint union of $\Gamma(u)$ and $\Gamma(v)$, hence $u \times v$ has coset-type $\rho \cup \sigma$. Now calculate:

$$\begin{aligned}\varphi_\rho * \varphi_\sigma &= \tilde{e}_{m+n}((\tilde{e}_m u \tilde{e}_m) \times (\tilde{e}_n v \tilde{e}_n))\tilde{e}_{m+n} \\ &= \tilde{e}_{m+n}(\tilde{e}_m \times \tilde{e}_n)(u \times v)(\tilde{e}_m \times \tilde{e}_n)\tilde{e}_{m+n} \\ &= \tilde{e}_{m+n}(u \times v)\tilde{e}_{m+n} = \varphi_{\rho \cup \sigma},\end{aligned}$$

because $\tilde{e}_{m+n}h = h\tilde{e}_{m+n} = \tilde{e}_{m+n}$ for $h \in H_{m+n}$, and hence $\tilde{e}_{m+n}(\tilde{e}_m \times \tilde{e}_n) = (\tilde{e}_m \times \tilde{e}_n)\tilde{e}_{m+n} = \tilde{e}_{m+n}$. |

From (2.8) it follows that the multiplication (2.7) makes C into a commutative and associative graded **C**-algebra.

For each $w \in S_{2n}$ let $\rho(w)$ denote the coset-type of w, and let

$$\psi'(w) = p_{\rho(w)}$$

in analogy with Chapter I, §7, cycle-type being replaced by coset-type. Note that $\psi'(w) = \psi'(w^{-1})$, by (2.1)(ii). Now define a **C**-linear mapping

$$\mathrm{ch}': C \to \Lambda_{\mathbf{C}}$$

as follows: if $f \in C_n$, then

$$\mathrm{ch}'(f) = \sum_{w \in S_{2n}} f(w)\psi'(w).$$

From this definition it is clear that

$$\mathrm{ch}'(\varphi_\rho) = p_\rho$$

for all partitions ρ, and hence it follows from (2.8) that

(2.9) *The mapping* $\mathrm{ch}': C \to \Lambda_{\mathbf{C}}$ *is an isomorphism of graded* **C***-algebras.* |

Next, we shall define a scalar product on C: if $f, g \in C$, say $f = \sum f_n$, $g = \sum g_n$ with $f_n, g_n \in C_n$, we define

$$(f, g) = \sum_{n \geq 0} (f_n, g_n)$$

where

(2.10) $$(f_n, g_n) = \sum_{w \in S_{2n}} f_n(w)\overline{g_n(w)}$$

(i.e. the usual scalar product, multiplied by $|S_{2n}| = (2n)!$). On the other hand, we have a scalar product (u, v) on $\Lambda_{\mathbf{C}}$, defined by

(2.11) $$(p_\rho, p_\sigma) = \delta_{\rho\sigma} z_{2\rho} = \delta_{\rho\sigma} 2^{l(\rho)} z_\rho$$

as in Chapter VI, (1.3). For these scalar products, the mapping ch′ is almost an isometry:

(2.12) *For each $n \geqslant 0$, the mapping*

$$f \mapsto |H_n|^{-1}\text{ch}'(f)$$

is an isometry of C_n onto $\Lambda^n_{\mathbf{C}}$.

Proof. Let ρ and σ be partitions of n. Clearly $(\varphi_\rho, \varphi_\sigma) = 0$ if $\rho \neq \sigma$, and

$$(\varphi_\rho, \varphi_\rho) = |H_\rho|^{-1} = |H_n|^{-2} z_{2\rho}$$

by (2.3). Hence

$$(\varphi_\rho, \varphi_\sigma) = |H_n|^{-2}(\text{ch}'\varphi_\rho, \text{ch}'\varphi_\sigma)$$

for all ρ, σ, which proves (2.12). |

Now define, for each $n \geqslant 0$ and each partition λ of n,

$$\begin{aligned} Z_\lambda &= |H_n|^{-1}\ \text{ch}'\omega^\lambda \\ &= |H_n|^{-1} \sum_{s \in S_{2n}} \omega^\lambda(s)\psi'(s) \end{aligned}$$

i.e.

(2.13) $$Z_\lambda = |H_n| \sum_{|\rho| = n} z_{2\rho}^{-1}\omega_\rho^\lambda p_\rho$$

where ω_ρ^λ is the value of ω^λ at elements of the double coset H_ρ. The symmetric functions Z_λ are called *zonal polynomials*.† Since the ω^λ form a **C**-basis of C, the Z_λ form a **C**-basis of $\Lambda_{\mathbf{C}}$.

(2.14) $Z_\lambda \in \mathbf{Z}[p_1, p_2, \ldots]$ *for all partitions* λ.

Proof. We have to show that

$$|H_n| z_{2\rho}^{-1}\omega_\rho^\lambda = |H_n|^{-1}|H_\rho|\omega_\rho^\lambda$$

is an integer for all λ, ρ. It is certainly a rational number, by (2.6), and it is an algebraic integer by (1.10). hence it is indeed an integer. |

† 'Zonal symmetric functions' would perhaps be a more logical name; but the present terminology is universally used, and I have therefore not sought to change it.

If λ, μ are partitions of n, it follows from (1.4)(iv) that

$$(\omega^\lambda, \omega^\mu) = \delta_{\lambda\mu}(2n)!/\chi^{2\lambda}(1)$$

or

(2.15) $$(\omega^\lambda, \omega^\mu) = \delta_{\lambda\mu} h(2\lambda)$$

where $h(2\lambda)$ is the product of the hook-lengths of the partition 2λ. Hence the matrix (ω_ρ^λ) satisfies the orthogonality relations

(2.15′) $$\sum_\rho z_{2\rho}^{-1} \omega_\rho^\lambda \omega_\rho^\mu = \frac{h(2\lambda)}{|H_n|^2} \delta_{\lambda\mu},$$

(2.15″) $$\sum_\lambda h(2\lambda)^{-1} \omega_\rho^\lambda \omega_\sigma^\lambda = \frac{z_{2\rho}}{|H_n|^2} \delta_{\rho\sigma},$$

from which and (2.12) we have

(2.16) $$p_\rho = |H_n| \sum_\lambda h(2\lambda)^{-1} \omega_\rho^\lambda Z_\lambda.$$

Hence if we define

$$C_\lambda = \frac{|H_n|}{h(2\lambda)} Z_\lambda$$

the relation (2.16) takes the form

(2.16′) $$p_\rho = \sum_{\lambda} \omega_\rho^\lambda C_\rho$$

and in particular

(2.16″) $$p_1^n = \sum_\rho C_\rho.$$

This renormalization of the zonal polynomials is often preferred by statisticians.

Next, from (2.12) and (2.15) it follows that

(2.17) $$(Z_\lambda, Z_\mu) = \delta_{\lambda\mu} h(2\lambda).$$

(2.18) *We have*

$$\prod_{i,j} (1 - x_i y_j)^{-1/2} = \sum_\lambda h(2\lambda)^{-1} Z_\lambda(x) Z_\lambda(y).$$

Proof. By (2.17), (Z_λ) and $(h(2\lambda)^{-1} Z_\lambda)$ are dual bases of $\Lambda_{\mathbf{Q}}$ for the scalar product (f, g), and so also are (p_ρ) and $(z_{2\rho}^{-1} p_\rho)$ by definition (2.11). It follows (cf. Chapter I, (4.6)) that

$$\sum_\lambda h(2\lambda)^{-1} Z_\lambda(x) Z_\lambda(y) = \sum_\rho z_{2\rho}^{-1} p_\rho(x) p_\rho(y)$$

$$= \prod_{r \geq 1} \exp \frac{p_r(xy)}{2r}$$

which in turn is equal to

$$\exp \tfrac{1}{2} \log \prod (1 - x_i y_j)^{-1} = \prod (1 - x_i y_j)^{-1/2}. \quad |$$

By considering the hook-lengths at $(i, 2j-1)$ and $(i, 2j)$ in the diagram of 2λ, where $(i, j) \in \lambda$, it follows that

$$h(2\lambda) = h_1(2\lambda) h_2(2\lambda)$$

where

$$h_i(2\lambda) = \prod_{s \in \lambda} (2a(s) + l(s) + i) \qquad (i = 1, 2) \tag{2.19}$$

in the notation of Chapter VI, (6.14). From the results of Chapter VI, §10 it follows that the Z_λ satisfy the same orthogonality relations as the Jack symmetric functions $J_\lambda^{(\alpha)}$ with parameter $\alpha = 2$. In fact, $Z_\lambda = J_\lambda^{(2)}$ for all λ, as we shall now show.

For each integer $n \geq 0$, let

$$\begin{aligned} g_n &= |H_n|^{-1} Z_{(n)} \\ &= \sum_{|\rho| = n} z_{2\rho}^{-1} p_\rho \end{aligned} \tag{2.20}$$

since $\omega_\rho^{(n)} = 1$ for all ρ. If now $\lambda = (\lambda_1, \lambda_2, \ldots)$ is any partition, let $g_\lambda = g_{\lambda_1} g_{\lambda_2} \ldots$. Then we have

$$(g_\lambda, m_\mu) = \delta_{\lambda\mu} \tag{2.21}$$

for all λ, μ.

Proof. The generating function for the g_n is

$$\sum_{n \geq 0} g_n(x) y^n = \sum_\rho z_{2\rho}^{-1} p_\rho(x) y^{|\rho|}$$

$$= \prod_{i \geq 1} (1 - x_i y)^{-1/2}$$

as in the proof of (2.18). It follows that

$$\prod_{i,j}(1-x_iy_j)^{-1/2}=\sum_\lambda g_\lambda(x)m_\lambda(y)$$

and hence that (g_λ), (m_λ) are dual bases for the scalar product (f,g). ∎

(2.22) *The transition matrix $M(Z,m)$ is strictly upper triangular.*

Proof. The coefficient of m_μ in the expansion of Z_λ as a linear combination of monomial symmetric functions is equal to (Z_λ, g_μ) by (2.21), hence we have to show that $(Z_\lambda, g_\mu)=0$ unless $\lambda \geqslant \mu$. From the definition of g_μ and from (2.9) it follows that

$$g_\mu = c_\mu \operatorname{ch}'(\theta^\mu)$$

where c_μ is a positive scalar, whose exact value need not concern us, and $\theta^\mu = \omega^{(\mu_1)} * \omega^{(\mu_2)} * \dots$. Hence by (2.12) it is enough to show that $(\omega^\lambda, \theta^\mu) = 0$ unless $\lambda \geqslant \mu$.

Now $\theta^\mu = \tilde{e}_n \eta \tilde{e}_n$, where $n=|\mu|$ and η is the trivial one-dimensional character of the subgroup $S_{2\mu} = S_{2\mu_1} \times S_{2\mu_2} \times \dots$ of S_{2n}. Since left and right multiplications by $\tilde{e}_n$ are self-adjoint for the scalar product (2.10), we have

$$\begin{aligned}(\omega^\lambda, \theta^\mu) &= \left(\tilde{e}_n \chi^{2\lambda}, \tilde{e}_n \eta \tilde{e}_n\right) = (\tilde{e}_n \chi^{2\lambda}, \eta) \\ &= (\tilde{e}_n \chi^{2\lambda}\eta)(1).\end{aligned}$$

Now the element $\chi^{2\lambda}\eta$ of the group algebra of S_{2n} generates the $\chi^{2\lambda}$-isotypic component of the induced representation $\eta^{S_{2n}}$, which, in the notation of Chapter I, §7, is $\eta_{2\mu}$. Hence $\chi^{2\lambda}\eta = 0$ unless $\chi^{2\lambda}$ occurs in $\eta_{2\mu}$; and since

$$\langle \chi^{2\lambda}, \eta_{2\mu}\rangle = \langle s_{2\lambda}, h_{2\mu}\rangle = K_{2\lambda,2\mu}$$

(Chapter I, §7) is zero unless $2\lambda \geqslant 2\mu$, i.e. unless $\lambda \geqslant \mu$, it follows that $\chi^{2\lambda}\eta = 0$ and hence $(\omega^\lambda, \theta^\mu) = 0$ unless $\lambda \geqslant \mu$. ∎

From (2.17) and (2.22) it follows (cf. Chapter VI, §1) that Z_λ is a scalar multiple of $J_\lambda^{(2)}$, for each partition λ. But by (2.13) and Chapter VI, (10.29), the coefficient of p_1^n in each of Z_λ and $J_\lambda^{(2)}$ (where $n=|\lambda|$) is equal to 1, and therefore

(2.23) $Z_\lambda = J_\lambda^{(2)}$ *for all partitions* λ. ∎

For each partition λ, let c_λ denote the polynomial

(2.24)
$$c_\lambda(X) = \prod_{s\in\lambda}(X + 2a'(s) - l'(s))$$
$$= \prod_{(i,j)\in\lambda}(X - i + 2j - 1)$$

Then if $Z_\lambda(1_n)$ denotes the value of $Z_\lambda(x_1, \dots, x_n)$ at $x_1 = \dots = x_n = 1$, we have

(2.25)
$$Z_\lambda(1_n) = c_\lambda(n)$$

by (2.23) and Chapter VI, (10.24). For an independent proof of (2.25), see (4.15) below.

(2.26) *The coefficient of x^λ in $Z_\lambda(x)$ is*
$$h_1(2\lambda) = \prod_{s\in\lambda}(2a(s) + l(s) + 1).$$

This follows from (2.23) and the definition (Chapter VI, (10.22)) of $J_\lambda^{(\alpha)}$. For another proof, see §3, Example 1 below.

(2.27) *If $\mu \leqslant \lambda$, the coefficient of x^μ in $Z_\lambda(x)$ is a positive integer divisible by $u_\mu = \prod_{i\geqslant 1} m_i(\mu)!$.*

Proof. The 'augmented' monomial symmetric functions $\tilde{m}_\mu = u_\mu m_\mu$ are a $\mathbf{Z}$-basis of the subring $\mathbf{Z}[p_1, p_2, \dots]$ of Λ (Chapter I, §6, Example 10). Hence it follows from (2.14) that the coefficient of x^μ (or of m_μ) in Z_λ is an integer divisible by u_μ; moreover it is positive by virtue of (2.23) and Chapter VI, (10.12). |

(2.28) *Let λ, μ, ν be three partitions. Then $(Z_\lambda, Z_\mu Z_\nu)$ is an integer $\geqslant 0$, and is positive only if $c^{2\lambda}_{2\mu,2\nu} = \langle s_{2\lambda}, s_{2\mu}s_{2\nu}\rangle$ (Chapter I, §9) is positive.*

Proof. The definition (2.11) of the scalar product shows that it is integer-valued on the subring $\mathbf{Z}[p_1, p_2, \dots]$ of Λ. Hence it follows from (2.14) that $(Z_\lambda, Z_\mu Z_\nu)$ is an integer. To show that it is non-negative, it is enough by (2.9) and (2.12) to show that $(\omega^\lambda, \omega^\mu * \omega^\nu) \geqslant 0$.

Let $|\mu| = m$, $|\nu| = n$, $|\lambda| = p$. We may assume that $m + n = p$, otherwise the scalar product is certainly zero. We have

$$(\omega^\lambda, \omega^\mu * \omega^\nu) = \left(\omega^\lambda, \tilde{e}_p(\omega^\mu \times \omega^\nu)\tilde{e}_p\right)$$
$$= (\omega^\lambda, \omega^\mu \times \omega^\nu)$$

which up to a positive scalar factor is the coefficient of $\omega^\mu \times \omega^\nu$ in the restriction of ω^λ to $S_{2m} \times S_{2n}$, and hence is non-negative by (1.8).

Finally, since

$$(\omega^\lambda, \omega^\mu \times \omega^\nu) = \left(\tilde{e}_p \chi^{2\lambda}, (\tilde{e}_m \times \tilde{e}_n)(\chi^{2\mu} \times \chi^{2\nu})\right)$$
$$= \tilde{e}_p \chi^{2\lambda}(\chi^{2\mu} \times \chi^{2\nu})(1),$$

for $(Z_\lambda, Z_\mu Z_\nu)$ to be positive we must have $\chi^{2\lambda}(\chi^{2\mu} \times \chi^{2\nu}) \neq 0$, which implies that $\chi^{2\lambda}$ occurs in the character obtained by inducing $\chi^{2\mu} \times \chi^{2\nu}$ from $S_{2m} \times S_{2n}$ to S_{2p}, i.e. that $\langle s_{2\lambda}, s_{2\mu} s_{2\nu} \rangle > 0$. ∎

Remark. It seems to be an open question whether $(Z_\lambda, Z_\mu Z_\nu) > 0$ if and only if $c^\lambda_{\mu\nu} = \langle s_\lambda, s_\mu s_\nu \rangle > 0$.

Examples

1. If $\lambda = (r|s) = (r+1, 1^s)$ is a hook, then Z_λ is equal to the determinant

$$D(1, 2, \ldots, s, -2r, -2r+2, \ldots, -2)$$

in the notation of Chapter VI, §10, Example 4(a).

2. There are explicit formulas for some of the entries in the matrix (ω^λ_ρ).

(a) $\omega^\lambda_\rho = 1$ if $\lambda = (n)$ or $\rho = (1^n)$.

(b) When $\lambda = (1^n)$,

$$Z_\lambda = n! e_n = n! \sum_\rho \varepsilon_\rho z_\rho^{-1} p_\rho$$

(Chapter I, (2.14′)) and hence

$$\omega^{(1^n)}_\rho = (-1)^{n - l(\rho)} / 2^{n - l(\rho)}.$$

(c) From (2.25) it follows that

$$|H_n| \sum_\rho z_{2\rho}^{-1} \omega^\lambda_\rho X^{l(\rho)} = c_\lambda(X)$$

and hence, on equating coefficients of X on either side,

$$\omega^\lambda_{(n)} = \frac{1}{|H_{n-1}|} \prod_{s \in \lambda}{}' (2a'(s) - l'(s))$$

where the product omits the square (1,1) (for which $a'(s) = l'(s) = 0$). Hence $\omega^\lambda_{(n)} = 0$ if $\lambda \supset (2^3)$.

(d)
$$\omega^\lambda_{(21^{n-2})} = \frac{2n(\lambda') - n(\lambda)}{n(n-1)}$$

where $n(\lambda) = \lambda_2 + 2\lambda_3 + \ldots = \Sigma_{s \in \lambda} l'(s)$. (Same method as in (c): equate coefficients of X^{n-1}.)

(e) Let φ: $\Lambda_{\mathbf{Q}} \to \Lambda_{\mathbf{Q}}$ be the automorphism defined by $\varphi(p_r) = \frac{1}{2}p_r$ for all $r \geqslant 1$. Use Example 1 to show that $Z_{(n-1,1)}$ is the image under φ of

$$2^{n-2}(\bar{s}_{(n-1,1)} + h_1 \bar{s}_{(n-1)})$$

where $\bar{s}_\mu = h(\mu)s_\mu$ for any partition μ, and deduce that

$$\omega_\rho^{(n-1,1)} = \frac{(2n-1)m_1(\rho) - n}{2n(n-1)}$$

where $m_1(\rho)$ is the number of parts of ρ equal to 1.

(f) Show that

$$\omega_\rho^{(21^{n-2})} = \omega_\rho^{(1^n)} \frac{(n+1)m_1(\rho) - 2n}{n(n-1)}.$$

(Same method as in (e).)

3. Let

$$\square_n = \sum_{i=1}^{n} x_i^2 D_i^2 + \sum_{i \neq j} (x_i - x_j)^{-1} x_i^2 D_i$$

where $D_i = \partial/\partial x_i$ (Laplace–Beltrami operator). Then

$$\square_n Z_\lambda(x_1, \ldots, x_n) = (2n(\lambda') - n(\lambda)) Z_\lambda(x_1, \ldots, x_n)$$

for partitions of λ of length $\leqslant n$. (Chapter VI, §3, Example 3(e).) The coefficients $a_{\lambda\mu}$ in

$$Z_\lambda = \sum_{\mu \leqslant \lambda} a_{\lambda\mu} m_\mu$$

may be computed recursively in terms of $a_{\lambda\lambda}$ by means of this equation (Chapter VI, §4, Example 3(d)).

4. Let $\Delta: \Lambda \to \Lambda \otimes \Lambda$ be the comultiplication defined in Chapter I, §5, Example 25. Show that

$$(f, g) = \langle \Delta f, \Delta g \rangle$$

for all $f, g \in \Lambda$, where the scalar product on the right is that defined *loc. cit.*

5. Let X denote the set of all graphs with $2n$ vertices $1, 2, \ldots, 2n$ and n edges, one through each vertex. If $x, y \in X$, the connected components of $x \cup y$ are cycles of even lengths $2\rho_1 \geqslant 2\rho_2 \geqslant \ldots$. Let $\rho(x, y)$ denote the partition $(\rho_1, \rho_2, \ldots)$ of n so defined.

Show that the eigenvalues of the matrix

$$(p_{\rho(x,y)})_{x, y \in X}$$

are the zonal polynomials Z_λ such that $|\lambda| = n$, and that the multiplicity of Z_λ as an eigenvalue is $\chi^{2\lambda}(1) = (2n)!/h(2\lambda)$.

6. Let ε denote the sign character of S_{2n}, and let ε_n denote its restriction to H_n. From §1, Example 11, the induced character $\varepsilon_n^{S_{2n}}$ is multiplicity-free, so that $(S_{2n}, H_n, \varepsilon_n)$ is a twisted Gelfand pair, and

$$\varepsilon_n^{S_{2n}} = \sum_{|\lambda|=n} \varepsilon\chi^{2\lambda} = \sum_{|\lambda|=n} \chi^{(2\lambda)'}.$$

Hence the irreducible characters that appear in $\varepsilon_n^{S_{2n}}$ are those corresponding to partitions of $2n$ with all columns of even length. The ε-spherical functions are

$$\pi^{\lambda'} = \chi^{(2\lambda)'}e_n'$$

where $e_n' = |H_n|^{-1}\varepsilon_n$ is the idempotent corresponding to ε_n.

(a) For each $f \in C_n = C(S_{2n}, H_n)$ let f^ε be the function defined by $f^\varepsilon(s) = \varepsilon(s)f(s)$ for $s \in S_{2n}$. Then the mapping $f \mapsto f^\varepsilon$ is an isometric isomorphism of C_n onto $C_n^\varepsilon = C^\varepsilon(S_{2n}, H_n)$, the algebra of functions f on S_{2n} that satisfy $f(hs) = f(sh) = \varepsilon(h)f(s)$ for all $s \in S_{2n}$ and $h \in H_n$. Under this isomorphism, the zonal spherical function $\omega^\lambda \in C_n$ is mapped to $\pi^{\lambda'} \in C_n^\varepsilon$.

(b) Let

$$C^\varepsilon = \bigoplus_{n \geq 0} C_n^\varepsilon$$

and define a bilinear multiplication on C^ε by the rule

$$f * g = e_{m+n}'(f \times g)e_{m+n}'$$

if $f \in C_m^\varepsilon$ and $g \in C_n^\varepsilon$. Then the mapping $f \mapsto f^\varepsilon$ defined above extends to a graded C-algebra isomorphism of C onto C^ε. The functions φ_ρ^ε form a C-basis of C^ε, and $\varphi_\rho^\varepsilon * \varphi_\sigma^\varepsilon = \varphi_{\rho\cup\sigma}^\varepsilon$ for any two partitions ρ, σ.

(c) Define an embedding $w \mapsto w^*$ (which is not a homomorphism) of S_n into S_{2n} as follows. For each cycle $c = (i_1, \dots, i_r)$ in the cycle-decomposition of $w \in S_n$, let c^* denote the $2r$-cycle $(2i_1 - 1, 2i_1, 2i_2 - 1, 2i_2, \dots, 2i_r - 1, 2i_r)$, and let w^* be the product of these c^*. If w has cycle-type ρ, then w^* has cycle-type 2ρ and coset-type ρ, and $\varepsilon(w^*) = (-1)^{l(\rho)} = (-1)^n\varepsilon(w)$. For each $f \in C_n^\varepsilon$, let f^* be the function on S_n defined by $f^*(w) = f(w^*)$, which is a class function on S_n.

Now define a C-linear mapping

$$\mathrm{ch}^\varepsilon : C^\varepsilon \to \Lambda_{\mathbf{C}}$$

as follows: if $f \in C_n^\varepsilon$, then $\mathrm{ch}^\varepsilon(f) = (-1)^n|H_n|^2\mathrm{ch}(f^*)$, where ch is the characteristic map of Chapter I, §7. We have $\mathrm{ch}^\varepsilon(\varphi_\rho^\varepsilon) = \varepsilon_\rho 2^{l(\rho)}p_\rho$, from which it follows that ch^ε is an isomorphism of graded C-algebras.

(d) Define a scalar product $(u, v)'$ on $\Lambda_{\mathbf{C}}$ by

$$(p_\rho, p_\sigma)' = \delta_{\rho\sigma}2^{-l(\rho)}z_\rho.$$

Then $f \mapsto |H_n|^{-1}\mathrm{ch}^\varepsilon(f)$ is an isometry of C_n^ε onto $\Lambda_{n,\mathbf{C}}$ for each $n \geq 0$.

7. In continuation of Example 6, define

$$Z_\lambda' = |H_n|^{-1}\mathrm{ch}^\varepsilon(\pi^\lambda)$$

if λ is a partition of n. Since

$$\pi^\lambda(w^*) = \varepsilon(w^*)\omega^{\lambda'}(w^*) = (-1)^n \varepsilon_\rho \omega_\rho^{\lambda'}$$

if $w \in S_n$ has cycle-type ρ, it follows that

$$Z'_\lambda = |H_n| \sum_\rho \varepsilon_\rho z_\rho^{-1} \omega_\rho^{\lambda'} p_\rho \tag{1}$$

which is the image of $Z_{\lambda'}$, under the automorphism of $\Lambda_{\mathbf{C}}$ which maps each power sum p_r to $(-1)^{r-1}2p_r$, and hence p_ρ to $\varepsilon_\rho 2^{l(\rho)} p_\rho$ for each partition ρ. From Chapter VI, (10.24) it follows that

$$Z'_\lambda = 2^n J_\lambda^{(1/2)} \tag{2}$$

if λ is a partition of n.

(a) From (1) and (2.15′) it follows that

$$(Z'_\lambda, Z'_\mu)' = h(2\lambda')\delta_{\lambda\mu}.$$

(b) From Chapter VI, (10.24) we have

$$Z'_\lambda(1_n) = \prod_{s \in \lambda} (2n + a'(s) - 2l'(s)).$$

(c) From Chapter VI, (10.22) the coefficient of x^λ in $Z'_\lambda(x)$ is

$$\prod_{s \in \lambda} (a(s) + 2l(s) + 2).$$

(d) We have (compare (2.18))

$$\prod_{i,j} (1 - x_i y_j)^{-2} = \sum_\lambda h(2\lambda')^{-1} Z'_\lambda(x) Z'_\lambda(y).$$

Notes and references

The zonal polynomials were introduced by L.-K. Hua [H11] and A. T. James [J4] independently, around 1960. There is a large literature devoted to them, mostly due to statisticians: see for example [F3], Chapters 12 and 13, and [M17]; also Takemura's monograph [T1] and the bibliography there.

The proof of (2.23) presented here is due to J. Stembridge [S29]. The formula (2.25) was first proved by A. G. Constantine [C4], and (2.26) is due to A. T. James [J5]. Proposition (2.28) is due to N. Bergeron and A. Garsia [B3].

Example 2. These formulas are due to P. Diaconis and E. Lander [D4].

Example 3. This method of calculating the zonal polynomials is due to A. T. James [J6].

Example 5. I learnt this combinatorial definition of the zonal polynomials from R. Stanley.

The zonal polynomials Z_λ have been calculated up to weight $|\lambda| = 12$: see [P1].

3. The Gelfand pair $(GL_n(\mathbf{R}), O(n))$

Let $V = \mathbf{R}^n$, with standard basis of unit vectors $e_1, \dots, e_n$, and let $G = GL(V) = GL_n(\mathbf{R})$, acting on V as follows:

$$ge_j = \sum_{i=1}^{n} g_{ij} e_i$$

if $g = (g_{ij})$. Let

$$P(G) = \bigoplus_{m \geqslant 0} P^m(G)$$

be the space of all polynomial functions on G, as in Chapter I, Appendix A, §8: $P^m(G)$ is the subspace of $P(G)$ consisting of the polynomial functions that are homogeneous of degree m. The group G acts on $P(G)$ by the rule

$$(3.1) \qquad (gp)(x) = p(xg)$$

for $p \in P(G)$ and $g, x \in G$, and under this action $P(G)$ decomposes as a direct sum

$$(3.2) \qquad P(G) = \bigoplus_{\mu} P_\mu,$$

where μ ranges over all partitions of length $\leqslant n$, and P_μ is the subspace of $P(G)$ spanned by the matrix coefficients R_{ij}^μ of an irreducible polynomial representation R^μ of G corresponding to the irreducible G-module $F_\mu(V)$, in the notation of Chapter I, Appendix A. Moreover we have

$$(3.3) \qquad P_\mu \cong F_\mu(V)^{d_\mu}$$

where d_μ is the dimension of $F_\mu(V)$.

Now let $K = O(n)$ be the compact subgroup of G consisting of the orthogonal matrices, and let e_K, e'_K be the operators on $P(G)$ defined by

$$(e_K p)(x) = \int_K p(kx)\,dk,$$

$$(e'_K p)(x) = \int_K p(xk)\,dk$$

where the integration is with respect to normalized Haar measure on the compact group K. Clearly e_K (but not e'_K) commutes with the action (3.1) of G. Since

$$e_K R^\mu_{ij}(x) = \int_K R^\mu_{ij}(kx)\,\mathrm{d}k$$
$$= \sum_r \left(\int_K R^\mu_{ir}(k)\,\mathrm{d}k\right) R^\mu_{rj}(x)$$

it follows that $e_K P_\mu \subset P_\mu$ for each μ, and a similar calculation shows that $e'_K P_\mu \subset P_\mu$.

Next, we have $e_K p(kx) = e_K p(x)$ for all $k \in K$, by the invariance of Haar measure, and therefore

$$(e_K^2 p)(x) = \int_K e_K p(kx)\,\mathrm{d}k = \int_K e_K p(x)\,\mathrm{d}k = e_K p(x)$$

which shows that e_K is idempotent, and likewise e'_K is idempotent. Moreover, the two operators e_K, e'_K are related by

(3.4) $$e'_K = \tau e_K \tau$$

where τ is the involution of $P(G)$ defined by $\tau p(x) = p(x')$, where x' is the transpose of the matrix $x \in G$.

Let

$$P(K\backslash G) = \bigoplus_{m \geqslant 0} P^m(K\backslash G)$$

(resp.
$$P(G/K) = \bigoplus_{m \geqslant 0} P^m(G/K),$$
$$P(G,K) = \bigoplus_{m \geqslant 0} P^m(G,K))$$

denote the subspace of $P(G)$ consisting of the functions constant on each coset Kx (resp. each coset xK, each double coset KxK) of K in G. In each case the superscript m denotes the subspace of functions that are homogeneous of degree m. Since $-1 \in K$ we have $p(x) = p(-x)$ for p in any of these spaces, and hence $P^m(K\backslash G)$, $P^m(G/K)$, and $P^m(G,K)$ are all zero if m is odd. Also, by (3.1), $P^m(G/K)$ is the same thing as $P^m(G)^K$.

(3.5) (i) *For each partition* μ, $e_K P_\mu$ *and* $e'_K P_\mu$ *are subspaces of* P_μ *of the same dimension.*
(ii) e_K (resp. e'_K) *projects* $P(G)$ *onto* $P(K\backslash G)$ (resp. $P(G/K)$).

Proof. (i) From (3.4) it follows that e_K and $e_{K'}$ are isomorphic projections,

so that their restrictions to each P_μ have the same rank, i.e. $\dim(e_K P_\mu) = \dim(e'_K P_\mu)$.
(ii) If $p \in e_K P(G)$ then clearly $p(kx) = p(x)$ for all $k \in K$ and $x \in G$. Conversely, if $p(kx) = p(x)$ for $k \in K$ and $x \in G$, we have $e_K p = p$ and hence $p \in e_K P(G)$. Hence $e_K P(G) = P(K\backslash G)$, and likewise $e'_K P(G) = P(G/K)$. |

Next we have

(3.6) $$\mathrm{End}_G P(K\backslash G) \cong P(G, K).$$

Proof. The proof is best expressed in terms of the Schur algebra $\mathfrak{S}^m = \mathrm{End}_{S_m} \mathbf{T}^m(V)$ (Chapter I, Appendix A, §8, Example 6), with G-action $g\alpha = \mathbf{T}^m(g)\alpha$ $(g \in G, \alpha \in \mathfrak{S}^m)$. The mapping $\alpha \mapsto p_\alpha$ defined by $p_\alpha(g) = \mathrm{trace}(g\alpha)$ is a G-isomorphism of $\mathfrak{S}^m$ onto $P^m(G)$, and since

$$\mathrm{trace}(kg\alpha) = \mathrm{trace}(\mathbf{T}^m(k)\mathbf{T}^m(g)\alpha) = \mathrm{trace}(g\alpha\mathbf{T}^m(k))$$

it follows that

$$\begin{aligned}(e_K p_\alpha)(g) &= \int_K \mathrm{trace}(kg\alpha)\,\mathrm{d}k \\ &= \mathrm{trace}(g\alpha\varepsilon_K)\end{aligned}$$

where

$$\varepsilon_K = \int_K \mathbf{T}^m(k)\,\mathrm{d}k \in \mathfrak{S}^m.$$

Hence $e_K p_\alpha = p_{\alpha\varepsilon_K}$, and therefore $P^m(K\backslash G) = e_K P^m(G)$ is the image of $\mathfrak{S}^m\varepsilon_K$ under the isomorphism $\alpha \mapsto p_\alpha$. Likewise $P^m(G/K)$ (resp. $P^m(G, K)$) is the image of $\varepsilon_K \mathfrak{S}^m$ (resp. $\varepsilon_K \mathfrak{S}^m \varepsilon_K$). Now ε_K is idempotent, and therefore

$$\begin{aligned}\mathrm{End}_G P^m(K\backslash G) &\cong \mathrm{End}_G(\mathfrak{S}^m\varepsilon_K) = \mathrm{End}_{\mathfrak{S}^m}(\mathfrak{S}^m\varepsilon_K) \\ &\cong \varepsilon_K \mathfrak{S}^m \varepsilon_K \cong P^m(G, K). \quad |\end{aligned}$$

Consider now the double cosets of K in G. We have

(3.7) *If $x \in G$, there exists a diagonal matrix $\xi = \mathrm{diag}(\xi_1, \dots, \xi_n)$ such that $x \in K\xi K$, and the eigenvalues of $x'x$ are $\xi_1^2, \dots, \xi_n^2$.*

Proof. The matrix $x'x$ is symmetric and positive definite, hence can be brought to diagonal form by conjugating with an orthogonal matrix, say $x'x = k'yk$ where $k \in K$ and y is a diagonal matrix whose diagonal entries y_i are the eigenvalues of $x'x$, and hence are positive real numbers. So if ξ_i is a square root of y_i for each i, and $\xi = \mathrm{diag}(\xi_1, \dots, \xi_n)$, we have

$x'x = k'\xi^2 k$, which shows that $k_1 = xk^{-1}\xi^{-1}$ is orthogonal and hence $x = k_1 \xi k \in K\xi K$. |

From (3.7) it follows that the polynomial functions $p \in P^{2m}(G, K)$ are determined by their values on the diagonal matrices, and that $p(x) = p^*(\xi_1^2, \ldots, \xi_n^2)$ where the ξ_i^2 are the eigenvalues of $x'x$, and p^* is a symmetric polynomial in n variables, homogeneous of degree m. We conclude that

(3.8) *The mapping $p \mapsto p^*$ is an isomorphism of $P^{2m}(G, K)$ onto $\Lambda^m_{n,\mathbf{R}}$ for each $m \geq 0$.* |

Next, let Σ denote the space of real symmetric $n \times n$ matrices, and let

$$P(\Sigma) = \bigoplus_{m \geq 0} P^m(\Sigma)$$

denote the space of polynomial functions on Σ, where $P^m(\Sigma)$ consists of the functions that are homogeneous of degree m. The group G acts on $P(\Sigma)$ by

$$(gp)(\sigma) = p(g'\sigma g)$$

for $g \in G$, $p \in P(\Sigma)$, and $\sigma \in \Sigma$.

Also let $\mathbf{T} = \mathbf{T}^{2m}(V)$ be the $2m$th tensor power of V. The group G acts diagonally on $\mathbf{T}$:

$$g(v_1 \otimes \ldots \otimes v_{2m}) = gv_1 \otimes \ldots \otimes gv_{2m}$$

and the symmetric group S_{2m} acts on $\mathbf{T}$ by permuting the factors:

$$w(v_1 \otimes \ldots \otimes v_{2m}) = v_{w^{-1}(1)} \otimes \ldots \otimes v_{w^{-1}(2m)}$$

where $g \in G$, $w \in S_{2m}$, and $v_1, \ldots, v_{2m} \in V$.

Let $\langle u, v \rangle$ be the standard inner product on V for which $\langle e_i, e_j \rangle = \delta_{ij}$, and let $\varphi: \mathbf{T} \to P^m(\Sigma)$ denote the linear mapping defined by

$$\varphi(v_1 \otimes \ldots \otimes v_{2m})(\sigma) = \prod_{i=1}^{m} \langle v_{2i-1}, \sigma v_{2i} \rangle .$$

Clearly φ commutes with the actions of G on $\mathbf{T}$ and on $P^m(\Sigma)$, and since $\langle u, \sigma v \rangle = \langle \sigma u, v \rangle = \langle v, \sigma u \rangle$ it follows that $\varphi(ht) = \varphi(t)$ for $t \in \mathbf{T}$ and h in the hyperoctahedral group $H = H_m$. Since $\langle e_i, \sigma e_j \rangle = \sigma_{ij}$ is the (i, j) element of $\sigma \in \Sigma$, it follows that φ is surjective. Hence if

$$\mathbf{T}^H = \{t \in \mathbf{T} : ht = t \text{ for all } h \in H\}$$

is the subspace of $\mathbf{T}$ fixed by H, we have

(3.9) *The mapping* φ *restricted to* $\mathbf{T}^H$ *is a G-isomorphism of* $\mathbf{T}^H$ *onto* $P^m(\Sigma)$. ∎

From Chapter I, Appendix A, §5, we know that $\mathbf{T}$ decomposes under the action of $S_{2m}\times G$ as follows:

(3.10) $$\mathbf{T}=\bigoplus_{\mu}\mathbf{T}_{\mu}$$

where

$$\mathbf{T}_{\mu}\cong M_{\mu}\otimes F_{\mu}(V).$$

Here μ ranges over the partitions of $2m$ of length $\leqslant n$, and M_{μ} is an irreducible S_{2m}-module corresponding to μ. Hence by (2.5) we have

(3.11) $$\mathbf{T}^H\cong\bigoplus_{\lambda}F_{2\lambda}(V)$$

as G-module, where λ ranges over the partitions of m of length $\leqslant n$.

If $p\in P(\Sigma)$, the function $\tilde{p}$ defined by

$$\tilde{p}(x)=p(x'x)$$

is a polynomial function on G, and the mapping $p\mapsto\tilde{p}$ commutes with the actions of G on $P(\Sigma)$ and $P(G)$. Moreover, it doubles degrees: if $p\in P^m(\Sigma)$ then $\tilde{p}\in P^{2m}(G)$.

(3.12) *The mapping* $p\mapsto\tilde{p}$ *defined above is a G-isomorphism of* $P^m(\Sigma)$ *onto* $P^{2m}(K\backslash G)$, *for each* $m\geqslant 0$.

Proof. Since $k'k=1$ for $k\in K$, we have $\tilde{p}(kx)=\tilde{p}(x)$ for all $k\in K$ and $x\in G$, and hence $\tilde{p}\in P^{2m}(K\backslash G)$. Next, each positive definite matrix σ can be written in the form $\sigma=x'x$ for some $x\in G$. Since the positive definite matrices form a non-empty open subset Σ^+ of Σ, a polynomial function $p\in P^m(\Sigma)$ that vanishes on Σ^+ must vanish identically, from which it follows that the mapping $p\mapsto\tilde{p}$ is injective.

Now from (3.9) and (3.11) we have

(1) $$P^m(\Sigma)\cong\bigoplus_{\lambda}F_{2\lambda}(V)$$

where λ runs through the partitions of m of length $\leqslant n$; and since e_K is a G-homomorphism it follows from (3.3) that $e_K P_{\mu}$ is isomorphic to the direct sum of say m_{μ} copies of $F_{\mu}(V)$, for each partition μ, so that

(2) $$P^{2m}(K\backslash G)=\bigoplus_{\mu}e_K P_{\mu}\cong\bigoplus_{\mu}F_{\mu}(V)^{m_{\mu}},$$

where μ runs through the partitions of $2m$ of length $\leqslant n$. We have already shown that $P^m(\Sigma)$ is isomorphic to a submodule of $P^{2m}(K\backslash G)$, and hence from (1) and (2) we have $m_\mu \geqslant 1$ if μ is even (i.e. $\mu = 2\lambda$).

From (2) we have

$$\sum_\mu m_\mu^2 = \dim \operatorname{End}_G P^{2m}(K\backslash G)$$

$$= \dim P^{2m}(G, K) \quad \text{by (3.6)},$$

which by (3.8) is equal to the number of partitions λ of m of length $\leqslant n$. It follows that $m_\mu = 1$ if $\mu = 2\lambda$ is even, and $m_\mu = 0$ otherwise. Hence (1) and (2) show that $P^m(\Sigma)$ and $P^{2m}(K\backslash G)$ have the same dimension, and the proof is complete. |

Remark. Let $v_1, \ldots, v_n \in V$ be the columns of the matrix $x \in G$, so that the (i, j) element of $x'x$ is the scalar product $\langle v_i, v_j \rangle = v_i' v_j$ of v_i and v_j. Then (3.12) says that any polynomial function $f(v_1, \ldots, v_n)$ such that $f(kv_1, \ldots, kv_n) = f(v_1, \ldots, v_n)$ for all $k \in K$—that is to say, any polynomial invariant of n vectors $v_1, \ldots, v_n$ under the orthogonal group—is a polynomial in the scalar products $\langle v_i, v_j \rangle$. In this form it is part of the 'first main theorem' of invariant theory for the orthogonal group ([W2], Chapter II, §9).

As a corollary of the proof of (3.12) we record

(3.13) (i) *The G-module $e_K P_\mu$ is isomorphic to $F_\mu(V)$ if μ is even, and is zero otherwise.*
(ii) *We have*

$$e_K P(G) = P(K\backslash G) \cong \bigoplus_\lambda F_{2\lambda}(V)$$

where λ ranges over all partitions of length $\leqslant n$. |

Hence the G-module $e_K P(G)$ is multiplicity-free. The situation is quite analogous to that of §1, except that the groups G, K involved are now infinite, and the space of all functions on a finite group is replaced by the space of polynomial functions on G. Thus (G, K) may be called a Gelfand pair.

(3.14) *Let μ be a partition of length $\leqslant n$. Then*

$$\dim F_\mu(V)^K = \begin{cases} 1 & \text{if } \mu \text{ is even,} \\ 0 & \text{otherwise.} \end{cases}$$

Proof. From (3.3) we have

$$\dim P_\mu^K = d_\mu \dim F_\mu(V)^K \tag{1}$$

and on the other hand, from (3.5) and (3.13),

$$\dim P_\mu^K = \dim e'_K P_\mu = \dim e_K P_\mu$$

$$(2) \qquad = \begin{cases} d_\mu & \text{if } \mu \text{ is even,} \\ 0 & \text{otherwise.} \end{cases}$$

The result follows from (1) and (2). |

For each partition μ of length $\leqslant n$, let

$$P_\mu(G,K) = P_\mu \cap P(G,K) = (e_K P_\mu)^K.$$

(3.15) *$P_\mu(G,K)$ is one-dimensional if μ is even, and is zero otherwise. Moreover*

$$P(G,K) = \bigoplus_\lambda P_{2\lambda}(G,K)$$

summed over all partitions λ of length $\leqslant n$.

This follows from (3.13) and (3.14). |

If $f \in \Lambda_{\mathbf{R}}$ we may regard f as a function on G:

$$f(x) = f(\xi_1, \ldots, \xi_n)$$

where $\xi_1, \ldots, \xi_n$ are the eigenvalues of $x \in G$. For example,

$$(3.16) \qquad p_r(x) = \operatorname{trace}(x^r) \qquad (r \geqslant 1)$$

since the eigenvalues of x^r are ξ_i^r, and

$$(3.17) \qquad e_r(x) = \operatorname{trace}(\Lambda^r x) \qquad (r \geqslant 1).$$

Since each $f \in \Lambda_{\mathbf{R}}$ is a polynomial in the e_r, it follows that f is a polynomial function on G that satisfies

$$(3.18) \qquad f(xy) = f(yx)$$

for all $x, y \in G$ (because xy and $yx = y(xy)y^{-1}$ have the same eigenvalues).

In particular (Chapter I, Appendix A, §8), the character of the irreducible representation R^μ of G is the Schur function s_μ:

$$s_\mu = \sum_i R_{ii}^\mu \in P_\mu$$

as a function on G. From (3.13) we have $e_K s_\mu = 0$ unless μ is an even partition:

$$(3.19) \qquad \int_K s_\mu(xk)\,dk = 0$$

for all $x \in G$, unless μ is an even partition.

Let $\Omega_\lambda = e_K s_{2\lambda}$, or explicitly

$$\Omega_\lambda(x) = \int_K s_{2\lambda}(xk)\, dk = \int_K s_{2\lambda}(kx)\, dk \tag{3.20}$$

for any partition λ of length $\leqslant n$.

(3.21) (i) $\Omega_\lambda(1) = 1$.
(ii) *$\Omega_\lambda \in P_{2\lambda}(G, K)$ and generates the irreducible G-submodule $e_K P_{2\lambda}$ of $P(K \backslash G)$.*

Moreover, Ω_λ is characterized by these two properties.

Proof. (i) We have

$$\Omega_\lambda(1) = \int_K s_{2\lambda}(k)\, dk = \dim F_{2\lambda}(V)^K = 1$$

by (3.14).
(ii) Since $s_{2\lambda} \in P_{2\lambda}$, it is clear from (3.20) that $\Omega_\lambda \in P_{2\lambda} \cap P(G, K) = P_{2\lambda}(G, K)$; and since $\Omega_\lambda \neq 0$ (by (i) above) it follows that Ω_λ generates the irreducible G-module $e_K P_{2\lambda}$.

Conversely, any $f \in P_{2\lambda}(G, K)$ must be a scalar multiple of Ω_λ, by (3.15); if also $f(1) = 1$, then $f = \Omega_\lambda$. |

The function Ω_λ is the *zonal spherical function* of (G, K) associated with the partition λ.

We return now to the $S_{2m} \times G$-module $\mathbf{T} = \mathbf{T}^{2m}(V)$, where $m \leqslant n = \dim V$. From (3.10) and (3.14) we have

$$\mathbf{T}^K = \bigoplus_\lambda \mathbf{T}_{2\lambda}^K \cong \bigoplus_\lambda M_{2\lambda}$$

as S_{2m}-module, where λ ranges over all partitions of m. From (2.4) it follows that $\mathbf{T}^K$ is isomorphic to the induced module $1_H^{S_{2m}}$. Now the tensor

$$\delta = \sum_{i=1}^n e_i \otimes e_i$$

is fixed by K, and hence $\delta^{\otimes m} = \delta \otimes \ldots \otimes \delta$ (m factors) lies in $\mathbf{T}^K$. Since the subgroup of S_{2m} that fixes $\delta^{\otimes m}$ is precisely H, it follows that $\delta^{\otimes m}$ generates $\mathbf{T}^K$ as S_{2m}-module. Hence by (1.5) the space

$$\mathbf{T}_{2\lambda}^{H \times K} = (\mathbf{T}_{2\lambda}^K)^H = (\mathbf{T}_{2\lambda}^H)^K$$

is one-dimensional, generated by

(3.22)
$$\omega^\lambda \delta^{\otimes m} = \sum_{w \in S_{2m}} \omega^\lambda(w) . w\delta^{\otimes m}.$$

We need now to compute the image of this element of $\mathbf{T}^H$ under the isomorphism $\varphi: \mathbf{T}^H \to P^m(\Sigma)$ (3.9). This is given by

(3.23)
$$\varphi(w\delta^{\otimes m}) = p_\rho$$

where ρ is the coset-type of w.

Proof. Let $f_w = \varphi(w\delta^{\otimes m})$. For any $h \in H$ we have $f_{wh} = f_w$, since $h\delta^{\otimes m} = \delta^{\otimes m}$; also $f_{hw} = f_w$, since $\varphi(ht) = \varphi(t)$ for all $t \in \mathbf{T}$. Hence f_w depends only on the double coset H_ρ to which w belongs. By taking w to be the product of consecutive cycles of lengths $2\rho_1, 2\rho_2, \ldots$, we reduce to the case where w is the $2m$-cycle $(1, 2, \ldots, 2m)$. Since

$$\delta^{\otimes m} = \sum e_{i_1} \otimes e_{i_1} \otimes \ldots \otimes e_{i_m} \otimes e_{i_m},$$

it follows that

$$w\delta^{\otimes m} = \sum e_{i_m} \otimes e_{i_1} \otimes e_{i_1} \otimes \ldots \otimes e_{i_{m-1}} \otimes e_{i_m}$$

and hence that

$$\begin{aligned} f_w(\sigma) &= \sum \sigma_{i_m i_1} \sigma_{i_1 i_2} \ldots \sigma_{i_{m-1} i_m} \\ &= \operatorname{trace}(\sigma^m) = p_m(\sigma). \quad | \end{aligned}$$

We can now express the zonal spherical function Ω_λ in terms of the zonal polynomial Z_λ defined in §2:

(3.24) *We have*

$$\Omega_\lambda(x) = Z_\lambda(x'x)/Z_\lambda(1_n)$$

for all $x \in G$.

Proof. Let $\varphi_\lambda = \varphi(\omega^\lambda \delta^{\otimes m})$, and define $\tilde{\varphi}_\lambda \in P(G)$ by $\tilde{\varphi}_\lambda(x) = \varphi_\lambda(x'x)$. It follows from (3.12) that $\tilde{\varphi}_\lambda$ is a generator of the one-dimensional space $P_{2\lambda}(G, K)$ that is the image of $\mathbf{T}_{2\lambda}^{H \times K}$ under the composite isomorphism

$$\mathbf{T}^H \to P^M(\Sigma) \to P^{2m}(K \backslash G).$$

Hence, by (3.21), Ω_λ is a scalar multiple of $\tilde{\varphi}_\lambda$, namely

(1)
$$\Omega_\lambda(x) = \varphi_\lambda(x'x)/\varphi_\lambda(1_n)$$

since $\Omega_\lambda(1) = 1$. On the other hand, it follows from (3.22) and (3.23) that

(2) $$\varphi_\lambda = \sum_\rho |H_\rho| \omega_\rho^\lambda p_\rho = |H_m| Z_\lambda$$

by (2.3) and (2.12). The result now follows from (1) and (2). |

Examples

1. (a) Let λ be a partition of length n and let $\mu = (\lambda_1 - 1, \dots, \lambda_n - 1)$. If $x \in G$ we have

$$\Omega_\lambda(x) = \int_K s_{2\lambda}(xk)\,dk$$

$$= \int_K (\det xk)^2 s_{2\mu}(xk)\,dk$$

$$= (\det x)^2 \Omega_\mu(x)$$

and hence if $\xi = \operatorname{diag}(\xi_1, \dots, \xi_n)$ is a diagonal matrix we have

(1) $$\frac{Z_\lambda(\xi)}{Z_\lambda(1_n)} = \xi_1 \dots \xi_n \frac{Z_\mu(\xi)}{Z_\mu(1_n)}.$$

(b) Let a_λ denote the coefficient of ξ^λ in $Z_\lambda(\xi)$. From (1) above we have

$$a_\lambda / a_\mu = Z_\lambda(1_n)/Z_\mu(1_n) = c_\lambda(n)/c_\mu(n)$$

by (2.25). If s is the square (i, λ_i) at the right-hand end of the ith row of λ, and t is the square $(i, 1)$ at the left-hand end of the same row, we have $a'(s) = a(t)$ and $n - l'(s) = l(t) + 1$, from which it follows that

$$c_\lambda(n)/c_\mu(n) = h_1(2\lambda)/h_1(2\mu)$$

and hence, by induction on $|\lambda|$, that $a_\lambda = h_1(2\lambda)$ (2.26).

2. Show that if $t \in \mathbf{R}$ and $x \in G$,

$$\int_K \exp(t \operatorname{trace} xk)\,dk = \sum_\lambda \frac{t^{|2\lambda|}}{h(2\lambda)} \Omega_\lambda(x).$$

(We have

$$p_1^m = \sum_{|\mu| = m} \frac{m!}{h(\mu)} s_\mu$$

(Chapter I, §7), hence

(1) $$\int_K (\operatorname{trace} xk)^m\,dk = \sum_\mu \frac{m!}{h(\mu)} \int_K s_\mu(xk)\,dk$$

summed over the partitions μ of m. The integral on the right is zero unless $\mu = 2\lambda$

for some partition λ, hence the integral (1) is zero if m is odd; and if $m = 2r$ is even it is equal to

$$\sum_{|\lambda|=r} \frac{(2r)!}{h(2\lambda)} \Omega_\lambda(x).)$$

3. For each partition μ of length $\leqslant n$, the irreducible G-module $F_\mu(V)$ has a basis indexed by the column-strict tableaux T of shape μ, such that relative to this basis a diagonal matrix $x = \operatorname{diag}(x_1, \dots, x_n) \in G$ is represented by a diagonal matrix $R^\mu(x)$ with diagonal entries x^T (Chapter I, Appendix A, §8, Example 2). Hence

$$\begin{aligned} s_\mu(xk) &= \operatorname{trace} R^\mu(x) R^\mu(k) \\ &= \sum_T x^T R^\mu_{T,T}(k) \end{aligned}$$

and therefore

$$\begin{aligned} \Omega_\lambda(x) &= \int_K s_{2\lambda}(xk)\, dk \\ &= \sum_T x^T \int_K R^{2\lambda}_{T,T}(k)\, dk \end{aligned}$$

summed over the column-strict tableaux T of shape 2λ that can be formed with the numbers $1, 2, \dots, n$. Each monomial x^T in this sum belongs to a monomial symmetric function $m_\nu(x_1, \dots, x_n)$, where $\nu \leqslant 2\lambda$, and hence $\Omega_\lambda(x)$ is a linear combination of these, since it is symmetric in $x_1, \dots, x_n$. But $\Omega_\lambda(x)$ is unaltered by replacing any x_i by $-x_i$, and hence is a symmetric polynomial in $x_1^2, \dots, x_n^2$. Consequently only the m_ν such that ν is even, say $\nu = 2\mu$, will occur, and so $\Omega_\lambda(x)$ is of the form

$$\Omega_\lambda(x) = \sum_{\mu \leqslant \lambda} a_{\lambda\mu} m_\mu(x_1^2, \dots, x_n^2),$$

since $2\mu \leqslant 2\lambda$ if and only if $\mu \leqslant \lambda$. This gives an independent proof of (2.22) and hence of (2.23).

Notes and references

The material in this section is due to A. T. James [J4].

4. Integral formulas

(4.1) *Let $f \in P(G)$, $f \neq 0$. Then f is a zonal spherical function of (G, K) if and only if*

$$\int_K f(xky)\, dk = f(x) f(y)$$

for all $x, y \in G$.

Proof. Let λ be a partition of length $\leqslant n$, and let

$$\varphi_y(x) = \int_K \Omega_\lambda(xky)\,dk$$

$$= \int_K \int_K s_{2\lambda}(xkyk_1)\,dk_1\,dk$$

by (3.16). Since $s_{2\lambda}$, as a function on G, lies in $P_{2\lambda}$ it is clear that $\varphi_y \in P_{2\lambda} \cap P(G,K) = P_{2\lambda}(G,K)$, and hence by (3.15) and (3.17) it follows that $\varphi_y = c\Omega_\lambda$ for some $c \in \mathbf{R}$. Since $\Omega_\lambda(1) = 1$, we have

$$c = \varphi_y(1) = \int_K \Omega_\lambda(ky)\,dk = \Omega_\lambda(y)$$

and hence

$$\int_K \Omega_\lambda(xky)\,dk = \Omega_\lambda(y)\Omega_\lambda(x) \tag{1}$$

for all $x, y \in G$.

Conversely, suppose that $f \in P(G)$ is such that $f \neq 0$ and

$$\int_K f(xky)\,dk = f(x)f(y)$$

for all $x, y \in G$. By choosing y such that $f(y) \neq 0$ we see that $f(xk_1) = f(x)$ for all $x \in G$ and $k_1 \in K$, and likewise that $f(k_1 y) = f(y)$ for all $y \in G$ and $k_1 \in K$. Hence $f \in P(G,K)$, say

$$f = \sum_\lambda a_\lambda \Omega_\lambda$$

by (3.15) and (3.21). We have then

$$f(x)f(y) = \int_K f(xky)\,dk = \sum_\lambda a_\lambda \int_K \Omega_\lambda(xky)\,dk$$

and hence by (1)

$$\sum a_\lambda \Omega_\lambda(x) f(y) = \sum_\lambda a_\lambda \Omega_\lambda(x)\Omega_\lambda(y).$$

Since the Ω_λ are linearly independent, it follows that $a_\lambda(f - \Omega_\lambda) = 0$ for all λ. Since not all the a_λ are zero, we conclude that $f = \Omega_\lambda$ for some λ. ∎

(4.2) *Let λ be a partition of length $\leqslant n$. Then*

$$\int_K Z_\lambda(\sigma k \tau k')\,dk = \frac{Z_\lambda(\sigma)Z_\lambda(\tau)}{Z_\lambda(1_n)}$$

for all $\sigma, \tau \in \Sigma$.

Proof. Both sides of (4.2) are polynomial functions of σ and τ, hence we may assume that σ and τ are positive definite, say $\sigma = x'x$ and $\tau = yy'$. Then $\sigma k \tau k' = x'xkyy'k'$, and hence by (3.18) and (3.24)

$$Z_\lambda(\sigma k \tau k') = Z_\lambda(1_n)\Omega_\lambda(xky).$$

(4.2) now follows from (4.1). |

Let $dx = \prod_{i,j=1}^n dx_{ij}$ be Lebesgue measure on $G = GL_n(\mathbf{R})$, and let

$$d\nu(x) = (2\pi)^{-n^2/2} \exp(-\tfrac{1}{2} \operatorname{trace} x'x)\, dx,$$

the 'normal distribution' of the statisticians. Since $\operatorname{trace} x'x = \Sigma_{i,j} x_{ij}^2$ and

$$\int_{-\infty}^{\infty} e^{-t^2/2}\, dt = (2\pi)^{1/2},$$

it follows that

$$\int_G d\nu(x) = 1.$$

The measure ν is K-invariant, that is to say

$$d\nu(k_1 x k_2) = d\nu(x)$$

for $k_1, k_2 \in K$. For if $y = k_1 x k_2$ we have $\operatorname{trace} y'y = \operatorname{trace} x'x$ and $dy = dx$.

(4.3) *For each integer* $m \geqslant 0$ *and all* $\sigma, \tau \in \Sigma$ *we have*

$$\int_G (\operatorname{trace} \sigma x \tau x')^m\, d\nu(x) = 2^m m! \sum_{|\lambda|=m} \frac{Z_\lambda(\sigma) Z_\lambda(\tau)}{h(2\lambda)}$$

where $h(2\lambda)$ *is the product of the hook-lengths of the partition* 2λ.

Proof. We may write

$$\sigma = k_1' \xi k_1, \qquad \tau = k_2' \eta k_2$$

where $\xi = \operatorname{diag}(\xi_1, \ldots, \xi_n)$ and $\eta = \operatorname{diag}(\eta_1, \ldots, \eta_n)$ are diagonal matrices, and $k_1, k_2 \in K$. Then

$$\begin{aligned} \operatorname{trace} \sigma x \tau x' &= \operatorname{trace} k_1' \xi k_1 x k_2' \eta k_2 x' \\ &= \operatorname{trace} \xi y \eta y' \end{aligned}$$

where $y = k_1 x k_2'$, and hence by K-invariance of ν

$$\int_G (\operatorname{trace} \sigma x \tau x')^m\, d\nu(x) = \int_G (\operatorname{trace} \xi x \eta x')^m\, d\nu(x)$$

which is the coefficient of $t^m/m!$ in

$$\int_G \exp(t \operatorname{trace} \xi x \eta x')\, \mathrm{d}\nu(x). \tag{1}$$

Now

$$t \operatorname{trace} \xi x \eta x' - \tfrac{1}{2} \operatorname{trace} x'x = -\tfrac{1}{2} \sum_{i,j} (1 - 2t\xi_i \eta_j) x_{ij}^2.$$

Hence the integral (1) is equal to

$$\prod_{i,j} (1 - 2t\xi_i\eta_j)^{-1/2}$$

which by (2.18) is equal to

$$\sum_\lambda (2t)^{|\lambda|} Z_\lambda(\sigma) Z_\lambda(\tau)/h(2\lambda) \tag{2}$$

since $Z_\lambda(\sigma) = Z_\lambda(\xi)$ and $Z_\lambda(\tau) = Z_\lambda(\eta)$. The result now follows from equating the coefficients of t^m in (1) and (2). |

(4.4) *Let $\sigma, \tau \in \Sigma$. Then*

$$\int_G Z_\lambda(\sigma x \tau x')\, \mathrm{d}\nu(x) = Z_\lambda(\sigma) Z_\lambda(\tau)$$

for all partitions λ of length $\leqslant n$.

Proof. Let $I_\lambda(\sigma, \tau)$ denote the integral on the left. Then we have

$$\begin{aligned} I_\lambda(\sigma, \tau) &= \int_K \left(\int_G Z_\lambda(\sigma k x \tau x' k')\, \mathrm{d}\nu(x) \right) \mathrm{d}k \\ &= \int_G \left(\int_K Z_\lambda(\sigma k x \tau x' k')\, \mathrm{d}k \right) \mathrm{d}\nu(x) \\ &= \int_G \frac{Z_\lambda(\sigma) Z_\lambda(x \tau x')}{Z_\lambda(1_n)}\, \mathrm{d}\nu(x) \end{aligned}$$

by (4.2). Hence

$$\begin{aligned} I_\lambda(\sigma, \tau) &= \frac{Z_\lambda(\sigma)}{Z_\lambda(1_n)} I_\lambda(1_n, \tau) \\ &= \frac{Z_\lambda(\sigma) Z_\lambda(\tau)}{Z_\lambda(1_n)^2} I_\lambda(1_n, 1_n) \end{aligned} \tag{1}$$

by repeating the argument.

On the other hand we have from (2.16)

$$p_1^m = 2^m m! \sum_{|\lambda|=m} h(2\lambda)^{-1} Z_\lambda$$

and therefore

$$\int_G (\text{trace}\ \sigma x \tau x')^m \, d\nu(x) = 2^m m! \sum_{|\lambda|=m} h(2\lambda)^{-1} I_\lambda(\sigma,\tau)$$

$$= 2^m m! \sum_{|\lambda|=m} \frac{Z_\lambda(\sigma) Z_\lambda(\tau)}{h(2\lambda)} \cdot \frac{I_\lambda(1_n,1_n)}{Z_\lambda(1_n)^2}.$$

It now follows from (4.3) that

$$I_\lambda(1_n,1_n) = Z_\lambda(1_n)^2$$

and hence from (1) that $I_\lambda(\sigma,\tau) = Z_\lambda(\sigma) Z_\lambda(\tau)$. |

By taking $\tau = 1_n$ in (4.4) we have

(4.5) $$\int_G Z_\lambda(x'\sigma x)\, d\nu(x) = Z_\lambda(1_n) Z_\lambda(\sigma).$$

In other words, the Z_λ are eigenfunctions of the linear operator $U_n: \Lambda_n \to \Lambda_n$ defined by

(4.6) $$(U_n f)(\sigma) = \int_G f(x'\sigma x)\, d\nu(x)$$

and the eigenvalue of U_n corresponding to Z_λ is $Z_\lambda(1_n)$.

As in Chapter I, Appendix A, §8, Example 5, for each partition μ of length $\leqslant n$ let Δ_μ be the polynomial function on G defined by

$$\Delta_\mu(x) = \prod_{i \geqslant 1} \det(x^{(\mu_i')})$$

for $x = (x_{ij}) \in G$, where $x^{(r)} = (x_{ij})_{1 \leqslant i,j \leqslant r}$ for $1 \leqslant r \leqslant n$. Then (*loc. cit.*) the G-submodule E_μ of $P(G)$ generated by Δ_μ is irreducible and isomorphic to $F_\mu(V)$. We have

(4.7) $$\Delta_\mu(xb) = \Delta_\mu(b'x) = \Delta_\mu(b)\Delta_\mu(x)$$

if $b \in G$ is upper triangular.

Now let B^+ be the subgroup of G consisting of the upper triangular matrices (b_{ij}) with positive diagonal elements b_{ii}. If $x \in G$, Gram–Schmidt orthogonalization applied to the successive columns of the matrix x shows that

$$x = kb \tag{4.8}$$

with $k \in K$ and $b \in B^+$; moreover, this factorization of x is unique, since $K \cap B^+ = \{1\}$.

(4.9) *We have*

$$e_K \Delta_{2\lambda}(x) = c_\lambda \Delta_\lambda(x'x)$$

where c_λ is a positive constant, independent of $x \in G$.

Proof. If $x = kb$ as above then

$$\begin{aligned} e_K \Delta_{2\lambda}(kb) &= \int_K \Delta_{2\lambda}(k_1 kb)\, dk_1 \\ &= \Delta_{2\lambda}(b) \int_K \Delta_{2\lambda}(k_1)\, dk_1 \\ &= c_\lambda \Delta_\lambda(b'b) = c_\lambda \Delta_\lambda(x'x) \end{aligned}$$

by use of (4.7), where

$$c_\lambda = \int_K \Delta_{2\lambda}(k)\, dk$$

is positive, since $\Delta_{2\lambda} = \Delta_\lambda^2$ is $\geqslant 0$ on G, and is > 0 on a dense open subset of K. |

(4.10) *We have*

$$\int_K \Delta_\lambda(k'\sigma k)\, dk = Z_\lambda(\sigma)/Z_\lambda(1_n)$$

for $\sigma \in \Sigma$.

Proof. We may assume that σ is positive definite, say $\sigma = x'x$ $(x \in G)$, since both sides of (4.10) are polynomial functions on Σ. From (4.9) it follows that the function $x \mapsto \Delta_\lambda(x'x)$ generates the irreducible G-module $P_{2\lambda}(K\backslash G) \cong F_{2\lambda}(V)$, and hence that

$$\begin{aligned} \int_K \Delta_\lambda(k'x'xk)\, dk &= \Omega_\lambda(x) \\ &= Z_\lambda(x'x)/Z_\lambda(1_n) \end{aligned}$$

by (3.24). |

(4.11) *We have*

$$\int_G \Delta_\lambda(x'\sigma x)\,d\nu(x) = Z_\lambda(\sigma)$$

for $\sigma \in \Sigma$.

Proof. Since the measure ν is K-invariant we have

$$\begin{aligned}\int_G \Delta_\lambda(x'\sigma x)\,d\nu(x) &= \int_K \left(\int_G \Delta_\lambda(k'x'\sigma xk)\,d\nu(x)\right) dk \\ &= \int_G \left(\int_K \Delta_\lambda(k'x'\sigma xk)\,dk\right) d\nu(x) \\ &= Z_\lambda(1_n)^{-1} \int_G Z_\lambda(x'\sigma x)\,d\nu(x) \\ &= Z_\lambda(\sigma)\end{aligned}$$

by Fubini's theorem, (4.10), and (4.5). |

In particular ($\sigma = 1_n$)

$$Z_\lambda(1_n) = \int_G \Delta_\lambda(x'x)\,d\nu(x). \tag{4.12}$$

We shall now calculate this integral directly, using the factorization $x = kb$ (4.8). Let

$$db = \prod_{i \leq j} db_{ij}$$

be Lebesgue measure on B^+, and let

$$d\nu(b) = (2\pi)^{-n(n+1)/4} \exp(-\tfrac{1}{2}\,\text{trace}\, b'b)\,db.$$

The measures dx, dk, db on G, K, B^+ respectively are related by

$$dx = c_n \Delta_\delta(b)\,dk\,db$$

where $\delta = (n-1, n-2, \ldots, 1, 0)$ and c_n is a positive real number depending only on n (see [B7], Chapter VII, §3, or Example 2 below). Since $x = kb$ we have $x'x = b'b$ and hence

$$d\nu(x) = c'_n \Delta_\delta(b)\,dk\,d\nu(b) \tag{4.13}$$

where $c'_n = (2\pi)^{-n(n-1)/4} c_n$.

From (4.7), (4.12), and (4.13) it follows that

$$Z_\lambda(1_n) = c'_n \int_{B^+} \Delta_\lambda(b)^2 \Delta_\delta(b)\, d\nu(b)$$

$$= c'_n \prod_{i=1}^{n} (2\pi)^{-1/2} \int_0^\infty b_{ii}^{2\lambda_i+n-i} e^{-b_{ii}^2/2}\, db_{ii}$$

(4.14) $$= (2\pi)^{-n/2} c'_n \prod_{i=1}^{n} 2^{\lambda_i+(n-i-1)/2} \Gamma(\lambda_i + \tfrac{1}{2}(n-i+1)).$$

Since $Z_0 = 1$ we obtain

$$Z_\lambda(1_n) = \frac{Z_\lambda(1_n)}{Z_0(1_n)} = \prod_{i=1}^{n} 2^{\lambda_i} \frac{\Gamma(\lambda_i + \frac{1}{2}(n-i+1))}{\Gamma(\frac{1}{2}(n-i+1))}$$

$$= 2^{|\lambda|} \prod_{i=1}^{n} \prod_{j=1}^{\lambda_i} (\tfrac{1}{2}(n-i+1) + j - 1)$$

or equivalently

(4.15) $$Z_\lambda(1_n) = \prod_{s\in\lambda} (n + 2a'(s) - l'(s)) = c_\lambda(n)$$

in agreement with (2.25).

Examples

1. If $k = (k_{ij}) \in K$, let

$$\omega_{ij}(k) = \sum_{r=1}^{n} k_{ri}\, dk_{rj},$$

a differential 1-form on K.

(a) Since $\Sigma_r k_{ri} k_{rj} = \delta_{ij}$, it follows that $\omega_{ij} + \omega_{ji} = 0$.

(b) If a is a fixed element of K then $\omega_{ij}(ak) = \omega_{ij}(k)$, i.e. the ω_{ij} are left-invariant differential forms on K.

(c) Hence

$$\omega_K = \bigwedge_{i>j} \omega_{ij}$$

is a left-invariant differential form of degree $\frac{1}{2}n(n-1) = \dim K$ (defined only up to sign, because we have not specified the order of the factors in the exterior product); moreover it is not identically zero, since $\omega_K(1_n) = \bigwedge_{i>j} dk_{ij}$. Hence (see e.g. [D5], (16.24.1)) ω_K determines a Haar measure d^*k on K. Since Haar measure is unique up to a positive scalar multiple, it follows that

$$d^*k = c_n dk$$

for some $c_n > 0$ depending only on n. The constant c_n is calculated in Example 2(d) below.

2. (a) For $x = (x_{ij}) \in G$ let

$$\omega_G(x) = \bigwedge_{i,j} \mathrm{d}x_{ij}$$

be the differential form on G (defined only up to sign, as in Example 1) that corresponds to Lebesgue measure $\mathrm{d}x$. If $a \in G$ we have

$$\omega_G(ax) = \pm\omega_G(xa) = \pm(\det a)^n \omega_G(x).$$

(b) For $b = (b_{ij}) \in B^+$, let

$$\omega_{B^+}(b) = \bigwedge_{i \leqslant j} \mathrm{d}b_{ij}$$

be the differential form on B^+ that corresponds to Lebesgue measure db. If $a \in B^+$ we have

$$\omega_{B^+}(ba) = \pm(\det a)^n \Delta_\delta(a)^{-1} \omega_{B^+}(b),$$

$$\omega_{B^+}(ab) = \pm(\det a)\Delta_\delta(a)\omega_{B^+}(b).$$

(c) From the factorization $x = kb$ (4.8) we have

$$(\mathrm{d}x) = (\mathrm{d}k)b + k(\mathrm{d}b)$$

where $(\mathrm{d}x)$ denotes the matrix of differentials $(\mathrm{d}x_{ij})$, and likewise for $(\mathrm{d}k)$ and $(\mathrm{d}b)$. Hence

$$k'(\mathrm{d}x)b^{-1} = k'(\mathrm{d}k) + (\mathrm{d}b)b^{-1},$$

in which the matrix $k'(\mathrm{d}k) = (\omega_{ij})$ (Example 1) is skew-symmetric, and the matrix $(\mathrm{d}b)b^{-1}$ is upper triangular. Deduce from (a) and (b) above that

$$\omega_G = \pm\omega_K \wedge \Delta_\delta \omega_{B^+}$$

and hence that

$$\mathrm{d}x = \mathrm{d}^*k \,.\, \Delta_\delta(b)\,\mathrm{d}b$$
$$= c_n \Delta_\delta(b)\,\mathrm{d}k\,\mathrm{d}b.$$

(d) Deduce from (4.14) (with $\lambda = 0$) that

$$c_n = v_1 v_2 \ldots v_n$$

where $v_r = 2\pi^{r/2}/\Gamma(\frac{1}{2}r)$ (which is the 'area' of the unit sphere in $\mathbf{R}^r$: $v_1 = 2$, $v_2 = 2\pi$, $v_3 = 4\pi, \ldots$).

3. Let A be the group of diagonal matrices $\xi = \mathrm{diag}(\xi_1, \ldots, \xi_n)$ with diagonal elements $\xi_i > 0$, and let $\Sigma^+ \subset \Sigma$ be the cone of positive definite symmetric matrices. Orthogonal reduction of a real symmetric matrix shows that each $\sigma \in \Sigma^+$ can be written in the form

(1) $$\sigma = k\xi k'$$

with $k \in K$ and $\xi \in A$. The ξ_i are the eigenvalues of σ, and hence the diagonal matrix ξ is unique up to conjugation by a permutation matrix. Moreover, if the ξ_i are all distinct, the columns of K are unique up to sign. From this it follows that the fibre over $\sigma \in \Sigma^+$ of the mapping $K \times A \to \Sigma^+$ defined by (1) is a finite set of cardinality $2^n n! = |H_n|$ whenever the discriminant of σ does not vanish, hence almost everywhere on Σ^+.

From (1) we have

$$(\mathrm{d}\sigma) = (\mathrm{d}k)\xi k' + k(\mathrm{d}\xi)k' + k\xi(\mathrm{d}k)'$$

where $(\mathrm{d}\sigma)$ denotes the matrix of differentials $(\mathrm{d}\sigma_{ij})$, and likewise for $(\mathrm{d}k)$, $(\mathrm{d}\xi)$. Hence

$$k'(\mathrm{d}\sigma)k = k'(\mathrm{d}k)\xi + (\mathrm{d}\xi) + \xi(\mathrm{d}k)'k,$$

the (i,j) element of which is the differential 1-form

$$\omega_{ij}\xi_j + \delta_{ij}\,\mathrm{d}\xi_i + \xi_i\omega_{ji} = (\xi_j - \xi_i)\,\omega_{ji} + \delta_{ij}\,\mathrm{d}\xi_i$$

in the notation of Example 1. Hence if

$$\omega_{\Sigma^+} = \bigwedge_{i \leqslant j} \mathrm{d}\sigma_{ij}, \qquad \omega_A = \bigwedge_{i=1}^{n} \mathrm{d}\xi_i$$

(defined only up to sign, as in Example 2) we have

$$\omega_{\Sigma^+} = \pm\,\omega_K \wedge a_\delta\,\omega_A \tag{2}$$

where $a_\delta(\xi) = \prod_{i<j}(\xi_i - \xi_j)$. It follows that

$$\mathrm{d}\sigma = \frac{1}{2^n . n!}\,|a_\delta(\xi)|\,\mathrm{d}k^*\,\mathrm{d}\xi$$

$$= \frac{c_n}{2^n . n!}\,|a_\delta(\xi)|\,\mathrm{d}k\,\mathrm{d}\xi. \tag{3}$$

4. Let $m \geqslant n$ and let $X = M_{m,n}(\mathbf{R})$ be the vector space of all real matrices $x = (x_{ij})$ with m rows and n columns. Also let Σ_m (resp. Σ_n) denote the space of real symmetric matrices of size $m \times m$ (resp. $n \times n$).

If $x, y \in X$ we have $\mathrm{trace}(xy') = \mathrm{trace}(y'x)$, and more generally

$$e_r(xy') = \mathrm{trace}\,\wedge^r(xy') = \mathrm{trace}\,\wedge^r(x)\wedge^r(y')$$

$$= \mathrm{trace}\,\wedge^r(y')\wedge^r(x) = e_r(y'x)$$

for all $r \geqslant 1$, so that

$$f(xy') = f(y'x) \tag{1}$$

for all $f \in \Lambda$ and $x, y \in X$.

Now let $\mathrm{d}x = \prod_{i,j}\mathrm{d}x_{ij}$ be Lebesgue measure on X, and let

$$\mathrm{d}\nu(x) = (2\pi)^{-mn/2}\exp(-\tfrac{1}{2}\,\mathrm{trace}\,x'x)\,\mathrm{d}x$$

be the normal distribution. Let $\sigma \in \Sigma_m$ and $\tau \in \Sigma_n$. Then

(a) We have

$$\int_X (\text{trace}\,\sigma x\tau x')^r \,d\nu(x) = 2^r r! \sum_{|\lambda|=r} \frac{Z_\lambda(\sigma)Z_\lambda(\tau)}{h(2\lambda)}$$

(same proof as (4.3)).

(b) We have

$$\int_X Z_\lambda(\sigma x\tau x')\,d\nu(x) = Z_\lambda(\sigma)Z_\lambda(\tau)$$

if $l(\lambda) \leqslant n$. (Same proof as (4.4).) In particular,

$$\int_X Z_\lambda(xx')\,d\nu(x) = \int_X Z_\lambda(x'x)\,d\nu(x) = Z_\lambda(1_m)Z_\lambda(1_n).$$

(c) Let $U_{m,n}: \Lambda_m \to \Lambda_m$ be the operator defined by

$$(U_{m,n}f)(\sigma) = \int_X f(x'\sigma x)\,d\nu(x)$$

where $\sigma \in \Sigma_m$. Then

$$U_{m,n}Z_\lambda = Z_\lambda(1_n)Z_\lambda$$

for all partitions λ of length $\leqslant n$. (Set $\tau = 1_n$ in (b) above.) Hence $\rho_{m,n} \circ U_{m,n} = U_n \circ \rho_{m,n}$, where $\rho_{m,n}: \Lambda_m \to \Lambda_n$ is the restriction homomorphism (Chapter I, §2).

5. Let $x, y \in G$. Then

(a)
$$\int_G \Omega_\lambda(xzy)\,d\nu(z) = \Omega_\lambda(x)\Omega_\lambda(y)Z_\lambda(1_n)$$

if $l(\lambda) \leqslant n$. (Use (3.24), (4.1), and (4.5).)

(b)
$$\int_G s_{2\lambda}(xzy)\,d\nu(z) = \Omega_\lambda(yx)Z_\lambda(1_n)$$

if $l(\lambda) \leqslant n$. (Use (3.20) and (a) above.)

6. Let μ be a positive K-invariant measure on the cone Σ^+ of positive definite symmetric $n \times n$ matrices, such that

$$\int_{\Sigma^+} f(\sigma)\,d\mu(\sigma) < \infty$$

for all $f \in \Lambda_n$. Define an operator $E: \Lambda_n \to \Lambda_n$ by

$$(Ef)(\sigma) = \int_{\Sigma^+} f(\sigma\tau)\,d\mu(\tau).$$

Then each Z_λ such that $l(\lambda) \leqslant n$ is an eigenfunction of E, with eigenvalue

$$Z_\lambda(1_n)^{-1}\int_{\Sigma^+} Z_\lambda(\tau)\,d\mu(\tau).$$

(Use (4.2).)

7. (a) Let $d\sigma = \prod_{i \leqslant j} d\sigma_{ij}$ be Lebesgue measure on Σ. If $x \in G$, the Jacobian of the linear transformation $\sigma \mapsto x'\sigma x$ is the symmetric square $\mathbf{S}^2(x)$ of x. If the eigenvalues of x are $\xi_1, \dots, \xi_n$, those of $\mathbf{S}^2(x)$ are $\xi_i \xi_j$ $(1 \leqslant i \leqslant j \leqslant n)$, whence $\det(\mathbf{S}^2 x) = (\det x)^{n+1}$. Hence

$$d(x'\sigma x) = |\det x|^{n+1}\, d\sigma$$

and therefore

$$d^*\sigma = (\det \sigma)^{-(n+1)/2}\, d\sigma$$

is a G-invariant measure on Σ (and on Σ^+).

(b) Let $m \geqslant n$ and let $X = M_{m,n}(\mathbf{R})$, as in Example 3. The group G acts on X by right multiplication, and we have

$$d(xg) = |\det g|^m\, dx$$

if $g \in G$. Hence

$$d^*x = (\det x'x)^{-m/2}\, dx$$

is a G-invariant measure on X.

(c) Hence the linear functional

$$f \mapsto \int_X f(x'x)\, d^*x$$

defines a G-invariant measure on Σ^+. Since G-invariant measures on homogeneous spaces (when they exist) are unique up to a positive scalar multiple, it follows from (a) that

$$\int_X f(x'x)\, d^*x = c_{m,n} \int_{\Sigma^+} f(\sigma)\, d^*\sigma \tag{1}$$

for some real number $c_{m,n} > 0$ independent of f.

If we take

$$f(\sigma) = (2\pi)^{-mn/2} \exp(-\tfrac{1}{2}\operatorname{trace}\sigma)(\det\sigma)^{m/2} g(\sigma)$$

in (1), we shall obtain

$$\int_X g(x'x)\, d\nu(x) = \int_{\Sigma^+} g(\sigma)\, d\mu_{m,n}(\sigma) \tag{2}$$

where

$$d\mu_{m,n}(\sigma) = c'_{m,n} \exp(-\tfrac{1}{2}\operatorname{trace}\sigma)(\det\sigma)^{(m-n-1)/2}\, d\sigma$$

and $c'_{m,n} = (2\pi)^{-mn/2} c_{m,n}$.

(d) Each matrix $\sigma \in \Sigma^+$ is uniquely of the form $\sigma = b'b$, with $b \in B^+$, and the mapping $b \mapsto \sigma$ is a diffeomorphism of B^+ onto Σ^+. The Jacobian matrix

$(\partial\sigma_{ij}/\partial b_{kl})$, with rows and columns arranged in lexicographical order, is lower triangular, so that its determinant is equal to

$$\prod_{i\leqslant j}(\partial\sigma_{ij}/\partial b_{ij}) = \prod_{i\leqslant j}(1+\delta_{ij})b_{ii}.$$

Hence

(3) $$d\sigma = 2^n(\det b)\Delta_8(b)\,db.$$

(e) By taking $f(\sigma) = \exp(-\pi \operatorname{trace}\sigma)(\det\sigma)^m$ in (1) and using (3), deduce that

$$c_{m,n} = 2^{-n}\prod_{i=1}^{n} v_{m-i+1}$$

where $v_r = 2\pi^{r/2}/\Gamma(\frac{1}{2}r)$ as in Example 2(d).

8. In this example and the following we shall use the abbreviations

$$\operatorname{etr}(\sigma) = \exp(\operatorname{trace}\sigma), \qquad |\sigma| = \det\sigma$$

for a square (not necessarily symmetric) matrix σ.

Let $f = f_{a,\lambda}$ be the function on Σ^+ defined by

$$f(\sigma) = |\sigma|^a Z_\lambda(\sigma)$$

where $a > \frac{1}{2}n$ and $l(\lambda) \leqslant n$.

(a) Show that

$$\int_{\Sigma^+} \operatorname{etr}(-\sigma)f(\sigma)\,d^*\sigma = Z_\lambda(1_n)\Gamma_n(a,\lambda)$$

where $d^*\sigma = |\sigma|^{-(n+1)/2}\,d\sigma$ as in Example 7, and

$$\Gamma_n(a,\lambda) = \int_{\Sigma^+} \operatorname{etr}(-\sigma)|\sigma|^a\Delta_{\lambda'}(\sigma)\,d^*\sigma$$

$$= \pi^{n(n-1)/4}\prod_{i=1}^{n}\Gamma(a+\lambda_i-\tfrac{1}{2}(i-1)).$$

(Use (4.10) and Example 7 above.)

(b) Let $\tau\in\Sigma^+$ and let

$$g(\tau) = \int_{\Sigma^+} \operatorname{etr}(-\sigma\tau)f(\sigma)\,d^*\sigma.$$

Show that

$$g(\tau) = \Gamma_n(a,\lambda)f(\tau^{-1}).$$

(Let $\tau^{1/2}$ be the positive definite square root of τ. Under the change of variables $\sigma\mapsto\tau^{1/2}\sigma\tau^{1/2}$ we obtain

$$g(\tau) = |\tau|^{-a}\int_{\Sigma^+} \operatorname{etr}(-\sigma)|\sigma|^a Z_\lambda(\sigma\tau^{-1})\,d^*\sigma.$$

Now replace σ by $k\sigma k'$ $(k\in K)$, integrate over K and use (4.2) and (a) above.)

This calculation shows that the Laplace transform of the function $f_{a-(n+1)/2,\lambda}$ is the function $\tau \mapsto \Gamma_n(a,\lambda) f_{a,\lambda}(\tau^{-1})$.

(c) More generally, if $\rho, \tau \in \Sigma^+$ show that

$$\int_{\Sigma^+} \operatorname{etr}(-\rho\sigma)|\sigma|^a Z_\lambda(\sigma\tau)\,d^*\sigma = \Gamma_n(a,\lambda)|\rho|^{-a} Z_\lambda(\rho^{-1}\tau).$$

9. (a) Let $a, b \geqslant \frac{1}{2}n$ and let λ be a partition of length $\leqslant n$. Show that

$$(1) \qquad \int Z_\lambda(\sigma\tau)|\sigma|^{a-p}|1_n - \sigma|^{b-p}\,d\sigma = \frac{\Gamma_n(a,\lambda)\Gamma_n(b,0)}{\Gamma_n(a+b,\lambda)} Z_\lambda(\tau)$$

for $\tau \in \Sigma^+$, where $p = \frac{1}{2}(n+1)$ and the integral is taken over $\Sigma^+ \cap (1_n - \Sigma^+)$.

(Let $f(\tau)$ denote the left-hand side of (1). By replacing σ by $k\sigma k'$, where $k \in K$, and then integrating over K and using (4.2), it follows that

$$(2) \qquad f(\tau) = f(1_n) Z_\lambda(\tau).$$

Next, from Example 8(a) we have

$$(3) \qquad \Gamma_n(a,\lambda)\Gamma_n(b,0)Z_\lambda(1_n) = \int \operatorname{etr}(-\sigma-\tau)|\sigma|^{a-p}|\tau|^{b-p} Z_\lambda(\sigma)\,d\sigma\,d\tau$$

integrated over $\Sigma^+ \times \Sigma^+$. Let

$$\tau_1 = \sigma + \tau, \qquad \sigma_1 = \tau_1^{-1/2}\sigma\tau_1^{-1/2},$$

where $\tau_1^{1/2}$ is the positive definite square root of τ_1; then $|\sigma| = |\sigma_1||\tau_1|$ and $|\tau| = |1 - \sigma_1||\tau_1|$, and $d\sigma_1 d\tau_1 = |\tau_1|^{-p} d\sigma\,d\tau$. Hence the integral (3) is equal to

$$\int_{\Sigma^+} \operatorname{etr}(-\tau_1)|\tau_1|^{a+b-p} f(\tau_1)\,d\tau_1$$

which by (2) above and Example 8(a) is equal to $\Gamma_n(a+b,\lambda)Z_\lambda(1_n)f_\lambda(1_n)$. Hence

$$f_\lambda(1_n) = \frac{\Gamma_n(a,\lambda)\Gamma_n(b,0)}{\Gamma_n(a+b,\lambda)}$$

which together with (2) completes the proof.)

(b) Show that

$$\int Z_\lambda(\xi)|\xi|^{a-p}|1_n - \xi|^{b-p}|a_\delta(\xi)|\,d\xi$$

$$= \frac{2^n n!}{c_n} \frac{\Gamma_n(a,\lambda)\Gamma_n(b,0)}{\Gamma_n(a+b,\lambda)} Z_\lambda(1_n)$$

where the integral is taken over $\xi = \operatorname{diag}(\xi_1, \dots, \xi_n)$ such that $0 \leqslant \xi_i \leqslant 1$ for $1 \leqslant i \leqslant n$. (Use (a) above and Example 3.)

This integral is the case $\alpha = 2$ (i.e. $k = \frac{1}{2}$) of the Selberg-type integral of Chapter VI, §10, Example 7.

10. Let λ be a partition of length $\leqslant n$, and let $m \geqslant n+2\lambda_1$. Then in the notation of Example 7(c) we have

$$\int_{\Sigma^+} Z_\lambda(\sigma^{-1}\tau)\,d\mu_{m,n}(\sigma) = \frac{Z_\lambda(-\tau)}{c_\lambda(n-m+1)}$$

for $\tau \in \Sigma$, where c_λ is the polynomial defined in (2.24) or (4.15). (Let $f(\tau)$ denote the integral above. By replacing σ by $k\sigma k'$, where $k \in K$, and then integrating over K, it follows from (4.2) that $f(\tau) = cZ_\lambda(\tau)$, where

$$c = \int_{\Sigma^+} \frac{Z_\lambda(\sigma^{-1})}{Z_\lambda(1_n)}\,d\mu_{m,n}(\sigma) = \int_{\Sigma^+} \Delta_\lambda(\sigma^{-1})\,d\mu_{m,n}(\sigma)$$

by (4.10). Evaluate the latter integral by setting $\sigma = bb'$, where $b \in B^+$.)

11. Let x and ξ be elements of G, where $\xi = \operatorname{diag}(\xi_1, \dots, \xi_n)$ is a diagonal matrix. Then the coefficient of $\xi_1 \dots \xi_r$ in $\det(x'\xi x)^{(r)}$ is $(\det x^{(r)})^2$, and hence if $l(\lambda) \leqslant n$ the coefficient of ξ^λ in $\Delta_\lambda(x'\xi x)$ is $\Delta_\lambda(x)^2 = \Delta_{2\lambda}(x)$. From (4.10) and (2.26) it follows that

$$\int_G \Delta_{2\lambda}(x)\,d\nu(x) = h_1(2\lambda),$$

the coefficient of ξ^λ in $Z_\lambda(\xi)$.

Deduce that

$$\int_K \Delta_{2\lambda}(k)\,dk = h_1(2\lambda)/Z_\lambda(1_n).$$

(Write $x = kb$ with $k \in K$ and $b \in B^+$, and use (4.13) and (4.14).)

12. Let λ be a partition of length $\leqslant n$. Show that

(a) $$\int_{K\times K} \Delta_{2\lambda}(k_1 x k_2)\,dk_1\,dk_2 = \frac{h_1(2\lambda)Z_\lambda(x'x)}{Z_\lambda(1_n)^2}.$$

(Let $f_\lambda(x)$ denote the integral on the left. Clearly $f_\lambda \in P(G,K)$, hence is a linear combination of zonal polynomials, and we may assume that $x = \xi = \operatorname{diag}(\xi_1, \dots, \xi_n)$ is a diagonal matrix. Now the leading term in $\det(k_1\xi k_2)^{(r)}$, as a polynomial in the ξ_i, is $(\det k_1^{(r)})(\det k_2^{(r)})\xi_1 \dots \xi_r$, and hence the leading term in $\Delta_{2\lambda}(k_1\xi k_2)$ as a polynomial in $\xi_1, \dots, \xi_n$ is $\Delta_{2\lambda}(k_1)\Delta_{2\lambda}(k_2)\xi^{2\lambda}$. Hence $f_\lambda(x)$ is a scalar multiple of $Z_\lambda(x'x)$, and the scalar multiple is given by Example 11.)

(b) Show that

$$\int_{K\times G} \Delta_{2\lambda}(kxy)\,dk\,d\nu(y) = \frac{h_1(2\lambda)Z_\lambda(x'x)}{Z_\lambda(1_n)},$$

$$\int_{G\times G} \Delta_{2\lambda}(yxz)\,d\nu(y)\,d\nu(z) = h_1(2\lambda)Z_\lambda(x'x).$$

(Use (a) above and (4.5).)

13. Let $x_1, \ldots, x_r \in G$. Then

$$\int_{G^{r+1}} \Delta_{2\lambda}(y_0 x_1 y_1 x_2 \ldots y_{r-1} x_r y_r)\, d\nu(y_0) \ldots d\nu(y_r) = h_1(2\lambda) \prod_{i=1}^{r} Z_\lambda(x_i' x_i)$$

if $l(\lambda) \leqslant n$, where $G^{r+1} = G \times \ldots \times G$ to $r+1$ factors. (Induction on r: the case $r = 1$ is Example 12(b). If we denote the integral above by $f_\lambda(x_1, \ldots, x_r)$ we have

$$f_\lambda(x_1, \ldots, x_{r+1}) = \int_G f_\lambda(x_1 y x_2, x_3, \ldots, x_{r+1})\, d\nu(y)$$

if $r \geqslant 1$, and the inductive step is completed by (4.5).)

14. Let μ be a positive measure on G such that $d\mu(kx) = d\mu(xk) = d\mu(x)$ for all $k \in K$, and such that

$$\int_G f(x)\, d\mu(x) < \infty$$

for all polynomial functions f on G. If $l(\lambda) \leqslant n$, show that

$$\int_G \Delta_\lambda(x' \sigma x)\, d\mu(x) = u_\lambda Z_\lambda(\sigma)$$

for all $\sigma \in \Sigma$, where

$$u_\lambda = Z_\lambda(1_n)^{-1} \int_G \Omega_\lambda(x)\, d\mu(x).$$

(Use (4.10) and (4.2).)

15. Let μ_1, μ_2 be positive measures on G satisfying the conditions of Example 14. Show that

$$\int_{G \times G} \Delta_{2\lambda}(yxz)\, d\mu_1(y)\, d\mu_2(z) = v_\lambda \Omega_\lambda(x),$$

where $v_\lambda = h_1(2\lambda)\mu_1(\Omega_\lambda)\mu_2(\Omega_\lambda)/Z_\lambda(1_n)$. (Replace y by $k_1 y k_3$ and z by $k_4 z k_2$, where $k_1, \ldots, k_4 \in K$; integrate first with respect to k_1 and k_2, and then with respect to k_3 and k_4, using Example 12(a) and (4.1).)

Notes and references

Most of the integral formulas in this section are to be found in Takemura's monograph [T1]. In particular, Proposition (4.1) is a classical criterion for zonal spherical functions, due originally to Gelfand [G7], and (4.2), (4.3) are due to James [J5]. Takemura (*loc. cit.*) essentially takes (4.6) as his definition of the zonal polynomials (the eigenvalues $Z_\lambda(1_n)$, where λ runs through the partitions of a fixed number r, will be all distinct provided n is large enough). Proposition (4.10) is a particular case of a well-known formula of Harish-Chandra (see, for example [H6], p. 418), and (4.11) is due to Kates [K7] (see [T1], p. 34). Example 7 deals with the 'Wishart density' (see e.g. [F3], p. 53), and Example 8 is due to Constantine [C4].

5. The complex case

In this section let $G = GL_n(\mathbf{C})$ and let K be the unitary group $U(n)$. We shall show that (G, K) is a Gelfand pair in the sense of §3, and that the associated zonal spherical functions may be identified, up to a scalar factor, with the Schur functions s_λ, where λ is a partition of length $\leqslant n$.

A *polynomial function* on G is by definition the restriction to G of a (complex-valued) polynomial function f on the space of complex $n \times n$ matrices, regarded as a *real* vector space of dimension $2n^2$. Thus f is a polynomial in the $2n^2$ algebraically independent functions X_{ij}, $\overline{X}_{ij}$, where $X_{ij}(g) = g_{ij}$ and $\overline{X}_{ij}(g) = \bar{g}_{ij}$ if $g = (g_{ij}) \in G$.

Let

$$P(G) = \bigoplus_{m \geqslant 0} P^m(G)$$

be the space of all polynomial functions on G, where $P^m(G)$ is the subspace of $P(G)$ consisting of the $p \in P(G)$ that are homogeneous of degree m (as polynomials in the X_{ij} and $\overline{X}_{ij}$). The group G acts on $P(G)$ by the rule

$$(5.1) \qquad (gp)(x) = p(xg)$$

for $p \in P(G)$ and $g, x \in G$. If $p \in P(G)$ then the function $\bar{p}: x \mapsto \overline{p(x)}$ is also in $P(G)$.

Let $Q(G) = \mathbf{C}[X_{ij} : 1 \leqslant i, j \leqslant n]$. Then we have

$$P(G) \cong Q(G) \otimes \overline{Q(G)}$$

and

$$Q(G) = \bigoplus_{\lambda} P_\lambda$$

where λ ranges over all partitions of length $\leqslant n$, and P_λ is the subspace of $Q(G)$ spanned by the matrix coefficients R^λ_{ij} of an irreducible polynomial representation R^λ of G (Chapter I, Appendix A). Hence

$$(5.2) \qquad P(G) = \bigoplus_{\lambda, \mu} P_{\lambda, \mu}$$

where λ, μ are partitions of length $\leqslant n$, and $P_{\lambda,\mu}$ is the subspace of $P(G)$ spanned by the products $R^\lambda_{ij} \overline{R}^\mu_{kl}$.

We can now define the operators e_K, e'_K as in §3 (K now being the unitary group), and propositions (3.3)–(3.8) will remain valid in the present context, provided that the transposed matrix x' is replaced by the conjugate transposed $x^* = \bar{x}'$. Correspondingly, Σ is now the space of hermitian $n \times n$ matrices, and

$$P(\Sigma) = \bigoplus_{m \geq 0} P^m(\Sigma)$$

the space of polynomial functions on Σ (regarded as a real vector space of dimension $n + n(n-1) = n^2$). The group G acts on $P(\Sigma)$ by the rule

$$(gp)(\sigma) = p(g^* \sigma g)$$

for $g \in G$, $p \in P(\Sigma)$, and $\sigma \in \Sigma$.

Let V (resp. $\bar{V}$) be a complex vector space of dimension n, with basis $e_1, \ldots, e_n$ (resp. $\bar{e}_1, \ldots, \bar{e}_n$). The group G acts on V and $\bar{V}$ by

$$ge_j = \sum_{i=1}^{n} g_{ij} e_i,$$

$$g\bar{e}_j = \sum_{i=1}^{n} \bar{g}_{ij} \bar{e}_i$$

if $g = (g_{ij}) \in G$.

The tensor space $\mathbf{T}^{2m}(V)$ of §3 is here replaced by $\mathbf{T} = \mathbf{T}^m(V) \otimes \mathbf{T}^m(\bar{V})$; the group G acts diagonally, as before, and (in place of the symmetric group S_{2m}) the group $S_m \times S_m$ acts by permuting the factors in $\mathbf{T}$.

Let $\langle u, \bar{v} \rangle$ be the bilinear form on $V \times \bar{V}$ for which $\langle e_i, \bar{e}_j \rangle = \delta_{ij}$. We have then

$$\langle gu, \bar{v} \rangle = \langle u, g^* \bar{v} \rangle$$

for $g \in G$, $u \in V$, and $\bar{v} \in \bar{V}$.

Let $\varphi: \mathbf{T} \to P^m(\Sigma)$ denote the linear mapping defined by

$$\varphi(u_1 \otimes \ldots \otimes u_m \otimes \bar{v}_1 \otimes \ldots \otimes \bar{v}_m)(\sigma) = \prod_{i=1}^{m} \langle u_i, \sigma \bar{v}_i \rangle.$$

From our definitions it follows that φ commutes with the actions of G on $\mathbf{T}$ and $P^m(\Sigma)$. The place of the hyperoctahedral group H_m in §3 is taken here by the diagonal subgroup $\Delta = \Delta_m = \{(w, w) : w \in S_m\}$ of $S_m \times S_m$, and it is clear from the definition of φ that $\varphi(ht) = \varphi(t)$ for $t \in \mathbf{T}$ and $h \in \Delta$. If $\mathbf{T}^\Delta$ is the subspace of $\mathbf{T}$ fixed by Δ, we have as in §3

(5.3) *The mapping φ is restricted to $\mathbf{T}^\Delta$ is a G-isomorphism of $\mathbf{T}^\Delta$ onto $P^m(\Sigma)$.* |

From Chapter I, Appendix A, §5 it follows that $\mathbf{T}$ decomposes under the action of $(S_m \times S_m) \times G$ as follows:

$$\mathbf{T} = \bigoplus_{\lambda,\mu} \mathbf{T}_{\lambda,\mu}$$

where

$$\mathbf{T}_{\lambda,\mu} \cong (M_\lambda \otimes M_\mu) \otimes \left(F_\lambda(V) \otimes F_\mu(\bar{V})\right).$$

Here λ and μ are partitions of m of length $\leqslant n$, and M_λ (resp. M_μ) is an irreducible S_m-module corresponding to λ (resp. μ). Since (Chapter I, §7)

$$\begin{aligned}\dim(M_\lambda \otimes M_\mu)^\Delta &= \langle \chi^\lambda \chi^\mu, 1 \rangle_{S_m} \\ &= \langle \chi^\lambda, \chi^\mu \rangle_{S_m} = \delta_{\lambda\mu}\end{aligned}$$

it follows that

(5.4) $$\mathbf{T}^\Delta \cong \bigoplus_\lambda \left(F_\lambda(V) \otimes F_\lambda(\bar{V})\right)$$

as G-module, where λ ranges after the partitions of m of length $\leqslant n$.

Now the G-modules $F_\lambda(V) \otimes F_\lambda(\bar{V})$, and more generally $F_\lambda(V) \otimes F_\mu(\bar{V})$ for any two partitions λ, μ of length $\leqslant n$, are irreducible. For the matrix coefficients $R_{ij}^\lambda \in Q(G)$ are linearly independent (Chapter I, Appendix A, (8.1)), and likewise the $\bar{R}_{kl}^\mu \in \overline{Q(G)}$ are linearly independent; hence the products $R_{ij}^\lambda \bar{R}_{kl}^\mu$ are linearly independent functions on G, which (*loc. cit.*) proves the assertion.

If $p \in P(\Sigma)$, the function $\tilde{p}$ defined by

$$\tilde{p}(x) = p(x^* x)$$

is a polynomial function on G, and the mapping $p \mapsto \tilde{p}$ commutes with the actions of G on $P(\Sigma)$ and $P(G)$. Moreover, it doubles degrees: if $p \in P^m(\Sigma)$ then $\tilde{p} \in P^{2m}(G)$.

(5.5) *The mapping $p \mapsto \tilde{p}$ defined above is a G-isomorphism of $P^m(\Sigma)$ onto $P^{2m}(K \backslash G)$, for all $m \geqslant 0$.*

The proof is the same as that of (3.12), and just as in the real case has the following consequences:

(5.6) (i) *The G-module $e_K P_{\lambda,\mu}$ is isomorphic to $F_\lambda(V) \otimes F_\lambda(\bar{V})$ if $\lambda = \mu$, and is zero otherwise.*

(ii) *We have*

$$e_K P(G) = P(K \backslash G) = \bigoplus_\lambda \left(F_\lambda(V) \otimes F_\lambda(\bar{V})\right)$$

where λ ranges over all partitions of length $\leqslant n$.

(5.7) *Let λ, μ be partitions of length $\leqslant n$. Then*

$$\dim\left(F_\lambda(V)\otimes F_\mu(\bar{V})\right)^K = \begin{cases} 1 & \text{if } \lambda=\mu, \\ 0 & \text{otherwise.} \end{cases}$$

(5.8) *$P_{\lambda,\mu}(G,K) = P_{\lambda,\mu}\cap P(G,K)$ is one-dimensional if $\lambda=\mu$, and is zero otherwise. Moreover,*

$$P(G,K) = \bigoplus_\lambda P_{\lambda,\lambda}(G,K)$$

summed over all partitions λ of length $\leqslant n$.

In these statements, $P(K\backslash G)$, $P(G,K)$ etc. are defined exactly as in the real case (§3). In particular, it follows from (5.6)(ii) that the G-module $P(K\backslash G)$ is multiplicity-free, so that (G,K) is a Gelfand pair.

As in §3, we regard each symmetric function $f\in\Lambda_{\mathbf{C}}$ as a polynomial function on G: if $x\in G$ has eigenvalues $\xi_1,\ldots,\xi_n$ then $f(x)=f(\xi_1,\ldots,\xi_n)$. Then the character of the irreducible G-module $F_\lambda(V)\otimes F_\mu(\bar{V})$ is $s_\lambda\bar{s}_\mu\in P_{\lambda,\mu}$, and hence by (5.6) we have $e_K(s_\lambda\bar{s}_\mu)=0$ unless $\lambda=\mu$:

$$\int_K (s_\lambda\bar{s}_\mu)(xk)\,dk = 0 \tag{5.9}$$

for all $x\in G$, unless $\lambda=\mu$. |

Let $\Omega_\lambda = e_K(s_\lambda\bar{s}_\lambda)$, or explicitly

$$\Omega_\lambda(x) = \int_K s_\lambda(xk)\bar{s}_\lambda(xk)\,dk \tag{5.10}$$

where λ is a partition of length $\leqslant n$. As in §3 we have

(5.11) (i) $\Omega_\lambda(1)=1$.
(ii) *$\Omega_\lambda\in P_{\lambda,\lambda}(G,K)$ and generates the irreducible G-submodule $e_K P_{\lambda,\lambda}$ of $P(K\backslash G)$.*

Moreover, Ω_λ is characterized by these two properties. |

These functions Ω_λ are the zonal spherical functions of the pair (G,K). In order to compute them we return to the $(S_m\times S_m)\times G$-module $\mathbf{T} = \mathbf{T}^m(V)\otimes\mathbf{T}^m(\bar{V})$, where $m\leqslant n$.

From (5.7) we have

$$\mathbf{T}^K = \bigoplus_\lambda \mathbf{T}^K_{\lambda,\lambda} \cong \bigoplus_\lambda (M_\lambda\otimes M_\lambda)$$

as $S_m \times S_m$-module. Hence $\mathbf{T}^K$ is isomorphic to the induced module $1_\Delta^{S_m \times S_m}$. Now the tensor

$$\delta = \sum_{i=1}^{n} e_i \otimes \bar{e}_i \in V \otimes \bar{V}$$

is fixed by K, and hence

$$\delta_m = \sum_{i_1, \ldots, i_m} e_{i_1} \otimes \ldots \otimes e_{i_m} \otimes \bar{e}_{i_1} \otimes \ldots \otimes \bar{e}_{i_m}$$

lies in $\mathbf{T}^K$. Since the subgroup of $S_m \times S_m$ that fixes δ_m is precisely Δ, it follows that δ_m generates $\mathbf{T}^K$ as $S_m \times S_m$ – module. Hence by (1.5) the space

$$\mathbf{T}_{\lambda,\lambda}^{\Delta \times K} = (\mathbf{T}_{\lambda,\lambda}^{K})^{\Delta} = (\mathbf{T}_{\lambda,\lambda}^{\Delta})^{K}$$

is one-dimensional, generated by $\omega^\lambda \delta_m$ where ω^λ is the zonal spherical function of the Gelfand pair $(S_m \times S_m, \Delta)$ corresponding to the partition λ. From §1, Example 9 we have

$$\omega^\lambda \delta_m = \chi^\lambda(1)^{-1} \sum_{v, w \in S_m} \chi^\lambda(vw^{-1}) . (v \times w) \delta_m$$

$$= \frac{n!}{\chi^\lambda(1)} \sum_{w \in S_m} \chi^\lambda(w) . (w \times 1) \delta_m$$

where χ^λ is the character of M_λ, since $(v \times w)\delta_m = (vw^{-1} \times 1)\delta_m$.

From the definitions of δ_m and the mapping $\varphi : \mathbf{T} \to P^m(\Sigma)$ (5.3) it follows easily that

$$\varphi((w \times 1)\delta_m) = p_\rho$$

where ρ is the cycle-type of $w \in S_m$, and hence

$$\varphi(\omega^\lambda \delta_m) = c_\lambda \sum_\rho z_\rho^{-1} \chi_\rho^\lambda p_\rho = c_\lambda s_\lambda$$

in the notation of Chapter I, §7, where $c_\lambda = (n!)^2 / \chi^\lambda(1)$. From this it follows, as in (3.24), that

(5.12) *We have*

$$\Omega_\lambda(x) = s_\lambda(x^* x) / s_\lambda(1_n)$$

for all $x \in G$.

Another proof of (5.12) is given in Example 1 below.

Examples

1. (a) Let $p \in Q(G)$. If $p|K = 0$ then $p = 0$. (Each $g \in G$ can be written as $g = k_1 x k_2$, where $k_1 k_2 \in K$ and x is a diagonal matrix, say $x = \operatorname{diag}(x_1, \ldots, x_n)$. For fixed $k_1, k_2 \in K$ let $q(x_1, \ldots, x_n) = p(k_1 x k_2)$; then q is a polynomial function of n complex variables $x_1, \ldots, x_n$ which vanishes whenever $|x_1| = \ldots = |x_n| = 1$. Deduce that $q = 0$ and hence that $p(g) = 0$ for all $g \in G$.)

(b) For each partition λ of length $\leqslant n$, the G-module $F_\lambda(V)$ remains irreducible as a K-module. (Consider the matrix coefficients of $F_\lambda(V)$, and use (a) above.)

(c) If $g \mapsto (R_{ij}^\lambda(g))$ is a matrix representation of G afforded by $F_\lambda(V)$, we have $(x \in G, k \in K)$

$$s_\lambda(xk) = \sum_{i,j} R_{ji}^\lambda(x) R_{ij}^\lambda(k) \tag{1}$$

and

$$\bar{s}_\lambda(xk) = s_\lambda(\bar{x}\bar{k}) = s_\lambda(k^{-1}x^*)$$

$$= \sum_{r,s} R_{rs}^\lambda(k^{-1}) R_{sr}^\lambda(x^*). \tag{2}$$

The matrix coefficients satisfy the orthogonality relations [S12]

$$\int_K R_{ij}^\lambda(k) R_{rs}^\lambda(k^{-1})\, \mathrm{d}k = d_\lambda^{-1} \delta_{is} \delta_{jr} \tag{3}$$

where $d_\lambda = \dim F_\lambda(V) = s_\lambda(1_n)$, since by (b) above the matrix representation $k \mapsto (R_{ij}^\lambda(k))$ of K is irreducible.

Deduce from (1), (2), and (3) that

$$\Omega_\lambda(x) = \int_K (s_\lambda \bar{s}_\lambda)(xk)\, \mathrm{d}k = d_\lambda^{-1} s_\lambda(xx^*).$$

2. Let $f \in P(G)$. Then f is a zonal spherical function of (G, K) if and only if $f \neq 0$ and

$$\int_K f(xky)\, \mathrm{d}k = f(x) f(y)$$

for all $x, y \in G$. (Same proof as for (4.1).)

3. Let $\sigma, \tau \in \Sigma$. Then

$$\int_K s_\lambda(\sigma k \tau k^{-1})\, \mathrm{d}k = s_\lambda(\sigma) s_\lambda(\tau) / s_\lambda(1_n).$$

In view of (5.12), this follows from Example 2.

4. Let $\mathrm{d}x$ be Lebesgue measure on G and let

$$\mathrm{d}\nu(x) = (2\pi)^{-n^2} \exp(-\tfrac{1}{2} \operatorname{trace} x^* x)\, \mathrm{d}x$$

so that $\mathrm{d}\nu(xk) = \mathrm{d}\nu(kx) = \mathrm{d}\nu(x)$ for $k \in K$, and $\int_G \mathrm{d}\nu(x) = 1$.

Let $\sigma, \tau \in \Sigma$. Then

$$\int_G (\text{trace}\, \sigma x \tau x^*)^m \, d\nu(x) = 2^m \sum_\lambda s_\lambda(\sigma) s_\lambda(\tau)$$

summed over partitions λ of m. (The proof follows that of (4.3).)

5. Let $\sigma, \tau \in \Sigma$. Then

$$\int_G s_\lambda(\sigma x \tau x^*) \, d\nu(x) = 2^{|\lambda|} h(\lambda) s_\lambda(\sigma) s_\lambda(\tau)$$

for all partitions λ of length $\leqslant n$, where $h(\lambda)$ is the product of the hook-lengths of λ. (The proof follows the same lines as that of (4.4).)

6. Let λ be a partition of length $\leqslant n$, and let Δ_λ be as in §4.

(a) Show that

$$e_K |\Delta_\lambda(x)|^2 = c_\lambda \Delta_\lambda(x^* x)$$

for $x \in G$, where $c_\lambda = \int_K |\Delta_\lambda(k)|^2 \, dk$.

(b) If $\sigma \in \Sigma$, show that

$$\int_K \Delta_\lambda(k^{-1} \sigma k) \, dk = s_\lambda(\sigma)/s_\lambda(1_n),$$

and that

$$\int_G \Delta_\lambda(x^* \sigma x) \, d\nu(x) = 2^{|\lambda|} h(\lambda) s_\lambda(\sigma).$$

In particular,

$$\int_G \Delta_\lambda(x^* x) \, d\nu(x) = 2^{|\lambda|} \prod_{s \in \lambda} (n + c(s))$$

where $c(s) = a'(s) - l'(s)$ is the content of $s \in \lambda$. (These formulas are the counterparts of (4.9)–(4.12), and may be proved in the same way.)

Notes and references

James [J5] was aware that the complex zonal polynomials are essentially the Schur functions (5.12). See also Farrell's paper [F3], and [T1], Chapter V.

All the integral formulas of §4 have their complex counterparts, and Examples 2–6 are merely a sample.

6. The quaternionic case

In this section let $G = GL_n(\mathbf{H})$, the group of non-singular $n \times n$ matrices over the division ring of quaternions, and let $K = U(n, \mathbf{H})$ be the quaternionic unitary group. We shall show that (G, K) is a Gelfand pair in the sense of §3, and that the associated zonal spherical functions may be identified, up to a scalar multiple, with the Jack symmetric functions (Chapter VI, §10) with parameter $\alpha = \frac{1}{2}$.

If $x = a + bi + cj + dk \in \mathbf{H}$ in the usual notation for quaternions ($i^2 = j^2 = k^2 = -1, ij = -ji = k$, etc.), where $a, b, c, d \in \mathbf{R}$, let

$$\theta(x) = \begin{pmatrix} a+bi & c+di \\ -c+di & a-bi \end{pmatrix} \tag{6.1}$$

so that $x \mapsto \theta(x)$ embeds $\mathbf{H}$ in the algebra of complex 2×2 matrices. More generally, if $g = (g_{rs}) \in G = GL_n(\mathbf{H})$, let

$$\theta(g) = (\theta(g_{rs}))_{1 \leqslant r,s \leqslant n} \tag{6.2}$$

so that θ is an injective homomorphism of G into the complex general linear group $G_{\mathbf{C}} = GL_{2n}(\mathbf{C})$. Observe that $\theta(g^*) = \theta(g)^*$, where the asterisks denote the conjugate transpose. We shall usually identify G with its image $\theta(G)$ in $G_{\mathbf{C}}$.

A polynomial function f on G is by definition the restriction to G of a (complex-valued) polynomial function on the matrix space $M_n(\mathbf{H})$, regarded as a real vector space of dimension $4n^2$. In view of (6.1) and (6.2), f is a polynomial in the $4n^2$ algebraically independent functions

$$\begin{cases} \dfrac{1}{2}(X_{2r-1,2s-1} + X_{2r,2s}), & \dfrac{1}{2i}(X_{2r-1,2s-1} - X_{2r,2s}), \\ \dfrac{1}{2}(X_{2r-1,2s} - X_{2r,2s-1}), & \dfrac{1}{2i}(X_{2r-1,2s} + X_{2r,2s-1}) \end{cases} \tag{6.3}$$

for $r, s = 1, 2, \dots, n$, where X_{pq} $(1 \leqslant p, q \leqslant 2n)$ are the coordinate functions on $G_{\mathbf{C}}$ (i.e. $X_{pq}(g) = g_{pq}$ for $g = (g_{pq}) \in G_{\mathbf{C}}$).

Let $P(G)$ denote the algebra of polynomial functions on G, generated over $\mathbf{C}$ by the functions (6.3), and let $P(G_{\mathbf{C}})$ denote the algebra $\mathbf{C}[X_{pq}: 1 \leqslant p, q \leqslant 2n]$ of polynomial functions on $G_{\mathbf{C}}$.

(6.4) (i) *The restriction map* $f \mapsto f|G$ *is an isomorphism of* $P(G_{\mathbf{C}})$ *onto* $P(G)$.
(ii) *Each irreducible polynomial representation of* $G_{\mathbf{C}}$ *remains irreducible on restriction to* G.

Proof. (i) Each $f \in P(G_{\mathbf{C}})$ is uniquely expressible as a polynomial in the $4n^2$ functions (6.3). If $f|G = 0$, then f vanishes for all real values of these $4n^2$ functions, hence is identically zero.

(ii) Suppose that $R: g \mapsto (R_{ij}(g))$ is a polynomial representation of $G_{\mathbf{C}}$ whose restriction to G is reducible. By replacing R by an equivalent representation, we may assume that for some r such that $1 \leqslant r < d$, where d is the degree of R, we have $R_{ij}(g) = 0$ for all (i, j) such that $i > r$ and

$j \leqslant r$, and all $g \in G$. By (i) above, it follows that this is so for all $g \in G_{\mathbf{C}}$, and therefore R is reducible. |

In view of (6.4), we may identify $P(G)$ and $P(G_{\mathbf{C}})$. As in §3, G acts on $P(G)$ by the rule

$$(gp)(x) = p(xg)$$

where $p \in P(G)$ and $g, x \in G$, and under this action $P(G) = P(G_{\mathbf{C}})$ decomposes as a direct sum

$$P(G) = \bigoplus_{\mu} P_{\mu},$$

where μ ranges over all partitions of length $\leqslant 2n$, and P_{μ} is the subspace of $P(G)$ spanned by the matrix coefficients R_{ij}^{μ} of an irreducible polynomial representation R^{μ} of G (or $G_{\mathbf{C}}$) corresponding to the irreducible G-module $F_{\mu}(\mathbf{C}^{2n})$, in the notation of Chapter I, Appendix A.

Now let

$$K = U(n, \mathbf{H}) = \{g \in G : gg^* = 1\}.$$

Since $\theta(K) = \theta(G) \cap U(2n)$, it follows that K is a bounded closed subgroup of G, and hence is compact. The set-up is now entirely analogous to that of §3, and we need only pay attention to those points in the development that differ from the real case. Propositions (3.3)–(3.8) remain valid in the present context, provided that the transpose x' $(x \in G)$ is replaced by x^*.

Next let Σ denote the space of quaternionic hermitian $n \times n$ matrices (i.e. matrices σ such that $\sigma^* = \sigma$), and let

$$P(\Sigma) = \bigoplus_{m \geqslant 0} P^m(\Sigma)$$

denote the space of polynomial functions on Σ (regarded as a real vector space of dimension $n + 2n(n-1)$), where $P^m(\Sigma)$ consists of the functions that are homogeneous of degree m. The group G acts on $P(\Sigma)$ by

$$(gp)(\sigma) = p(g^* \sigma g)$$

for $g \in G$, $p \in P(\Sigma)$, and $\sigma \in \Sigma$.

Also let $V = \mathbf{C}^{2n}$, the elements of V being regarded as column vectors of length $2n$. The group G acts on V via $\theta : gv = \theta(g)v$ for $g \in G$ and $v \in V$. Let $J = \theta(j1_n)$, so that J is the diagonal sum of n copies of the matrix $\begin{pmatrix} 0 & 1 \\ -1 & 0 \end{pmatrix}$, and let

$$\langle u, v \rangle = u'Jv \tag{6.5}$$

for $u, v \in V$. The bilinear form so defined on V is skew-symmetric, and we have

(6.6) $$\langle gu, v\rangle = \langle u, g^*v\rangle$$

for all $g \in G$. For it follows from (6.1) and (6.2) that $\theta(g)'J = J\theta(g^*)$, and hence that

$$\langle gu, v\rangle = u'\theta(g)'Jv = u'J\theta(g^*)v$$
$$= \langle u, g^*v\rangle.$$

Let $\mathbf{T} = \mathbf{T}^{2m}(V)$ be the $2m$th tensor power of V. As in §3, the group G acts diagonally on $\mathbf{T}$, and the symmetric group S_{2m} acts by permuting the factors. Let $\varphi: \mathbf{T} \to P^m(\Sigma)$ denote the linear mapping defined by

$$\varphi(v_1 \otimes \ldots \otimes v_{2m})(\sigma) = \prod_{i=1}^{m} \langle v_{2i-1}, \sigma v_{2i}\rangle.$$

Clearly φ commutes with the actions of G on $\mathbf{T}$ and $P^m(\Sigma)$, and since by (6.6) we have

$$\langle u, \sigma v\rangle = \langle \sigma u, v\rangle = -\langle v, \sigma u\rangle$$

for $\sigma \in \Sigma$, it follows that

$$\varphi(ht) = \varepsilon(h)\varphi(t)$$

for $t \in \mathbf{T}$ and $h \in H = H_m$, where ε is the sign character of S_{2m}. Finally, if $e_1, \ldots, e_{2n}$ is the standard basis of $V = \mathbf{C}^{2n}$ and $e_1^*, \ldots, e_{2n}^*$ is the dual basis, so that $\langle e_i^*, e_j\rangle = \delta_{ij}$, then $\langle e_p^*, \sigma e_q\rangle$ is the (p,q) element of $\theta(\sigma)$, from which it follows easily that φ is surjective. Hence if

$$\mathbf{T}^{\varepsilon H} = \{t \in \mathbf{T}: ht = \varepsilon(h)t \text{ for all } h \in H\}$$

we have (compare (3.9)):

(6.7) *The mapping φ restricted to $\mathbf{T}^{\varepsilon H}$ is a G-isomorphism of $\mathbf{T}^{\varepsilon H}$ onto $P^m(\Sigma)$.* |

Under the action of $S_{2m} \times G_{\mathbf{C}}$, T decomposes as

$$\mathbf{T} = \bigoplus_{\mu} \mathbf{T}_\mu$$

where

$$\mathbf{T}_\mu \cong M_\mu \otimes F_\mu(V).$$

Here μ ranges over the partitions of $2m$ of length $\leqslant 2n$, and M_μ is an irreducible S_{2m}-module corresponding to μ. By (6.4)(ii), $F_\mu(V)$ remains irreducible as a G-module. Since (§2, Example 6) $M_\mu^{\varepsilon H}$ is one-dimensional if μ' is an even partition (or equivalently if $\mu = \lambda \cup \lambda$ for some partition λ of m) and is zero otherwise, it follows that

$$\text{(6.8)} \qquad \mathsf{T}^{\varepsilon H} \cong \bigoplus_\lambda F_{\lambda \cup \lambda}(V)$$

as G-module, where λ ranges over the partitions of m of length $\leqslant n$.

If $p \in P(\Sigma)$, the function $\tilde{p}$ defined by

$$\tilde{p}(x) = p(x^* x)$$

is a polynomial function on G, and the mapping $p \mapsto \tilde{p}$ commutes with the actions of G on $P(\Sigma)$ and $P(G)$. Moreover, it doubles degrees: if $p \in P^m(\Sigma)$ then $\tilde{p} \in P^{2m}(G)$.

(6.9) *The mapping $p \mapsto \tilde{p}$ defined above is a G-isomorphism of $P^m(\Sigma)$ onto $P^{2m}(K \backslash G)$, for all $m \geqslant 0$.*

The proof is the same as that of (3.12), and just as in the real case has the following consequences:

(6.10) (i) *The G-module $e_K P_\mu$ is isomorphic to $F_\mu(V)$ if μ' is even, and is zero otherwise.*
(ii) *We have*

$$e_K P(G) = P(K \backslash G) \cong \bigoplus_\lambda F_{\lambda \cup \lambda}(V)$$

where λ ranges over all partitions of length $\leqslant n$.

(6.11) *Let μ be a partition of length $\leqslant 2n$. Then*

$$\dim F_\mu(V)^K = \begin{cases} 1 & \text{if } \mu' \text{ is even,} \\ 0 & \text{otherwise.} \end{cases}$$

(6.12) *$P_\mu(G, K)$ is one-dimensional if μ' is even, and is zero otherwise. Moreover,*

$$P(G, K) = \bigoplus_\lambda P_{\lambda \cup \lambda}(G, K)$$

summed over all partitions λ of length $\leqslant n$.

In these statements, e_K, $P(K \backslash G)$ etc. are defined exactly as in the real case.

As in §3, we may regard each $f \in \Lambda_{\mathbf{C}}$ as a polynomial function on $G_{\mathbf{C}}$: if

$x \in G_{\mathbf{C}}$ has eigenvalues $\xi_1, \ldots, \xi_{2n}$, then $f(x) = f(\xi_1, \ldots, \xi_{2n})$. With this understood, we have as in (3.19)

(6.13) $$\int_K s_\mu(\theta(xk))\,dk = 0$$

unless μ' is an even partition.

Let λ be a partition of length $\leqslant n$, and define $\Omega_\lambda \in P(G)$ by

(6.14) $$\Omega_\lambda(x) = \int_K s_{\lambda \cup \lambda}(\theta(xk))\,dk$$

for $x \in G$. Then as in (3.21) we have

(6.15) (i) $\Omega_\lambda(1) = 1$.
(ii) $\Omega_\lambda \in P_{\lambda \cup \lambda}(G, K)$ *and generates the irreducible G-submodule $e_K P_{\lambda \cup \lambda}$ of $P(K \backslash G)$.*
Moreover, Ω_λ is characterized by these two properties.

The function Ω_λ is the *zonal spherical function* of (G, K) associated with the partition λ.

We return now to the $S_{2m} \times G$-module $\mathbf{T} = \mathbf{T}^{2m}(V)$, where $m \leqslant n$. We have (compare §3)

$$\mathbf{T}^K = \bigoplus_\lambda \mathbf{T}^K_{\lambda \cup \lambda} \cong \bigoplus_\lambda M_{\lambda \cup \lambda}$$

as S_{2m}-module, where λ ranges over all partitions of m. From §2, Example 6 it follows that $\mathbf{T}^K$ is isomorphic to the induced module $\varepsilon_H^{S_{2m}}$, where ε_H is the restriction to H of the sign character of S_{2m}. As before, let $e_1, \ldots, e_{2n}$ be the standard basis of $V = \mathbf{C}^{2n}$, and let $e_1^*, \ldots, e_{2n}^*$ be the dual basis, such that $\langle e_p^*, e_q \rangle = \delta_{pq}$ for $1 \leqslant p, q \leqslant 2n$. We have $e_{2r-1}^* = -e_{2r}$ and $e_{2r}^* = e_{2r-1}$ for $1 \leqslant r \leqslant n$.

The tensor

(6.16) $$\delta = \sum_{p=1}^{2n} e_p^* \otimes e_p = -\sum_{p=1}^{2n} e_p \otimes e_p^*$$

is fixed by K, by virtue of (6.6), and hence $\delta^{\otimes m} = \delta \otimes \ldots \otimes \delta$ (m factors) lies in $\mathbf{T}^K$. From (6.16) it follows that $h\delta^{\otimes m} = \varepsilon(h)\delta^{\otimes m}$ for $h \in H$, and hence just as in the real case that $\delta^{\otimes m}$ generates $\mathbf{T}^K$ as S_{2m}-module. Hence (§2, Example 6) the space

$$\mathbf{T}^{\varepsilon H \times K}_{\lambda \cup \lambda} = (\mathbf{T}^K_{\lambda \cup \lambda})^{\varepsilon H} = (\mathbf{T}^{\varepsilon H}_{\lambda \cup \lambda})^K$$

is one-dimensional, generated by

(6.17)
$$\pi^\lambda \delta^{\otimes m} = \sum_{w \in S_{2m}} \varepsilon(w)\omega^{\lambda'}(w).w\delta^{\otimes m}.$$

We have now to compute $\varphi(\pi^\lambda \delta^{\otimes m}) \in P^m(\Sigma)$. For this purpose we need to interpret symmetric functions as polynomial functions on Σ. If $\sigma \in \Sigma$, then $\mathrm{trace}(\sigma) \in \mathbf{R}$, because the diagonal elements of σ are real. More generally, $\mathrm{trace}(\sigma^r) \in \mathbf{R}$ for each $r \geqslant 1$, since $\sigma^r \in \Sigma$. If now $\rho = (\rho_1, \rho_2, \ldots)$ is any partition, we define

$$p_\rho(\sigma) = \prod_{i \geqslant 1} \mathrm{trace}(\sigma^{\rho_i})$$

thus defining p_ρ, and hence by linearity any symmetric function, as a polynomial function on Σ. Since by (6.1) and (6.2) $\mathrm{trace}(\theta(\sigma)) = 2\,\mathrm{trace}(\sigma)$, it follows that

(6.18)
$$p_\rho(\theta(\sigma)) = 2^{l(\rho)} p_\rho(\sigma).$$

In place of (3.23) we now have

(6.19)
$$\varepsilon(w)\varphi(w\delta^{\otimes m}) = \varepsilon_\rho 2^{l(\rho)} p_\rho$$

as functions on Σ, *where* ρ *is the coset-type* (§2) *of* $w \in S_{2m}$, *and* $\varepsilon_\rho = (-1)^{m-l(\rho)}$.

Proof. Let $f_w = \varepsilon(w)\varphi(w\delta^{\otimes m})$. For any $h \in H$ we have $f_{wh} = f_w$, since $h\delta^{\otimes m} = \varepsilon(h)\delta^{\otimes m}$; also $f_{hw} = f_w$, since $\varphi(ht) = \varepsilon(h)\varphi(t)$ for any $t \in \mathbf{T}$. Hence f_w depends only on the double coset $H_\rho = HwH$. As in the proof of (3.23), we are reduced to the case where w is the $2m$-cycle $(1, 2, \ldots, 2m)$. Since

$$\delta^{\otimes m} = \sum e_{i_1}^* \otimes e_{i_1} \otimes \ldots \otimes e_{i_m}^* \otimes e_{i_m}$$

summed over $i_1, \ldots, i_m$ running independently from 1 to $2n$, we have

$$w\delta^{\otimes m} = \sum e_{i_m} \otimes e_{i_1}^* \otimes e_{i_1} \otimes \ldots \otimes e_{i_{m-1}} \otimes e_{i_m}^*$$

and hence (since $\varepsilon(w) = -1$)

(1)
$$f_w(\sigma) = -\sum \langle e_{i_m}, \sigma e_{i_1}^* \rangle \ldots \langle e_{i_{m-1}}, \sigma e_{i_m}^* \rangle.$$

Since

$$\langle e_i, \sigma e_j^* \rangle = \langle \sigma e_i, e_j^* \rangle = -\langle e_j^*, \sigma e_i \rangle = -\theta(\sigma)_{ji}$$

it follows from (1) that

$$f_w(\sigma) = (-1)^{m-1} \sum \theta(\sigma)_{i_1, i_m} \cdots \theta(\sigma)_{i_m i_{m-1}}$$
$$= (-1)^{m-1} \operatorname{trace}(\theta(\sigma)^m)$$
$$= (-1)^{m-1} 2 \operatorname{trace}(\sigma^m)$$

and hence that $f_w = (-1)^{m-1} 2 p_m$, which completes the proof. |

We can now express the zonal spherical functions Ω_λ in terms of the symmetric functions Z'_λ introduced in §2, Example 7:

(6.20) *We have*

$$\Omega_\lambda(x) = Z'_\lambda(x^*x)/Z'_\lambda(1_n)$$

for all $x \in G$.

Proof. Let $\varphi_\lambda = \varphi(\pi^\lambda \delta^{\otimes m})$, and define $\tilde{\varphi}_\lambda \in P(G)$ by $\tilde{\varphi}_\lambda(x) = \varphi_\lambda(x^*x)$. It follows from (6.9) that $\tilde{\varphi}_\lambda$ is a generator of the one-dimensional space $P_{\lambda \cup \lambda}(G, K)$ that is the image of $\mathrm{T}^{\varepsilon H \times K}$ under the composite isomorphism $\mathrm{T}^{\varepsilon H} \to P^m(\Sigma) \to P^{2m}(K \backslash G)$. Hence by (6.15)(ii) Ω_λ is a scalar multiple of $\tilde{\varphi}_\lambda$, namely

(1)
$$\Omega_\lambda(x) = \varphi_\lambda(x^*x)/\varphi_\lambda(1_n)$$

since $\Omega_\lambda(1_n) = 1$.

On the other hand, from (6.17) and (6.19) we have

$$\varphi_\lambda = \sum_\rho |H_\rho| \varepsilon_\rho \omega_\rho^{\lambda'} 2^{l(\rho)} p_\rho$$
$$= |H_m|^2 \sum_\rho \varepsilon_\rho z_\rho^{-1} \omega_\rho^{\lambda'} p_\rho$$

and therefore (§2, Example 7)

(2)
$$\varphi_\lambda = |H_m| Z'_\lambda.$$

The result now follows from (1) and (2). |

Examples

1. Let $f \in P(G)$. Then f is a zonal spherical function of (G, K) if and only if $f \neq 0$ and

$$\int_K f(xky)\, dk = f(x) f(y)$$

for all $x, y \in G$. (Same proof as for (4.1).)

2. In this and the following Examples we define p_ρ, for each partition ρ, as a function on G by the rule (6.18)

$$p_\rho(x) = 2^{-l(\rho)} p_\rho(\theta(x)).$$

By linearity this defines each symmetric function $f \in \Lambda$ as a polynomial function on G, such that $f(xy) = f(yx)$ for all $x, y \in G$. With this definition we have

$$\int_K Z'_\lambda(\sigma k \tau k^*)\, dk = \frac{Z'_\lambda(\sigma) Z'_\lambda(\tau)}{Z'_\lambda(1_n)}$$

for all $\sigma, \tau \in \Sigma$. In view of (6.20), this follows from Example 1.

3. Let dx be Lebesgue measure on G and let

$$d\nu(x) = (2\pi)^{-2n^2} \exp(-\tfrac{1}{2} \operatorname{trace} x^* x)\, dx,$$

so that $d\nu(kx) = d\nu(xk) = d\nu(x)$ for $k \in K$, and $\int_G d\nu(x) = 1$. Let $\xi, \eta \in G$ be real diagonal matrices. Then trace$(\xi x \eta x^*)$ is a real number for all $x \in G$, and we have

$$\int_G (\operatorname{trace} \xi x \eta x^*)^m\, d\nu(x) = 2^m m! \sum_{|\lambda| = m} \frac{Z'_\lambda(\xi) Z'_\lambda(\eta)}{h(2\lambda')}.$$

(The proof follows that of (4.3), using §2, Example 7(d).)

4. Let $\sigma, \tau \in \Sigma$. Then

$$\int_G Z'_\lambda(\sigma x \tau x^*)\, d\nu(x) = 2^{|\lambda|} Z'_\lambda(\sigma) Z'_\lambda(\tau). \tag{1}$$

In particular,

$$\int_G Z'_\lambda(x^* \sigma x)\, d\nu(x) = 2^{|\lambda|} Z'_\lambda(1_n) Z'_\lambda(\sigma). \tag{2}$$

(Let $I_\lambda(\sigma, \tau)$ denote the integral on the left-hand side of (1). We may assume without loss of generality that σ and τ are real diagonal matrices ξ, η. Since

$$p_1^m = m! \sum_{|\lambda| = m} h(2\lambda')^{-1} Z'_\lambda$$

(§2, Example 7) it follows that

$$\int_G (\operatorname{trace} \xi x \eta x^*)^m\, d\nu(x) = m! \sum_{|\lambda| = m} h(2\lambda')^{-1} I_\lambda(\xi, \eta).$$

On the other hand, Example 2 shows that $I_\lambda(\xi, \eta)$ is a scalar multiple of $Z'_\lambda(\xi) Z'_\lambda(\eta)$, and hence the result follows from Example 3.)

5. (a) If $x \in G$, then det $\theta(x)$ is real and positive. (Let $x = vb$ where $b = (b_{ij})$ is upper triangular and v is lower unitriangular. Then det $\theta(v) = 1$ and det $\theta(b) = \prod_{i=1}^n |b_{ii}|^2$. Since elements of the form $x = vb$ are dense in G, the result follows.)

Define det x to be the positive square root of det $\theta(x)$.

(b) If $x=(x_{ij})\in G$ (resp. $G_{\mathbf{C}}$), let $x^{(r)}=(x_{ij})_{1\leqslant i,j\leqslant r}$ for $1\leqslant r\leqslant n$ (resp. $1\leqslant r\leqslant 2n$). Then $\theta(x^{(r)})=\theta(x)^{(2r)}$ and hence

$$\det\theta(x)^{(2r)}=(\det x^{(r)})^2.$$

(c) If $x\in G$ (resp. $G_{\mathbf{C}}$) and μ is a partition of length $\leqslant n$ (resp. $2n$) let

$$\Delta_\mu(x)=\prod_{i\geqslant 1}(\det x^{(\mu_i')}).$$

Then we have

$$\Delta_{\lambda\cup\lambda}(\theta(x))=\Delta_{2\lambda}(x)$$

for $x\in G$ and $l(\lambda)\leqslant n$.

(d) Show (as in (4.9)) that

$$e_K\Delta_{2\lambda}(x)=c_\lambda\Delta_\lambda(x^*x)$$

where c_λ is a positive constant, and deduce that

$$\int_K\Delta_\lambda(k^*\sigma k)\,\mathrm{d}k=Z_\lambda'(\sigma)/Z_\lambda'(1_n)$$

for $\sigma\in\Sigma$.

6. (a) Let λ be a partition of length $\leqslant n$. Show that

$$\int_G\Delta_\lambda(x^*\sigma x)\,\mathrm{d}\nu(x)=2^{|\lambda|}Z_\lambda'(\sigma)$$

for $\sigma\in\Sigma$. (Use Examples 5(d) and 4.) In particular,

$$Z_\lambda'(1_n)=2^{-|\lambda|}\int_G\Delta_\lambda(x^*x)\,\mathrm{d}\nu(x). \tag{1}$$

(b) Let B^+ be the subgroup of G consisting of upper triangular matrices (b_{ij}) with positive real diagonal elements b_{ii}. Each $x\in G$ factorizes uniquely as $x=kb$, where $k\in K$ and $b\in B^+$, and we have

$$\mathrm{d}x=c_n\Delta_\delta(b)\,\mathrm{d}k\,\mathrm{d}b, \tag{2}$$

where $\mathrm{d}k$ is normalized Haar measure on K, $\mathrm{d}b$ is Lebesgue measure on B^+, and $\delta=(4n-1,4n-5,\ldots,3)$, and c_n is a positive constant. By evaluating the integral (1) show that

$$Z_\lambda'(1_n)=\prod_{s\in\lambda}(2n+a'(s)-2l'(s))$$

in agreement with §2, Example 7(b).

(c) Show that the constant c_n in (2) is given by

$$c_n=v_4v_8\ldots v_{4n}$$

where $v_r=2\pi^{r/2}/\Gamma(\frac{1}{2}r)$ as in §4, Example 2(d). (Use (2) to evaluate $\int_G\mathrm{d}\nu(x)=1$.)

Notes and references

As far as I am aware, the quaternionic analogue of James's theory (§3), and in particular the identification of the quaternionic zonal polynomials with the Jack symmetric functions with parameter $\alpha = \frac{1}{2}$, announced in [M5], has not been worked out in print before. See however [G16], where the three cases (real, complex, quaternionic) are handled simultaneously.

As in the complex case, all the integral formulas of §4 have their quaternionic analogues, and Examples 1–6 provide a sample of these.

BIBLIOGRAPHY

A1. Aitken, A. C. (1931). Note on dual symmetric functions. *Proc. Edin. Math. Soc.* (2) **2**, 164–7.

A2. Akin, K., Buchsbaum, D. A., and Weyman, J. (1982). Schur functions and Schur complexes. *Advances in Math.* **44**, 207–78.

A3. Andrews, G. E. (1976). The theory of partitions. *Encyclopaedia of mathematics and its applications*, Vol. 2. Addison–Wesley, Reading, Massachusetts.

A4. Andrews, G. E. (1977). MacMahon's conjecture on symmetric plane partitions. *Proc. Natl. Acad. Sci. USA*, **74**, 426–9.

A5. Andrews, G. E. (1979). Plane partitions (III): the weak Macdonald conjecture. *Inv. Math.*, **53**, 193–225.

A6. Aomoto, K. (1987). Jacobi polynomials associated with Selberg integrals. *SIAM J. Math. Anal.*, **18**, 545–9.

A7. Askey, R. (1980). Some basic hypergeometric extensions of integrals of Selberg and Andrews. *SIAM J. Math. Anal.*, **11**, 938–51.

B1. Bannai, E. and Ito, T. (1984). *Algebraic combinatorics I: Association schemes*. Benjamin–Cummings, Menlo Park, California.

B2. Berele, A. and Regev, A. (1987). Hook Young diagrams with applications to combinatorics and to representations of Lie superalgebras. *Advances in Math.*, **64**, 118–75.

B3. Bergeron, N. and Garsia, A. M. (1988). Zonal polynomials and domino tableaux. Preprint.

B4. Bergeron, N. and Garsia, A. M. (1990). Sergeev's formula and the Littlewood–Richardson rule. *Linear and Multilinear Alg.*, **27**, 79–100.

B5. Berthelot, P. (1971). Généralités sur les λ-anneaux. In SGA6 (Séminaire de Géométrie Algèbrique du Bois-Marie, 1966/7). *Springer Lecture Notes*, **225**.

B6. Borho, W. and MacPherson, R. (1981). Representations of Weyl groups and intersection homology of nilpotent varieties. *C.R. Acad. Sci. Paris*, **292**, 707–10.

B7. Bourbaki, N. (1967). *Intégration*, Chapters VII and VIII. Hermann, Paris.

B8. Bourbaki, N. (1968). *Groupes et algèbres de Lie*, Chapters IV, V, and VI. Hermann, Paris.

B9. Burge, W. H. (1974). Four correspondences between graphs and generalized Young tableaux. *J. Comb. Theory* (A) **17**, 12–30.

B10. Butler, L. M. (1986). Combinatorial properties of partially ordered sets associated with partitions and finite abelian groups. Thesis (M.I.T.).

C1. Cartan, H. and Eilenberg, S. (1956). *Homological algebra*. Princeton University Press.

C2. Chen, Y., Garsia, A. M., and Remmel, J. (1984). Algorithms for plethysm, in *Combinatorics and Algebra, Contemporary Math.*, **34**, 109–53.

C3. Comtet, L. (1970). *Analyse combinatoire* (2 vols.). Presses Universitaires de France, Paris.

C4. Constantine, A. G. (1963). Some non-central distribution problems in multivariate analysis. *Ann. Math. Stat.*, **34**, 1270–85.

C5. Curtis, C. W. and Reiner, I. (1962). *Representation theory of finite groups and associative algebras*. Interscience, New York.

D1. Debiard, A. (1983). Polynômes de Tchébychev et de Jacobi dans un espace euclidien de dimension *p*. *C.R. Acad. Sci. Paris* (*sér. I*), **296**, 529–32.

D2. Deligne, P. (1972). Les constantes des equations fonctionelles des fonctions *L*. *Springer Lecture Notes*, **349**, 501–97.

D3. Diaconis, P. (1988). *Group representations in probability and statistics*. Inst. of Math. Statistics, Hayward, California.

D4. Diaconis, P. and Lander, E. (1992). Some formulas for zonal polynomials. In preparation.

D5. Dieudonné, J. (1972). *Treatise on Analysis*, Vol. III. Academic Press, New York.

D6. Doubilet, P. (1972). On the foundations of combinatorial theory, VII: Symmetric functions through the theory of distribution and occupancy. *Studies in Appl. Math.*, **51**, 377–96.

E1. Eğecioğlu, O. and Remmel, J. (1990). A combinatorial interpretation of the inverse Kostka matrix. *Linear and Multilinear Alg.*, **26**, 59–84.

E2. Eğecioğlu, O. and Remmel, J. (1991). Brick tabloids and the connection matrices between bases of symmetric functions. *Discrete Appl. Math.*, **34**, 107–20.

F1. Farahat, H. K. (1956). On the blocks of characters of symmetric groups. *Proc. London Math. Soc.*, (3) **6**, 501–17.

F2. Farahat, H. K. and Higman, G. (1959). The centres of symmetric group rings. *Proc. Roy. Soc.* (A), **250**, 212–21.

F3. Farrell, R. H. (1980). Calculation of complex zonal polynomials, in *Multivariate Analysis* V (ed. Krishnaiah). North Holland, Amsterdam.

F4. Farrell, R. H. (1985). *Multivariate calculation: use of the continuous groups*. Springer-Verlag, New York.

F5. Foulkes, H. O. (1949). Differential operators associated with *S*-functions. *J. London Math. Soc.*, **24**, 136–43.

F6. Foulkes, H. O. (1954). Plethysm of *S*-functions. *Philos. Trans. Roy. Soc.* (A), **246**, 555–91.

F7. Foulkes, H. O. (1974). A survey of some combinatorial aspects of symmetric functions, in *Permutations*. Gauthier–Villars, Paris.

F8. Frame, J. S., Robinson, G. de B., and Thrall, R. M. (1954). The hook graphs of S_n. *Can. J. Math.*, **6**, 316–24.

F9. Franzblau, D. S. and Zeilberger, D. (1982). A bijective proof of the hook-length formula. *J. Algorithms*, **3**, 317–43.

F10. Frobenius, F. G. (1900). Über die Charaktere der symmetrischen Gruppe. *Sitzungsberichte der Königlich Preussischen Akademie der Wissenschaften zu Berlin* (1900) 516–34 (*Ges. Abhandlungen*, **3**, 148–66).

F11. Frobenius, F. G. (1904). Über die Charaktere der mehrfach transitiven Gruppen. *Sitzungsberichte der Königlich Preussischen Akademie der Wissenschaften zu Berlin* (1904) 558–71 (*Ges. Abhandlungen*, **3**, 335–48).

G1. Gansner, E. R. (1978). Matrix correspondences and the enumeration of plane partitions. Thesis, M.I.T.

G2. Garsia, A. M. (1985). Gelfand pairs in finite groups. Preprint.

G3. Garsia, A. M. (1992). Orthogonality of Milne's polynomials and raising operators. *Discrete Math.*, **99**, 247–64.

G4. Garsia, A. M. and Haiman, M. (1996). Some natural bigraded S_n-modules and q,t-Kostka coefficients. *Electronic J. Combinatorics*, **3**, R24, 60; also *The Fuata Festschrift*, Impruinene Louis-Jean, Gap, France, pp. 561–620.

G5. Gasper, G. and Rahman, M. (1990). Basic hypergeometric series. *Encyclopaedia of mathematics and its applications*, Vol. 35. Cambridge University Press.

G6. Geissinger, L. (1977). Hopf algebras of symmetric functions and class functions. *Springer Lecture Notes*, **579**, 168–81.

G7. Gelfand, I. M. (1950). Spherical functions on symmetric spaces. *Dokl. Akad. Nauk USSR*, **70**, 5–8.

G8. Giambelli, G. Z. (1903). Alcune proprietà delle funzioni simmetriche caratteristiche. *Atti Torino*, **38**, 823–44.

G9. Goulden, I. P. and Jackson, D. M. (1994). Symmetric functions and Macdonald's result for top connection coefficients in the symmetric group. *J. Alg.*, **166**, 364–78.

G10. Gow, R. (1983). Properties of characters of the general linear group related to the transpose-inverse involution, *Proc. London Math. Soc.*, (3) **47**, 493–506.

G11. Green, J. A. (1955). The characters of the finite general linear groups. *Trans. Am. Math. Soc.*, **80**, 402–47.

G12. Green, J. A. (1956). Les polynômes de Hall et les caractères des groupes $GL(n, q)$. *Colloque d'algèbre supérieure*, 207–15 (Brussels).

G13. Green, J. A. (1961). Symmetric functions and p-modules (lecture notes, Manchester).

G14. Green, J. A. (1980). Polynomial representations of GL_n. *Springer Lecture Notes*, **830**.

G15. Greene, C., Nijenhuis, A., and Wilf, H. S. (1979). A probabilistic proof of a formula for the number of Young tableaux of a given shape. *Advances in Math.*, **31**, 104–9.

G16. Gross, K. I. and Richards, D. St. P. (1987). Spherical functions of matrix argument I: algebraic induction, zonal polynomials and hypergeometric functions. *Trans. Am. Math. Soc.*, **301**, 781–811.

H1. Hall, P. (1938). A partition formula connected with abelian groups. *Comm. Math. Helv.*, **11**, 126–9.

H2. Hall, P. (1940). On groups of automorphisms. *J. reine angew. Math.*, **182**, 194–204.

H3. Hall, P. (1959). The algebra of partitions. *Proc. 4th Canadian Math. Congress*, 147–159. Banff.

H4. Hammond, J. (1883). On the use of certain differential operators in the theory of equations. *Proc. London Math. Soc.*, **14**, 119–29.

H5. Hanlon, P. (1988). Jack symmetric functions and some combinatorial properties of Young symmetrizers, *J. Comb. Theory* (A), **47**, 37–70.

H6. Helgason, S. (1984). *Groups and geometric analysis*. Academic Press, New York.

H7. Hoffman, P. and Humphreys, J. F. (1992). Projective representations of symmetric groups. *Oxford Mathematical Monographs*.

H8. Hotta, R. and Shimomura, N. (1979). The fixed point subvarieties of unipotent transformations on generalized flag varieties and the Green functions. *Math. Annalen*, **241**, 193–208.

H9. Hotta, R. and Springer, T. A. (1977). A specialization theorem for certain

Weyl group representations and an application to the Green polynomials of unitary groups. *Inv. Math.*, **41**, 113–27.

H10. Howe, R. (1987). (GL_n, GL_m)-duality and symmetric plethysm. *Proc. Indian Acad. Sci.* (*Math. Sci.*), (1987), 85–109.

H11. Hua, L.-K. (1963). Harmonic analysis of functions of several complex variables in the classical domains, *AMS Translations*, Vol. 6.

H12. Humphreys, J. F. (1986). Blocks of projective representations of the symmetric groups. *J. London Math. Soc.*, (2) **33**, 441–52.

J1. Jack, H. (1970). A class of symmetric polynomials with a parameter. *Proc. R. Soc. Edinburgh* (A), **69**, 1–18.

J2. Jack, H. (1972). A surface integral and symmetric functions. *Proc. R. Soc. Edinburgh* (A), **69**, 347–63.

J3. Jacobi, C. G. (1841). De functionibus alternantibus.... *Crelle's Journal*, **22**, 360–71 (*Werke*, **3**, 439–52).

J4. James, A. T. (1961). Zonal polynomials of the real positive definite matrices. *Annals of Math.*, **74**, 456–69.

J5. James, A. T. (1964). Distributions of matrix variates and latent roots derived from normal samples. *Ann. Math. Stat.*, **35**, 475–501.

J6. James, A. T. (1968). Calculation of zonal polynomial coefficients by use of the Laplace–Beltrami operator. *Ann. Math. Stat.*, **39**, 1711–18.

J7. James, G. D. (1978). The representation theory of the symmetric groups. *Springer Lecture Notes*, **682**.

J8. James, G. D. (1987). The representation theory of the symmetric groups. *Proc. of Symposia in Pure Math.*, **47** part 1, 111–26.

J9. James, G. D. and Kerber, A. (1981). The representation theory of the symmetric groups. *Encyclopaedia of mathematics and its applications*, Vol. 16. Addison–Wesley, Reading, Massachusetts.

J10. James, G. D. and Peel, M. (1979). Specht series for skew representations of symmetric groups. *J. Algebra*, **56**, 343–64.

J11. Jing, N. (1991). Vertex operators, symmetric functions and the spin group Γ_n. *J. Algebra*, **138**, 340–98.

J12. Jing, N. (1991). Vertex operators and Hall–Littlewood symmetric functions. *Advances in Math.*, **87**, 226–48.

J13. Johnsen, K. (1982). On a forgotten note by Ernst Steinitz. *Bull. London Math. Soc.*, **14**, 353–5.

J14. Józefiak, T. (1988). Semisimple superalgebras. *Springer Lecture Notes*, **1352**, 96–113.

J15. Józefiak, T. (1989). Characters of projective representations of symmetric groups. *Expositiones Math.*, **7**, 193–247.

J16. Józefiak, T. (1991). Schur Q-functions and cohomology of isotropic Grassmannians. *Math. Proc. Camb. Phil. Soc.*, **109**, 471–8.

J17. Józefiak, T. and Pragacz, P. (1991). A determinantal formula for skew Schur Q-functions. *J. London Math. Soc.*, (2) **43**, 76–90.

K1. Kac, V. G. (1980). Simple Lie groups and the Legendre symbol. *Springer Lecture Notes*, **848**, 110–23.

K2. Kac, V. G. (1983). *Infinite-dimensional Lie algebras*. Birkhäuser, Boston.

K3. Kadell, K. W. J. (1988). A proof of some analogues of Selberg's integral for $k = 1$. *SIAM J. Math. Anal.*, **19**, 944–68.

K4. Kadell, K. W. J. (1988). The q-Selberg polynomials for $n = 2$, *Trans. Amer. Math. Soc.*, **310**, 535–53.
K5. Kadell, K. W. J. (1992). The Selberg–Jack symmetric functions. Preprint.
K6. Kadell, K. W. J. (1996). An integral for the product of two Selberg–Jack symmetric polynomials. *Comp. Math.*, **87**, 5–43.
K7. Kates, L. K. (1980). Zonal polynomials. Thesis, Princeton Univ.
K8. Kerov, S. V. (1984). On the Littlewood–Richardson rule and the Robinson–Schensted–Knuth correspondence (in Russian). *Uspekhi Mat. Nauk*, **39**, 161–2.
K9. Kirillov, A. N. and Reshetikhin, N. Yu. (1988). The Bethe Ansatz and the combinatorics of Young tableaux. *J. Soviet Math.*, **41**, 925–55.
K10. Klein, T. (1969). The Hall polynomial. *J. Algebra*, **12**, 61–78.
K11. Kljačko, A. A. (1981). Models for complex representations of $GL(n,q)$ and Weyl groups. *Soviet Math. Dokl.*, **24**, 496–9.
K12. Knuth, D. E. (1970). Permutations, matrices and generalized Young tableaux. *Pacific J. Math.*, **34**, 709–27.
K13. Knutson, D. (1973). λ-rings and the representation theory of the symmetric group. *Springer Lecture Notes*, **308**.
K14. Kondo, T. (1963). On Gaussian sums attached to the general linear groups over finite fields. *J. Math. Soc. Japan*, **15**, 244–55.
K15. Koornwinder, T. (1988). Self-duality for q-ultraspherical polynomials associated with the root system A_n. (Unpublished manuscript.)
K16. Kostka, C. (1882). Über den Zusammenhang zwischen einigen Formen von symmetrischen Funktionen. *Crelle's Journal*, **93**, 89–123.
K17. Kostka, C. (1908). Tafeln für symmetrische Funktionen bis zur elften Dimension. *Wissenschaftliche Beilage zum Programm des königl. Gymnasiums und Realgymnasiums zu Insterberg.*
L1. Lascoux, A. (1978). Classes de Chern d'un produit tensoriel. *C.R. Acad. Sci. Paris*, **286A**, 385–7.
L2. Lascoux, A. and Pragacz, P. (1984). Equerres et fonctions de Schur. *C.R. Acad. Sci. Paris (série I)*, **299**, 955–8.
L3. Lascoux, A. and Pragacz, P. (1988). Ribbon Schur functions. *Eur. J. Combinatories*, **9**, 561–74.
L4. Lascoux, A. and Schützenberger, M. P. (1978). Sur une conjecture de H. O. Foulkes. *C.R. Acad. Sci. Paris*, **286A**, 323–4.
L5. Lascoux, A. and Schützenberger, M. P. (1985). Formulaire raisonné de fonctions symétriques. *Publ. Math. Univ. Paris*, VII.
L6. Lascoux, A., Leclerc, B., and Thibon, J.-Y. (1992). *Fonctions de Hall–Littlewood et polynômes de Kostka–Foulkes aux racines de l'unité*. Institut Blaise Pascal, Paris.
L7. Littelmann, P. (1990). A generalization of the Littlewood–Richardson rule. *J. Algebra*, **130**, 328–68.
L8. Littelmann, P. (1992). A Littlewood–Richardson rule for symmetrizable Kac–Moody algebras. Preprint.
L9. Littlewood, D. E. (1950). The theory of group characters (2nd edn). Oxford University Press.
L10. Littlewood, D. E. (1951). Modular representations of symmetric groups. *Proc. R. Soc.*, A, **209**, 333–53.

L11. Littlewood, D. E. (1956). The Kronecker product of symmetric group representations. *J. London Math. Soc.*, **31**, 89–93.
L12. Littlewood, D. E. (1961). On certain symmetric functions, *Proc. London Math. Soc.*, **43**, 485–98.
L13. Littlewood, D. E. and Richardson, A. R. (1934). Group characters and algebra. *Philos. Trans. R. Soc.*, A, **233**, 99–141.
L14. Luks, E. M. (1966). Spherical functions on GL_n over p-adic fields. Thesis, M.I.T.
L15. Lusztig, G. (1981). Green polynomials and singularities of unipotent classes, *Advances in Math.*, **42**, 169–78.
M1. Macdonald, I. G. (1971). Spherical functions on a group of p-adic type. *Publ. Ramanujan Inst. No.* 2, Madras.
M2. Macdonald, I. G. (1980). Zeta functions attached to finite general linear groups. *Math. Annalen*, **249**, 1–15.
M3. Macdonald, I. G. (1980). Polynomial functors and wreath products, *J. Pure Appl. Algebra*, **18**, 173–204.
M4. Macdonald, I. G. (1984). The algebra of partitions, in *Group Theory: essays for Philip Hall* (ed. Gruenberg and Roseblade) pp. 315–33. Academic Press, London.
M5. Macdonald, I. G. (1987). Commuting differential operators and zonal spherical functions. *Springer Lecture Notes*, **1271**, 189–200.
M6. Macdonald, I. G. (1988). A new class of symmetric functions. *Publ. I.R.M.A. Strasbourg, Actes* 20e *Séminaire Lotharingien*, 131–71.
M7. Macdonald, I. G. (1991). *Notes on Schubert polynomials*. Publications du LACIM, Montreal.
M8. Macdonald, I. G. (1992). Schur functions: theme and variations. *Publ. I.R.M.A. Strasbourg, Actes* 28e *Séminaire Lotharingien*, 5–39.
M9. MacMahon, P. A. (1915, 1916). *Combinatory Analysis* I, II. Cambridge University Press. (Reprinted by Chelsea, New York, 1960.)
M10. Mead, D. G. (1993). Generators for the algebra of symmetric polynomials. *Am. Math. Monthly*, **100**, 386–8.
M11. Mills, W. H., Robbins, D. P., and Rumsey, H. (1982). Proof of the Macdonald conjecture. *Inv. Math.*, **66**, 73–87.
M12. Morris, A. O. (1963). The characters of the groups $GL(n, q)$. *Math. Zeitschrift*, **81**, 112–23.
M13. Morris, A. O. (1964). A note on the multiplication of Hall functions. *J. London Math. Soc.*, **39**, 481–8.
M14. Morris, A. O. (1965). The spin representation of the symmetric group. *Can. J. Math.*, **17**, 543–9.
M15. Morris, A. O. (1971). Generalizations of the Cauchy and Schur identities. *J. Comb. Theory (A)*, **11**, 163–9.
M16. Morris, A. O. and Yaseen, A. K. (1986). Some combinatorial results involving shifted Young diagrams. *Math. Proc. Camb. Phil. Soc.*, **99**, 23–31.
M17. Muirhead, R. J. (1982). *Aspects of multivariate statistical theory*. Wiley, New York.
M18. Murnaghan, F. D. (1937). The characters of the symmetric group. *Am. J. Math.*, **59**, 739–53.
N1. Nakayama, T. (1940). On some modular properties of irreducible representations of a symmetric group I, II. *Jap. J. Math.*, **17**, 165–84 and 411–23.

N2. Nimmo, J. J. C. (1990). Hall–Littlewood symmetric functions and the BKP equation. *J. Physics (A)*, **23**, 751–60.

P1. Parkhurst, A. M. and James, A. T. (1974). Zonal polynomials of order 1 through 12. *Selected tables in math. statistics*, Vol. 2. American Math. Society, Providence, Rhode Island.

P2. Pragacz, P. (1991). Algebro-geometric applications of Schur S- and Q-polynomials, *Séminaire d'Algèbre Dubreil–Malliavin* 1989–90. *Springer Lecture Notes*, **1478**, 130–91.

P3. Pragacz, P. and Thorup, A. (1992). On a Jacobi–Trudi formula for supersymmetric polynomials. *Advances in Math.*, **95**, 8–17.

P4. Puttaswamaiah, B. M. and Dixon, J. D. (1977). *Modular representations of finite groups*. Academic Press, New York.

R1. Redfield, J. H. (1927). The theory of group reduced distributions, *Am. J. Math.*, **49**, 433–55.

R2. Ringel, C. M. (1990). Hall polynomials for the representation-finite hereditary algebras. *Advances in Math.*, **84**, 137–78.

R3. Ringel, C. M. (1990). Hall algebras and quantum groups. *Inv. Math.*, **101**, 583–91.

R4. Ringel, C. M. (1991). Hall Algebras, in *Topics in Algebra*, *Banach Centre Publ. 26*, Warsaw.

R5. Robinson, G. de B. (1938). On the representations of S_n, I. *Am. J. Math.*, **60**, 745–60.

R6. Robinson, G. de B. (1948). On the representations of the symmetric group, III. *Am. J. Math.*, **70**, 277–94.

R7. Robinson, G. de B. (1961). *Representation theory of the symmetric group*. Edinburgh University Press.

R8. Rota, G.-C. (1964). On the foundations of combinatorial theory I: Theory of Möbius functions. *Z. Wahrscheinlichkeitstheorie*, **2**, 340–68.

R9. Ryser, H. J. (1963). *Combinatorial mathematics*. Wiley, New York.

S1. Sagan, B. E. (1987). Shifted tableaux, Schur Q-functions and a conjecture of R. Stanley. *J. Comb. Theory* (A), **45**, 62–103.

S2. Sagan, B. E. (1991). *The symmetric group*. Wadsworth and Brooks, Pacific Grove, California.

S3. Satake, I. (1963). Theory of spherical functions on reductive algebraic groups over p-adic fields. *Publ. Math. IHES*, **18**, 5–70.

S4. Schur, I. (1901). Über ein Klasse von Matrizen die sich einer gegebenen Matrix zuordnen lassen. Dissertation, Berlin (*Ges. Abhandlungen* **1**, 1–72).

S5. Schur, I. (1911). Über die Darstellung der symmetrischen und der alternierenden Gruppe durch gebrochene lineare Substitutionen. *Crelle's Journal*, **139**, 155–250 (*Ges. Abhandlungen*, **1**, 346–441).

S6. Schur, I. (1927). Über die rationalen Darstellungen der allgemeinen linearen Gruppe. *Sitzungsberichte der Preussischen Akademie der Wissenschaften*, (1927), 58–75 (*Ges. Abhandlungen*, **3**, 68–85).

S7. Schützenberger, M. P. (1977). La correspondance de Robinson, in *Combinatoire et représentations du groupe symétrique*, *Strasbourg* 1976. *Springer Lecture Notes*, **579**, 59–135.

S8. Schützenberger, M. P. (1978). Propriétés nouvelles des tableaux de Young.

Séminaire Delange–Pisot–Poitou, 19e année, 1977 / 8, no. 26. Secrétariat Mathématique, Paris.

S9. Sekiguchi, J. (1977). Zonal spherical functions on some symmetric spaces. *Publ. RIMS, Kyoto Univ.*, **12**, 455–9.

S10. Selberg, A. (1944). Bemerkninger om et multipelt integral. *Norsk Mat. Tidsskrift*, **26**, 71–8.

S11. Sergeev, A. N. (1985). The tensor algebra of the identity representation as a module over the Lie superalgebras $gl(n, m)$ and $Q(n)$. *Math. USSR Sbornik*, **51**, 419–27.

S12. Serre, J.-P. (1967). *Représentations linéaires des groupes finis*. Hermann, Paris.

S13. Serre, J.-P. (1970). *Cours d'arithmétique*. Presses Universitaires de France, Paris.

S14. Shimomura, N. (1980). A theorem on the fixed point set of a unipotent transformation on the flag manifold. *J. Math. Soc. Japan*, **32**, 55–64.

S15. Shimura, G. (1971). Introduction to the arithmetic theory of automorphic functions. *Publ. Math. Soc. Japan*, No. 11, Princeton Univ. Press.

S16. Spaltenstein, N. (1976). On the fixed point set of a unipotent transformation on the flag manifold. *Proc. Kon. Akad. v. Wetenschappen*, **79**, 452–6.

S17. Specht, W. (1932). Eine Verallgemeinerung der symmetrischen Gruppe. *Schriften Math. Seminar Berlin*, **1**, 1–32.

S18. Specht, W. (1935). Die irreduziblen Darstellungen der symmetrischen Gruppe. *Math. Z.*, **39**, 696–711.

S19. Specht, W. (1960). Die Charaktere der symmetrischen Gruppe. *Math. Z.*, **73**, 312–29.

S20. Springer, T. A. (1970). Characters of special groups. *Springer Lecture Notes*, **131**, 121–56.

S21. Springer, T. A. and Zelevinsky, A. V. (1984). Characters of $GL(n, \mathbf{F}_q)$ and Hopf algebras. *J. London Math. Soc.*, (2) **30**, 27–43.

S22. Srinivasan, B. (1992). On Macdonald's symmetric functions. *Bull. London Math. Soc.*, **24**, 519–25.

S23. Stanley, R. P. (1971). Theory and application of plane partitions I, II. *Studies in Appl. Math.*, **50**, 167–88 and 259–79.

S24. Stanley, R. P. (1984). The q-Dyson conjecture, generalized exponents, and the internal product of Schur functions. *Contemporary Math.*, **34**, 81–93.

S25. Stanley, R. P. (1989). Some combinatorial properties of Jack symmetric functions. *Advances in Math.*, **77**, 76–115.

S26. Steinitz, E. (1901). Zur Theorie der Abel'schen Gruppen. *Jahresbericht der DMV*, **9**, 80–5.

S27. Stembridge, J. R. (1985). A characterization of supersymmetric polynomials. *J. Algebra*, **95**, 439–44.

S28. Stembridge, J. R. (1989). Shifted tableaux and the projective representations of symmetric groups. *Advances in Math.*, **74**, 87–134.

S29. Stembridge, J. R. (1992). On Schur's Q-functions and the primitive idempotents of a commutative Hecke algebra. *J. Alg. Combinatorics*, **1**, 71–95.

S30. Stembridge, J. R. (1995). The enumeration of totally symmetric plane partitions. *Advances in Mathematics*, **111**, 227–43.

S31. Stembridge, J. R. (1994). Some hidden relations involving the ten symmetry classes of plane partitions. *J. Comb. Theory Series A*, **68**, 372–409.

S32. Stembridge, J. R. (1994). Some particular entries of the two-parameter Kostka matrix. *Proc. Amer. Math. Soc.*, **121**, 367–73.
T1. Takemura, A. (1984). Zonal polynomials. *Inst. of Math. Statistics Lecture Notes —Monograph Series*, Vol. 4. Hayward, California.
T2. Tamagawa, T. (1963). On the ζ-functions of a division algebra. *Ann. Math.*, **77**, 121–56.
T3. Thoma, E. (1964). Die unzerlegbaren, positiv-definiten Klassenfunktionen der abzählbar unendlichen symmetrischen Gruppe. *Math. Z.*, **85**, 40–61.
T4. Thomas, G. P. (1974). Baxter algebras and Schur functions. Thesis, Swansea.
T5. Thrall, R. M. (1942). On symmetrized Kronecker powers and the structure of the free Lie ring. *Am. J. Math.*, **64**, 371–8.
T6. Thrall, R. M. (1952). A combinatorial problem. *Michigan Math. J.*, **1**, 81–8.
W1. Weil, A. (1949). Number of solutions of equations in finite fields. *Bull. Am. Math. Soc.*, **55**, 497–508.
W2. Weyl, H. (1946). *The classical groups*. Princeton Univ. Press.
W3. White, D. (1981). Some connections between the Littlewood–Richardson rule and a construction of Schensted. *J. Comb. Theory* (A), **30**, 237–47.
W4. Worley, D. R. (1984). A theory of shifted Young tableaux. Thesis, M.I.T.
Y1. You, Y. (1989). Polynomial solutions of the BKP hierarchy and projective representations of symmetric groups, in *Infinite-dimensional Lie algebras and groups*, pp. 449–464, *Adv. Ser. Math. Phys.*, 7. World Science Publishing, Teaneck, New Jersey.
Y2. Young, A. (1901–1952). Quantitative substitutional analysis I–IX. *Proc. London Math. Soc.*
Z1. Zeilberger, D. and Bressoud, D. (1985). A proof of Andrews' q-Dyson conjecture. *Discrete Math.*, **54**, 201–24.
Z2. Zelevinsky, A. V. (1981). Representations of finite classical groups: a Hopf algebra approach. *Springer Lecture Notes*, **869**.
Z3. Zelevinsky, A. V. (1981). A generalization of the Littlewood–Richardson rule and the Robinson–Schensted–Knuth correspondence. *J. Algebra*, **69**, 82–94.

Notation

Chapter I

$l(\lambda)$	length of a partition λ: I, 1
$\|\lambda\|$	weight (= sum of parts) of a partition λ: I, 1
$\mathscr{P}_n$	the set of partitions of n: I, 1
$\mathscr{P}$	the set of all partitions: I, 1
$m_i(\lambda)$	multiplicity of i as a part of λ: I, 1
λ'	conjugate of a partition λ: I, 1
$n(\lambda)$	$\sum(i-1)\lambda_i = \sum \binom{\lambda'_i}{2}$: I, 1
$(\alpha \mid \beta)$	Frobenius' notation for a partition: I, 1
$\lambda \supset \mu$	$\lambda_i \geqslant \mu_i$ for all $i \geqslant 1$: I, 1
$\lambda - \mu$	skew diagram: I, 1
$\lambda + \mu$	partition with parts $\lambda_i + \mu_i$: I, 1
$\lambda \cup \mu$	partition whose parts are those of λ and of μ: I, 1
$\lambda\mu$	partition with parts $\lambda_i \mu_i$: I, 1
$\lambda \times \mu$	partition with parts $\min(\lambda_i, \mu_j)$: I, 1
$\lambda \geqslant \mu$	$\lambda_1 + \ldots + \lambda_i \geqslant \mu_1 + \ldots + \mu_i$ for all $i \geqslant 1$: I, 1
R_{ij} $(i<j)$	raising operator: I, 1
S_n	symmetric group on n symbols: I, 1
δ	$(n-1, n-2, \ldots, 1, 0)$: I, 1
$h(x)$	hook length at $x \in \lambda$: I, 1, Example 1
$c(x)$	content of $x \in \lambda$: I, 1, Example 3
$\varphi_r(t)$	$(1-t)(1-t^2)\ldots(1-t^r)$: I, 1, Example 3
λ^*	p-quotient of λ: I, 1, Example 8
$\tilde{\lambda}$	p-core of λ: I, 1, Example 8
$\lambda \sim_p \mu$	λ, μ have the same p-core: I, 1, Example 8
$h(\lambda)$	product of the hook-lengths of λ: I, 1, Example 10
$c_\lambda(X)$	content polynomial of λ: I, 1, Example 11
Λ_n	$\mathbf{Z}[x_1, \ldots, x_n]^{S_n}$: I, 2
Λ	ring of symmetric functions: I, 2
Λ_A	$\Lambda \otimes_{\mathbf{Z}} A$ (A a commutative ring): I, 2
x^α	$x_1^{\alpha_1} x_2^{\alpha_2} \ldots$: I, 2
m_λ	monomial symmetric function generated by x^λ: I, 2
e_r	rth elementary symmetric function: I, 2
$E(t)$	$\sum e_r t^r = \prod(1 + x_i t)$: I, 2
e_λ	$e_{\lambda_1} e_{\lambda_2} \ldots$: I, 2
h_r	rth complete symmetric function: I, 2
$H(t)$	$\sum h_r t^r = \prod(1 - x_i t)^{-1}$: I, 2
h_λ	$h_{\lambda_1} h_{\lambda_2} \ldots$: I, 2
ω	involution of Λ which interchanges e_r and h_r: I, 2

f_λ	'forgotten' symmetric function $\omega(m_\lambda)$: I, 2
p_r	rth power sum: I, 2
$P(t)$	$\Sigma p_r t^{r-1}$: I, 2
p_λ	$p_{\lambda_1} p_{\lambda_2} \dots$: I, 2
ε_λ	$(-1)^{\lvert\lambda\rvert - l(\lambda)}$: I, 2
z_λ	$\prod_{i \geq 1} i^{m_i(\lambda)} . m_i(\lambda)!$: I, 2
$\begin{bmatrix} n \\ r \end{bmatrix}$	q-binomial coefficient $\varphi_n(q)/\varphi_r(q)\varphi_{n-r}(q)$: I, 2, Example 3
$c(G)$	cycle indicator of a subgroup G of S_n: I, 2, Example 9
$\varepsilon(w)$	sign of $w \in S_n$: I, 3
a_α	skew-symmetric polynomial generated by x^α: I, 3
s_λ	Schur function: I, 3
$\begin{bmatrix} n \\ \lambda \end{bmatrix}$	generalized q-binomial coefficient: I, 3, Example 1
$H_\lambda(q)$	hook polynomial $\prod_{s \in \lambda}(1 - q^{h(s)})$: I, 3, Example 2
$\binom{X}{\lambda}$	generalized binomial coefficient: I, 3, Example 4
$s_\lambda(x/y)$	'supersymmetric' Schur function: I, 3, Example 23
$s_{\lambda/\mu}$	skew Schur function: I, 5
$c^\lambda_{\mu\nu}$	coefficient of s_λ in $s_\mu s_\nu$: I, 5
$K_{\lambda-\mu,\nu}$	$\langle s_{\lambda/\mu}, h_\nu \rangle$: I, 5
$f^\perp$	adjoint of multiplication by $f \in \Lambda$: I, 5, Example 3
Δ	diagonal map $\Lambda \to \Lambda \otimes \Lambda$: I, 5, Example 25
$M(u, v)$	transition matrix: I, 6
K	Kostka matrix $M(s, m)$: I, 6
L	$M(p, m)$: I, 6
$\lambda \leq_R \mu$	λ is a refinement of μ: I, 6
χ^λ	irreducible character of S_n: I, 7
χ^λ_ρ	value of χ^λ at elements of cycle-type ρ: I, 7
$f * g$	internal product of $f, g \in \Lambda$: I, 7
M_λ	Specht module: I, 7, Example 15
$f \circ g$	plethysm $(f, g \in \Lambda)$: I, 8
F_λ	irreducible polynomial functor: I, Appendix A, 5

Chapter II

$\mathfrak{o}$	a discrete valuation ring: II, 1
$\mathfrak{p}$	maximal ideal of $\mathfrak{o}$: II, 1
k	residue field $\mathfrak{o}/\mathfrak{p}$: II, 1
$l(M)$	length of a finite $\mathfrak{o}$-module M: II, 1
$a_\lambda(q)$	number of automorphisms of a module of type λ: II, 1
$G^\lambda_{\mu\nu}(\mathfrak{o})$	number of submodules N of an $\mathfrak{o}$-module M of type λ, such that N has type ν and M/N has type μ: II, 2
$H(\mathfrak{o})$	Hall algebra of $\mathfrak{o}$: II, 2
$g_S(t), g^\lambda_{\mu\nu}(t)$	Hall polynomials: II, 4

Chapter III

$R_\lambda(x_1,\dots,x_n;t)$	symmetric polynomials defined in III, 1
$v_m(t)$	$\varphi_m(t)/(1-t)^m$: III, 1
$v_\lambda(t)$	$\prod_{i\geq 0} v_{m_i(\lambda)}(t)$: III, 1
$P_\lambda(x;t)$	Hall–Littlewood symmetric function: III, 2
$q_r(x;t)$	$(1-t)P_{(r)}(x;t)$ $(r\geq 1)$: III, 2
$b_\lambda(t)$	$\prod_{i\geq 1} \varphi_{m_i(\lambda)}(t)$: III, 2
$Q_\lambda(x;t)$	$b_\lambda(t)P_\lambda(x;t))$: III, 2
$q_\lambda(x;t)$	$q_{\lambda_1}(x;t)q_{\lambda_2}(x;t)\dots$: III, 2
$f^\lambda_{\mu\nu}(t)$	coefficient of P_λ in $P_\mu P_\nu$: III, 3
$z_\lambda(t)$	$z_\lambda \prod(1-t^{\lambda_i})^{-1}$: III, 4
$S_\lambda(x;t)$	$\det(q_{\lambda_i-i+j}(x;t))$: III, 4
$Q_{\lambda/\mu}, P_{\lambda/\mu}$	skew Hall–Littlewood functions: III, 5
$\varphi_{\lambda/\mu}(t), \psi_{\lambda/\mu}(t), \varphi_T(t), \psi_T(t)$:	III, 5
$K(t)$	transition matrix $M(s,P)$: III, 6
$X^\lambda_\rho(t)$	coefficient of $P_\lambda(x;t)$ in $p_\rho(x)$: III, 7
$Q^\lambda_\rho(q)$	$q^{n(\lambda)}X^\lambda_\rho(q^{-1})$ (Green's polynomials): III, 7
Γ	$\mathbb{Z}[q_1,q_2,q_3,\dots]$: III, 8
$\mathrm{Pf}(A)$	Pfaffian of a skew-symmetric matrix A: III, 8

Chapter IV

k	finite field: IV, 1
q	number of elements in k: IV, 1
$\bar{k}$	algebraic closure of k: IV, 1
F	Frobenius automorphism $x\mapsto x^q$: IV, 1
k_n	extension of k of degree n contained in $\bar{k}$: IV, 1
M	multiplicative group of $\bar{k}$: IV, 1
M_n	multiplicative group of k_n: IV, 1
$N_{n,m}$	norm homomorphism $M_n\to M_m$ (where m divides n): IV, 1
L	$\varinjlim \hat{M}_n$: IV, 1
L_n	subgroup of L fixed by F^n: IV, 1
$\langle \xi, x\rangle_n$	pairing between L_n and M_n: IV, 1
Φ	set of F-orbits in M, identified with the set of monic irreducible polynomials (other than t) in $k[t]$: IV, 1
$d(f)$	degree of $f\in\Phi$: IV, 1
G_n	$GL_n(k)$: IV, 2
μ	partition-valued function on Φ: IV, 2
$\lVert\mu\rVert$	$\Sigma_{f\in\Phi}\, d(f)\lvert\mu(f)\rvert$: IV, 2
c_μ	conjugacy class parametrized by μ: IV, 2
q_f	$q^{d(f)}$: IV, 2
a_μ	$\prod_{f\in\Phi} a_{\mu(f)}(q_f)$ = order of centralizer of any element of c_μ: IV, 2
$u_1\circ\dots\circ u_r$	induction product of class functions: IV, 3
π_μ	characteristic function of c_μ: IV, 3

A_n	space of class functions on G_n: IV, 3
A	$\bigoplus_{n\geq 0} A_n$: IV, 3
R_n	**Z**-module generated by the characters of G_n: IV, 3
R	$\bigoplus_{n\geq 0} R_n$: IV, 3
B	polynomial algebra over **C** generated by independent variables $e_n(f)$ $(n\geq 1, f\in\Phi)$: IV, 4
$\tilde{P}_\lambda(f)$	$q_f^{-n(\lambda)}P_\lambda(X_f;q_f^{-1})$: IV, 4
$\tilde{Q}_\lambda(f)$	$a_\lambda(q_f)\tilde{P}_\lambda(f)$: IV, 4
$\tilde{P}_{\boldsymbol{\mu}}$	$\prod_{f\in\Phi}\tilde{P}_{\boldsymbol{\mu}(f)}(f)$: IV, 4
$\tilde{Q}_{\boldsymbol{\mu}}$	$\prod_{f\in\Phi}\tilde{Q}_{\boldsymbol{\mu}(f)}(f)=a_{\boldsymbol{\mu}}\tilde{P}_{\boldsymbol{\mu}}$: IV, 4
ch	characteristic map: IV, 4
$\tilde{p}_n(x)$	$p_{n/d}(f)$ $(x\in f$, n a multiple of $d=d(f))$: IV, 4
$\tilde{p}_n(\xi)$	$(-1)^{n-1}\sum_{x\in M_n}\langle\xi,x\rangle_n\tilde{p}_n(x)$: IV, 4
Θ	set of F-orbits in L: IV, 4
$d(\varphi)$	number of elements of $\varphi\in\Theta$: IV, 4
$p_r(\varphi)$	$\tilde{p}_{rd}(\xi)$ $(\xi\in\varphi,\ d=d(\varphi))$: IV, 4
$s_\lambda(\varphi)$	Schur function in the φ-variables: IV, 4
$\boldsymbol{\lambda}$	partition-valued function on Θ: IV, 4
$\Vert\boldsymbol{\lambda}\Vert$	$\sum_{\varphi\in\Theta}d(\varphi)\lvert\boldsymbol{\lambda}(\varphi)\rvert$: IV, 4
$S_{\boldsymbol{\lambda}}$	$\prod_{\varphi\in\Theta}s_{\boldsymbol{\lambda}(\varphi)}(\varphi)$: IV, 4
S	**Z**-submodule of B generated by the $S_{\boldsymbol{\lambda}}$: IV, 4
$\chi^{\boldsymbol{\lambda}}$	irreducible character of G_n: IV, 6
$\chi^{\boldsymbol{\lambda}}_{\boldsymbol{\mu}}$	value of $\chi^{\boldsymbol{\lambda}}$ at the class $c_{\boldsymbol{\mu}}$: IV, 6

Chapter V

F	non-archimedean local field: V, 1
$\lvert a\rvert$	absolute value of $a\in F$: V, 1
$\mathfrak{o}$	ring of integers of F: V, 1
$\mathfrak{p}$	maximal ideal of $\mathfrak{o}$: V, 1
k	(finite) residue field $\mathfrak{o}/\mathfrak{p}$: V, 1
q	number of elements in k: V, 1
π	generator of $\mathfrak{p}$: V, 1
v	normalized valuation on F^*: V, 1
G	$GL_n(F)$: V, 2
G^+	$G\cap M_n(\mathfrak{o})$: V, 2
K	$GL_n(\mathfrak{o})$: V, 2
$L(G,K)$ (resp. $L(G^+,K)$)	space of complex-valued continuous functions of compact support on $K\backslash G/K$ (resp. $K\backslash G^+/K$): V, 2
$f*g$	convolution product in $L(G,K)$: V, 2
$H(G,K)$ (resp. $H(G^+,K)$)	Hecke ring of (G,K) (resp. (G^+,K)): V, 2
π^λ	$\mathrm{diag}(\pi^{\lambda_1},\dots,\pi^{\lambda_n})$: V, 2
c_λ	characteristic function of $K\pi^\lambda K$: V, 2
ρ	$\frac{1}{2}(n-1,n-3,\dots,1-n)$: V, 2
ω	spherical function on G relative to K: V, 3
$\hat{\omega}(f), \hat{f}(\omega)$	Fourier transform of $f\in L(G,K)$ by ω: V, 3.

ω_s	spherical function with parameter $s=(s_1,\dots,s_n)$: V, 3
G_m^+	$\{x\in G^+ : v(\det x)=m\}$: V, 4
τ_m	characteristic function of G_m^+: V, 4
$\tau(X)$	Hecke series $\Sigma_m \tau_m X^m$: V, 4
$\zeta(s,\omega)$	zeta function defined by ω: V, 4

Chapter VI

F	the field $\mathbf{Q}(q,t)$: VI, 2
$z_\lambda(q,t)$	$z_\lambda \Pi(1-q^{\lambda_i})/(1-t^{\lambda_i})$: VI, 2
$(a;q)_\infty$	$\Pi_0^\infty(1-aq^r)$: VI, 2
$\Pi(x,y;q,t)$	$\Pi_{i,j}(tx_iy_j;q)_\infty/(x_iy_j;q)_\infty$: VI, 2
$g_n(x;q,t)$	coefficient of y^n in $\Pi(tx_iy;q)_\infty/(x_iy;q)_\infty$: VI, 2
$g_\lambda(x;q,t)$	$\Pi\, g_{\lambda_i}(x;q,t)$: VI, 2
$\omega_{u,v}$	the endomorphism of Λ_F defined by $p_r \mapsto (-1)^{r-1}p_r(1-u^r)/(1-v^r)$ $(r\geqslant 1)$: VI, 2
T_{u,x_i}	$f(x_1,\dots,x_n)\mapsto f(x_1,\dots,ux_i,\dots,x_n)$: VI, 3
$D_n(X;q,t)$	see VI, (3.2)
D_n^r	coefficient of X^r in $D_n(X;q,t)$: VI, 3
$A_i(x;t)$	$\Pi_{j\neq i}(tx_i-x_j)/(x_i-x_j)$: VI, 3
$D_n(X;\alpha)$	see VI, 3, Example 3
$\square_n^\alpha$	Laplace–Beltrami operator: VI, 3, Example 3(e)
$P_\lambda(x;q,t)$	symmetric functions defined by VI, (4.7)
$b_\lambda(q,t)$	$\langle P_\lambda, P_\lambda\rangle^{-1}$: VI, 4
$Q_\lambda(x;q,t)$	$b_\lambda(q,t)P_\lambda(x;q,t)$: VI, 4
ω_α	the automorphism of Λ_F defined by $p_r\mapsto(-1)^{r-1}\alpha^r p_r$: IV, 5, Example 3
$\varphi_{\lambda/\mu}, \varphi'_{\lambda/\mu}, \psi_{\lambda/\mu}, \psi'_{\lambda/\mu}$	coefficients in Pieri formulas: VI, (6.24)
$a(s), a'(s)$	arm length and colength of $s\in\lambda$: VI, 6
$l(s), l'(s)$	leg length and colength of $s\in\lambda$: VI, 6
$\varepsilon_{u,t}$	specialization $p_r\mapsto(1-u^r)/(1-t^r)$: VI, 6
$b_\lambda(s;q,t)$	$(1-q^{a(s)}t^{l(s)+1})/(1-q^{a(s)+1}t^{l(s)})$: VI, 6
$f_{\mu\nu}^\lambda(q,t)$	coefficient of P_λ in $P_\mu P_\nu$: VI, 7
$P_{\lambda/\mu}(q,t)$, $Q_{\lambda/\mu}(q,t)$	skew functions: VI, 7
$c_\lambda(q,t)$	$\Pi_{s\in\lambda}(1-q^{a(s)}t^{l(s)+1})$: VI, 8
$c'_\lambda(q,t)$	$\Pi_{s\in\lambda}(1-q^{a(s)+1}t^{l(s)})$: VI, 8
$J_\lambda(x;q,t)$	$c_\lambda(q,t)P_\lambda(x;q,t)$: VI, 8
$K_{\lambda\mu}(q,t)$	coefficient of $S_\lambda(x;t)$ in $J_\mu(x;q,t)$: VI, 8
$p_\rho(x;t)$	$p_\rho(x)\Pi(1-t^{\rho_i})$: VI, 8
$X_\rho^\lambda(q,t)$	coefficient of $z_\rho^{-1}p_\rho(x;t)$ in $J_\lambda(x;q,t)$: VI, 8
L_n	$F[x_1^{\pm1},\dots,x_n^{\pm1}]$: VI, 9
$\bar f$	$f(x_1^{-1},\dots,x_n^{-1})$ if $f=f(x_1,\dots,x_n)$: VI, 9
$[f]_1$	constant term in $f\in L_n$: VI, 9
$\Delta(x;q,t)$	$\Pi_{i\neq j}(x_ix_j^{-1};q)_\infty/(tx_ix_j^{-1};q)_\infty$: VI, 9
$(q,t)\to_\alpha(1,1)$	$(q,t)\to(1,1)$ and $(1-q)/(1-t)\to\alpha$: VI, 10
$g_n^{(\alpha)}(x)$	coefficient of y^n in $\Pi(1-x_iy)^{-1/\alpha}$: VI, 10

$g_\lambda^{(\alpha)}(x)$	$\prod g_{\lambda_i}^{(\alpha)}(x)$: VI, 10
$b_\lambda^{(\alpha)}(s)$	$(a(s)\alpha + l(s) + 1)/(a(s)\alpha + l(s) + \alpha)$: VI, 10
$\psi_{\lambda/\mu}^{(\alpha)}$	limit of $\psi_{\lambda/\mu}(q,t)$ as $(q,t) \to_\alpha (1,1)$: VI, 10
$P_\lambda^{(\alpha)}$	Jack symmetric function: VI, 10
$b_\lambda^{(\alpha)}$	$\prod_{s\in\lambda} b_\lambda^{(\alpha)}(s)$: VI, 10
$Q_\lambda^{(\alpha)}$	$b_\lambda^{(\alpha)} P_\lambda^{(\alpha)}$: VI, 10
$P_{\lambda/\mu}^{(\alpha)}, Q_{\lambda/\mu}^{(\alpha)}$	skew Jack functions: VI, 10
$f_{\mu\nu}^\lambda(\alpha)$	coefficient of $P_\lambda^{(\alpha)}$ in $P_\mu^{(\alpha)} P_\nu^{(\alpha)}$: VI, 10
ε_X	F-algebra homomorphism $\Lambda_F \to F[X]$ defined by $p_r \mapsto X$ $(r \geqslant 1)$: VI, 10
$c_\lambda(\alpha)$	$\prod_{s\in\lambda}(a(s)\alpha + l(s) + 1)$: VI, 10
$c_\lambda'(\alpha)$	$\prod_{s\in\lambda}(a(s)\alpha + l(s) + \alpha)$: VI, 10
$J_\lambda^{(\alpha)}$	$c_\lambda(\alpha) P_\lambda^{(\alpha)}$: VI, 10
$\theta_\rho^\lambda(\alpha)$	coefficient of p_ρ in $J_\lambda^{(\alpha)}$: VI, 10

Chapter VII

$C(G,K)$	space of functions on G constant on the double cosets of K: VII, 1
1_K^G	permutation representation of G on G/K: VII, 1
V^K	subspace of vectors $v \in V$ fixed by K: VII, 1
H_n	hyperoctahedral subgroup of S_{2n}: VII, 2
$\Gamma(w)$	graph attached to $w \in S_{2n}$: VII, 2
H_ρ	double coset of H_n in S_{2n}: VII, 2
ω^λ	zonal spherical function of (S_{2n}, H_n): VII, 2
ω_ρ^λ	value of ω^λ at permutations of coset-type ρ: VII, 2
Z_λ	zonal polynomial, equal to $J_\lambda^{(2)}$: VII, 2
Z_λ'	$2^{\lvert\lambda\rvert} J_\lambda^{(1/2)}$: VII, 2, Example 7
G	$GL_n(\mathbf{R})$: VII, 3
K	$O(n)$: VII, 3
$\mathrm{d}k$	normalized Haar measure on K: VII, 3
$P(G)$	space of polynomial functions on G: VII, 3
$P(K\backslash G), P(G/K), P(G,K)$	subspaces of $P(G)$: VII, 3
Σ	space of real symmetric $n \times n$ matrices: VII, 3
Σ^+	cone of positive definite symmetric $n \times n$ matrices: VII, 3
$P(\Sigma)$	space of polynomial functions on Σ: VII, 3
$\mathbf{T}$	$2m$th tensor power of $V = \mathbf{R}^n$: VII, 3
Ω_λ	zonal spherical function of (G,K) indexed by λ: VII, 3
$\mathrm{d}x$	Lebesgue measure on G: VII, 4
$\mathrm{d}\nu(x)$	$(2\pi)^{-n^2/2}\exp(-\frac{1}{2}\operatorname{trace} x'x)\,\mathrm{d}x$: VII, 4
$x^{(r)}$	the $r \times r$ matrix $(x_{ij})_{1\leqslant i,j\leqslant r}$ $(x \in G)$: VII, 4
Δ_μ	polynomial function on G defined by $x \mapsto \prod(\det x^{(\mu_i')})$, μ a partition of length $\leqslant n$: VII, 4
B^+	subgroup of G consisting of upper triangular matrices with positive diagonal elements: VII, 4
$\Gamma_n(a,\lambda)$	$\pi^{n(n-1)/4}\prod_{i=1}^n \Gamma(a + \lambda_i - \frac{1}{2}(i-1))$: VII, 4, Example 8

INDEX

abacus: I, 1, Ex. 8
Adams operations: I, 2
addition of partitions: I, 1
antisymmetrization: I, 3
arm length, colength: VI, 6
array: II, Appendix
augmented monomial symmetric function: I, 6, Ex. 9
augmented Schur function: I, 7, Ex. 17

basic characters of $GL_n(k)$: IV, 6, Ex. 1
Bell polynomials: I, 2, Ex. 11
bitableau: I, 5, Ex. 23
border of a partition: I, 1
border strip: I, 1

Cauchy's determinant: I, 4, Ex. 6
character polynomial: I, 7, Ex. 14
characteristic map: I, 7 and IV, 4
characters of S_n: I, 7
charge of a word or tableau: III, 6
column-strict plane partition: I, 5, Ex. 13
column-strict tableau: I, 1
complement of a partition: I, 1, Ex. 17
complete symmetric functions: I, 2
completely symmetric plane partition: I, 5, Ex. 18
composition (= plethysm): I, 8 and I, Appendix A, 6
conjugate diagram: I, 1
conjugate partition: I, 1
connected components of a skew diagram: I, 1
content: I, 1, Ex. 3
content polynomial: I, 1, Ex. 11
coset-type of a permutation: VII, 2
cotype: II, 1
cycle indicator: I, 2, Ex. 9
cycle-product: I, Appendix B, 3
cycle-type of a permutation: I, 7
cyclic $\mathfrak{o}$-module: I, 2, Ex. 9
cyclically symmetric plane partitions: I, 5, Ex. 18

diagonal map: I, 5, Ex. 25
diagram of a partition: I, 1
Dickson's theorem: I, 2, Ex. 27
dominance partial order on partitions: I, 1
domino: I, 6, Ex. 5
domino tableau: I, 6, Ex. 7
domino tabloid: I, 6, Ex. 5
double diagram of a strict partition: I, 1, Ex. 9
double strip: III, 8, Ex. 11
doubly stochastic matrix: I, 1, Ex. 13
dual of a finite $\mathfrak{o}$-module: II, 1

elementary $\mathfrak{o}$-module: II, 1
elementary symmetric functions: I, 2
even partition: I, 5, Ex. 5 and VII, 2
exterior powers: I, 2

finite $\mathfrak{o}$-module: II, 1
flag manifold: II, 3, Ex.
forgotten symmetric functions: I, 2
Fourier transform: V, 3 and VII, 1, Ex. 4
Frobenius automorphism: IV, 1
Frobenius notation for partitions: I, 1
fundamental theorem on symmetric functions: I, 2

Gale–Ryser theorem: I, 7, Ex. 9
Gaussian polynomials: I, 2, Ex. 3
Gelfand pair: VII, 1
generalized binomial coefficients: I, 3, Ex. 4
generalized tableau: I, 8, Ex. 8
Green's polynomials: III, 7

Hall algebra: II, 2
Hall polynomial: II, 4
Hall–Littlewood functions: III, 2
Hecke ring: V, 2 and V, 5
Hecke series: V, 4 and V, 5
height of a border strip: I, 1
homogeneous polynomial functor: I, Appendix A, 1
hook-length: I, 1, Ex 1 and Ex. 14
hook polynomial: I, 3, Ex. 2
horizontal strip: I, 1
hyperoctahedral group: VII, 2

induction product: I, Appendix A, 6 and IV, 3
internal product: I, 7

Jack's symmetric functions: VI, 10
Jacobi–Trudi formula: I, 3
Jordan canonical form: IV, 2, Ex. 1

Kostka numbers: I, 6

Laplace–Beltrami operator: VI, 3, Ex. 3 and VII, 2, Ex. 3
lattice permutation: I, 9
lattice property: III, 8
leg length, colength: VI, 6
length of a finite o-module: II, 1
length of a partition: I, 1
linearization of a polynomial functor: I, Appendix A, 3
Littlewood–Richardson rule: I, 9
local field: V, 1
LR-sequence: II, 3

marked shifted tableau: III, 8
matrix coefficients of a representation: I, Appendix A, 8
maximal tori: IV, 2, Ex. 4
metric tensor: VI, 1, Ex.
modified cycle-type: I, 7, Ex. 24
monomial symmetric functions: I, 2
Muirhead's inequalities: I, 2, Ex. 18
multiplication of partitions: I, 1
multiplicative family: VI, 1, Ex.
multiplicity-free representation: VII, 1
Murnaghan–Nakayama rule: I, 7, Ex. 5

Nakayama's conjecture: I, 7, Ex. 19
natural ordering of partitions: I, 1
Newton's formulas: I, 2
normal distribution: VII, 4

odd partition: III, 8
orthogonality relations: III, 7 and VI, 8

parts of a partition: I, 1
parts of a plane partition: I, 5, Ex. 13
p-bar core: I, 1, Ex. 9
p-core of a partition: I, 1, Ex. 8
Pfaffian: III, 8
Pieri's formula: I, 5 and VI, 6
plane partition: I, 5, Ex. 13–19
plethysm: I, 8 and I, Appendix A, 6
Polya's theorem: I, 2, Ex. 9
polynomial functor: I, Appendix A, 1
polynomial representation: I, Appendix A, 8
positive type: VII, 1
power sums: I, 2
p-quotient of a partition: I, 1, Ex. 8
p-residue of a partition: I, 1, Ex. 8
p-tableau: I, 5, Ex. 2

q-binomial coefficient: I, 2, Ex. 3
quadratic reciprocity: I, 3, Ex. 17

raising operator: I, 1
refinement of a partition: I, 6
reverse lexicographical order on partitions: I, 1
ribbon: I, 1
rim of a partition: I, 1
ring of integers of a local field: V, 1
ring of symmetric functions: I, 2
row-strict array: II, Appendix

scalar product: I, 4; III, 4; VI, 1
Schur algebra: I, Appendix A, 8, Ex. 6
Schur functions: I, 3
Schur operation: I, 3
Schur *Q*-functions: III, 8
Sergeev–Pragacz formula: I, 3, Ex. 24
shape of a tableau: I, 1
shifted diagram: I, 1, Ex. 9
shifted standard tableau: III, 8, Ex. 12
shifted tableau: III, 8
skew diagram: I, 1
skew Hall–Littlewood functions: III, 5
skew hook: I, 1
skew Schur functions: I, 5
Specht module: I, 7, Ex. 15
special border strip: I, 6, Ex. 4
spherical functions: V, 3
spherical transform: VII, 1, Ex. 4
staircase partition: III, 8, Ex. 3
standard tableau: I, 1
Stirling numbers: I, 2, Ex. 11
strict partition: I, 1, Ex. 9
strictly lower (upper) (uni) triangular: I, 6
symmetric function: I, 2
symmetric polynomial: I, 2
symmetric powers: I, 2
symmetrical plane partitions: I, 5, Ex. 15

tableau: I, 1
transition matrix: I, 6
twisted Gelfand pair: VII, 1, Ex. 10
type of a finite o-module: II, 1

unimodal polynomial: I, 3, Ex. 1 and I, 8, Ex. 4

Vandermonde determinant: I, 3
vertical strip: I, 1

weight of a marked shifted tableau: III, 8
weight of a partition: I, 1
weight of a plane partition: I, 5, Ex. 13
weight of a tableau: I, 1
Weyl's identity: I, 5, Ex. 9
Wishart density: VII, 4, Ex. 7
wreath product: I, Appendix A, 6 and I, Appendix B

zeta function: V, 4, and V, 5
zonal polynomials: VII, 2
zonal spherical functions: VII, 1

Printed in the USA/Agawam, MA
September 27, 2021